VISUALIZING

GEOLOGY

THIRD EDITION

VISUALIZING
GEOLOGY

THIRD EDITION

BARBARA MURCK, PhD
UNIVERSITY OF TORONTO

BRIAN SKINNER, PhD
YALE UNIVERSITY

WILEY
VISUALIZING™

WILEY

VICE PRESIDENT AND EXECUTIVE PUBLISHER   Jay O'Callaghan
EXECUTIVE EDITOR   Ryan Flahive
DIRECTOR OF DEVELOPMENT   Barbara Heaney
MANAGER, PRODUCT DEVELOPMENT   Nancy Perry
PRODUCT DESIGNER   Beth Tripmacher
EDITORIAL OPERATIONS MANAGER   Lynn Cohen
EDITORIAL ASSISTANTS   Darnell Sessoms, Brittany Cheetham
DIRECTOR, MARKETING COMMUNICATIONS   Jeffrey Rucker
SENIOR MARKETING MANAGER   Margaret Barrett
SENIOR CONTENT MANAGER   Micheline Frederick
SENIOR PRODUCTION EDITOR   William Murray
CREATIVE DIRECTOR   Harry Nolan
COVER DESIGN   Harry Nolan
INTERIOR DESIGN   Jim O'Shea
PHOTO RESEARCHER   Ellinor Wagner

COVER CREDITS
Main cover photo: Chris Hill/NGS/© Corbis
Bottom inset photos (left to right): Greg Vaughn/Alamy; Doug Millar/Photo Researchers, Inc.; Jerry Dorill/
Getty Images, Inc.; Woods Wheatcroft/Getty Images, Inc.; Michael S. Yamashita/NG Image Collection
Back cover inset photo: Michael S. Yamashita/NG Image Collection

This book was set in New Baskerville by Preparé and printed and bound by Quad/Graphics.
The cover was printed by Quad/Graphics.

ISBN 13: 978-1-118-12986-9
BRV ISBN: 978-1-118-25281-9

Printed in the United States of America
10 9 8 7 6 5 4 3 2

How Is Wiley Visualizing Different?

Wiley Visualizing is based on decades of research on the use of visuals in learning (Mayer, 2005).[1] The visuals teach key concepts and are pedagogically designed to **explain, present,** and **organize** new information. The figures are tightly integrated with accompanying text; the visuals are conceived with the text in ways that clarify and reinforce major concepts while allowing students to understand the details. This commitment to distinctive and consistent visual pedagogy sets Wiley Visualizing apart from other textbooks.

The texts offer an array of remarkable photographs, maps, media, and film from photo collections around the world, including those of National Geographic. Wiley Visualizing's images are not decorative; such images can be distracting to students. Instead, they are purposeful and the primary driver of the content. These authentic materials immerse the student in real-life issues and experiences and support thinking, comprehension, and application.

Together these elements deliver a level of rigor in ways that maximize student learning and involvement. Wiley Visualizing has been proven to increase student learning through its unique combination of text, photographs, and illustrations, with online video, animations, simulations, and assessments.

1. Visual Pedagogy. Using the Cognitive Theory of Multimedia Learning, which is backed up by hundreds of empirical research studies, Wiley's authors create visualizations for their texts that specifically support students' thinking and learning—for example, the selection of relevant materials, the organization of the new information, or the integration of the new knowledge with prior knowledge.

2. Authentic Situations and Problems. *Visualizing Geology 3e* benefits from National Geographic's more than century-long recording of the world and offers an array of remarkable photographs, maps, media, and film. These authentic materials immerse the student in real-life issues in geology, thereby enhancing motivation, learning, and retention (Donovan & Bransford, 2005).[2]

3. Designed with Interactive Multimedia. *Visualizing Geology 3e* is tightly integrated with *WileyPLUS*, our online learning environment that provides interactive multimedia activities in which learners can actively engage with the materials. The combination of textbook and *WileyPLUS* provides learners with multiple entry points to the content, giving them greater opportunity to explore concepts and assess their understanding as they progress through the course. *WileyPLUS* is a key component of the Wiley Visualizing learning and problem-solving experience, setting it apart from other textbooks whose online component is mere drill-and-practice.

Wiley Visualizing and the *WileyPLUS* Learning Environment are designed as a natural extension of how we learn

To understand why the Visualizing approach is effective, it is first helpful to understand how we learn.

1. Our brain processes information using two main channels: visual and verbal. Our *working memory* holds information that our minds process as we learn. This "mental workbench" helps us with decisions, problem-solving, and making sense of words and pictures by building verbal and visual models of the information.

2. When the verbal and visual models of corresponding information are integrated in working memory, we form more comprehensive and lasting mental models.

3. When we link these integrated mental models to our prior knowledge, stored in our *long-term memory,* we build even stronger mental models. When an integrated (visual plus verbal) mental model is formed and stored in long-term memory, real learning begins.

The effort our brains put forth to make sense of instructional information is called *cognitive load*. There are two kinds of cognitive load: productive cognitive load, such as when we're engaged in learning or exert positive effort to create mental models; and unproductive cognitive load, which occurs when the brain is trying to make sense of needlessly complex content or when information is not presented well. The learning process can be impaired when the information to be processed exceeds the capacity of working memory. Well-designed visuals and text with effective pedagogical guidance can reduce the unproductive cognitive load in our working memory.

---

[1] Mayer, R.E. (Ed.) (2005). *The Cambridge Handbook of Multimedia Learning.* Cambridge University Press.
[2] Donovan, M.S., & Bransford, J. (Eds.) (2005). *How Students Learn: Science in the Classroom.* The National Academy Press. Available online at http://www.nap.edu/openbook.php?record_id=11102&page=1.

Wiley Visualizing is designed for engaging and effective learning

The visuals and text in *Visualizing Geology 3e* are specially integrated to present complex processes in clear steps and with clear representations, organize related pieces of information, and integrate related information. This approach, along with the use of interactive multimedia, minimizes unproductive cognitive load and helps students engage with the content. When students are engaged, they're reading and learning, which can lead to greater knowledge and academic success.

Research shows that well-designed visuals, integrated with comprehensive text, can improve the efficiency with which a learner processes information. In this regard, SEG Research, an independent research firm, conducted a national, multisite study evaluating the effectiveness of Wiley Visualizing. Its findings indicate that students using Wiley Visualizing products (both print and multimedia) were more engaged in the course, exhibited greater retention throughout the course, and made significantly greater gains in content area knowledge and skills, as compared to students in similar classes that did not use Wiley Visualizing.[3]

The use of *WileyPLUS* can also increase learning. According to a white paper titled "Leveraging Blended Learning for More Effective Course Management and Enhanced Student Outcomes" by Peggy Wyllie of Evince Market Research & Communications, studies show that effective use of online resources can increase learning outcomes. Pairing supportive online resources with face-to-face instruction can help students to learn and reflect on material, and deploying multimodal learning methods can help students to engage with the material and retain their acquired knowledge.

[3] SEG Research (2009). Improving Student-Learning with Graphically Enhanced Textbooks: A Study of the Effectiveness of the Wiley Visualizing Series.

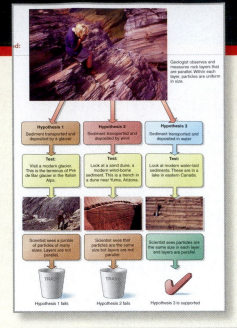

Using the scientific method (Figure 1.4) This matrix visually organizes abstract information to reduce cognitive load.

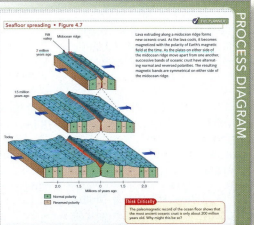

Seafloor spreading (Figure 4.7) Through a logical progression of graphics, this illustration directs learners' attention to the underlying concept. Textual and visual elements are physically integrated. This eliminates split attention—when too many sources of information divide attention.

Earth and lunar "soil"—Not the same! (Figure 7.11) Photos are paired so that students can compare and contrast them, thereby grasping the underlying concept. Adjacent caption eliminates split attention.

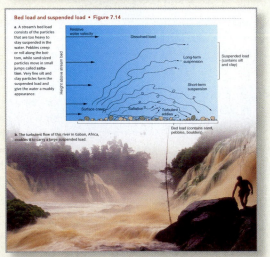

Bed load and suspended load (Figure 7.14) From abstraction to reality: Linking the graph to a photo illustrates how data on the graph relates to an actual river.

How Are the Wiley Visualizing Chapters Organized?

Student engagement is more than just exciting videos or interesting animations—engagement means keeping students motivated to keep going. It is easy to get bored or lose focus when presented with large amounts of information, and it is easy to lose motivation when the relevance of the information is unclear. The design of *WileyPLUS* is based on cognitive science, instructional design, and extensive research into user experience. It transforms learning into an interactive, engaging, and outcomes-oriented experience for students.

Each Wiley Visualizing chapter engages students from the start

Chapter opening text and visuals introduce the subject and connect the student with the material that follows.

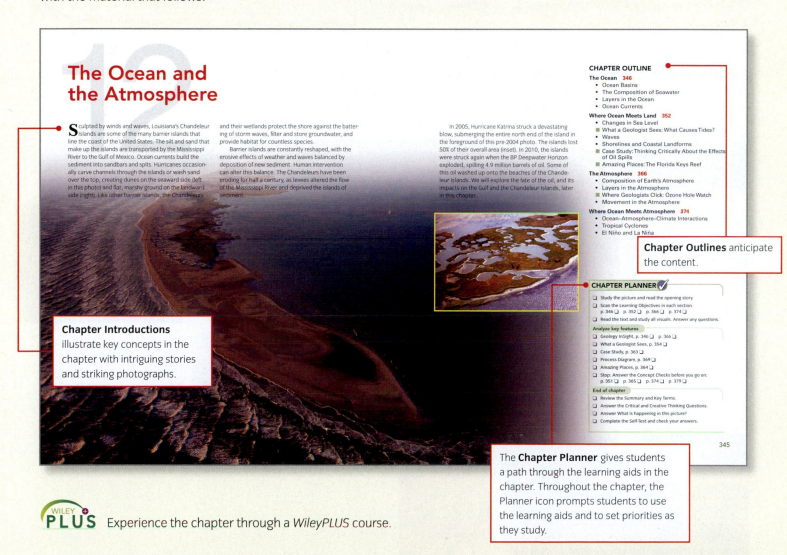

Chapter Introductions illustrate key concepts in the chapter with intriguing stories and striking photographs.

Chapter Outlines anticipate the content.

The **Chapter Planner** gives students a path through the learning aids in the chapter. Throughout the chapter, the Planner icon prompts students to use the learning aids and to set priorities as they study.

WILEY PLUS Experience the chapter through a *WileyPLUS* course.

Wiley Visualizing guides students through the chapter

The content of Wiley Visualizing gives students a variety of approaches—visuals, words, interactions, video, and assessments—that work together to provide a guided path through the content.

Learning Objectives at the start of each section indicate in behavioral terms the concepts that students are expected to master while reading the section.

Metamorphic Processes

LEARNING OBJECTIVES

1. **Identify** two types of physical changes that occur in rock during metamorphism.
2. **Describe** contact, burial, and regional metamorphism.
3. **Identify** the tectonic settings where the different types of metamorphism are likely to occur.
4. **Define** metasomatism and explain how it differs from metamorphism.

Process Diagrams provide in-depth coverage of processes correlated with clear, step-by-step narrative, enabling students to grasp important topics with less effort.

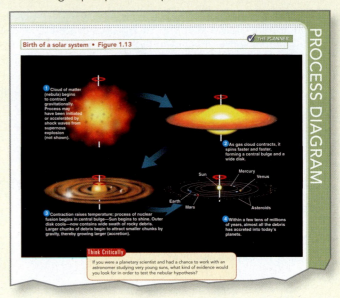

Birth of a solar system • Figure 1.13

PROCESS DIAGRAM

THE PLANNER

1. Cloud of matter (nebula) begins to contract gravitationally. Process may have been initiated or accelerated by shock waves from supernova explosion (not shown).

2. As gas cloud contracts, it spins faster and faster, forming a central bulge and a wide disk.

3. Contraction raises temperature; process of nuclear fusion begins in central bulge—Sun begins to shine. Outer disk cools—now contains wide swath of rocky debris. Larger chunks of debris begin to attract smaller chunks by gravity, thereby growing larger (accretion).

4. Within a few tens of millions of years, almost all the debris has accreted into today's planets.

Sun Mercury Venus Earth Mars Asteroids

Think Critically
If you were a planetary scientist and had a chance to work with an astronomer studying very young suns, what kind of evidence would you look for in order to test the nebular hypothesis?

Geology InSight features are multipart visual sections that focus on a key concept or topic in the chapter, exploring it in detail or in broader context using a combination of photos, diagrams, maps, and data.

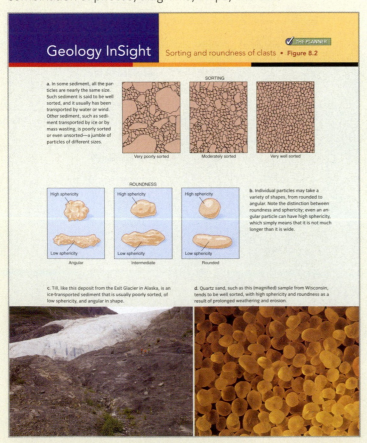

Geology InSight

Sorting and roundness of clasts • Figure 8.2

THE PLANNER

a. In some sediment, all the particles are nearly the same size. Such sediment is said to be well sorted, and it usually has been transported by water or wind. Other sediment, such as sediment transported by ice or by mass wasting, is poorly sorted or even unsorted—a jumble of particles of different sizes.

SORTING

Very poorly sorted   Moderately sorted   Very well sorted

ROUNDNESS

High sphericity   High sphericity   High sphericity

Low sphericity   Low sphericity   Low sphericity

Angular   Intermediate   Rounded

b. Individual particles may take a variety of shapes, from rounded to angular. Note the distinction between roundness and sphericity; even an angular particle can have high sphericity, which simply means that it is not much longer than it is wide.

c. Till, like this deposit from the Exit Glacier in Alaska, is an ice-transported sediment that is usually poorly sorted, of low sphericity, and angular in shape.

d. Quartz sand, such as this (magnified) sample from Wisconsin, tends to be well sorted, with high sphericity and roundness as a result of prolonged weathering and erosion.

Case Studies are in-depth examinations of fascinating and important issues in geology.

CASE STUDY

Global Locator

Providence Canyon

THE PLANNER

Bad and Good Soil Management

Providence Canyon in Georgia is a gorgeous example of deeply weathered soil, but it is also a dreadful example of poor soil management. In Figure a, in the canyon wall you can readily spot the dark brown A horizon, the bright red B horizon that is full of clay, and the paler E horizon. This is a good, productive soil. Some people have called Providence Canyon the "Little Grand Canyon," but in reality, the two canyons are very different. While the layered appearance of the Grand Canyon is due to the strata of bedrock, the appearance of Providence Canyon is caused by soil alone.

Another difference is age. Providence Canyon, believe it or not, is less than 200 years old. There was no canyon here when settlers from Europe began farming in the early 1800s. But the farmers plowed straight up and down the hills, and the furrows rapidly developed into gullies. By 1850, the gullies were 1 to 2 m deep. The farmers had to abandon their fields, but by then, erosion in the gullies was running amok. The canyon is now more

a.

A horizon   E horizon

B horizon

NATIONAL GEOGRAPHIC

b.

than 50 m deep. Unfortunately, there are many such locations in North America.

In the early 20th century, scientists involved with erosion studies pointed out that water flowing in plowed land needed to be controlled. To fight erosion, farmers should use contour plowing (Figure b). Instead of going in straight lines, the furrows follow the contour of the land, slowing runoff and inhibiting the formation of gullies. Crop rotation can also significantly help prevent erosion.

Despite measures such as contour plowing, the erosion of farmland soil is a massive worldwide problem. In the United States, the amount of agricultural soil eroded each year exceeds the amount of replenished soil by about 1 billion tons. For every kilogram of food we eat, the land loses 6 kilograms of soil. Although there is a small "sustainable farming" movement, we are very far from consuming only as much as we can put back. A lot more critical thinking and action are needed in order to control soil erosion.

WHAT A GEOLOGIST SEES

✓ THE PLANNER

The Red Sea and the Gulf of Aden

This spectacular photo, taken by astronauts on the *Gemini 11* mission, shows the southern end of the Red Sea (left top, center) and the Gulf of Aden (right), separating the southern tip of the Arabian Peninsula (right) from the northeastern corner of Africa. A geologist looking at this scene sees parallel coastlines and realizes that two lithospheric plates (the Arabian and African plates; see inset map) are splitting apart. A divergent plate margin runs down the center of the Gulf of Aden and joins another divergent margin that runs down the center of the Red Sea (both underwater). The Red Sea is widening at a rate of about 1 centimeter per year. If the process continues for several million years, the long narrow sea will become a wide gulf. The red dots on the map (inset) show the locations of recent earthquakes and volcanic activity.

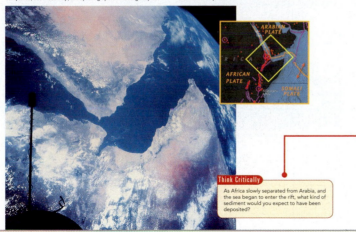

Think Critically

As Africa slowly separated from Arabia, and the sea began to enter the rift, what kind of sediment would you expect to have been deposited?

What a Geologist Sees highlights a concept or phenomenon that would stand out to a geologist. Photos and figures are used to improve students' understanding of the usefulness of a geology perspective and to develop their observational skills.

Think Critically questions let students analyze the material and develop insights into essential concepts.

Where Geologists Click showcases a Web site that professionals use and encourages students to try out its tools.

Where Geologists CLICK
The Geological Society of London

www.geolsoc.org.uk/gsl/education/resources/rockcycle/pid/3660

The Geological Society Web site is a fine source of information of interest to geologists and students of geology. An example of what you will find on the site is the article and video called "How do turbidity currents work?" To view turbidity currents happening in the ocean deeps would be far too dangerous, so geologists wishing to view the process would go to this site to see a turbidity current under laboratory conditions.

228

AMAZING PLACES

✓ THE PLANNER

Lechuguilla Cave

WILEY PLUS Video

Global Locator

NATIONAL GEOGRAPHIC

In 1986, cavers (also known as spelunkers) discovered the deepest-known cave in the United States, called Lechuguilla Cave, in Carlsbad Caverns National Park. Its entrance had been known for decades, but it had been considered a dead end until cavers dug through the floor to a huge network of passages on the other side.

Lechuguilla Cave has now been explored to a depth of 475 m, and it has almost 160 km of mapped passages. It is as spectacular as it is deep, but it is closed to the public to preserve its unusual formations.

a. A "bush" made of fragile aragonite pokes out of a stalagmite made of calcite. Aragonite and calcite are polymorphs of calcium carbonate (see Chapter 2).

b. "Soda straws" reach down from the ceiling in this small chamber. Water flows down through the center of each straw. If water starts flowing down the outside, it will build a stalactite.

c. The origin of this rare formation, called "pool fingers," is a mystery, perhaps related to bacterial activity. The "fingers" crystallized in a pool of water and were left behind when the water retreated.

d. Gypsum crystals provide a clue to this cave's unusual history. Unlike most other limestone caves, which are formed by carbonic acid in rainwater, Lechuguilla formed from the bottom up,

The **Amazing Places** sections take the student to a unique place that provides a vivid illustration of a concept in the chapter. Students could easily visit most of the Amazing Places someday and so continue their geologic education.

In concert with the visual approach of the book, **www.ConceptCaching.com** is an online collection of photographs that explores places, regions, people, and their activities. Photographs, GPS coordinates, and explanations of core geographic concepts are "cached" for viewing by professors and students alike. Professors can access the images or submit their own by visiting the Web site. Caches on the Web site are integrated in the *WileyPLUS* course as examples to help students understand the concepts.

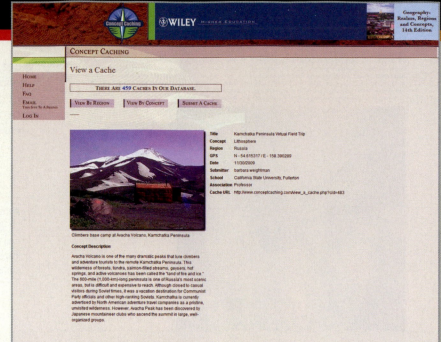

GeoDiscoveries Media Library is an interactive media source of animations, simulations, and interactivities allowing instructors to visually demonstrate key concepts in greater depth.

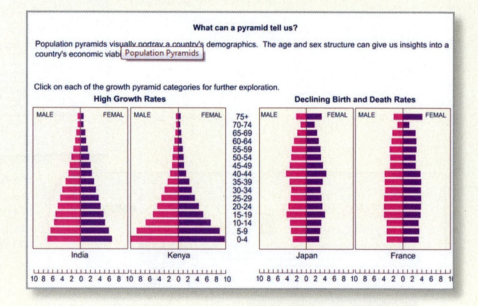

CONCEPT CHECK STOP

1. **What** physical and chemical changes happen in rock undergoing metamorphism?
2. **What** distinguishes burial metamorphism from regional metamorphism?
3. **How** does regional metamorphism in a subduction zone differ from regional metamorphism in a collision zone?
4. **What** process changes the chemical composition of a rock, rather than just its texture or mineral assemblage?

Coordinated with the section-opening **Learning Objectives**, at the end of each section **Concept Check** questions allow students to test their comprehension of the learning objectives.

Student understanding is assessed at different levels

Wiley Visualizing with *WileyPLUS* offers students lots of practice material for assessing their understanding of each study objective. Students know exactly what they are getting out of each study session through immediate feedback and coaching.

The **Summary** revisits each major section, with informative images taken from the chapter. These visuals reinforce important concepts.

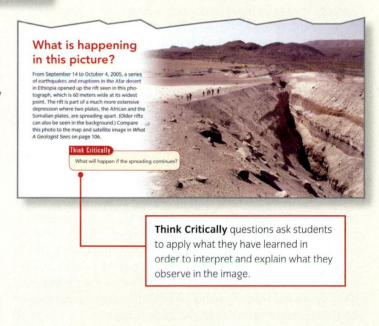

✓ THE PLANNER

Summary

1 Relative Age 66

- Geologists study the chronologic sequence of geologic events—that is, their **relative age**. Relative age is derived from **stratigraphy**, the study of rock layers and how those layers are formed.
- There are three basic principles of stratigraphy. (1) Strata, or sedimentary rock layers, are horizontal when they are deposited as water-laid sediment (Principle of Original Horizontality). (2) Strata accumulate in sequence, from the oldest on the bottom to the youngest on top (Principle of Stratigraphic Superposition). (3) Strata extend outward horizontally; they may thin or pinch out at their farthest edges, but they generally do not terminate abruptly unless cut by a younger rock unit (Principle of Lateral Continuity). In addition, the Principle of Cross-cutting Relationships states that a rock stratum is always older than any geologic feature, such as a fracture, that cuts across it.
- **Numerical age**, the exact number of years of a geologic feature, is more difficult to determine than relative age. The sequence of strata in any particular location is not necessarily continuous in time. An **unconformity** is a break or gap in the normal stratigraphic sequence. It usually marks a period during which sedimentation ceased and erosion removed some of the previously laid strata. The three common types of unconformities are nonconformities, angular unconformities, and disconformities as shown in the diagram.

Unconformities • Figure 3.3

Cycle continues, new rock forms.

STAGE 5

Ocean returns

2 The Geologic Column 73

- The **geologic column**, a stratigraphic time scale, is a composite diagram that shows the succession of all known strata, arranged in chronological order of formation, based on fossils and other age criteria as shown in the diagram.

The Geologic column in words and pictures • Figure 3.8

No rocks yet discovered. No formal name.

Precambrian time

88% of Earth's history

- **Correlation** of strata is the establishment of the time equivalence of strata in different places. Fossil assemblages, usually consisting of hard shells, bones, and wood, have

- The geologic column is divided into several different units of time, called eons, eras, periods, and epochs. The majority of Earth's history is divided into three eons,

Critical and Creative Thinking Questions

1. Do you think there may be life similar to ours on a planet outside our solar system? What would the atmosphere of that planet be like? Must it have a hydrosphere? Why or why not?

2. Why is the systems approach so useful in studying both natural and artificial processes? Can you think of examples of artificial (i.e., human-built) systems other than those given in the text? Are they open systems or closed systems? (Think about the materials and energy in them.)

3. How do you think the principle of uniformitarianism accounts for occasional catastrophic events such as meteorite impacts, huge volcanic eruptions, or great earthquakes?

4. In this chapter we have suggested that Earth is a close approximation of a natural closed system, and we have hinted at some of the ways that living in a closed system affects each of us. Can you think of some other ways?

5. In what ways do geologic processes affect your daily life?

Critical and Creative Thinking Questions challenge students to think more broadly about chapter concepts. The level of these questions ranges from simple to advanced; they encourage students to think critically and develop an analytical understanding of the ideas discussed in the chapter.

What is happening in this picture? presents a photograph that is relevant to a chapter topic and illustrates a situation students are not likely to have encountered previously.

What is happening in this picture?

From September 14 to October 4, 2005, a series of earthquakes and eruptions in the Afar desert in Ethiopia opened up the rift seen in this photograph, which is 60 meters wide at its widest point. The rift is part of a much more extensive depression where two plates, the African and the Somalian plates, are spreading apart. (Older rifts can also be seen in the background.) Compare this photo to the map and satellite image in *What A Geologist Sees* on page 106.

Think Critically
What will happen if the spreading continues?

Think Critically questions ask students to apply what they have learned in order to interpret and explain what they observe in the image.

Self-Test
(Check your answers in Appendix D.)

1. A(n) _____ is an atom that has gained or lost one or more electrons and has a net electric charge.
   a. molecule   d. ion
   b. isotope   e. compound
   c. element

2. On this diagram, locate and label the following parts of the atom:
   proton
   nucleus
   electron
   first energy-level electron shell
   neutron
   second energy-level electron shell

   ● positively charged
   ● uncharged
   ○ negatively charged

6. Volcanic glass is not considered a mineral because _____.
   a. is amorphous (i.e., it lacks a crystal structure)
   b. is not naturally occurring
   c. does not have enough silicon or oxygen in its composition
   d. All of the above answers are correct.

7. The photograph shows natural samples of the corundum. What best explains the striking difference in color between the red (ruby) and the blue samples?
   a. polymorphism
   b. small variations in composition
   c. differences in crystalline structure
   d. polymerization

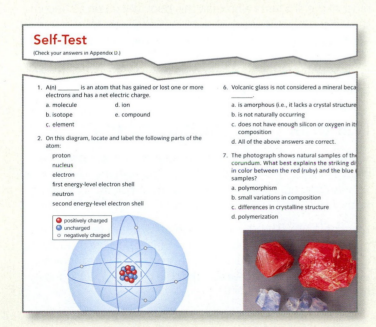

Visual end-of-chapter **Self-Tests** pose review questions that ask students to demonstrate their understanding of key concepts.

Why *Visualizing Geology 3e?*

The goal of *Visualizing Geology 3e* is to introduce students to geology and Earth system science through the distinctive mode of visual learning that is the hallmark of the Wiley Visualizing series. Students will learn that the geologic features we see and experience result from the interaction of three grand cycles, which extend from Earth's core to the fringes of our atmosphere: the tectonic cycle, the rock cycle, and the water cycle. We place special emphasis on plate tectonics because it is an organizing principle and a framework that unites many disparate observations into a coherent pattern of geologic activity on our planet. In this third edition we give added attention to the biosphere and its influence on the chemistry of the atmosphere and hydrosphere. We also add a new chapter on "Climate: Past, Present, and Future."

Case studies throughout the book and a final chapter devoted to Earth's resources bring the science of geology into focus in students' everyday lives. We fit current events, such as the March 2011 Tohoku earthquake and tsunami in Japan, the Haiti earthquake of 2010, and global warming, into a larger picture that explains how Earth works and why such events happen. Students will also learn about how human actions affect Earth systems and vice versa. The unique format of Wiley Visualizing allows us to reinforce the textual content with arresting images that are, in many cases, the next best thing to being there. Geology invites us to travel outside our familiar environment to distant parts of the world. Who better to take us on this journey than the photographers of National Geographic, who have been traveling the world and recording it visually for more than a century? The authors of *Visualizing Geology 3e* were given access to National Geographic's vast photo archive. With such photos, and with features such as Amazing Places in every chapter, we seek to instill what words sometimes cannot: a sense of wonder about the planet we call home.

Organization

Visualizing Geology 3e is organized as follows. In Chapters 1 through 4 we outline Earth system science as an approach to the study of our planet and our environment. We describe the various kinds of rocks and minerals, explain the ways in which geologists learn about Earth's changes over time, and present the unifying theory of plate tectonics. In Chapters 5 and 6 we discuss the hazards of earthquakes and volcanoes and explain how they relate to the tectonic cycle. Chapters 7 through 10 describe the major processes of the rock cycle—weathering, erosion, sedimentation, and metamorphism. In addition, students will learn the basics of geologic maps and structural geology. In Chapters 11 through 13 we turn our attention to the water cycle and explain the ubiquitous effects of water on Earth's surface, underground, and in the atmosphere. We devote a full chapter to deserts and glaciers, the two extreme environments that in recent years have become bellwethers of climate change, and in Chapter 14 we address the record of climate changes. Finally, Chapters 15 and 16 reintegrate the various parts of the Earth system to draw conclusions about two topics of great interest to students and to society as a whole: the history of life on Earth and the future of the natural resources on which humanity depends.

Changes in the New Edition

This third edition has been updated and modified in response to user suggestions:

- Discussion of climate has been removed from Chapter 13 on deserts and glaciers, and Chapter 14 is now devoted to climates, past, present, and future.

- Chapter 15 has been revised and expanded to bring it up to date with developments in the field of paleontology.

- Critical thinking questions called "Think Critically" have been added to many of the illustrated features, including Process Diagrams and What a Geologist Sees. These questions are not necessarily ones to which students are expected to know the answers; instead, they are intended to stimulate discussion, further thinking, and perhaps further reading.

To encourage students to explore the information sources that professionals use, each chapter lists a Web site under the heading Where Geologists Click.

This book is intended as a textbook for an introductory college-level course in geology. Because our emphasis is on physical processes, it could be used as well for an introductory physical geology or physical geography course. We do not expect that most of the students who read this book will go on to become geologists, but we hope that all will come to have a better understanding of, and appreciation for, their home planet. For those students who want to take further courses in the field—and we hope there are many—we provide a solid, sufficient, and challenging background to do so with confidence.

National Geographic Society

Visualizing Geology 3e offers an array of remarkable photographs, maps, media, and film from the National Geographic Society collections. Students using the book benefit from the long history and rich, fascinating resources of National Geographic.

Fact-checking: The National Geographic Society has also performed an invaluable service in fact-checking *Visualizing Geology 3e*. They have verified every fact in the book with two outside sources to ensure that the text is accurate and up-to-date.

Also available

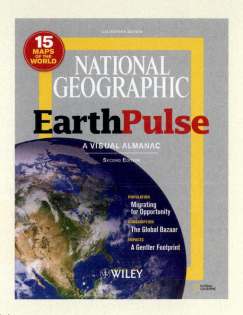

Earth Pulse 2e. Utilizing full-color imagery and National Geographic photographs, *EarthPulse* takes you on a journey of discovery covering topics such as *The Human Condition, Our Relationship with Nature,* and *Our Connected World*. Illustrated by specific examples, each section focuses on trends affecting our world today. Included are extensive full-color world and regional maps for reference. *EarthPulse* is available only in a package with *Visualizing Geology 3e.* Contact your Wiley representative for more information or visit www.wiley.com/college/earthpulse.

How Does Wiley Visualizing Support Instructors?

Wiley Visualizing Site

The Wiley Visualizing site hosts a wealth of information for instructors using Wiley Visualizing, including ways to maximize the visual approach in the classroom and a white paper titled "How Visuals Can Help Students Learn," by Matt Leavitt, instructional design consultant. Visit Wiley Visualizing at www.wiley.com/college/visualizing.

Wiley Custom Select

Wiley Custom Select gives you the freedom to build your course materials exactly the way you want them. Offer your students a cost-efficient alternative to traditional texts. In a simple three-step process create a solution containing the content you want, in the sequence you want, delivered how you want. Visit Wiley Custom Select at http://customselect.wiley.com.

Videos
(available in *WileyPLUS*)

Wherever the video icon appears in the text, **Video Explorations** provide visual context for key concepts, ideas, and terms through more than 30 National Geographic videos from their award-winning collection. Streaming videos are available to students in the context of *WileyPLUS* and accompanying assignments that can be graded online and added to the instructor gradebook.

Book Companion Site www.wiley.com/college/murck

All instructor resources (the Test Bank, Instructor's Manual, PowerPoint presentations, and all textbook illustrations and photos in jpeg format) are housed on the book companion site (www.wiley.com/college/murck). Student resources include self quizzes and flashcards.

PowerPoint Presentations
(available in *WileyPLUS* and on the book companion site)

A complete set of highly visual PowerPoint presentations—one per chapter—by Karen Savage, California State University, Northridge, is available online and in *WileyPLUS* to enhance classroom presentations. Tailored to the text's topical coverage and learning objectives, these presentations are designed to convey key text concepts, illustrated by embedded text art. We offer three different types of PowerPoint presentations for each chapter: PowerPoints with just the text art, PowerPoints with text art and presentation notes, and Media-Integrated PowerPoints with links to videos and animations.

Test Bank (available in *WileyPLUS* and on the book companion site)

The visuals from the textbook are also included in the Test Bank by Richard Josephs, Plymouth State University, who also authored the Pre-Lecture Clicker questions. The Test Bank has a diverse selection of test items, including multiple-choice and essay questions, with at least 20 percent of them incorporating visuals from the book. The test bank is available online in MS Word files as a Computerized Test Bank, and within *WileyPLUS*. The easy-to-use test-generation program fully supports graphics, print tests, student answer sheets, and answer keys. The software's advanced features allow you to produce an exam to your exact specifications.

Instructor's Manual (available in *WileyPLUS* and on the book companion site)

The Instructor's Manual includes creative ideas for in-class activities, discussion questions, and lecture transitions by Michael Farabee, Estrella Mountain Community College. It also includes answers to Critical and Creative Thinking questions and Concept Check questions.

Guidance is also provided on how to maximize the effectiveness of visuals in the classroom.

1. **Use visuals during class discussions or presentations.** Point out important information as the students look at the visuals, to help them integrate separate visual and verbal mental models.
2. **Use visuals for assignments and to assess learning.** For example, learners could be asked to identify samples of concepts portrayed in visuals.
3. **Use visuals to encourage group activities.** Students can study together, make sense of, discuss, hypothesize, or make decisions about the content. Students can work together to interpret and describe the diagram, or use the diagram to solve problems, conduct related research, or work through a case study activity.
4. **Use visuals during reviews.** Students can review key vocabulary, concepts, principles, processes, and relationships displayed visually. This recall helps link prior knowledge to new information in working memory, building integrated mental models.
5. **Use visuals for assignments and to assess learning.** For example, learners could be asked to identify samples of concepts portrayed in visuals.
6. **Use visuals to apply facts or concepts to realistic situations or examples.** For example, a familiar photograph, such as the Grand Canyon, can illustrate key information about the stratification of rock, linking this new concept to prior knowledge.

Image Gallery

All photographs, figures, maps, and other visuals from the text are online and in *WileyPLUS* and can be used as you wish in the classroom. These online electronic files allow you to easily incorporate images into your PowerPoint presentations as you choose, or to create your own handouts.

In addition to the text images, you also have access to **ConceptCaching**, an online database of photographs that explores what a region looks like and what it feels like to live in that region. Photographs and GPS coordinates are "cached" for viewing by core geographical concept and by region. Professors can access the images or submit their own by visiting www.conceptcaching.com.

Wiley Faculty Network

The Wiley Faculty Network (WFN) is a global community of faculty, connected by a passion for teaching and a drive to learn, share, and collaborate. Their mission is to promote the effective use of technology and enrich the teaching experience. Connect with the Wiley Faculty Network to collaborate with your colleagues, find a mentor, attend virtual and live events, and view a wealth of resources all designed to help you grow as an educator. Visit the Wiley Faculty Network at www.wherefacultyconnect.com.

How Has Wiley Visualizing Been Shaped by Contributors?

Wiley Visualizing and the *WileyPLUS* learning environment would not have come about without lots of people, each of whom played a part in sharing their research and contributing to this new approach.

Academic Research Consultants

Richard Mayer, Professor of Psychology, UC Santa Barbara. His *Cognitive Theory of Multimedia Learning* provided the basis on which we designed our program. He continues to provide guidance to our author and editorial teams on how to develop and implement strong, pedagogically effective visuals and use them in the classroom.

Jan L. Plass, Professor of Educational Communication and Technology in the Steinhardt School of Culture, Education, and Human Development at New York University. He co-directs the NYU Games for Learning Institute and is the founding director of the CREATE Consortium for Research and Evaluation of Advanced Technology in Education.

Matthew Leavitt, Instructional Design Consultant. He advises the Visualizing team on the effective design and use of visuals in instruction and has made virtual and live presentations to university faculty around the country regarding effective design and use of instructional visuals.

Independent Research Studies

SEG Research, an independent research and assessment firm, conducted a national, multisite effectiveness study of students enrolled in entry-level college Psychology and Geology courses. The study was designed to evaluate the effectiveness of Wiley Visualizing. You can view the full research paper at www.wiley.com/college/visualizing/huffman/efficacy.html.

Instructor and Student Contributions

Throughout the process of developing the concept of guided visual pedagogy for Wiley Visualizing, we benefited from the comments and constructive criticism provided by the instructors and colleagues listed below. We offer our sincere appreciation to these individuals for their helpful reviews and general feedback:

Visualizing Reviewers, Focus Group Participants, and Survey Respondents

James Abbott, Temple University
Melissa Acevedo, Westchester Community College
Shiva Achet, Roosevelt University
Denise Addorisio, Westchester Community College
Dave Alan, University of Phoenix
Sue Allen-Long, Indiana University Purdue
Robert Amey, Bridgewater State College
Nancy Bain, Ohio University
Corinne Balducci, Westchester Community College
Steve Barnhart, Middlesex County Community College
Stefan Becker, University of Washington—Oshkosh
Callan Bentley, NVCC Annandale

Valerie Bergeron, Delaware Technical & Community College
Andrew Berns, Milwaukee Area Technical College
Gregory Bishop, Orange Coast College
Rebecca Boger, Brooklyn College
Scott Brame, Clemson University
Joan Brandt, Central Piedmont Community College
Richard Brinn, Florida International University
Jim Bruno, University of Phoenix
William Chamberlin, Fullerton College
Oiyin Pauline Chow, Harrisburg Area Community College
Laurie Corey, Westchester Community College
Ozeas Costas, Ohio State University at Mansfield

Christopher Di Leonardo, Foothill College
Dani Ducharme, Waubonsee Community College
Mark Eastman, Diablo Valley College
Ben Elman, Baruch College
Staussa Ervin, Tarrant County College
Michael Farabee, Estrella Mountain Community College
Laurie Flaherty, Eastern Washington University
Susan Fuhr, Maryville College
Peter Galvin, Indiana University at Southeast
Andrew Getzfeld, New Jersey City University
Janet Gingold, Prince George's Community College
Donald Glassman, Des Moines Area Community College
Richard Goode, Porterville College
Peggy Green, Broward Community College
Stelian Grigoras, Northwood University
Paul Grogger, University of Colorado
Michael Hackett, Westchester Community College
Duane Hampton, Western Michigan University
Thomas Hancock, Eastern Washington University
Gregory Harris, Polk State College
John Haworth, Chattanooga State Technical Community College
James Hayes-Bohanan, Bridgewater State College
Peter Ingmire, San Francisco State University
Mark Jackson, Central Connecticut State University
Heather Jennings, Mercer County Community College
Eric Jerde, Morehead State University
Jennifer Johnson, Ferris State University
Richard Kandus, Mt. San Jacinto College District
Christopher Kent, Spokane Community College
Gerald Ketterling, North Dakota State University
Lynnel Kiely, Harold Washington College
Eryn Klosko, Westchester Community College
Cary T. Komoto, University of Wisconsin—Barron County
John Kupfer, University of South Carolina
Nicole Lafleur, University of Phoenix
Arthur Lee, Roane State Community College
Mary Lynam, Margrove College
Heidi Marcum, Baylor University
Beth Marshall, Washington State University
Dr. Theresa Martin, Eastern Washington University
Charles Mason, Morehead State University
Susan Massey, Art Institute of Philadelphia
Linda McCollum, Eastern Washington University
Mary L. Meiners, San Diego Miramar College
Shawn Mikulay, Elgin Community College
Cassandra Moe, Century Community College
Lynn Hanson Mooney, Art Institute of Charlotte
Kristy Moreno, University of Phoenix
Jacob Napieralski, University of Michigan—Dearborn

Gisele Nasar, Brevard Community College, Cocoa Campus
Daria Nikitina, West Chester University
Robin O'Quinn, Eastern Washington University
Richard Orndorff, Eastern Washington University
Sharen Orndorff, Eastern Washington University
Clair Ossian, Tarrant County College
Debra Parish, North Harris Montgomery Community College District
Linda Peters, Holyoke Community College
Robin Popp, Chattanooga State Technical Community College
Michael Priano, Westchester Community College
Alan "Paul" Price, University of Wisconsin—Washington County
Max Reams, Olivet Nazarene University
Mary Celeste Reese, Mississippi State University
Bruce Rengers, Metropolitan State College of Denver
Guillermo Rocha, Brooklyn College
Penny Sadler, College of William and Mary
Shamili Sandiford, College of DuPage
Thomas Sasek, University of Louisiana at Monroe
Donna Seagle, Chattanooga State Technical Community College
Diane Shakes, College of William and Mary
Jennie Silva, Louisiana State University
Michael Siola, Chicago State University
Morgan Slusher, Community College of Baltimore County
Julia Smith, Eastern Washington University
Darlene Smucny, University of Maryland University College
Jeff Snyder, Bowling Green State University
Alice Stefaniak, St. Xavier University
Alicia Steinhardt, Hartnell Community College
Kurt Stellwagen, Eastern Washington University
Charlotte Stromfors, University of Phoenix
Shane Strup, University of Phoenix
Donald Thieme, Georgia Perimeter College
Pamela Thinesen, Century Community College
Chad Thompson, SUNY Westchester Community College
Lensyl Urbano, University of Memphis
Gopal Venugopal, Roosevelt University
Daniel Vogt, University of Washington—College
  of Forest Resources
Dr. Laura J. Vosejpka, Northwood University
Brenda L. Walker, Kirkwood Community College
Stephen Wareham, Cal State Fullerton
Fred William Whitford, Montana State University
Katie Wiedman, University of St. Francis
Harry Williams, University of North Texas
Emily Williamson, Mississippi State University
Bridget Wyatt, San Francisco State University
Van Youngman, Art Institute of Philadelphia
Alexander Zemcov, Westchester Community College

Student Participants

Karl Beall, Eastern Washington University
Jessica Bryant, Eastern Washington University
Pia Chawla, Westchester Community College
Channel DeWitt, Eastern Washington University
Lucy DiAroscia, Westchester Community College

Heather Gregg, Eastern Washington University
Lindsey Harris, Eastern Washington University
Brenden Hayden, Eastern Washington University
Patty Hosner, Eastern Washington University
Tonya Karunartue, Eastern Washington University

Sydney Lindgren, Eastern Washington University
Michael Maczuga, Westchester Community College
Melissa Michael, Eastern Washington University
Estelle Rizzin, Westchester Community College

Andrew Rowley, Eastern Washington University
Eric Torres, Westchester Community College
Joshua Watson, Eastern Washington University

Acknowledgments

Our sincere appreciation to the following professionals who provided valuable feedback and suggestions
for the first, second, and third editions of *Visualizing Geology*:

Laura Sue Allen-Long
*Indiana University–Purdue University,
Indianapolis*
Sylvester Allred
Northern Arizona University
Laurie Anderson
Louisiana State University
Jake Armour
University of North Carolina, Charlotte
Jerry Bartholomew
University of Memphis
Jay D. Bass
University of Illinois at Urbana-Champaign
David Basterdo
San Bernardino Valley College
Barbara Bekken
Virginia Polytechnic and State University
Gregory Bishop
Orange Coast College
Ross A. Black
University of Kansas
Rebecca Boger
Brooklyn College
Theodore J. Bornhorst
Michigan Technological University
Michael Bradley
Eastern Michigan University
Ann Brandt-Williams
Glendale Community College
Natalie Bursztyn
Bakersfield College
Michael Canestaro
Sinclair Community College
Richard L. Carlson
Texas A&M University
Victor V. Cavatroc
*Northern Carolina State
University*
Stan Celestian
Glendale Community College
Chu-Yen Chen
University of Illinois at Urbana-Champaign
Nehru Cherukupalli
Brooklyn College

O. Pauline Chow
Harrisburg Area Community College
Diane Clemens-Knott
California State University, Fullerton
Mitchell Colgan
College of Charleston
Constantin Cranganu
Brooklyn CUNY
Dee Cooper
University of Texas
Cathy Connor
University of Alaska Southeast
Peter Copeland
University of Houston
Linda Crow
Montgomery College
Michael Dalman
Blinn College
John Dassinger
Chandler-Gilbert Community College
Smruti Desai
Cy-Fair College
Chris DiLeonardo
Foothill College
Charles Dick
Pasco-Hernando Community College
W. Crawford Elliott
Georgia State University
Robert Eves
Southern Utah University
Mike Farabee
Estrella Mountain Community College
Mark Feigenson
Rutgers University
Lynn Fielding
El Camino College
David Foster
University of Florida
Carol D. Frost
University of Wyoming
Tracy Furutani
North Seattle Community College
Yongli Gao
East Tennessee State University

William Garcia
*University of North Carolina at
  Charlotte*
Donald Glassman
Des Moines Area Community College
Richard Goode
Porterville College
Pamela Gore
Georgia Perimeter College
Mark Grobner
California State University, Stanislaus
Paul Grogger
University of Colorado, Colorado Springs
Erich Guy
Ohio University
Daniel Habib
Queens College
Michael Hackett
Westchester Community College
Duane Hampton
Western Michigan University
Gale Haigh
McNeese State University
Duane Hampton
Western Michigan University
Roger Hangarter
*Indiana University–Purdue University,
  Indianapolis*
Michael Harman
North Harris College
Frederika Harmsen
California State University, Fresno
Michael J. Harrison
Tennessee Technological University
Terry Harrison
Arapahoe Community College
Javier Hasbun
University of West Georgia
Michael J. Harrison
Tennessee Tech University
Stephen Hasiotis
University of Kansas
Adam Hayashi
Central Florida Community College

Dan Hembree
Ohio University
Mary Anne Holmes
University of Nebraska, Lincoln
William Hoyt
University of Northern Colorado
Laura Hubbard
University of California, Berkeley
James Hutcheon
Georgia Southern University
Scott Jeffrey
Community College of Baltimore County,
Catonsville Campus
Eric Jerde
Morehead State University
Verner Johnson
Mesa State College
Marie Johnson
U.S. Military Academy
Richard Josephs
University of North Dakota
Amanda Julson
Blinn College
Matthew Kapell
Wayne State University
Arnold Karpoff
University of Louisville
Alan Kehew
Western Michigan University
Dale Lambert
Tarrant County College,
Northeast
Arthur Lee
Roane State Community College
Harvey Liftin
Broward Community College
Walter Little
University at Albany, SUNY
Steven Lower
Ohio State University
Ntungwa Maasha
Coastal Georgia Community College
Ronald Martino
Marshall University

Anthony Martorana
Chandler-Gilbert Community College
Charles Mason
Morehead State University
Ryan Mathur
Juniata College
Brendan McNulty
California State University, Dominguez Hills
Joseph Meert
University of Florida
Mary Anne Meiners
San Diego Miramar College
Erik Melchiorre
California State University, San Bernardino
Ken Miller
Rutgers University
Scott Miller
Penn State University
Katherine Miller
Florida A&M University
Keith Montgomery
University of Wisconsin—Marathon
David Morris
Valdosta State University
Jane Murphy
Virginia College Online
Bethany Myers
Wichita State University
Jacob Napieralski
University of Michigan, Dearborn
Pamela Nelson
Glendale Community College
Terri Oltman
Westwood College
Keith Prufer
Wichita State University
Steve Ralser
University of Wisconsin—Madison
Kenneth Rasmussen
Northern Virginia Community College
Guillermo Rocha
Brooklyn College
Gary D. Rosenburg
Indiana University–Purdue University

Ian Saginor
Keystone College
Karen Savage
California State University, Northridge
Steve Schimmrich
SUNY Ulster County Community College
Laura Sherrod
Kutztown University
Bruce Simonson
Oberlin College
Jay Simms
University of Arkansas at Little Rock
Jeff Snyder
Bowling Green State University
Ann Somers
University of North Carolina,
Greensboro
Debra Stakes
Cuesta College
Alycia Stigall
Ohio University
Donald Thieme
Georgia Perimeter College
Carol Thompson
Tarleton State University
Kip Thompson
Ozarks Technical Community
College
Heyo Van Iten
Hanover College
Judy Voelker
Northern Kentucky University
Arthur Washington
Florida A&M University
Karen When
Buffalo State University
Harry Williams
University of North Texas
Stephen Williams
Glendale Community College
Feranda Williamson
Capella University
Thomas C. Wynn
Lock Haven University

About the Authors

Barbara Murck is a geologist and senior lecturer in environmental science at the University of Toronto Mississauga. She completed her undergraduate degree in Geological and Geophysical Sciences at Princeton University and then spent two years in the Peace Corps in West Africa, before returning to Ph.D. studies at the University of Toronto. Her subsequent teaching and research has involved an interesting combination of geology, natural hazards, environmental science, and environmental issues in the developing world, primarily in Africa and Asia. She also carries out practical research on pedagogy. She is an award-winning lecturer who has co-authored a number of books, including several with Brian Skinner.

Brian Skinner was born and raised in Australia, studied at the University of Adelaide in South Australia, worked in the mining industry in Tasmania, and in 1951 entered the Graduate School of Arts and Sciences, Harvard University, from which he obtained his Ph.D. in 1954. Following a period as a research scientist in the United States Geological Survey in Washington D.C., he joined the faculty at Yale in 1966, where he continues his teaching and research as the Eugene Higgins Professor of Geology and Geophysics. Brian Skinner has been president of the Geochemical Society, the Geological Society of America, and the Society of Economic Geologists. He holds an honorary Doctor of Science from Toronto University and an honorary Doctor of Engineering from the Colorado School of Mines.

Thanks for Participating in a Special Project

Visualizing Geology was one of the "pioneers" in the Visualizing series; it was one of the first to go into a second edition, and now it is leading the pack into a third edition. Working on the book this time around was just as exciting as it was the previous times. The great thing about doing a third edition of a book that you are proud of is the chance to make improvements based on the thoughtful suggestions of users and to make the book even better than its predecessors.

With *Visualizing Geology 3e* we are aiming for a book that incorporates many new design features, that is even more visual, more meaningful for students, more useful for instructors, and more faithful to the science we love.

As with previous editions, it was a privilege to work with the staff at John Wiley & Sons; they are professional, efficient, and fun to work with. At every step along the way, and especially those steps where we stumbled, these thoughtful, sensitive, and very professional experts steered us and cheered us on, and led us in the right direction. Murck and Skinner, full-time geologists and long-time collaborators, revised this third edition without the aid of Dana Mackenzie, a science writer with a mathematics background, who collaborated on the previous editions. It's hard for us to imagine a mix working

better together, and that remains as true for the third edition as it was for the first and second.

Once the revision started we were in frequent contact with Charity Robey, who set up conference calls, was in ever-ready contact by email, and, with her voice-of-reason, kept us working smoothly together. Ellinor Wagner, long-suffering photo researcher, was a crucial help with the many photo replacements for this new edition. Nancy Perry waited in the wings to make decisions when issues about the new design arose, and to provide help when needed. If ever editors should be acknowledged as essential parts of a team, they are Charity and Nancy. Behind the scenes were members of the skilled team that make a project work; our Executive Editor Ryan Flahive, Jay O'Callaghan, the Executive Publisher, Barbara Heaney, Director of Product and Market Development, and William Murray, Senior Production Editor.

We say special thanks to all of those professional colleagues who read, thought about, and offered advice on our proposal, plans, writing, and production. They are thanked and named elsewhere, but we authors want to acknowledge the essential part they played in the production of the book. Their input has made the book stronger.

Contents in Brief

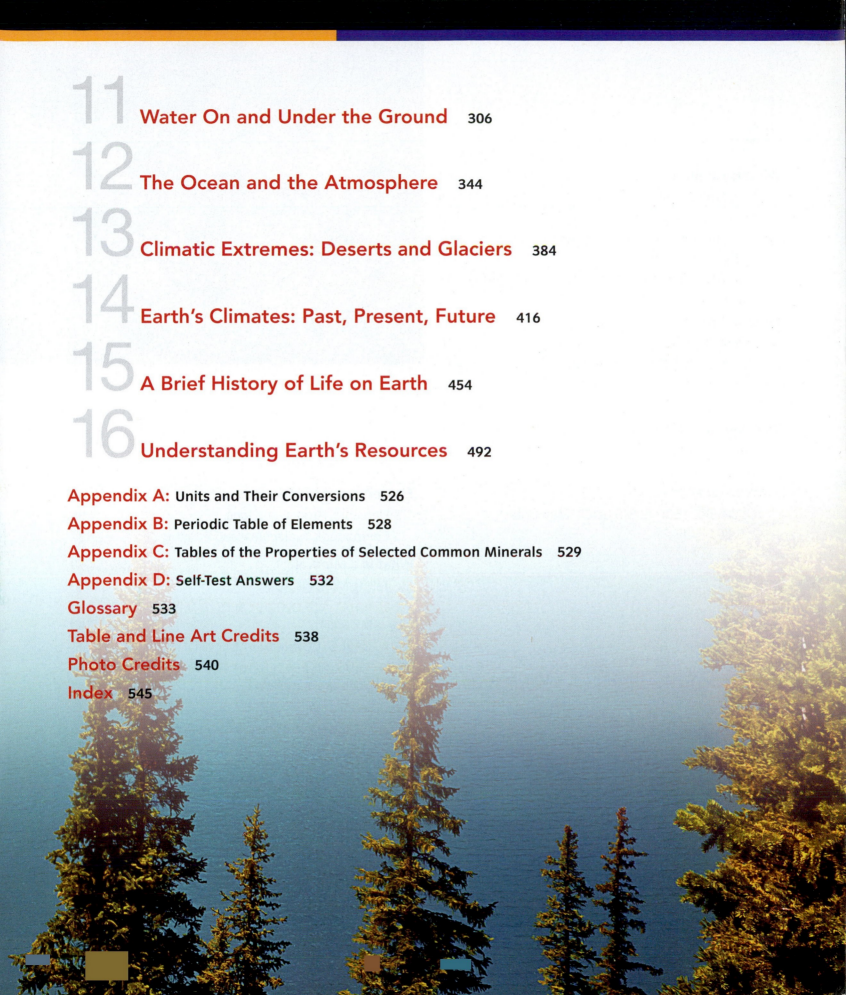

Table of Contents

3 How Old Is Old? The Rock Record and Geologic Time

4 Plate Tectonics

7 Weathering and Erosion

8 From Sediment to Sedimentary Rock

9 Folds, Faults, and Geological Maps

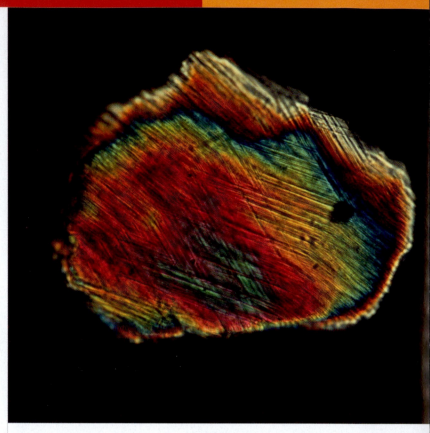

10 Metamorphism: New Rock from Old

13 Climatic Extremes: Deserts and Glaciers

14 Earth's Climates: Past, Present, Future

15 A Brief History of Life on Earth

16 Understanding Earth's Resources

Amazing Places

In every chapter in this book, we take you to an "Amazing Place" that is both beautiful and of geologic interest.

Here is our itinerary:

Chapter 1: Meteorite Impact Craters, to see evidence of the way Earth was assembled
Chapter 2: Naica Mine in Chihuahua, Mexico, for a look at the world's largest crystals
Chapter 3: Famous Unconformities, to see evidence of ancient uplift and erosion
Chapter 4: The Hawaiian Islands, to see plate tectonics and volcanoes in action
Chapter 5: Point Reyes, California, to explore the most famous fault in America, the San Andreas fault
Chapter 6: Mt. St. Helens, to witness the most famous eruption in U.S. history
Chapter 7: Mt. Monadnock—and Monadnocks, for a look at the power of erosion
Chapter 8: The Navajo Sandstone, for its beautiful sedimentary rock formations

Chapter 9: The Canadian Rockies, for spectacular examples of folding and thrusting
Chapter 10: The source of Olmec jade, for evidence of high-pressure metamorphism
Chapter 11: Lechuguilla Cave, in New Mexico, for incredible shapes made by groundwater
Chapter 12: The Florida Keys Reef, to visit a geologic formation that is also alive
Chapter 13: Glacier Bay National Park and Wilderness, Alaska, to see evidence of glacier melting in recent historic times
Chapter 14: Fossil Forests of the High Arctic, to see remarkable preservation of forests that grew in the now-frozen north
Chapter 15: The Burgess Shale, for its fossil record of the first animals on Earth
Chapter 16: Saugus Iron Works, Massachusetts, to see the first iron-smelting operation in North America

The most amazing place of all, however, is Earth itself—the only world in the universe where we *know* that life exists.

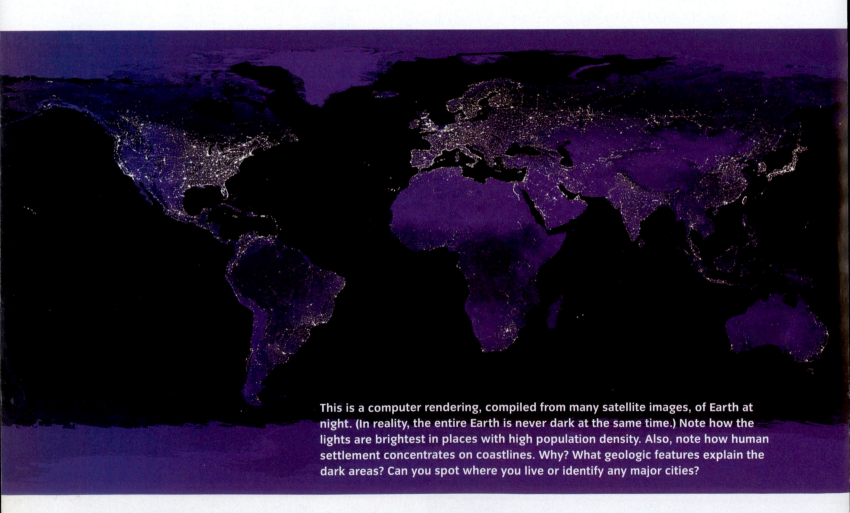

This is a computer rendering, compiled from many satellite images, of Earth at night. (In reality, the entire Earth is never dark at the same time.) Note how the lights are brightest in places with high population density. Also, note how human settlement concentrates on coastlines. Why? What geologic features explain the dark areas? Can you spot where you live or identify any major cities?

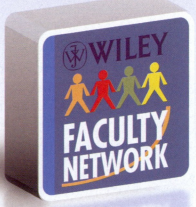

VISUALIZING

GEOLOGY

THIRD EDITION

1. Earth as a Planet

P hotographs of Earth from space have profoundly influenced our thoughts about Earth.
Getting a whole-Earth photograph is difficult—
planes don't do the job; you must be out in space,
and the Sun has to be directly behind the camera
so that Earth is shadow free. A striking whole-Earth
photograph was taken by the *Apollo 17* astronauts
on their way to the Moon in 1972. When they
were 45,000 kilometers out, they looked back,
and there was the fully illuminated Earth,
like a blue marble suspended in space. The
name stuck. NASA now has a stunning
group of whole-Earth images called the
Blue Marble series.

This particular version of a blue
marble image was obtained in 1997.
It is one of the most detailed images
ever made of Earth. There is a huge
storm off the west coast of North
America—it is Hurricane Linda—
and the Moon is rising over Earth
in the upper left. Hurricane Linda
reminds us that the different parts
of Earth—rocks, water, atmosphere, living things—all interact.
The Moon reminds us that Earth is
a member of the solar system. To
know Earth, we must understand its
parts and also the system to which it
belongs.

CHAPTER OUTLINE

CHAPTER PLANNER ✔

- ❑ Study the picture and read the opening story.
- ❑ Scan the Learning Objectives in each section:
  p. 4 ❑   p. 16 ❑   p. 24 ❑
- ❑ Read the text and study all visuals.
  Answer any questions.

Analyze key features

- ❑ What a Geologist Sees, p. 10 ❑
- ❑ Amazing Places, p. 11 ❑
- ❑ Case Study, p. 17 ❑
- ❑ Process Diagram, p. 19 ❑   p. 20 ❑
- ❑ Geology InSight, p. 26 ❑
- ❑ Stop: Answer the Concept Checks before you go on:
  p. 15 ❑   p. 23 ❑   p. 29 ❑

End of chapter

- ❑ Review the Summary and Key Terms.
- ❑ Answer the Critical and Creative Thinking Questions.
- ❑ Answer What is happening in this picture?
- ❑ Complete the Self-Test and check your answers.

What Is Geology?

LEARNING OBJECTIVES

1. **Describe** several of the many branches of geology.
2. **Explain** how scientists use the scientific method.
3. **Explain** what is meant by a systems approach to geology.
4. **Identify** four major subsystems of the Earth system.
5. **Explain** how these subsystems interact, using the concept of cycles.

The word **geology** comes from two Greek roots: *geo-*, meaning "Earth," and *logis*, meaning "study" or "science." The science called geology encompasses the study of our planet: how it formed; the nature of its interior; the materials of which it is composed; its water, glaciers, mountains, and deserts; its earthquakes and volcanoes; its resources; and its history—physical, chemical, and biological. Scientists who make a career of geology are geologists. Geology, like all other sciences, is based on factual observations, testable hypotheses, reproducible procedures, and open communication of information.

> **geology** The scientific study of Earth.

The study of geology is traditionally divided into two broad subject areas: physical geology and historical geology. **Physical geology** is concerned with understanding the *processes* that operate at or beneath the surface of Earth and the *materials* on which those processes operate. Some examples of geologic processes are mountain building, volcanic eruptions, earthquakes, river flooding, and the formation of ore deposits. Some examples of materials are minerals, soils, rocks, air, and water.

Historical geology, on the other hand, is concerned with the sequence of *events* that have occurred in the past. These events can be inferred from the evidence left in Earth's rocks (**Figure 1.1**). Through the findings of historical geology, scientists seek to resolve questions such as When did the oceans form? Why did the dinosaurs die out? When did the Rocky Mountains rise? and When and where did the first trees appear? Historical geology gives us a perspective on the past. It also establishes a context for thinking about present-day changes in our natural environment. This book is concerned mainly with physical geology, but it includes many lessons we can learn from historical geology.

Within the traditional domains of physical and historical geology are many specialized disciplines, some of which are illustrated in **Figure 1.2**. Economic geology, for example, is concerned with the formation and occurrence of, and the search for, valuable mineral deposits. Environmental geology focuses on how materials and processes in the natural geologic environment affect—and are affected by—human activities. Volcanologists study volcanoes and eruptions, past and present; seismologists study earthquakes; mineralogists undertake the microscopic study of minerals

Digging the past • Figure 1.1

The light layer of clay seen in this rock from Clear Creek, Colorado, provided geologists with crucial evidence concerning one of the great mysteries of the past: Why did the dinosaurs die out 65 million years ago? Layers of clay like this are found at many places in rocks that are 65 million years old. Many geologists hypothesize that the clay represents the debris from an enormous meteorite impact. Above the clay layer is a layer that contains fragments of "shocked" or fractured quartz, also indicative of a very violent event, which may have contributed to, or even caused the extinction of the dinosaurs.

Shocked quartz fragments

Clay

Impact debris

Faces and places of geology • Figure 1.2

Geologists are privileged to work in some of the most exotic places on Earth—and beyond.

a. Harrison (Jack) Schmitt, a planetary geologist, is the only scientist (so far) to walk on the Moon. Schmitt flew on the *Apollo 17* mission in 1972. Here, he is collecting a lunar sample to take back to Earth.

b. Volcanologists get uncomfortably close to the 2002 eruption of Mt. Etna in Sicily, Italy, to record the sounds of the eruption.

c. A paleontologist dives into the waters off The Bahama Islands to study stromatolites, a living algal formation reminiscent of Earth's oldest fossils.

d. Climatologists collect an ice core from a floe off the coast of Antarctica. Ice samples can tell us about the changes in Earth's climate over hundreds or thousands of years.

e. A speleologist (expert on caves) collects a water sample from a sinkhole in Oman. Groundwater is the force that sculpts most of the world's caves, and it can be found even in the most arid environments.

f. A seismologist (expert on earthquakes) inspects one of the fissures that opened up in the Santa Cruz Mountains in California during the Loma Prieta earthquake of 1989.

and crystals; paleontologists study fossils and the history of life on Earth; structural geologists study how rocks break and bend. These specialties are needed because geology encompasses a broad range of topics.

To a certain extent we are all geologists, even though only a few of us make a career out of geology. Everyone living on this planet relies on geologic resources: water, soil, building stones, metals, fossil fuels, gemstones, plastics (from petroleum), ceramics (from clay minerals), glass (from silica sand), salt (a mineral called halite), and many others. Geologic processes affect us every day. We also influence the geologic environment through our daily activities, whether we are drinking water that came from an aquifer or planting trees to control soil erosion. This book will help you to become better informed and more mindful of these interactions. As a result, you will be better equipped to make decisions about Earth materials and processes that affect your life.

Using the Scientific Method

scientific method
The way a scientist approaches a problem; steps include observing, formulating a hypothesis, testing, and evaluating results.

Like all other scientists, geologists use a logical research strategy, called the **scientific method**, that has developed through trial and error over many years. The scientific method is based on observations and the collection of evidence that can be seen and tested by anyone who cares to do so. Although it varies in details, the scientific method includes the basic steps outlined in **Figure 1.3**.

Let's consider how the scientific method might be applied in a real geologic situation.

Step 1. *Observe and gather data.* Scientists start with a question and acquire trustworthy evidence about it, especially measurements. In **Figure 1.4**, a geologist asks the question "How did this group of rocks form?" She observes and measures the sequence of layered rocks in question. She sees that the layers are *parallel*—An important clue. Further, each layer consists of innumerable *small* grains, and the size of the grains *varies* from layer to layer but is approximately the same within each layer.

Step 2. *Formulate a hypothesis.* Scientists explain their observations by developing a **hypothesis**. The geologist in our example develops three hypotheses. She hypothesizes that the rocks were formed from material that was transported and deposited where she has found it; but how was it transported? Hypothesis 1 is that a *glacier* was the transporting agent. Hypothesis 2 is that *wind* did the transporting. Hypothesis 3 is that *water* did the transporting.

hypothesis
A plausible but yet-to-be-proved explanation for how something happens.

Using the scientific method: Part 1 • Figure 1.3

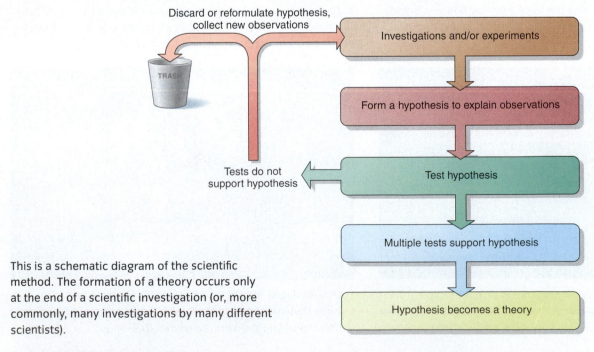

Discard or reformulate hypothesis, collect new observations

TRASH

Investigations and/or experiments

Form a hypothesis to explain observations

Test hypothesis

Tests do not support hypothesis

Multiple tests support hypothesis

Hypothesis becomes a theory

This is a schematic diagram of the scientific method. The formation of a theory occurs only at the end of a scientific investigation (or, more commonly, many investigations by many different scientists).

Using the scientific method: Part 2 • Figure 1.4

Top, the geologist observes and measures rock layers that are parallel. She also sees that the particles within each layer have a uniform size. Middle, the geologist formulates three hypotheses about how the layers might have formed. She tests the hypotheses by visiting three geologic sites. Bottom, she concludes that the third hypothesis best explains the observations.

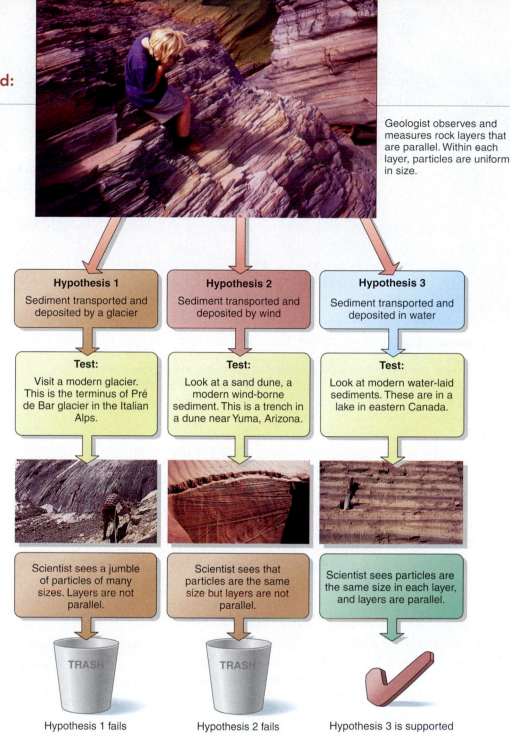

Geologist observes and measures rock layers that are parallel. Within each layer, particles are uniform in size.

Hypothesis 1
Sediment transported and deposited by a glacier

Hypothesis 2
Sediment transported and deposited by wind

Hypothesis 3
Sediment transported and deposited in water

Test:
Visit a modern glacier. This is the terminus of Pré de Bar glacier in the Italian Alps.

Test:
Look at a sand dune, a modern wind-borne sediment. This is a trench in a dune near Yuma, Arizona.

Test:
Look at modern water-laid sediments. These are in a lake in eastern Canada.

Scientist sees a jumble of particles of many sizes. Layers are not parallel.

Scientist sees that particles are the same size but layers are not parallel.

Scientist sees particles are the same size in each layer, and layers are parallel.

TRASH

TRASH

Hypothesis 1 fails

Hypothesis 2 fails

Hypothesis 3 is supported

A scientist's hypotheses are often influenced by prior experience or knowledge. Chapters 7 and 8 explain why the three hypotheses in our current example are reasonable explanations. Note that the scientist does not have to select one hypothesis at this point. In fact, choosing a "leading candidate" too early may prejudice the scientist and cause her to overlook some relevant clues. T.C. Chamberlin, a 19th-century geologist, argued that scientists should consider *all* reasonable explanations—an approach he called the "method of multiple working hypotheses."

Step 3. *Test the hypotheses.* Scientists use a hypothesis—or multiple hypotheses—to make predictions and to develop tests. The tests may involve controlled experiments in a laboratory, further observations and measurements, and possibly the development of a mathematical model. Geologists in particular like to test their hypotheses against real observations. Here's how our geologist tests the hypotheses:

- Our scientist travels to a modern glacier and studies the jumble of debris it deposits. She notes

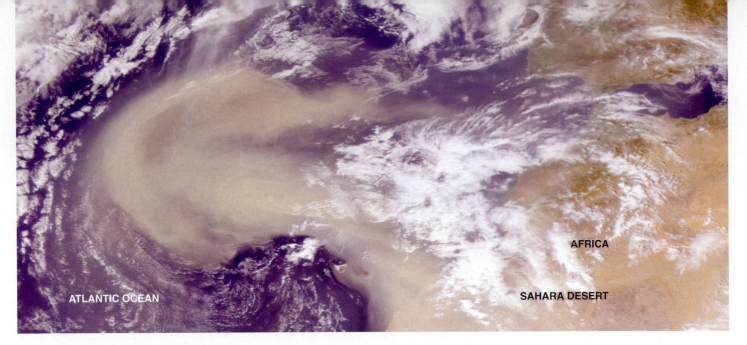

AFRICA

SAHARA DESERT

ATLANTIC OCEAN

Earth from orbit • Figure 1.5

Satellite images can reveal interactions within Earth systems. In this photo, dust storms from the Sahara Desert blow far out into the Atlantic. In fact, geologists have found African dust all the way across the Atlantic Ocean, and some think that it might contribute to the death of coral reefs off the coast of Florida.

that the grains are different sizes, all mixed up, and not in neatly defined layers. So, Hypothesis 1 fails.

- Then she goes to a desert region where she sees wind-transported material deposited in dunes. She observes that particle sizes are approximately the same, but they aren't in parallel layers; the layers are at odd angles. So Hypothesis 2 fails.

- Finally, our scientist visits a lake and observes materials transported by a river and deposited in lake water. Now she sees layers that are parallel, and the particles in each layer are approximately the same size. Hypothesis 3 has potential, but more testing is needed. Our scientist notes that plants are growing in the lake. She goes a step further and hypothesizes that if the material that formed the rocks really was deposited in a lake, the remains of aquatic plants might still be present. If, on further observation, she finds fossilized freshwater plant remains in the layered rocks, she will be even more confident that she is on the right track.

This is how the scientific method tests and retests hypotheses.

Step 4. *Formulate a theory.* Once a hypothesis has withstood numerous tests, scientists become more confident in its validity. The hypothesis is elevated to a **theory**. It is not the final word, however, and a theory is always open to further testing. (Note: In everyday speech, people often misuse the term "theory" to mean "hypothesis" by saying dismissively, "That's just a theory." What they really mean is, "That's just a hypothesis." In science, by the time a statement attains the stature of a theory, it is very substantial and must be taken seriously.)

> **theory** A hypothesis that has been tested and is strongly supported by experimentation, observation, and scientific evidence.

Step 5. *Formulate a law or principle.* Ultimately, a theory or group of theories may be formulated into a **law** or **principle**. Laws and principles are statements that some natural phenomenon invariably is observed to happen in the same way, and no deviations ever have been observed. For example, in geology the *Law of Original Horizontality* states that sediment deposited in water is always in horizontal layers (or nearly always so, because a lake or sea-floor might have slight irregularities) and the layers are parallel to Earth's surface (or nearly always so). No exceptions have ever been observed.

Earth System Science

Traditionally, scientists have studied Earth by focusing on separate units—the atmosphere, the oceans, or a single mountain range—in isolation from the other units. However, the first photographs of Earth taken from space (like the

The system concept • Figure 1.6

This figure shows a variety of systems. The entire diagram—mountains, river, and lake—illustrates one kind of system: a coastal watershed. The individual pieces enclosed by boxes, such as the river, are also systems. Even small volumes of water or lake sediment (foreground boxes) are systems in their own right.

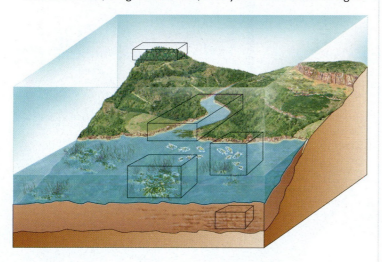

Two kinds of systems • Figure 1.7

A closed system is one that does not allow matter but does allow energy, to pass through its boundaries. An open system allows both matter and energy to pass through its boundaries. Most systems in geology are open.

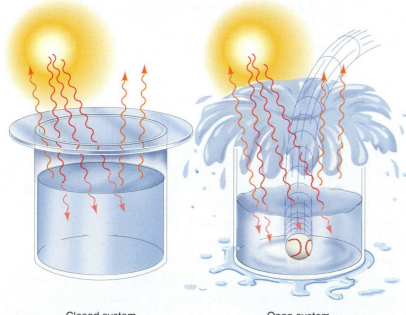

Closed system   Open system

chapter-opening photograph) caused a dramatic rethinking of this traditional view. For the first time, it was possible to see the whole planet in one sweeping view. We could see everything at a glance—clouds, oceans, polar ice caps, and continents—all at the same time, and in their proper scale. The astronauts, like the rest of us, marveled at Earth's "overwhelming beauty . . . the stark contrast between bright colorful home and stark black infinity" (Rusty Schweikart, *Apollo 9*). Yet from space it was also clear how small Earth is—just a dust speck compared to the vastness of the solar system and the universe. On such a small planet, it no longer made sense to study all the pieces separately. There was only one geology that mattered, not the geology of America or the Atlantic Ocean but the geology of the whole Earth.

Instruments carried by satellites in space have also given us new ways to study the relationships of the parts on a global scale, as we never could before (**Figure 1.5**). This new, more all-inclusive view of geology is called **Earth system science**.

The system concept A systems approach is a helpful way to break down a large, complex problem into smaller pieces that are easier to study without losing sight of the connections between those pieces. We start by defining a **system** as any portion of the universe that can be isolated for the purpose of observing and measuring change. The definition allows the observer to choose the system to be whatever he or she wants it to be. That is why a system is only a concept; you choose its limits for the convenience of your study. A system may be large or small, simple or complex (**Figure 1.6**). It could be the contents of the beaker in a laboratory experiment or the contents of an ocean. A leaf is a system, but it is also part of a larger system (a tree), which is part of a still larger system (a forest).

The fact that we distinguish a system from the rest of the universe for specific study does not mean that we ignore its surroundings. In fact, the nature of a system's boundaries is one of its most important defining characteristics. **Figure 1.7** illustrates two basic kinds of systems. A **closed system** has boundaries that do not allow any matter to enter or escape the system. The boundaries may (and in the real world, always do) allow energy, such as sunlight, to pass through. An example of a closed system would be a perfectly sealed oven, which would allow the material inside to be heated but would not allow any of that material to escape. (Note that in real life, ovens do allow some vapor to escape, so they are not perfect examples of closed systems.)

WHAT A GEOLOGIST SEES

An Island Is Not a Closed System

To a casual tourist, the island of Bora Bora (shown here) may seem like a closed system, isolated from the rest of the world, a great place to "get away from it all." But how isolated is it, really?

A geologist would look at this volcanic island and see the forested slopes as evidence of abundant precipitation; the flat area between the mountain and the sea as evidence of erosion transferring material from the mountain toward the sea; and, in the foreground, coral reefs growing on the shallow, submerged part of the island as evidence of warm waters and plentiful nutrients. A geologist, like all other scientists, would conclude that the island is an open system. (Remember from Figure 1.7 that an open system allows both matter and energy to cross its boundaries.)

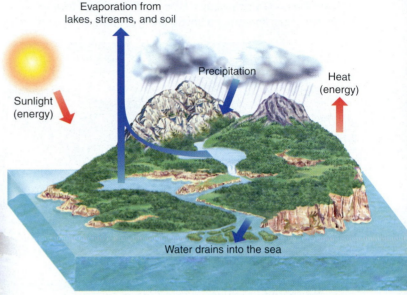

Evaporation from lakes, streams, and soil

Precipitation

Heat (energy)

Sunlight (energy)

Water drains into the sea

Energy (in the form of sunlight) and matter (in the form of precipitation) reach the island from outside sources. Energy leaves the island as heat. Matter in the form of water either evaporates or drains into the sea. Birds may fly into and out of the system, and, in the modern era, humans may also bring materials into and out of the system by importing and exporting resources.

Think Critically

Bora Bora is a volcanic island. The volcano is considered to be extinct, but suppose it were just dormant and waiting to erupt again. How would you modify the diagram and the caption to describe this system, a volcanic island, during an eruption?

A second kind of system, **an open system**, can exchange *both* matter and energy across its boundaries. An island offers a simple example (see *What a Geologist Sees*). The system concept can also be applied to artificial environments. For example, urban geographers and land use planners sometimes use a systems approach to the study of cities. Enormous flows of energy and materials occur across city borders.

The Earth system Earth itself is a very close approximation of a closed system. Energy enters the Earth system as solar radiation. The energy is used in various biologic and geologic processes and then departs in the form of heat. Very little matter crosses the boundaries of the Earth system. We do lose some hydrogen and helium atoms from the outer atmosphere, and we gain some material in the form of meteorites. However, for most purposes, especially over the short term, we can treat Earth as a closed system. Over the long term, geologically speaking, we cannot ignore the effects of meteorites, as illustrated by *Amazing Places*.

AMASING PLACES

Meteorite Impact Craters

Not far from Alice Springs in central Australia, a ring of hills called Gosses Bluff juts up from the endlessly flat landscape. The origin of this curious formation was a puzzle until the 1960s, when it was shown to be the relic of a great meteorite that crashed to Earth 142 million years ago. The impact and explosion blasted out a crater 24 kilometers in diameter, surrounded by a ring of debris (**Figure a**). What we see today is the eroded remnant of the center of the impact site. You can see a ghostly, highly eroded remnant of the outer rim in a photograph from space (**Figure b**).

Evidence for the origin of Gosses Bluff came from Meteor Crater, in Arizona (**Figure c**). G.K. Gilbert, a distinguished government geologist, hypothesized in 1891 that Meteor Crater was formed by a great volcanic steam explosion. D.M. Barringer, a mining engineer, disagreed. Barringer thought it was the impact site of a great iron meteorite, and he was given the right to prospect and mine the iron object. He never found it because, as we now know, the meteorite mostly vaporized as it raced through the atmosphere, and what was left disintegrated into small fragments on impact. Convincing evidence for the impact origin of the crater finally came in 1960, when E.M. Shoemaker, another government geologist, discovered a mineral called

stishovite, a form of silica that is formed only at super-high pressures, such as those caused by an impacting meteorite.

Meteor Crater, about 70 kilometers east of Flagstaff, Arizona, is well preserved. Formed 50,000 years ago, the crater is 1200 meters in diameter, much smaller than Gosses Bluff, and 170 meters deep. It is surrounded by a 45-meter-high raised rim of rocky debris blasted out of the crater by the impact. Scientists estimate that the impacting body was a metallic object about 40 meters in diameter. The speed of the meteorite at the moment of impact has been estimated at 46,000 km/hr.

Our planet formed about 4.56 billion years ago out of innumerable rocks very much like the meteorites that left Gosses Bluff and Meteor Crater. Though impacts of large meteorites are rare in the present era, scars from past impacts remind us that we are not isolated in space but, rather, are still an integral part of an active solar system.

b.

Outer rim of crater

Gosses Bluff

a.

c.

Living in a closed system The fact that Earth is a closed system (on the scale of the time that humans have existed) has three important consequences. First, *because the amount of matter in a closed system is fixed and finite*, the mineral and fossil fuel resources on this planet are all we have and all we will ever have until we learn to mine other planets. Second, *all the waste materials we develop remain within the confines of the Earth system;* or, as environmentalists say, "There is no away to throw things to." Third, *if changes are made in one part of a closed system, the results of those changes eventually will affect other parts of the system.* For instance, when we divert a river to provide drinking water for a city, we may deplete the water resources somewhere else (**Figure 1.8**).

A river no longer runs through it • Figure 1.8

a. The Colorado River and its tributaries provide drinking water to 25 million people in California, Nevada, Arizona, Utah, New Mexico, Colorado, and Wyoming. It also irrigates 3.5 million acres (1.4 million hectares) of fields. Because of the massive diversion of water away from the river in those states, little water reaches the river's historic terminus, the Gulf of California (or Sea of Cortez) in Mexico. This river provides an example of how changes made in one part of a system eventually affect other parts of the system.

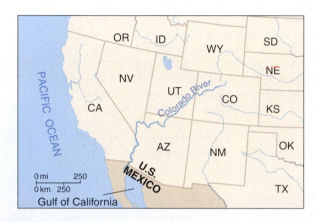

b. Instead of a broad river mouth, the northern end of the Gulf of California is now a vast mud flat. Most of the land on both sides of Isla Montague would have been underwater in the 19th century, before water management began.

Earth's subsystems The Earth system can be divided into four very large subsystems: the **geosphere**, **biosphere**, **atmosphere**, and **hydrosphere** (**Figure 1.9**). These can be further subdivided into many subsystems of interest to geologists; for example, the hydrosphere consists of oceans, glacial ice, streams, lakes, groundwater, and so on.

The geosphere may come to mind first, but in fact all four spheres play important

> **geosphere** The solid Earth, as a whole.
>
> **biosphere** The system consisting of all living and recently dead organisms on Earth.
>
> **atmosphere** The envelope of gases that surrounds Earth.
>
> **hydrosphere** The system comprising all of Earth's bodies of water and ice, both on the surface and underground.

roles in geology. Plants draw nutrients from the lithosphere and incorporate them into the biosphere when they die and decompose. The material they contain may enter the atmosphere or return to the geosphere. Rocks erode, and the minerals they contain become salts in the hydrosphere; evaporation returns these salts to the geosphere. The exchanges of materials between spheres never stop.

Earth's subsystems: The four "spheres" • Figure 1.9

This figure illustrates Earth's four principal subsystems: geosphere, biosphere, atmosphere, and hydrosphere. Materials and energy cycle among these subsystems, as shown by the arrows, making them open systems.

Atmosphere

Biosphere

Geosphere

Hydrosphere

The life zone The four major Earth reservoirs interact most intensively in a narrow life zone, a region that extends to about 10 kilometers above and 10 kilometers below the Earth's surface (**Figure 1.10**). In this narrow zone, all known forms of life exist. It is only here that conditions favorable for life are created by interactions among the geosphere, hydrosphere, atmosphere, and biosphere.

Cycles and Interactions

Figure 1.9 shows the interactions among Earth's reservoirs in a simplistic fashion, but we can be a good deal more precise about the nature of the flows by focusing on the way materials move, or cycle, among the reservoirs. In this book, we will discuss the Earth system as a series of three interrelated cycles that facilitate the movement of materials and energy among the reservoirs.

These are the **water cycle**, or **hydrologic cycle**, the **rock cycle**, and the **tectonic cycle**. They are sketched in **Figure 1.11**. It is not necessary for you to understand the details of this diagram yet; we will return to this figure repeatedly in later chapters and label each of the processes that are illustrated with icons and arrows here. The most important points to understand now are that the interactions form cycles—that is, processes without beginning or end—and that they are closely interconnected.

Because this is a book about physical geology, we will often

> **hydrologic cycle** A model that describes the movement of water through the reservoirs of the Earth system; the water cycle.
>
> **rock cycle** The set of crustal processes that form new rock, modify it, transport it, and break it down.
>
> **tectonic cycle** Movements and interactions in the geosphere and the internal Earth processes that drive them.

The life zone • Figure 1.10

All life on Earth inhabits a zone no wider than 20 kilometers, and the majority lives in a region no wider than 1 kilometer.

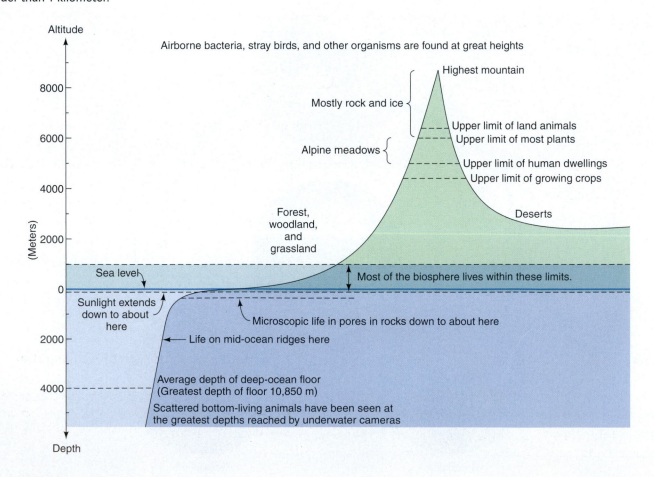

Interconnected cycles • Figure 1.11

In the hydrologic cycle (right), water circulates through various reservoirs: ocean, atmosphere, geosphere (as surface water and groundwater), and biosphere (in plants and animals), eventually returning to the ocean. The rock cycle (center) describes crustal processes through which rock is uplifted into mountains, then eroded and weathered. The resulting sediment may be modified, transformed, or reformed into rock, and thrust up into mountains again. The tectonic cycle (left) explains where igneous rock comes from, and how new crust is formed and re-cycled by large-scale motions of Earth's surface and interior. Energy from the Sun powers the hydrologic cycle. Heat from Earth's interior powers the tectonic cycle, and both sources power the rock cycle.

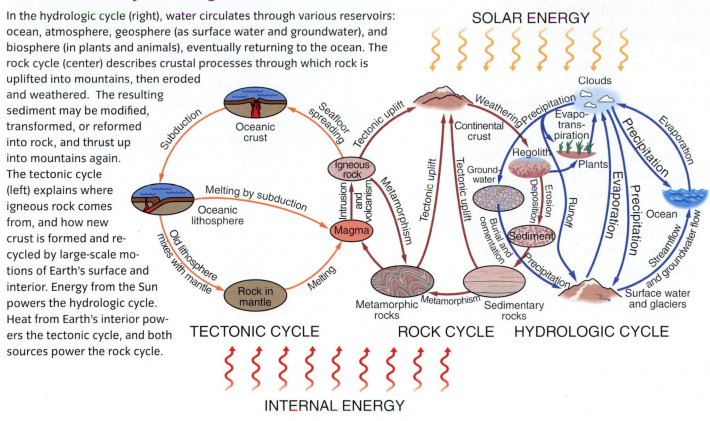

focus primarily on the **lithosphere**—the outermost part of the geosphere. However, the systems approach tells us that it is unreasonable—even impossible—to consider one part of the Earth system in isolation from the rest. We cannot fully understand where rocks come from without also understanding the hydrologic

> **lithosphere** Earth's rocky outermost layer.

cycle and the tectonic cycle. Nor can we understand the lithosphere without learning something about the hydrosphere, atmosphere, and biosphere, as well as Earth's deep interior (which is distinct from the lithosphere). Thus, in this course, you will study not only geology but also a little bit of oceanography, hydrology, meteorology, physics, chemistry, biology, and astronomy.

CONCEPT CHECK

1. **What** is the difference between historical and physical geology, and what are some specialized disciplines within geology?

2. **What** distinguishes a theory from a hypothesis?

3. **Why** is the system concept a key part of modern geology?

4. **How** does the geosphere differ from the hydrosphere?

5. **What** are the three main cycles that describe the exchange of materials and energy among Earth's subsystems?

Earth in Space

LEARNING OBJECTIVES

1. **Meet** the different members of the solar system.

2. **Outline** the difference between the rocky planets and the gaseous planets.

3. **Describe** how planetary accretion and meteorite impacts have shaped the solar system.

4. **Explain** why the inner planets are chemically differentiated.

5. **Identify** several similarities and differences among Earth and the other inner planets.

A s geologists, we mainly study the processes that occur on Earth, in isolation from the rest of the solar system. How did Earth become a unique and special place? We must remember that Earth is a planet that is part of a larger system, the solar system. The characteristics of the planet we live on today are linked to the origins of both the planet and the solar system as a whole, as well as what has happened to the planet over the past 4.56 billion years.

The Solar System

Earth is one of eight large members of the **solar system**, traditionally called **planets** (from *planetai*, Greek for "wanderers" because they seem to wander across the sky). The solar system consists of the Sun and the group of objects in orbit around it. In addition to the Sun and the planets, the solar system includes 143 known moons, a vast number of asteroids, millions of comets, and innumerable fragments of rock and dust. All the objects in our solar system move through space in smooth, regular orbits, held in place by gravitational attraction. The planets, asteroids, comets, and meteoroids orbit the Sun, and the moons orbit the planets.

We can separate the planets into two groups on the basis of their physical characteristics and distances from the Sun (**Figure 1.12**). The innermost planets—Mercury, Venus, Earth, and Mars—are small, rocky, and relatively dense. They are similar in size and chemical composition and are called terrestrial planets because they resemble Earth (*Terra* in Latin). Among the **terrestrial planets**, only Earth is the right size and has the right temperature because of its distance from the Sun to support life. Whether there are other planets in the Milky Way Galaxy that can support life remains an open question, though (see *Case Study*).

Family portrait of the solar system • Figure 1.12

Our solar system's eight recognized planets are shown here to scale against the Sun. (Note, however, that the distances between planets are much greater than shown, and the planets never line up neatly like this.) Between the terrestrial and jovian planets lies the asteroid belt, consisting of more than 100,000 small pieces of rock that have never coalesced into a planet. Beyond Neptune lies Pluto, formerly considered a ninth planet, and a large number of other icy objects, mostly like Pluto—some larger but mostly smaller—which collectively comprise the Kuiper Belt.

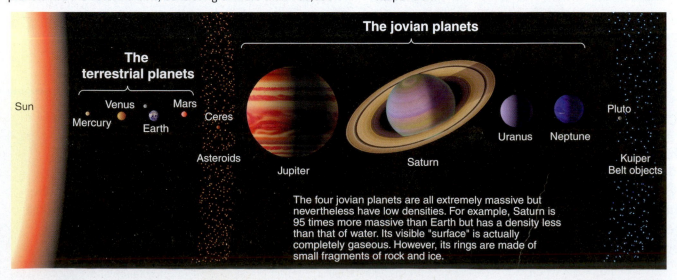

The terrestrial planets — Sun, Mercury, Venus, Earth, Mars, Ceres, Asteroids

The jovian planets — Jupiter, Saturn, Uranus, Neptune, Pluto, Kuiper Belt objects

The four jovian planets are all extremely massive but nevertheless have low densities. For example, Saturn is 95 times more massive than Earth but has a density less than that of water. Its visible "surface" is actually completely gaseous. However, its rings are made of small fragments of rock and ice.

Are We Alone?

Are we humans alone in space? This has been a question of great interest for centuries, but we still don't know the answer. We can break the question down into three parts:

1. Do other suns have planetary systems?
2. If there are other planetary systems, do any of them support life?
3. If life is present on planets around other suns, has intelligent life emerged?

As recently as 25 years ago, the answer to each of the questions was "We don't know," but the development of more sensitive telescopes has provided a "yes" for the first question. Slight fluctuations in light due to a planet moving in front of a star and/or slight wobbles in the position of a star due to the gravitational pull of a large planet provide evidence that other planetary systems exist. By October 2011, 695 extraterrestrial planets (exoplanets) had been identified around suns in the Milky Way Galaxy, and an additional 1200 light and wobble anomalies were being further tested. It is clear that there are plenty of planets around suns in our galaxy.

The second question remains unanswered. Even in our own solar system, there is no evidence that life is now, or has ever been, present beyond life on Earth, though much research remains to be done before we write off the possibility completely. So what chance do we have of detecting the presence of life outside the solar system? Remember that materials flow between the different parts of the Earth system. Our atmosphere has the composition it does in large part because it has been changed by the biosphere. Possibly that may be so on other planets, too, so careful measurement of atmospheres around suitable exoplanets may someday provide an answer to this question.

Now for the toughest but most fascinating question of all: If the answer to question 2 is "yes," do any planets out there support *intelligent* life? The clearest proof would be detection of radio signals or possible laser signals, so the "search for extraterrestrial intelligence" (SETI) has been focused on detecting possible radio signals. So far, nothing has been confirmed.

One of the first researchers involved in SETI was Dr. Frank Drake. In 1961, Drake wrote down an equation for the number of detectable extraterrestrial civilizations in the Milky Way Galaxy. The equation combines all the things we would need to know in order to estimate the existence of other civilizations. Drake was not concerned just with the detection of life; he was interested in intelligent life that could send and receive radio waves.

The Drake formula is:

$$N = R^* \times f_p \times n_e \times f_t \times f_i \times f_c \times L$$

In this formula, N is the number of detectable civilizations; R^* the average rate at which stars form in our galaxy; f_p is the fraction of stars that have planets; n_e is the average number of planets/stars that can potentially support life; f_t is the fraction of n_e planets that go on to develop life; f_i is the fraction of f_t planets that go on to develop intelligent life; f_c is the fraction of f_t planets where intelligent life develops technology that releases detectable signs into space; and L is the length of time a civilization lasts and releases detectable signals into space.

The only items in the Drake equation for which we can make an estimate based on evidence, no matter how tenuous, are R^*, f_p, n_e, and L. For the other items, we can only guess. You should try making the calculation yourself. Drake's estimates in 1961 yielded an estimate of 10 intelligent civilizations within our galaxy with which we might hope to communicate. More recent estimates have yielded numbers between 2 and 5000.

So far, all investigations have been confined to the Milky Way Galaxy. Astronomers have now discovered that there are billions of other galaxies out in space. Most are incredibly distant, so that detection of planets and life seems highly unlikely. But there are several galaxies close by; the closest is the Canis Major Dwarf Galaxy, about 25,000 light-years from Earth, followed by the Sagittarius Dwarf Galaxy, at 70,000 light-years, and the two Magellanic Cloud Galaxies, at 160,000 and 200,000 light-years. Future research with new generations of telescopes may well find stars with planets in these galaxies.

This is an artist's rendition of the five exoplanets around the star 55 Cancri, the brightest dot. The large blue planet is about the size of Neptune and is thought to be orbiting within the habitable zone of the star (just the right distance to have a temperature that allows H_2O to be present on the surface in liquid, solid, and vapor forms).

This photo shows the two Magellanic Cloud Galaxies. They are both smaller than the Milky Way Galaxy of which our solar system is a part, but they are close enough in space so that they may be the first galaxies outside of the Milky Way in which exoplanets will someday be detected.

The outer planets of the solar system are much larger and more massive than the terrestrial planets, but they are much less dense. These **jovian planets**—Jupiter, Saturn, Uranus, and Neptune—take their name from *Jove*, the name for Jupiter in Roman mythology. (Jupiter was the king of the gods, and the god of light and weather, among other things. In size, Jupiter is certainly the king of the planets.) The jovian planets probably have small solid centers that may resemble terrestrial planets, but much of their planetary mass is contained in thick atmospheres of hydrogen, helium, and other gases. The atmospheres are what we actually see when we observe these planets.

Pluto, until recently considered to be a ninth planet, doesn't fit into either of these planetary groups. It is much smaller and denser than the jovian planets but much less dense than the terrestrial planets. Recent discoveries suggest that Pluto is actually the nearest of a sizeable population of Pluto-like bodies beyond the orbit of Neptune called **Kuiper Belt objects**, some of which are actually larger than Pluto. In 2006, the International Astronomical Union (IAU) adopted a new designation of **dwarf planet** for an object that orbits the sun, is large enough that its own gravity has pulled it into a spherical shape, but is small enough that it has not "cleared the neighborhood around its orbit" by gravitational attraction of surrounding debris. Under the new definition, Pluto is a dwarf planet, and so is the large asteroid Ceres. Although the decision aroused controversy (especially among fans of Pluto), it is a normal part of science to adopt new terminology when the old nomenclature has been based on assumptions that later turn out to be flawed.

The Origin of the Solar System

How did the solar system form? The answer to this ancient question is still incomplete, and it is an excellent example of a carefully researched scientific hypothesis. The **nebular hypothesis**, originally formulated by the German philosopher Immanuel Kant in 1755 and now widely accepted as the best description of planetary formation, hypothesizes that the solar system coalesced out of a swirling cloud of interstellar dust and gas called a **nebula** (**Figure 1.13**). When the gases at the center of the cloud became sufficiently hot and dense, an energy-producing nuclear reaction called **fusion** began, and our Sun was born. The process of fusion, which combines hydrogen atoms to form helium, has powered the sun for 4.56 billion years, and will continue to power it for billions of years into the future.

The nebular hypothesis explains very well why the inner planets are rocky, while the outer planets contain a higher proportion of ice and gas. The temperature was higher in the portion of the nebula that was to become the innermost part of the solar system. Elements with a high melting point (iron, silicon, and so on—the chemicals that make up rock) would have been early to condense from the cloud. Meanwhile, a strong solar wind stripped much of the lighter gases, such as helium and hydrogen, from the inner planets. But the solar wind was not strong enough to do the same to the outer planets, which grew into gas giants. Volatile compounds such as water and methane condense only at lower temperatures. Therefore, water ice, methane ice, and other ices are abundant on the moons and smaller bodies of the outer solar system, where the primordial nebula was cooler.

Planetary accretion, a 20th-century supplement to the nebular hypothesis, accounts for the existence of meteoroids and asteroids. According to the accretion hypothesis, the outer portions of the solar nebula cooled as it flattened out into a disk of rocky, metallic, and icy debris before any of the planets were formed. Through random collisions, the debris began to form clumps. Eventually some clumps grew large enough to pull in the remaining debris by the force of gravity. Thus, the process of accretion accounts for two parts of the IAU's definition of a planet. Planets are round because their gravity overcomes the material strength of the rocks that form the planet. They are isolated because they long ago swept up virtually all the material that was formerly in or near their orbit.

Even so, some stray rocks never managed to be swept up by any planet; these are called **meteoroids**. Sometimes pieces of this debris happen to fall to Earth; they are called **meteors** if they burn up in the atmosphere or **meteorites** if they reach Earth's surface. Meteorites are fascinating relics of the early days of the solar system.

> **meteorite** A fragment of extraterrestrial material that falls to Earth.

A major consequence of the planetary accretion hypothesis is the great importance of violent collisions in the history of the solar system. Every crater on our Moon, as far as we know, was formed by a meteorite impact. (Astronauts and scientists have searched for volcanic craters

Birth of a solar system • Figure 1.13

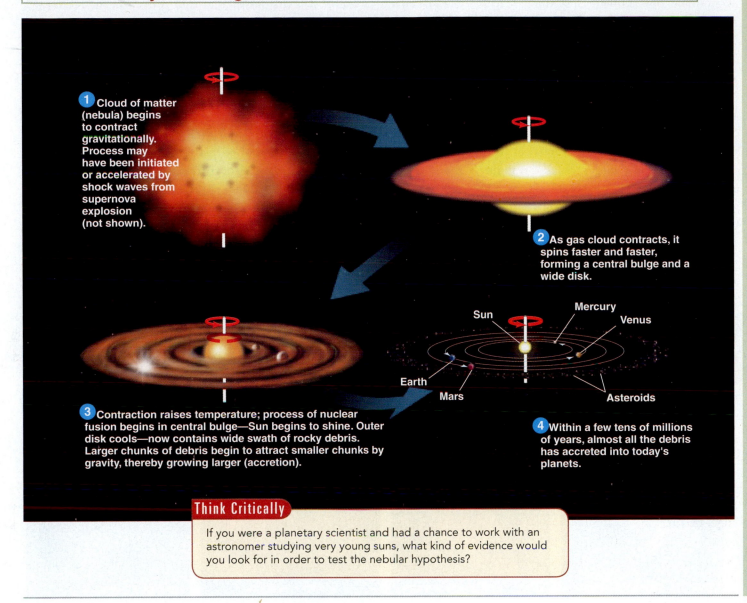

1 Cloud of matter (nebula) begins to contract gravitationally. Process may have been initiated or accelerated by shock waves from supernova explosion (not shown).

2 As gas cloud contracts, it spins faster and faster, forming a central bulge and a wide disk.

3 Contraction raises temperature; process of nuclear fusion begins in central bulge—Sun begins to shine. Outer disk cools—now contains wide swath of rocky debris. Larger chunks of debris begin to attract smaller chunks by gravity, thereby growing larger (accretion).

4 Within a few tens of millions of years, almost all the debris has accreted into today's planets.

Sun · Mercury · Venus · Earth · Mars · Asteroids

Think Critically

If you were a planetary scientist and had a chance to work with an astronomer studying very young suns, what kind of evidence would you look for in order to test the nebular hypothesis?

on the Moon, but none has ever been found.) Scientists recognize impact craters on Earth, too, although they are harder to find because erosion and other geologic processes cover them up or erase them over time. A meteorite impact in the Yucatán region of Mexico may have been responsible for the catastrophe that killed the dinosaurs more than 65 million years ago.

Even larger impacts than this were commonplace in the early solar system. Most planetary scientists now think

it is likely that Earth collided with another planetary body, roughly the size of Mars, around 4.5 billion years ago. The impact tilted Earth's axis of rotation at an angle to the plane of its orbit around the Sun, and that is why we have seasons. The impact must also have melted most of Earth's surface due to the tremendous amount of energy released. The collision completely destroyed the other planet, and it blasted so much debris into orbit that for a little while, Earth had rings much denser than Saturn's. Eventually the ring of debris coalesced into the most familiar of as-tronomical objects—Earth's Moon (**Figure 1.14**). This hypothesis explains the existence of a magma ocean early in lunar history (shown by rocks retrieved from the Moon). It also explains our Moon's relatively large size in contrast to other moons, which are many times smaller than their parent bodies, and accounts for certain chemical discrep-ancies and similarities between Earth and its Moon.

Such giant collisions were the inevitable final stage of planetary accretion, when most of the debris has been swept up and only larger objects remain. Signs of giant impacts abound in the solar system. One such impact probably caused Uranus's axis to tip over on its side. Pluto's moon, Charon, was probably created by an impact because it is unusually large compared to Pluto. Perhaps

Venus experienced a giant impact, too. Although it lacks a moon, it is the only planet that rotates in the retrograde (east-to-west) direction, an effect that could have been produced by a large glancing blow, essentially tipping the planet upside down.

Why is the accretion history of the planets impor-tant to geologists? Because of the heat generated by all of these collisions, every rocky planet probably started out hot enough to melt either partially or completely. During the period of partial melting, terrestrial planets separated into layers of differing chemical composition, a process called **differentiation**. Earth's geosphere differentiated into three layers: a relatively thin, low-density, rocky **crust**;

> **crust** The outermost compositional layer of the solid Earth.

Formation of the Moon • Figure 1.14

THE PLANNER

Impact

Some 4550 million years ago, the still-forming Earth (unrecognizable because it probably did not yet have oceans) runs into another growing planet, which scientists have named Theia.

Impact + 8 hours

Theia is obliterated, and its remnants—along with a good chunk of Earth's mantle—are blasted into orbit around Earth. The off-center impact has knocked Earth's axis of rotation askew.

Impact + 24 hours

The debris spreads itself into a ring and begins to clump together.

Impact + 1 year

The largest clump starts to attract other fragments and is well on its way to becoming the Moon. Its surface is initially molten. Earth has recovered its shape, leaving no trace of the most violent event in its history.

Think Critically

What differences would you notice on Earth today if Theia had been a smaller body and, as a result, the Moon were in about the same place as it is today but only one-tenth as large?

Inside the terrestrial planets • Figure 1.15

This figure shows a comparison of the sizes of the terrestrial planets, and the relative amount of total planetary volume represented by the core in each case.

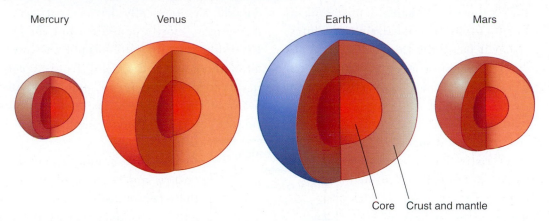

Mercury   Venus   Earth   Mars

Core   Crust and mantle

a rocky, intermediate-density **mantle**; and a metallic, high-density **core**. Similar layers are present in Mercury, Venus, Mars, and the Moon, although they have different proportional sizes and compositions (**Figure 1.15**).

> **mantle** The middle compositional layer of Earth, between the core and the crust.
>
> **core** Earth's innermost compositional layer.

There are other important similarities among the terrestrial planets. All of them have experienced volcanic activity, which means they have or once had internal heat sources. (We will explain where the heat originates in Chapter 3.) The volcanism is dominated by the eruption of lava that cools to form a volcanic rock called **basalt** (**Figure 1.16**). All of the planets have also been through intense cratering processes, although the signs are well hidden on Venus and Earth. Finally, all have lost their primordial atmospheres. The three that ended up with new atmospheres (Earth, Mars, and Venus) evolved them over time, from material that leaked from their interiors via volcanoes.

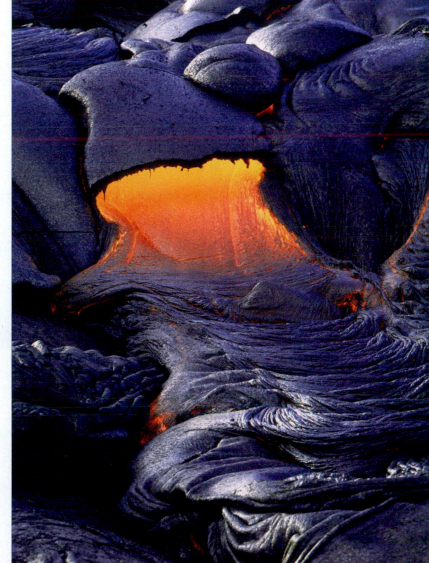

Basalt: The most common rock in the solar system • Figure 1.16

Lava from Mauna Loa volcano in Hawaii cools to form the volcanic rock basalt, a rock type that also comprises most of the surfaces of Venus and Mars, and the "seas" (dark-colored spots) of Earth's Moon. Though taken only a few years ago, this picture could represent the surface of any of these planets 4.56 billion years ago.

What Makes Earth Unique?

Venus and Mars, Earth's nearest neighbors, are in some ways very similar to our planet. In terms of size, Venus is nearly Earth's twin. Yet there is no chance of mistaking either of them for Earth (**Table 1.1**). Earth's blues, whites, and greens attest that it has three things neither Venus nor Mars, nor any other planet or moon in our solar system, possesses: an oxygen-rich atmosphere; a hydrosphere that contains water in solid, liquid, and vapor forms; and a biosphere full of living organisms.

The nature of Earth's solid surface is another special characteristic. Earth is covered by an irregular blanket of loose debris formed as a result of **weathering**—the chemical alteration and mechanical breakdown of rock caused by exposure to water, air, and living organisms. This layer is called **regolith** (from the Greek words for "blanket" and "stone"). It includes soil, river mud, desert sand, rock fragments, and other unconsolidated debris. Earth's regolith is unique because it teems with life. When material from the biosphere becomes incorporated with rock material, the result is a special type of regolith unique to Earth: **soil**.

Other planets and moons with rocky surfaces have regolith, too (**Figure 1.17**). In the case of the Moon, the regolith has formed from endless pounding by impacting meteorites. Both Venus and Mars have atmospheres that contain water vapor and carbon dioxide, so chemically driven weathering occurs in addition to meteorite pounding, but the key ingredient of Earth-like regolith—life—is not known to be present. In all of these cases, the regoliths are not true soils. (You may read about "lunar soil" and "Martian soil," but such usage of the word "soil" is geologically incorrect. The correct terms are "lunar regolith" and "Martian regolith.")

Another unique property of Earth is the nature and extent of its tectonic activity. **Plate tectonics** has shaped Earth's continents and oceans and governs, to a large extent, the location of Earth's volcanoes and the occurrence of earthquakes. Tectonic activity has given Earth two different kinds of crust—thin, basaltic **oceanic crust** and thick, granitic **continental crust**. The latter seems to be unique to Earth—at least Mars and Venus do not appear to have it. Because the location of continents affects oceanic and atmospheric currents, plate tectonics exerts a powerful influence on Earth's climate, which in turn affects the evolution of life. We will have much more to say about plate tectonics in Chapter 4 and later chapters.

---

plate tectonics
The movement and interactions of large fragments of Earth's lithosphere, called plates.

oceanic crust
The thinner, denser, and younger part of Earth's crust, underlying the ocean basins.

continental crust
The older, thicker, and less dense part of Earth's crust; the bulk of Earth's land masses.

Goldilocks planets: Too much, too little, just right   Table 1.1

In spite of their similar origins, Venus, Earth, and Mars have profound geologic differences that have made Earth the only one that is hospitable for life.

Visible Light / Cloud-penetrating radar — Venus

Earth

Mars

	Venus	Earth	Mars
Atmosphere	White in photos, 97% carbon dioxide; temperature averages a blistering 480°C (hot enough to melt lead)	78% nitrogen, 21% oxygen; average temperature 14.6°C	Thin and insufficient to retain much heat; average temperature -63°C; temperatures usually too low to melt water ice (0°C)
Hydrosphere	Exists only as vapor in the atmosphere due to high temperatures	Contains water as solid, liquid, and vapor	Water cannot exist in liquid form on the surface due to low temperatures and pressure
Biosphere	None	Only known biosphere	None known

Regolith on Mars • Figure 1.17

This photo shows a trench dug on the Martian surface by the Phoenix lander. The regolith is red, indicating that it has been oxidized—a sign of chemical weathering. The white material is thought to be water ice.

CONCEPT CHECK STOP

1. **What** are the principal members of the solar system?

2. **How** do the terrestrial planets differ from the jovian planets?

3. **What** are the nebular hypothesis and the planetary accretion model?

4. **How** did the inner planets develop cores?

5. **How** are Venus, Mars, and Earth similar, and how do they differ?

The Ever-Changing Earth

LEARNING OBJECTIVES

1. **Explain** Hutton's principle of uniformitarianism.
2. **Describe** why the Earth system can be so dynamic and yet appear so stable.
3. **Identify** several key benefits of studying geology.

Uniformitarianism

Since material is constantly being transferred from one open system on Earth to another, you may wonder why these systems seem so stable. Why doesn't the sea become saltier or fresher? Why doesn't all the water in the world flow into the sea and stay there? Why should the chemical composition of the atmosphere be mainly nitrogen and oxygen, as it has been for millions of years? How can rock that is 2 billion years old have the same composition as rock that is being formed today? If mountains are constantly being worn down by erosion, why are there still high mountains?

The answers to these questions are all basically the same: Materials cycle from one reservoir to another, but the reservoirs themselves don't change noticeably because the different parts of the cycle balance each other: The amounts added approximately equal the amounts removed. While a mountain is worn down in one part of the cycle, a new mountain is being built up in another part. This cycling of materials has been going on since Earth was formed, and it continues today.

A fundamental principle of geology, attributed to an 18th-century Scottish geologist named James Hutton, is based on this idea. This principle was given the name **uniformitarianism** by geologists who followed Hutton, and one way to express it is to state that "the present is the key to the past." We can examine any rock, no matter how old, and compare its characteristics with those of similar rocks forming today. We can then infer that the ancient rock likely formed in a similar environment through similar processes and on a comparable time scale. For example, in many deserts today, we can see gigantic dunes formed from sand

> **uniformitarianism**
> The concept that the processes governing the Earth system today have operated in a similar manner throughout geologic time.

grains transported by the wind. Because of the way they form, dunes have a distinctive internal structure (**Figure 1.18**). Using the **principle of uniformitarianism**, we can infer that a rock composed of cemented grains of sand and having the same distinctive internal structure as modern dunes is the remains of an ancient dune.

Time and Change

The **principle of uniformitarianism** provides the first step toward understanding Earth's history. Geologists have used this principle to explain Earth's features in a logical manner. In so doing, they have discovered that Earth is incredibly old. An enormously long time is needed to erode a mountain range or for huge quantities of sand and mud to be transported by streams, deposited in the ocean, and cemented into rocks, and for the rocks to be uplifted to form a mountain. Yet the rock record tells us that the cycle of erosion, formation of new rock, uplift, and more erosion has been repeated many times during Earth's long history.

One ongoing process that Hutton and those who followed him did not know about is the slow motion of Earth's plates. In a nutshell, plate tectonics involves the motion of about a half dozen large, curved fragments of Earth's lithosphere, plus a large number of smaller ones. Like so much else about Earth, we can now observe the motion of plates from space: Global Positioning System (GPS) satellites can measure the shifting of plates in centimeters per year. By the **principle of uniformitarianism**, this process should also have operated in the past. When we extrapolate these imperceptibly slow motions over millions of years, we discover a stunning result, which is supported by many decades of scientific observation: Earth's continents were in very different positions in the past (**Figure 1.19**).

This observation leads us to a more sophisticated understanding of Hutton's principle. The physical processes that occur on Earth have not changed over time, but the physical conditions of Earth have changed dramatically. Sea levels drop and rise; the chemical composition of Earth's atmosphere fluctuates, albeit ever so slowly. The cycles maintain a balance, but in doing so the sizes of the reservoirs of the Earth system may change and the speed of cycles and processes may increase or decrease. This

a. A distinctive pattern of wind-deposited sand grains can be seen in a trench dug in this modern sand dune near Yuma, Arizona.

NATIONAL GEOGRAPHIC

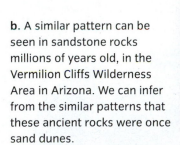

b. A similar pattern can be seen in sandstone rocks millions of years old, in the Vermilion Cliffs Wilderness Area in Arizona. We can infer from the similar patterns that these ancient rocks were once sand dunes.

is an especially important lesson today, when it appears that our planet has entered a period of human-mediated climatic change.

Throughout this book we will explore the process of plate tectonics in much greater depth. We will see how virtually every aspect of the Earth system owes its essen-tial character to the existence of plate tectonics. Because plate tectonics has become the overarching theme in ge-ology, it is viewed as a unifying theory. Practically every aspect of geologic science, including the histories of the atmosphere, hydrosphere, and biosphere, is connected to the motion of lithospheric plates.

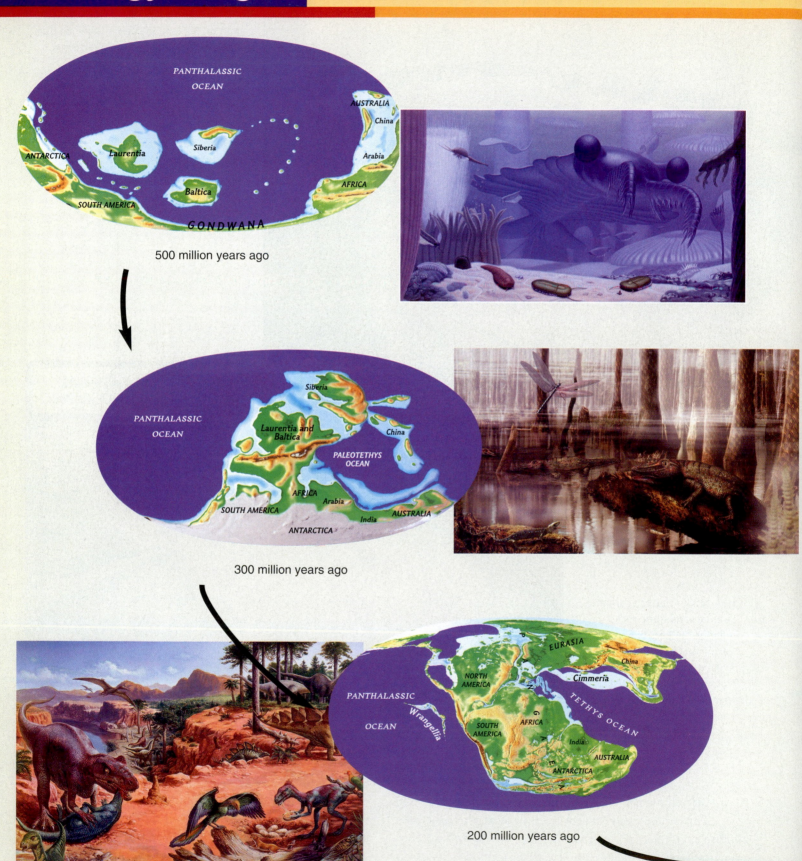

500 million years ago

300 million years ago

200 million years ago

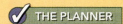

These figures show geologists' best reconstruction of the way Earth's landmasses have changed position over the last 500 million years. Note that Earth had one contiguous supercontinent from about 300 million years ago to about 200 million years ago—a point we return to in Chapter 4.

Present day

50 million years ago

100 million years ago

Why Study Geology?

With this brief introduction to geology and the Earth system, you have probably deduced some of the reasons it is important, as well as interesting, to study geology. We need to understand Earth materials because we depend on them for all of our material resources—the minerals, rocks, and metals with which we construct our built environment; the energy with which we run it; the soil that supports agriculture and other plant life; and the air and water that sustain life. Many of Earth's resources are limited and require knowledgeable and thoughtful management. The materials of Earth also have physical and chemical properties that affect us, such as their tendency to flow or fail during a landslide, their capacity to hold or transmit fluids such as water or oil, their ability to absorb waste or prevent it from migrating, or their ability to affect human health if released into the environment in a dangerous manner.

You have learned that Earth is essentially a closed system, which means that all materials remain within the system. It is important to understand how materials move from one reservoir to another. It is also important to understand the time scales that govern these processes in order to gain some perspective on the changes that we see occurring in the natural environment. Some Earth processes are hazardous—that is, damaging to human interests. These geologic hazards include earthquakes, tsunamis, volcanic eruptions, landslides, hurricanes, floods, and meteorite impacts (**Figure 1.20**). Though impacts of large meteorites are rare in the present era, scars from past impacts remind us that we are not isolated in space but, rather, are still an integral part of an active solar

Lethal eruption • Figure 1.20

Mt. Pinatubo in the Philippines erupted in 1991. Volcanologists predicted the eruption, making it possible to evacuate residents from the area and prevent thousands of deaths. When it erupted, the volcano sent this lethal cloud of searing, dust-laden gases rolling down its flanks, to spread rapidly across the surrounding plains. This particular car and driver escaped, but many houses, trees, and fields were smothered with volcanic ash.

Global Locator

Philippines

NATIONAL GEOGRAPHIC

Where Geologists CLICK

Meteorite Impact Sites

http://www.passc.net/EarthImpactDatabase/Worldmap.html

The fifty most easily recognizable meteorite impact sites around the world appear on the interactive map at the Earth Impact Database at the Planetary and Space Science Centre of the University of New Brunswick (Canada). These impact sites include Meteor Crater in Arizona, Chicxulub Crater in Mexico, and Manicouagan Crater in Québec. The site provides photos and other information about confirmed impact structures around the world.

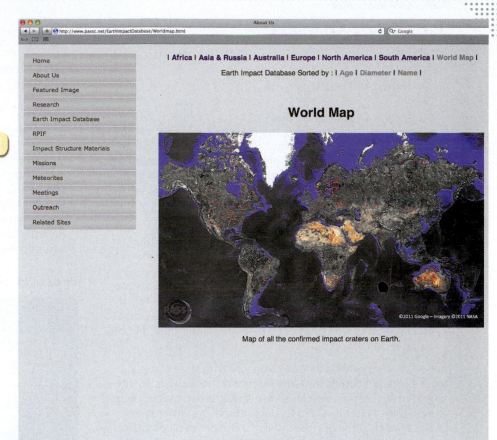

Map of all the confirmed impact craters on Earth.

system. You can explore some of Earth's most impressive meteorite impact craters using an interactive map online (see *Where Geologists Click*). The more we know about these hazardous processes, the more successful we will be in protecting ourselves from future natural disasters.

From its beginnings a couple of centuries ago, geology has been an interdisciplinary science because Earth operates through the interactions of biologic, physical, and chemical processes. Yet we are discovering that the interactions are more complex and dynamic than we would have believed only decades ago. We are still learning about the complexities and interrelationships of subsystems such as climate, ocean currents, and shifting continents. We are now beginning to appreciate our own role in geologic change—for example, rivers, lakes, the atmosphere, soil erosion, and the decline of certain animal species—as well as the need to study the Earth system as a whole rather than in separate fragments.

If you are planning to become a geologist, this book will be an introduction to some of the many fascinating possibilities that await you in your career. If you are taking this course out of personal interest or to fulfill a degree re-

quirement, you will emerge more aware of the geologic nature of our planet and better prepared to make informed decisions about the natural processes that affect your life on a daily basis.

Earth is our home planet. The features that make Earth unique and the powerful geologic processes that characterize the Earth system are a constant source of awe and fascination to those who study them. One of the most important things we can do is to deepen and refine our understanding of our home planet.

CONCEPT CHECK

1. **Why** do geologists consider the present to be the "key to the past"?

2. **What** are some geologic processes that take place over long periods of time? Can you think of some geologic processes that happen very quickly?

3. **What** are some benefits of understanding geologic processes and principles?

Summary

1 What Is Geology?   4

- **Geology** is the scientific study of Earth, including its formation and internal structure; the materials it is composed of and the properties of those materials; its chemical and physical processes; and its physical, chemical, and biologic history. **Physical geology** focuses on the materials and processes of the Earth system. **Historical geology** seeks to establish the chronology of geologic events in Earth's history.

- Geology, like all other sciences, employs the **scientific method**, with four major steps: observation and data collection; formulation of a **hypothesis** (or several hypotheses); testing of the hypotheses; and formulation of a **theory** after the hypotheses have been sufficiently tested and revised. A theory is not merely a guess or an opinion but an idea that is well supported by experiment, observation, and logical reasoning.

- Geologists study Earth today using Earth system science. This concept comes from the discovery that Earth is an integrated **system** of interconnected and interdependent parts. Individual systems within the larger Earth system may be classified as *open or closed*. Earth is, for practical purposes, a closed system, even though some small quantities of matter, such as meteorites, do cross its boundary. The Earth system consists of four principal open subsystems, the **atmosphere**, **hydrosphere**, **biosphere**, and the **geosphere**. Materials and energy are stored for varying lengths of time in each of these reservoirs and can move from one to another.

- Three major processes that describe the movement of materials among the subsystems are the **hydrologic cycle**, the **rock cycle**, and the **tectonic cycle**. The term cycle indicates that these processes never end as shown in the diagram.

Interconnected cycles • Figure 1.11

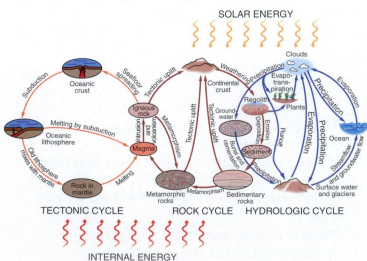

2 Earth in Space   16

- Earth is one of eight bodies in the **solar system** recognized as **planets**. The four inner planets, Mercury, Venus, Earth, and Mars, are called the terrestrial planets; they are all small, rocky, and relatively dense, and they have similar sizes and chemical composition. The four outer planets, Jupiter, Saturn, Uranus, and Neptune, are the jovian planets; they consist of huge gaseous atmospheres with small solid cores, giving them very low densities overall. Pluto, until recently considered to be a ninth planet, has been reclassified as a **dwarf planet** as shown in the diagram.

Birth of a solar system • Figure 1.13

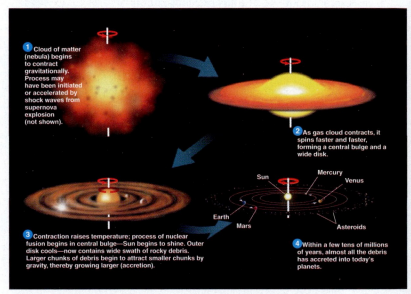

- The nebular hypothesis is the generally accepted hypothesis that the solar system formed from the coalescence and condensation of a nebula, a cloud of interstellar gas and dust. This theory is supplemented by the planetary accretion hypothesis, which states that all of the planets assembled themselves by the accretion of debris within the nebula and that today's meteoroids and asteroids are fragments that were never swept into the forming planets.

- The hypothesis of planetary accretion and intense early heating is important to geology because it helps explain Earth's beginnings. Early in its history, Earth underwent compositional **differentiation** into a dense, metallic **core**, a rocky **mantle**, and a brittle, rocky outer **crust**.

- Earth is unique in the solar system in that it possesses an oxygen-rich atmosphere. Earth is also the only planet in the solar system with a hydrosphere in which water exists near the surface in solid, liquid, and gaseous forms, and a biosphere with living organisms. Finally, Earth is the only planet where true soil is formed from regolith by interactions among physical, chemical, and biologic processes, and where life as we know it could exist.

- **Plate tectonics** is a unifying theory in geology. It describes the motion and interaction of large segments of the **lithosphere**. It is because of plate tectonics that Earth has two fundamentally different types of crust: the relatively thin, dense **oceanic crust** of basaltic composition, and the thicker, less dense **continental crust** of granitic composition.

3 The Ever-Changing Earth 24

- The principle of **uniformitarianism** states that the processes we see operating in the Earth system today have operated in a similar manner throughout much of geologic time. In other words, "the present is the key to the past."

- From a human standpoint, most geologic processes are incredibly slow. Earth is roughly 4.56 billion years old, and the rock cycle has been continuous throughout this long history. Though the processes that occur on Earth have not changed, the rates of the different cycles, such as the rock cycle and plate tectonics, have differed over time. The physical conditions on Earth—such as the tempera-

ture and composition of the atmosphere, the level of the oceans, and the location of the continents—have also been dramatically different at times in the past as shown in the diagram.

Earth's changing face • Figure 1.19

- Earth materials and processes affect our lives through our dependence on Earth resources; through geologic hazards such as volcanic eruptions, floods, and earthquakes; and through the physical properties of the natural environment.

Key Terms

- atmosphere   13
- biosphere   13
- continental crust   22
- core   21
- crust   21
- geology   4
- geosphere   13

- hydrologic cycle   14
- hydrosphere   13
- hypothesis   6
- lithosphere   15
- mantle   21
- meteorite   19
- oceanic crust   22

- plate tectonics   22
- rock cycle   14
- scientific method   5
- tectonic cycle   14
- theory   8
- uniformitarianism   24

Critical and Creative Thinking Questions

1. Do you think there may be life similar to ours on a planet outside our solar system? What would the atmosphere of that planet be like? Must it have a hydrosphere? Why or why not?

2. Why is the systems approach so useful in studying both natural and artificial processes? Can you think of examples of artificial (i.e., human-built) systems other than those given in the text? Are they open systems or closed systems? (Think about the materials and energy in them.)

3. How do you think the principle of uniformitarianism accounts for occasional catastrophic events such as meteorite impacts, huge volcanic eruptions, or great earthquakes?

4. In this chapter we have suggested that Earth is a close approximation of a natural closed system, and we have hinted at some of the ways that living in a closed system affects each of us. Can you think of some other ways?

5. In what ways do geologic processes affect your daily life?

What is happening in this picture?

NATIONAL GEOGRAPHIC

This rock, photographed in Saudi Arabia's Rub' al Khali ("Empty Quarter"), was discovered in 1965. It is believed to be the largest fragment of a meteorite that fell to Earth sometime before 1863 (when the first piece of it was discovered).

Think Critically

1. How do you think these scientists can tell it is a meteorite?
2. Why did it break up into pieces?
3. Why is the desert a good place to look for meteorites? (Hint: Think about what would have happened to this rock if it had fallen in a jungle or a mountain range.)

Self-Test

(Check your answers in Appendix D.)

1. _____ is fundamentally concerned with understanding the processes that operate at or beneath the surface of Earth and the materials on which those processes operate.
   - a. Economic geology
   - b. Physical geology
   - c. Historical geology
   - d. Environmental geology
   - e. Planetary science

2. In the scientific method, a theory is _____.
   - a. an assumption that cannot be either proven or refuted.
   - b. a plausible, but yet to be proved, explanation of a phenomenon
   - c. a plausible explanation that has been tested and is strongly supported by experimental or observational evidence
   - d. the same thing as a hypothesis
   - e. a guess that scientists make when they cannot find enough evidence to determine the facts

3. In the scientific method, suppose that an experimental test fails to confirm a certain hypothesis. Which of the following is *never* an appropriate step for a scientist to take?
   - a. Discard the hypothesis.
   - b. Alter the experimental data to agree with the hypothesis.
   - c. Repeat the experiment more carefully.
   - d. Formulate and test an alternative hypothesis.

4. The island depicted in the diagram acts as _____.
   - a. an intermittent system
   - b. a closed system
   - c. a solar system
   - d. an open system
   - e. a isolated system

Evaporation from lakes, streams, and soil

Precipitation

Heat (energy)

Sunlight (energy)

Water drains into the sea

5. On the time scale of a human lifetime, Earth acts as _____.
   - a. an intermittent system
   - b. a closed system
   - c. a solar system
   - d. an open system
   - e. an isolated system

6. The _____ is a subset of the Earth system that comprises all of its bodies of water and ice, both on the surface and underground.
   - a. atmosphere
   - b. hydrosphere
   - c. lithosphere
   - d. ionosphere
   - e. biosphere

7. The _____ explains how new crust is formed and recycled by large-scale motions of Earth's surface and interior.
   a. tectonic cycle
   b. rock cycle
   c. water (hydrologic) cycle
   d. nebular hypothesis

8. According to the planetary accretion model, _____.
   a. planets evolved from molten material condensing from an early solar nebula
   b. planets assembled themselves from meteorite-like debris
   c. most of today's meteorites are the debris that never managed to be swept up by any planet
   d. Both b and c are correct.
   e. Answers a through c are all correct.

9. The inner planets of the solar system are rocky, whereas the outer planets contain a higher proportion of ice and gas. This differentiation occurred in the early solar system because _____.
   a. the rocky and metallic components, which have higher melting points, would have condensed at an early stage and in the innermost region of the solar system
   b. the solar wind was strong enough in the inner solar system to push the lighter gases, such as helium and hydrogen, to the outer solar system
   c. volatile compounds, such as water and methane, condense at lower temperatures common to the outer solar system, where the primordial nebula was cooler
   d. All of the above statements are correct.

10. The photograph is of a basalt flow on the island of Hawaii. Which one of the following statements is correct?
    a. Basaltic lava flows have occurred only on Earth.
    b. These types of flows have been common only on Earth and in the early history of the Moon.
    c. Basalt is the most common volcanic rock known in our solar system.
    d. Basaltic lava flows would have been common in the early history of Earth, but in modern times they have largely ceased.
    e. None of the above statements is correct.

11. Earth, Mars, and Venus all have _____.
    a. an oxygen- and nitrogen-rich atmosphere
    b. a core, mantle, and crust
    c. a hydrosphere with liquid water
    d. a biosphere
    e. All of the above answers are correct.

12. These two photographs compare distinctive patterns in windblown sand in a modern dune (a) with similar structures inside an ancient sandstone (b). Our ability to infer that the structures inside the sandstone were formed by the same processes that formed them in a modern sand dune is an application of _____.
    a. the accretionary hypothesis
    b. plate tectonic theory
    c. Hutton's principle of uniformitarianism
    d. the nebular hypothesis
    e. planetary differentiation

a.
b.

13. Hutton's principle of uniformitarianism _____.
    a. states that the physical processes that act on Earth have not changed over time, even though the physical conditions of Earth have changed dramatically
    b. states that the physical processes that act on Earth and the physical conditions of Earth have not changed over time
    c. cannot be used to explain rapid fluctuations in the climate system
    d. has been rendered obsolete by the modern unifying concept of plate tectonics

14. The rates of some processes involved in the cycles of the Earth system, such as the rock cycle and tectonic cycle, _____.
    a. have been constant over time
    b. have varied over time
    c. have been steadily increasing with time
    d. have been steadily decreasing with time

15. The study of geology is important because _____.
    a. it helps us understand the processes that govern the Earth system
    b. it helps us understand the time scales that govern Earth processes
    c. it helps us understand and mitigate the potential threats of geologic hazards, such as earthquakes, volcanic eruptions, landslides, floods, and meteorite impact
    d. All of the above answers are correct.

THE PLANNER ✓

Review your Chapter Planner on the chapter opener and check off your completed work.

Earth Materials

These diamonds (inset) come from Point Lake, Northwest Territories, Canada, where geologists Charles Fipke and Stewart Blusson discovered diamonds in 1991. They hypothesized that diamonds found in Wisconsin had been pushed from northern Canada by glaciers during the last ice age. They looked for the source, and eventually found it. Since 1998, when the mine opened at Point Lake, Canada has become the third largest producer of diamonds in the world.

Apart from its rarity and value, diamond is a remarkable mineral. It is pure elemental carbon, the same chemical composition as graphite and charcoal. Unlike graphite and charcoal, natural diamond forms at extraordinarily high pressures and temperatures, deep under Earth's surface. It is the hardest mineral known, and an excellent conductor of heat. Its properties make it valuable for industrial purposes.

Artificial diamonds might one day replace silicon in computers; a computer with diamond chips could run at much higher temperatures.

In this chapter you will learn about rocks and minerals. Some are beautiful, some are important economically, and some are vital as nutrients; others are hazardous to human health. By studying rocks and minerals, we can learn to balance their positive and negative effects in our lives.

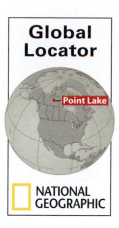

CHAPTER OUTLINE

CHAPTER PLANNER ✓

- ☐ Study the picture and read the opening story.
- ☐ Scan the Learning Objectives in each section:
  p. 36 ☐   p. 40 ☐   p. 50 ☐   p. 56 ☐
- ☐ Read the text and study all visuals. Answer any questions.

Analyze key features

- ☐ Process Diagram, p. 38 ☐
- ☐ Geology InSight, p. 39 ☐
- ☐ What a Geologist Sees, p. 40 ☐
- ☐ Case Study, p. 41 ☐   p. 49 ☐
- ☐ Amazing Places, p. 45 ☐
- ☐ Stop: Answer the Concept Checks before you go on:
  p. 38 ☐   p. 49 ☐   p. 55 ☐   p. 58 ☐

End of chapter

- ☐ Review the Summary and Key Terms.
- ☐ Answer the Critical and Creative Thinking Questions.
- ☐ Answer What is happening in this picture?
- ☐ Complete the Self-Test and check your answers.

Elements and Compounds

LEARNING OBJECTIVES

1. **Define** element, atom, compound, molecule, and ion.
2. **Explain** the difference between an atom and a molecule.
3. **Describe** the internal structure of an atom.
4. **Identify** four kinds of chemical bonding.
5. **Explain** how the kinds of bond in a material affect its physical properties.

All matter on and in Earth, including the page of text you are reading and the eyes you are reading it with, consists of one or more chemical elements. All of the chemical reactions that control our lives and make life on Earth possible depend on the ways in which these elements interact.

Elements, Atoms, and Ions

element The most fundamental substance into which matter can be separated using chemical means.

Chemical **elements** are the most fundamental substances into which matter can be separated and analyzed by ordinary chemical methods. Ninety-two naturally occurring elements are known, and atomic scientists have synthesized a number of other elements. Each element is identified by a symbol, such as H for hydrogen and Si for silicon. Some of the symbols come from the names of the elements in languages other than English, such as Fe for iron, from the Latin *ferrum*, and Na for sodium, from the Latin *natrium*. Others are named in honor of famous scientists, such as element 99, Es, einsteinium. A table of the elements listed in alphabetical order is shown in Appendix B. The table also gives the percentage of each element in the continental crust.

Even the tiniest grain of dust is made of innumerable particles, called **atoms**, which are much too small for the eye to see (**Figure 2.1**). In a pure sample of an element, every atom would be the same kind. Atoms are the building blocks of chemistry; they account for all of an element's chemical properties and many of its physical ones, such as density and color. They are so tiny, about one-billionth of a millimeter, that they cannot be seen at all with a conventional optical

atom The smallest individual particle that retains the distinctive chemical properties of an element.

microscope; only specially designed microscopes have succeeded in imaging atoms.

Chemical reactions produce the rocks, minerals, liquids, and gases of which Earth is made and that geologists study. For this reason, a quick review of chemical elements and atoms is a good place to begin our study of Earth materials. Atoms themselves are composed of even smaller particles, which have no independent chemical properties. The **nucleus** (plural **nuclei**) of an atom contains **protons**, with positive electric charges, and **neutrons**, which are electrically neutral. The number of protons in an atom—its **atomic number**—determines which element the atom is and gives the atom its chemical characteristics. Atomic numbers range from 1 for the lightest element, hydrogen, to 92 for the heaviest naturally occurring element, uranium. Every element from atomic number 1 to 92 has either been synthesized in the laboratory or found in nature, so there are no new elements to be discovered in that range. However, some laboratories are still working on synthesizing heavier elements, and they have gotten up to element 118 (ununoctium).

The number of protons plus the number of neutrons in the nucleus of an atom is the **mass number**. Atoms of a given element always have the same atomic number, but they can have different mass numbers. For example, there are three naturally occurring **isotopes** of carbon: carbon-12, carbon-13, and carbon-14. Each of the isotopes of carbon has 6 protons and thus an atomic number of 6. However, the three isotopes contain different numbers of neutrons: 6, 7, and 8 per atom, respectively (thus different mass numbers: 12, 13, and 14).

isotopes Atoms with the same atomic number but different mass numbers.

The third small component of an atom is called an *electron*. Electron interactions help determine the make-up of ions and compounds. Electrons orbit the nucleus in complex patterns, shown schematically in Figure 2.1

as circles of different sizes. An electron has a negative charge that is equal in magnitude but opposite in sign to the positive charge of a proton. In its ideal state, an atom has an equal number of protons and electrons and thus is electrically neutral. Under certain circumstances, however, an atom may gain or lose an electron during a chemical reaction. Although atoms can exchange electrons, they never exchange protons or neutrons as a result of chemical processes.

An atom that has lost or gained one or more electrons has a net electric charge and is called an **ion**. If the charge is positive, meaning that the atom has lost one or more electrons, the ion is called a **cation**. If the charge is negative, meaning that the atom has gained one or more electrons, the ion is called an **anion**. A convenient way to indicate ionic charges is to record them as superscripts. For example, Na^+ is the symbol for an atom of sodium that has given up an electron; Cl^- is the symbol for an atom of chlorine that has accepted an electron; and Fe^{2+} is the symbol for an atom of iron that has given up two electrons.

Compounds, Molecules, and Bonding

Chemical **compounds** form when atoms of one or more elements combine with atoms of another element in a specific ratio. For example, sodium and chlorine combine to form sodium chloride (a mineral called halite, also known as table salt), which is written NaCl. For every Na atom in this compound,

> **compound** A combination of atoms of one or more elements in a specific ratio.

there is one Cl atom. The element that tends to form cations is written first; the element that tends to form anions is written second; and the relative numbers of atoms are indicated by subscripts. For example, water forms when hydrogen (a cation, H^+) combines with oxygen (an anion, O^{2-}) in the ratio of two atoms of hydrogen to one atom of oxygen. Thus, for water, we write H_2O.

Inside an atom • Figure 2.1

A single atom of carbon-12

As shown in this schematic diagram of an atom of carbon-12, six electrons orbit the nucleus in two complex paths called *orbitals*, rendered here (unrealistically) as circles. The orbitals arrange themselves in energy-level shells, which are more stable when completely filled. Besides being the sole component of diamond, graphite, and coal, carbon is the most important atom in living beings.

Three materials made of carbon

Diamond

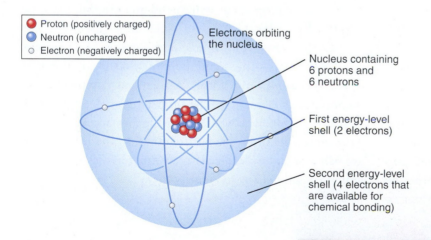

- ● Proton (positively charged)
- ● Neutron (uncharged)
- ○ Electron (negatively charged)

Electrons orbiting the nucleus

Nucleus containing 6 protons and 6 neutrons

First energy-level shell (2 electrons)

Second energy-level shell (4 electrons that are available for chemical bonding)

Graphite

Coal

How ions and compounds form • Figure 2.2

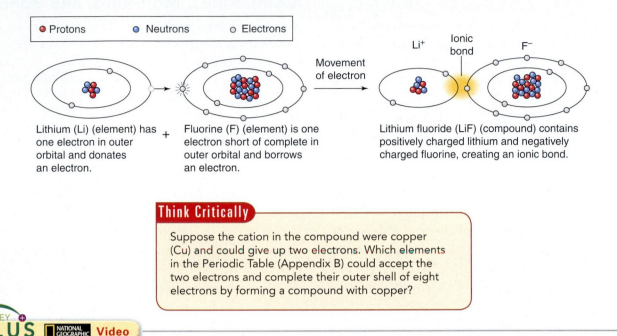

● Protons   ● Neutrons   ○ Electrons

Movement of electron

Li⁺   Ionic bond   F⁻

Lithium (Li) (element) has one electron in outer orbital and donates an electron.

Fluorine (F) (element) is one electron short of complete in outer orbital and borrows an electron.

Lithium fluoride (LiF) (compound) contains positively charged lithium and negatively charged fluorine, creating an ionic bond.

Think Critically

Suppose the cation in the compound were copper (Cu) and could give up two electrons. Which elements in the Periodic Table (Appendix B) could accept the two electrons and complete their outer shell of eight electrons by forming a compound with copper?

WILEY PLUS NATIONAL GEOGRAPHIC Video

Figure 2.2 shows how lithium and fluorine combine to form an ionic compound, lithium fluoride, abbreviated as LiF.

Clearly, the properties of compounds are not the same as the properties of their constituent elements. For example, hydrogen and oxygen are both gases at surface temperature and pressure, whereas water is a liquid. Similarly, elemental sodium and elemental chlorine are both highly toxic, whereas their compound, salt, is essential for life.

The smallest unit that has the properties of a given compound is a **molecule**. Do not confuse a molecule and an atom; the definitions seem similar, but a molecular compound always consists of two or more atoms; the compound H_2O is an example. Molecules are held together by

molecule The smallest chemical unit that has all the properties of a particular compound.

bond The force that holds together the atoms in a chemical compound.

electromagnetic forces known as **bonds**. Bonding involves the transfer of electrons from one atom to another or in some cases the sharing of electrons. There are four principal kinds of bonds, as illustrated in **Figure 2.3**. You will see that different bonding explains the difference in properties of diamond and graphite.

We have spent some time discussing elements, compounds, and bonding because minerals—the main building blocks of the geosphere—occur in the form of chemical compounds (or sometimes as elements). The properties of these minerals depend very much on what chemicals they are made of and how those chemicals are put together. In the next section we will look more closely at the characteristics that define minerals and help us to identify them.

CONCEPT CHECK   STOP

1. **How** does an atom differ from an ion?

2. **Why** can water be separated into two chemically distinct substances (hydrogen and oxygen), but gold cannot?

3. **What** determines an element's atomic number?

4. **What** are the four main types of bonds in minerals?

5. **Why** are metals good conductors of electricity?

Geology InSight

Thinking critically about the four types of bonding • **Figure 2.3**

Bonding type	Crystal structure	Occurence	Uses

Ionic bonding: What happens when one atom transfers an electron to another? As illustrated in Figure 2.2, an attractive force is set up between the donor and the receiver that creates an ionic bond. In a crystal such as table salt (NaCl), the positive sodium ions (red) are attracted to all of the negative neighboring chlorine ions (gray), not just to one of them. Thus the ionic bonds form a cubic lattice. Compounds with ionic bonds tend to have moderate strength and hardness.

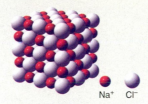

In table salt (sodium chloride, NaCl), each sodium cation is surrounded by chlorine anions.

Crystals of sodium chloride are rectangular, with straight edges.

Salt is a moderately hard solid that dissolves easily in water.

Covalent bonding: What happens when electrons from different atoms "pair up"? The force of sharing is called a covalent bond. Note that electron sharing does not produce ions. These are the strongest chemical bonds, and elements and compounds with covalent bonds (e.g., diamond) tend to be strong and hard.

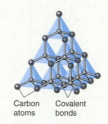

Carbon atoms / Covalent bonds

Diamond consists of carbon atoms connected in a network of covalent bonds. Each atom is connected to four others.

Diamond crystals appear in a rock called kimberlite. Covalent compounds are often strong and hard; diamond is one of the hardest substances known.

Cut and polished diamonds are prized gems. Tiny diamonds are used in industry for cutting and grinding instruments.

Metallic bonding: Metals are different. Why? In metals, atoms are so tightly packed that electrons can be shared among several atoms. In fact, the outermost electrons are so loosely held that they can readily drift from one atom to another. This mobility of electrons explains why metals are good at conducting electricity and heat.

Atoms of gold are packed in the densest possible manner. Each atom is surrounded by, and in contact with, 12 other gold atoms.

This 2.3-kg nugget of gold was once embedded in rock, but weathering and erosion have removed most of the rock.

Gold is durable as well as malleable; it has been used as currency since ancient times.

Van der Waals bonding: Are there any weak kinds of bonds? Yes, a weak attraction occurs between electrically neutral molecules that have an asymmetrical charge distribution. The positive end of one molecule is attracted to the negative end of another molecule, but the attraction is weak. For example, the carbon atoms in graphite form sheets in which each carbon atom has strong covalent bonds with three neighbors. The sheets are held together by weak Van der Waals bonds. This is why graphite feels slippery when you rub it between your fingers—you are breaking the Van der Waals bonds.

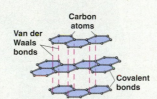

Carbon atoms / Van der Waals bonds / Covalent bonds

In graphite, carbon atoms form layers connected by covalent bonds. The layers are weakly held together by Van der Waals bonds.

Graphite is not a strong material and can be easily crumbled into small particles.

The "lead" in pencils is not lead, it's really graphite. When you write, the pressure of your hand breaks off a trail of carbon particles.

What Is a Mineral?

LEARNING OBJECTIVES

1. **Identify** four requirements for a solid material to be classified as a mineral.

2. **Explain** the principle of atomic substitution.

3. **Explain** why crystals tend to have flat faces with specific angles between them.

4. **Explain** why color is one of the least dependable ways to identify a mineral.

To be classified as a **mineral**, a substance must satisfy four criteria; first, it must be a naturally occurring solid, second it must be formed by inorganic processes, third, it must have a characteristic crystal structure and fourth, it must have a specific chemical composition. The *Case Study* discusses some of the ways of classifying a material as a mineral.

> **mineral** A naturally formed solid, inorganic substance with a characteristic crystal structure and a specific chemical composition.

Telling Minerals Apart

The compositions and crystal structures of minerals influence their physical properties and characteristics. If we have an unidentified mineral sample (as in *What a Geologist Sees*), we can apply a few simple tests to determine what mineral it is without taking it to a laboratory or using expensive equipment. The properties most often used to identify minerals are the quality and intensity of light reflected from the mineral, the crystal form and habit of the mineral, its hardness, its tendency to break in preferred directions, its color, and its specific gravity or density. Color, perhaps the most obvious characteristic, is often the least reliable identifier. Let's look at the properties that are used to identify different minerals and follow the reasoning of a geologist as she identifies the mystery mineral in *What a Geologist Sees*.

WHAT A GEOLOGIST SEES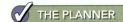

A "Mystery Mineral"

A geologist would first note the metallic luster of the specimen and note that the mineral has grown as a group of cubes—a geologist would make a note that says "cubic habit." Picking up the specimen, the geologist would discover that it is very heavy—that is, it has a high density. We will discuss other things that a geologist would see in this mineral sample as we continue with the text.

Mineral or Not?

Ice is a mineral, though you may not usually think of it this way. It occurs in nature in the form of hexagonal crystals and has a specific chemical formula (H_2O).

Water is *not* a mineral because it is not a solid. This criterion also means that naturally occurring substances such as oil and natural gas cannot be considered minerals.

Coal fails the second of the four tests for a mineral because it is derived from the remains of plant material and was formed as a result of organic processes. Coal also fails the composition test because it is not a single compound; rather, it is composed of many different compounds.

Steel (being produced in the background) fails the first of the four tests for a mineral because it does not occur naturally. It is formed by extensive human processing of naturally occurring ores, which are minerals.

Bones are a tricky case. They do contain the same chemical compound found in a common mineral called apatite, but they are *not* minerals because they form by organic processes. Thus the bone in this modern crocodile skull is not a mineral. Bones and other solids formed by organic processes are often called **biominerals**.

This **fossil**, a crocodile skull in the Kenyan National Museum, was once made of bone but is now composed of minerals. During fossilization, the materials of the original bone were replaced in an inorganic process called **mineralization**.

Quartz is an easily recognizable mineral. Its chemical formula is SiO_2. Note that some minerals have very complex formulas. For example, phlogopite, a form of mica, is $KMg_3AlSi_3O_{10}(OH)_2$. The important thing is that the elements combine in specific ratios. Phlogopite is a mineral even though its chemical formula is complicated.

Opal, a gemstone, is typically included in books about minerals, but it is not a true mineral. Opals do not have a specific composition, and they lack a crystalline structure (they are amorphous). Opal is an example of a **mineraloid**.

Think Critically

Which of the following substances fulfill all of the defining characteristics of minerals? Asphalt; a kidney stone; a synthetic diamond; quartz that has been broken apart by being transported in a stream; glass in a lightbulb; a nugget of gold panned from a stream; volcanic glass.

Composition of Minerals A complication to the rule that a mineral must have a specific chemical composition arises from a phenomenon called **atomic substitution**. In some cases, the ions of two elements can be similar enough in size and in bonding properties that they can substitute for each other in a mineral. For example, magnesium and iron ions (Mg^{2+} and Fe^{2+}) often can take the place of one another in a mineral lattice because they are similar in size and have identical electrical charge. The mineral olivine (an important component of Earth's mantle) can occur as pure Fe_2SiO_4 or pure Mg_2SiO_4 or an intermediate mixture in which some of the Fe^{2+} cations are replaced by Mg^{2+} cations. The formula for olivine, therefore, becomes $(Mg,Fe)_2SiO_4$, indicating that Mg and Fe can substitute for one another in this mineral. Note that the ratio of cations to anions is not changed by atomic substitution, so the specific composition rule is not violated.

The composition requirements for minerals specifically rule out materials whose composition varies so much that it cannot be expressed by an exact chemical formula. An example of such a material is glass, which is a mixture of many elements and can have a wide range of compositions.

Glass—even naturally formed volcanic glass—fails the test of having a characteristic **crystal structure**. In a crystal, the atoms are arranged in regular, highly repetitive geometric patterns, as shown in **Figure 2.4a**. By contrast, the atoms in a liquid or in an **amorphous** solid like glass are mixed up or randomly jumbled.

All specimens of a given mineral have an identical crystal structure. Extremely sensitive scanning tunneling microscopes enable scientists to look at the crystal structures of minerals and actually see the orderly arrangement of atoms in the mineral. As you can see in **Figure 2.4b**, the atoms in a crystalline material resemble the regular, orderly rows in an egg carton.

Sometimes two minerals can have the same chemical formula but different crystal structures. A very common example is calcium carbonate ($CaCO_3$), which forms two different minerals called calcite and aragonite. These different forms are called **polymorphs**. Likewise, graphite and diamond are polymorphs of carbon, as explained in Figure 2.3. Remarkably, water ice has 14 different polymorphs that have been discovered so far, only 1 of which occurs naturally on Earth. Some of the other polymorphs of ice are likely to turn up on the ice-covered moons of other planets, such as Jupiter.

Note that some materials occur in nature as **native elements**; that is, they are not combined in compounds with other elements. Minerals that can occur in this form include some metals, such as gold (Au) and silver (Ag), and some nonmetals, such as sulfur (S), graphite (C), and diamond (also C).

> **crystal structure**
> An arrangement of atoms or molecules into a regular geometric lattice. Materials that possess a crystal structure are said to be crystalline.

Atomic structures of crystals • Figure 2.4

Sulfur
(S)

Lead
(Pb)

Lead atom

Sulfur atom

a. The atoms of all crystalline materials are arranged in orderly lattices, like the cubical lattice illustrated here for a mineral called galena (PbS), the main source of lead. The atoms are so small that a cube of galena 1 centimeter on an edge would contain 10^{22} atoms (that's 1 followed by 22 zeros). The inset shows an "exploded view" of the packing arrangement of atoms in a galena crystal. The atoms are shown pulled apart along the black lines to demonstrate how they fit together. Compare the arrangement of atoms in NaCl (Figure 2.3); it is the same as the arrangement of lead and sulfur atoms in PbS.

b. Atoms are too small to see with an optical microscope, but a scanning tunneling microscope can show the atoms in a crystal. This is an image of a galena crystal; the sulfur atoms look like large bumps and the lead atoms like small ones.

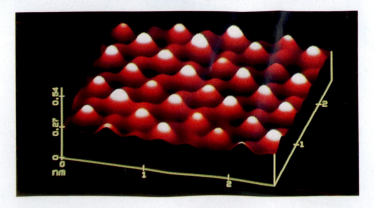

Luster Minerals reflect light in different ways; we call this property the **luster** of a mineral. The luster of the mystery mineral in *What a Geologist Sees* is **metallic**, meaning that it looks like a polished metal surface. As illustrated in **Figure 2.5**, other kinds of luster you might encounter are **vitreous**, like glass; **resinous**; **pearly**; or **greasy**, as if the surface were covered by a film of oil.

Crystal form and habit

Ancient Greeks were intrigued by the smooth, planar surfaces of ice needles. The Greeks called ice *krystallos*. Eventually, the word **crystal** came to be applied to any solid body that has grown with planar faces. The planar surfaces that bound a crystal are called **crystal faces**.

During the 17th century, scientists investigated crystal form as a possible way to identify minerals. But the sizes of crystal faces vary widely from one sample to another. It is apparent in **Figure 2.6** that crystal size and the relative sizes of crystal faces are not the same for these two crystals of quartz. In fact, the relative sizes of crystal faces are not definitive for any mineral.

In 1669 a Danish physician, Nils Stensen (known by his Latin name, Nicolaus Steno) demonstrated that the key property is not the size of faces but rather the angles between them. According to **Steno's Law**, the angle between any corresponding pair of crystal faces of a given mineral is constant, no matter what the shape or size of the crystal may be (see Figure 2.6).

Crystal faces and angles • Figure 2.6

Crystals of the same mineral may differ widely in shape, but the angles between faces will remain the same in all specimens. In the two quartz crystals pictured, numbers identify equivalent faces. According to Steno's Law, the angle between faces 1 and 3 (for example) is the same in both specimens.

Steno suspected that a mineral must have some kind of internal order that predisposes it to form crystals with constant interfacial angles. Proof of a crystal's internal atomic structure finally arrived in 1912, when German scientist Max von Lauë beamed X-rays at crystals and showed that the diffracted rays matched the patterns a geometric array would generate. More recently, scanning tunneling microscopes have given more direct proof of atomic lattice structure, as shown in Figure 2.4b.

Crystals can form either by **crystallization**—essentially, freezing—which occurs when a molten mineral cools to a temperature where it cannot stay in the liquid state, or by **precipitation**, which occurs when a dissolved mineral becomes sufficiently concentrated that it cannot remain dissolved. Both types of crystals get started by nucleation: Molecules of the mineral adhere either to each other, or to

Mineral luster • Figure 2.5

Here are examples that show four common types of luster.

Metallic
Pyrite (a form of iron sulfide and a common iron mineral) has a metallic luster.

Vitreous
Quartz has a glassy luster.

Resinous
Sphalerite (ZnS, a source of zinc) has a resinous luster, like dried tree resin.

Pearly
Talc has a pearly luster.

any solid object that happens to be present in the liquid, to establish a tiny crystal nucleus. (This is just a tiny "starter crystal," not a nucleus in the same sense as the nucleus of an atom.) Once the nucleus is established, the crystal grows by attracting more molecules of the same type, as a result of one of the intermolecular forces mentioned earlier (i.e., ionic bonding, covalent bonding, metallic bonding, or Van der Waals bonds). As molecules are added, they pack together, following the regular geometric lattice characteristic of the mineral that is forming.

Crystals with planar faces form most easily when mineral grains grow freely in an open space, such as in a cave. You can see a truly remarkable example of this at the Naica Mine in Chihuahua, Mexico, profiled in *Amazing Places*. Because most mineral grains do not form in open, unobstructed spaces, nicely formed crystals are uncommon in nature. Usually other mineral grains get in the way. As a result, most mineral grains have an irregular shape. However, in both a crystal and an irregularly shaped grain of the same mineral, all the atoms present are packed in the same strict geometric pattern. This is why we use the term crystal structure rather than just crystal in defining the characteristics of minerals.

habit The distinctive shape of a particular mineral.

Some minerals grow in such distinctive ways that their shape—called the **habit**—can be used as an identification tool. The geologist noted that the mystery sample in *What a Geologist Sees* clearly has a cubic

Fibers of asbestos • Figure 2.7

Some minerals have distinctive growth habits even though they do not develop well-formed crystal faces. The mineral chrysotile $[Mg_3Si_2O_5(OH_4)]$ sometimes grows as fine, cotton-like threads that can be separated and woven into fireproof fabric. When the mineral occurs like this, it is said to have an **asbestiform** habit. Many different minerals can grow with asbestiform habits, and several are mined and commercially sold as asbestos.

habit because it looks like a collection of interlocked cubes. A very different example is the mineral chrysotile, shown in **Figure 2.7**, which takes the form of fine fibers or threads. This **fibrous** habit is characteristic of asbestiform minerals such as chrysotile.

hardness A mineral's resistance to scratching.

Hardness The **hardness** of a mineral, like habit and crystal form, is governed by crystal structure and by the strength of the bonds between atoms. The stronger the bonds, the harder the mineral. Relative hardness values can be assigned by determining whether one mineral will scratch another, using the **Mohs relative hardness scale**. The scale is divided into 10 steps, each marked by a common mineral listed in **Table 2.1**.

The Mohs scale* of relative hardness of minerals   Table 2.1

These are the 10 reference minerals of the Mohs scale, starting with the softest, talc, in the upper-left corner, and proceeding across two rows to diamond, the hardest mineral in the lower-right corner.

*Named for Friedrich Mohs, a German mineralogist, who chose the 10 minerals of the scale.

Relative Hardness	Number	Reference Mineral	Hardness of Common Objects
Softest	1	Talc	
	2	Gypsum	
			Fingernail
	3	Calcite	
			Copper penny
	4	Fluorite	
	5	Apatite	
			Pocketknife; glass
	6	Potassium feldspar	
	7	Quartz	
	8	Topaz	
	9	Corundum	
Hardest	10	Diamond	

AMAZING PLACES

The Naica Mine, Chihuahua, Mexico

In 2000, two miners in one of Mexico's largest lead and silver mines, the Naica mine near Chihuahua, unexpectedly broke through to a new cave 290 meters below the surface. A similar opening, called the Cave of Swords, had been discovered in 1912, with walls covered by large crystals of gypsum (calcium sulfate). But no one could have expected what they found in the newly discovered cave: crystals, mainly of selenite (a variety of gypsum), that are the largest known crystals of any kind, anywhere in the world.

The giant crystals grew from saline solutions rich in calcium sulfate at a temperature just below 58°C. They grew undisturbed over a period of 600,000 years. When the water was pumped out of the adjacent lead mine, this cave was exposed to the air for the first time, and therefore the crystals are no longer growing.

▲ The largest crystals in the Cave of Crystals can be seen in the foreground. They exceed 11 meters in length and weigh an estimated 55 tons.

▲ The Cave of Swords, discovered in 1912, is still in its original condition because it has been protected from vandalism by the mining company.

◄ The Cave of Crystals, discovered in 2000, contains the most striking collection of giant crystals ever discovered. The sizes of the crystals can be judged in comparison to the size of the miner.

What Is a Mineral? 45

Talc, the basic ingredient of "talcum" powders, is the softest mineral known and therefore is assigned a value of 1 on the relative hardness scale. Diamond, the hardest mineral, has a value of 10. The 10 steps of the hardness scale do not represent equal intervals of hardness; the important thing is that any mineral on the scale will scratch all minerals below it. For convenience, we often test relative hardness by using a common object such as a penny or a penknife as the scratching instrument, or glass as the object to be scratched. The hardness values of these objects are also shown in Table 2.1.

Our geologist investigating the mystery mineral in *What a Geologist Sees* would find that it is relatively soft and scratches easily with a copper penny or a piece of fluorite. It can even be scratched by a fingernail, but just barely. This puts its hardness somewhere around 2.5 on the Mohs hardness scale.

Cleavage If you break a mineral with a hammer or drop it on the floor so that it shatters, some of the broken fragments will be bounded by surfaces that are smooth and planar; these are called **cleavage** surfaces. In **Figure 2.8**, the muscovite breaks or cleaves easily into thin sheets but does not cleave at all in any other direction. In certain minerals, several cleavage directions are present. Potassium feldspar ($KAlSi_3O_8$) has two cleavages, and halite (NaCl) has three directions, so that all the fragments of halite shown in Figure 2.8 are bounded by smooth planar surfaces. Don't confuse crystal faces and cleavage surfaces, even though the two often look alike; a cleavage surface is a breakage surface, whereas a crystal face is a growth surface. Also, note the difference between hardness and cleavage: A mineral might be quite hard—that is, resistant to scratching—but it may still cleave easily in one or more directions.

Mineral cleavage • Figure 2.8

a. Muscovite (a potassium alumino-silicate mineral), a form of mica, cleaves so easily in one direction that it can be split by hand into flakes that are suggestive of the pages of a book.

b. Potassium feldspar (another potassium alumino-silicate), one of the most common minerals on Earth, breaks along two directions at right angles.

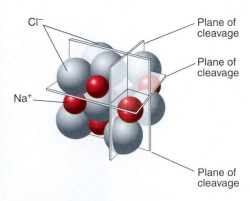

c. Halite (NaCl, or table salt) has three distinct cleavage directions. No matter how small the bits you break it into, they will always have perpendicular faces.

Diamond, the hardest mineral, has four directions of cleavage.

The directions in which cleavage occurs are governed by crystal structure. Cleavage takes place along planes where the bonds between atoms are relatively weak, as in the case of muscovite, or where there are fewer bonds per unit area, as in the case of diamond. Because cleavage directions are directly related to crystal structure, the angles between equivalent pairs of cleavage directions are the same for all grains of a given mineral. Thus, analogously to Steno's Law for crystals, the angles between cleavage planes are constant. A small hand lens is usually enough for a geologist to spot these distinctive angles. As we remarked earlier, crystals and crystal faces are rare; but almost every mineral grain you see in a rock shows one or more breakage surfaces. That is why cleavage is such a useful aid in the identification of minerals.

Our geologist would carefully examine the mystery mineral in *What a Geologist Sees* and discover a place where rough handling has broken off a fragment, revealing three cleavages at right angles.

Color and streak The color of a mineral, though often striking, is not a reliable means of identification (**Figure 2.9**). A mineral's color is determined by several factors, but the main determinant is chemical composition. Some elements can create strong color effects, even when they are present only as trace impurities. For example, the mineral corundum (Al_2O_3) is commonly white or grayish, but when small amounts of chromium are present as a result of atomic substitution of Cr^{3+} for Al^{3+}, corundum is blood red and is given the gem name ruby. Similarly, when small amounts of iron and titanium are present, the corundum is deep blue, producing another gem, sapphire. The *Case Study* on the next page discusses a number of other gemstones. Many other colors, such as green, gray, white, black, and pink, are not particularly useful in identifying minerals because there are so many minerals that occur in these colors.

Same mineral, different colors • Figure 2.9

These cut gems are all synthetic sapphires, the mineral corundum. They are all the same type of material, but slight differences in chemical composition give them very different colors.

Uncut natural specimens of the same material, corundum (Al_2O_3), also have very different colors. The red crystals, rubies, are from Tanzania, and the blue crystals, sapphires, come from Newton, New Jersey. The ruby in the upper-right corner is about 2.5 centimeters across.

Color can be particularly confusing in opaque minerals that have metallic luster. This is because the color is partly a property of the size of the mineral grains. One way to reduce error is to prepare a **streak** by rubbing a specimen of a metallic luster mineral on an unglazed fragment of porcelain called a **streak plate**. The color of a streak is reliable because all the grains in the streak are very small, and the effect of grain size is reduced. For example, hematite (Fe_2O_3), an important iron-bearing mineral, produces a reddish-brown streak, even though a specimen may look black and metallic (**Figure 2.10**). Our geologist would discover that the streak of the mystery mineral from *What a Geologist Sees* is gray.

> **streak** A thin layer of powdered mineral made by rubbing a specimen on an unglazed fragment of porcelain.

Density Another important physical property of a mineral is how light or heavy it feels; this is an indication of its **density**. Two equal-sized baskets have different weights when one is filled with feathers and the other with rocks because the rocks have greater density than the feathers. Minerals that have a high density, such as gold, have closely packed atoms. A sample of gold feels distinctly heavy when you pick it up. Minerals with a low density, such as ice, have less closely packed atoms.

It is not easy to measure the density of a mineral in a laboratory because doing so requires dividing the mass by the volume, and the volume of an irregularly shaped object is difficult to determine. This was the same problem that Archimedes faced more than 2000 years ago, when he was asked to determine whether the gold in a crown was pure or alloyed with a less dense material. Geologists use a modern version of Archimedes' method. First, they measure the weight of the mineral grain in air (W_A). Then they submerge it in water and measure the weight again (W_W). The weight in water is less than that in air, and the difference ($W_A - W_W$) is the weight of water displaced by the mineral grain. The ratio of W_A to ($W_A - W_W$) gives the **specific gravity** of the mineral. Water has a density of 1.0 grams per cubic centimeter, which means that its specific gravity is numerically equal to its density.

Many common minerals, such as quartz, have specific gravities in the range of 2.5 to 3.0 g/cm^3. Fortunately, fancy equipment is usually not needed to determine the specific gravity of an unknown mineral sample. If a mineral is substantially lighter than 2.5 g/cm^3 or substantially heavier than 3.0 g/cm^3, a geologist can immediately tell this by lifting the sample. Metallic minerals generally feel heavy, whereas minerals with vitreous luster tend to feel light. Our geologist testing the mystery mineral in *What a Geologist Sees* would discover that it feels quite a bit heavier than most other rocks and minerals of the same size; if she took it to a lab, she would find that its specific gravity is 7.5 g/cm^3.

Mineral streak • Figure 2.10

Although hematite (Fe_2O_3) is shiny, black or gray, and metallic looking, it makes a distinctive red-brown smear when rubbed on a porcelain streak plate.

Other Mineral Properties

We have described the most commonly used properties for identifying a mineral. However, other properties can also be helpful. Some minerals are translucent or transparent, while others are opaque. Certain minerals, such as calcite, are **birefringent**; when you look at an object through them, you can see a double image because the mineral's crystal structure splits light beams in two. If you can afford to destroy some of your sample, dissolving it in acid may provide useful information about its chemical composition. For instance, carbonate minerals such as calcite dissolve readily in dilute hydrochloric acid or even in vinegar; a droplet of the acidic liquid onto the surface of the sample will yield tiny bubbles as the min-

CASE **STUDY**

Minerals for Adornment

Real or synthetic? All gems can be synthesized, so how can you be sure you are buying a natural stone? Trusting reliable jewelers is a good idea, but you can also make your own test. Synthetic stones are usually free of flaws and inclusions, while natural stones almost always have some tiny defects. Diamond, for example, often contains tiny specks of black graphite. A 10x magnifier is a good way to look for imperfections and make sure your sample is natural.

Think Critically

Why do you think natural minerals tend to have inclusions and impurities, whereas synthetic ones don't?

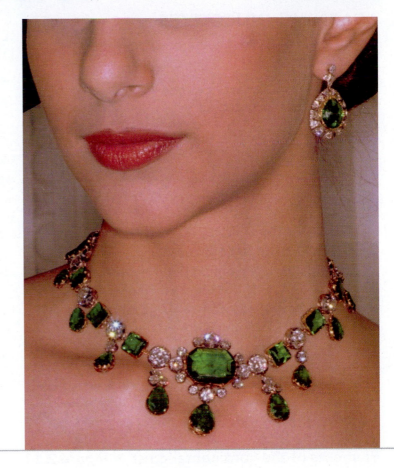

eral dissolves and releases carbon dioxide gas. Certain minerals, particularly those that contain iron in their chemical composition, may be **magnetic**; that is, they are attracted by a magnet. Still other minerals, such as scheelite and calcite, have an unusual property—called **luminescence** or **fluorescence**—that causes them to glow when exposed to an ultraviolet light. This property can be easily tested with a handheld "black light" source.

Mystery mineral clues What is the identity of the mystery mineral in *What a Geologist Sees?* Our geologist discovered several clues:

1. It has a metallic luster.
2. It has a cubic habit.
3. It is fairly soft, with a hardness of about 2.5 on the Mohs scale.
4. Both the color and the streak are gray.
5. It has three perpendicular cleavage directions.

6. It feels heavy and has an unusually high specific gravity of 7.5 g/cm^3.

Can you figure out the identity of the mineral? Refer to the table of minerals in Appendix C and see if your identification agrees with that of the geologist. (The answer is given at the end of the chapter.)

CONCEPT CHECK STOP

1. **What** requirements must be satisfied if a substance is to be called a mineral?
2. **Why** is it possible for an ion of Fe to substitute for an ion of Mg in minerals such as olivine?
3. **What** is the difference between a cleavage surface and a crystal face?
4. **What** is the difference between luster, color, and streak, and how are they related?

Mineral Families

LEARNING OBJECTIVES

1. **Name** the 12 most common chemical elements in Earth's crust.

2. **Identify** the 2 most common mineral families and 4 common accessory mineral families.

3. **Describe** the various molecular structures of silicate minerals.

4. **Give** examples of the economic uses of minerals.

5. **Discuss** some of the challenges and impacts of mining.

G eologists have identified approximately 4000 mineral species. This number may seem large, but it is tiny compared with the number of synthetic materials, such as ceramics, concrete, drugs like aspirin, and solid chemical reagents. The reason for the disparity between the number of minerals and the millions of solids that have been synthesized in laboratories becomes clear when we consider the relative abundances of the chemical elements in nature. Out of every kilogram of material in Earth's continental crust, only 12 elements are present in quantities greater than 1 gram, as shown in **Figure 2.11**. The abundant 12 account for 992.3 of the 1000 grams in a kilogram of Earth's continental crust; all common minerals have compositions based on 1 or more of these abundant elements. The remaining 80

elements, combined, make up less than 1% of the crust by weight and less than 2% by volume. Minerals made of the scarcer elements occur only in small amounts, and ore deposits of scarce elements such as gold, uranium, and tin are rare and hard to find. However, these scarce elements can be extracted and used to synthesize a wide range of materials in the laboratory and in manufacturing for everyday use.

To see some of the great variety of minerals that have been discovered, and to explore a fantastic collection of mineral photographs, see *Where Geologists Click*.

Minerals of Earth's Crust

Mineral families are groups of minerals that are similar to one another in terms of chemistry, atomic structure, or (more commonly) both. Two elements—oxygen and

Elements of the continental crust • Figure 2.11

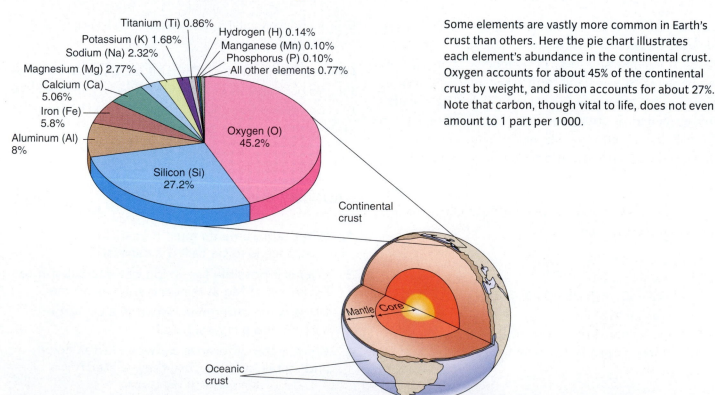

Titanium (Ti) 0.86%
Potassium (K) 1.68%
Sodium (Na) 2.32%
Magnesium (Mg) 2.77%
Calcium (Ca) 5.06%
Iron (Fe) 5.8%
Aluminum (Al) 8%

Hydrogen (H) 0.14%
Manganese (Mn) 0.10%
Phosphorus (P) 0.10%
All other elements 0.77%

Oxygen (O) 45.2%
Silicon (Si) 27.2%

Continental crust

Mantle Core

Oceanic crust

Some elements are vastly more common in Earth's crust than others. Here the pie chart illustrates each element's abundance in the continental crust. Oxygen accounts for about 45% of the continental crust by weight, and silicon accounts for about 27%. Note that carbon, though vital to life, does not even amount to 1 part per 1000.

silicon—make up more than 80% of the atoms in Earth's crust and comprise more than 70% of its mass. Thus it should be no surprise that the great majority of minerals contain one or both of these elements. The most common family of minerals, the **silicate minerals**, contain a strongly bonded complex anion called the silica anion that contains both silicon and oxygen, $(SiO_4)^{4-}$. The bonding in most silicate minerals is a mixture of ionic and covalent; as a result, silicates tend to be hard, tough minerals (**Figure 2.12**). The next most abundant family is the **oxide minerals**, which contain the simple oxide anion O^{2-}.

Rock-forming and accessory minerals

A few silicate minerals and a few oxide minerals, together with calcium sulfate and calcium carbonate, comprise the bulk of Earth's crust—an estimated 99.2% by volume. These common minerals, of which there are about 30, are called **rock-forming minerals**. Rock-forming minerals are everywhere—not only in rocks but also in soils and sediments, and even in the dust that we breathe. Rock-forming minerals are common and inexpensive but nevertheless economically vital; you rely on them every day when you drive on a paved road or enter a concrete building.

Less common minerals are called **accessory minerals**. These are widely present in common rocks but usually in such small amounts that they do not determine the properties of the rocks. Though they are less common and typically present in lesser abundance than the principal rock-formers, many are important economically. Some are ore minerals; as mentioned, galena (PbS) is the main source of lead, and chalcopyrite ($CuFeS_2$) is the principal source of copper. Others have significant biological

Where Geologists CLICK

Photo Gallery: Minerals

http://science.nationalgeographic.com/science/photos/minerals/

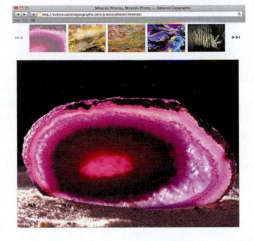

The National Geographic Society maintains an extensive and beautiful collection of mineral photographs that geologists refer to and enjoy. Elsewhere in National Geographic's Photo Gallery you will find all sorts of information on geologic processes. Try clicking on "Earth" for photos of lava flows, rocks, gems, earthquake damage, and lots more. Under "Prehistoric World" you will find photos of all sorts of fossils.

functions; the accessory mineral apatite $[Ca_5(PO_4)_3(F,OH)]$ supplies the phosphorus for agricultural fertilizers and is also the fundamental compound in our bones and teeth.

Silicate links • Figure 2.12

Silicate minerals consist of silicon and oxygen groupings linked together, with other elements in the spaces between them.

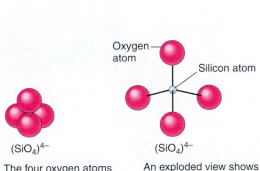

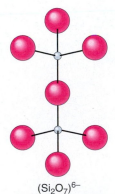

$(SiO_4)^{4-}$
The four oxygen atoms of a silicate ion form a regular tetrahedron.

$(SiO_4)^{4-}$
An exploded view shows the smaller silicon atom in the center.

$(Si_2O_7)^{6-}$
Two tetrahedra can link together by sharing an oxygen atom.

$(Si_2O_7)^{6-}$
Exploded view of the same configuration.

Oxygen atom

Silicon atom

Silicates: The most important rock-formers

Not only are silicates the most common minerals and the main rock-formers, they also have an unusual diversity of atomic structures. To explain this phenomenon, we begin by looking at the silica anion itself, in which four oxygen atoms are tightly bonded to the single silicon atom. As you saw in Figure 2.12, if we draw a stick figure, joining the centers of the oxygen atoms with lines, we get a regular **tetrahedron** (i.e., a pyramid with a triangular base). The small, ionically bonded silicon atom occupies the space at the center of the tetrahedron, and the oxygen atoms occupy the corners. This is the building block of silicate mineral structures.

Two silica tetrahedra can bond by sharing an oxygen atom. This process can be repeated over and over, with the silica anions assembling themselves into large, com-

Polymerization and the common silicate minerals • Figure 2.13

This chart summarizes the ways in which silica ions can polymerize to form minerals. Typical examples of each type are shown in the photographs. There are many other silicate materials in each category, but all of the principal categories that occur in nature are illustrated here.

Silicate Structure	Mineral/Formula	Cleavage	Example of a Specimen
 Single tetrahedron	Olivine Mg_2SiO_4	None	
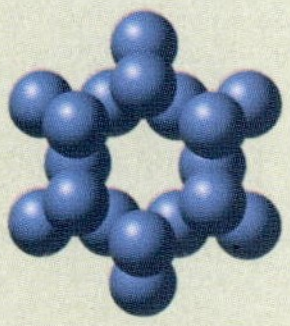 Hexagonal ring	Beryl (Gem form is emerald) $Be_3Al_2Si_6O_{18}$	One direction	
 Single chain	Pyroxene group $CaMg(SiO_3)_2$ (variety: diopside)	Two directions at 90°	

plex linked structures called **polymers**. Silica anions only polymerize by joining at corners, never along edges or faces. **Figure 2.13** illustrates the different mineral structures, such as rings, chains, sheets, or three-dimensional frameworks that can be formed out of silica ions through **polymerization**. Different common cations, such as calcium (Ca^{2+}), aluminum (Al^{3+}), magnesium (Mg^{2+}), iron (Fe^{2+}), sodium (Na^+), and others, can fit into the spaces between the polymerized silica tetrahedra.

The identity of a silicate mineral is determined by three factors: (1) how the silica tetrahedra occur within the mineral (i.e., whether they are single or

> **polymerization**
> The formation of a complex molecule by the joining of repeated simpler units.

Silicate Structure	Mineral/Formula	Cleavage	Example of a Specimen
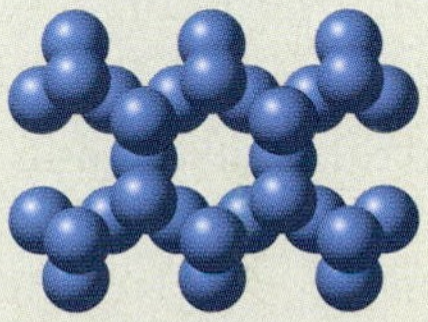 Double chain	Amphibole group $Ca_2Mg_5(Si_4O_{11})_2(OH)_2$ (variety: tremolite)	Two directions at 120°	
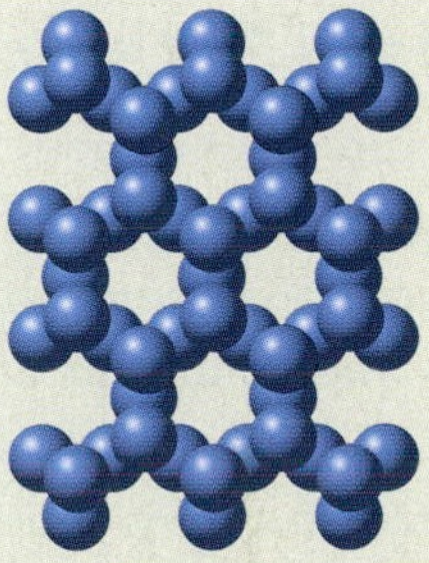 Sheet	Mica $KAl_2(AlSi_3O_{10})(OH)_2$ (variety: muscovite) $K(Mg,Fe)_3(AlSi_3O_{10})(OH)_2$ (variety: biotite)	One direction	
See Figure 2.14 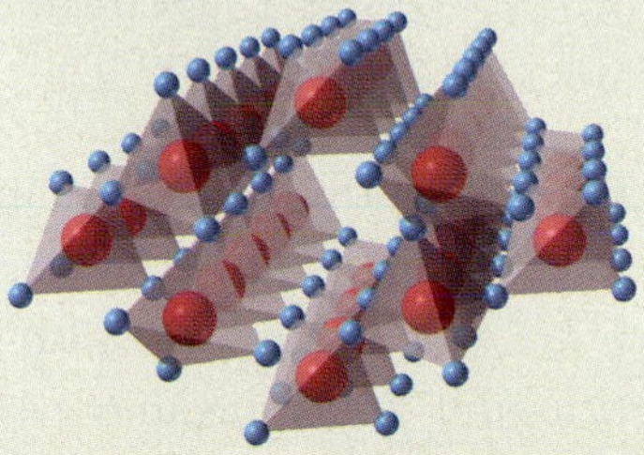	Feldspar $KAlSi_3O_8$ (variety: orthoclase)	Two directions at 90°	
	Quartz SiO_2	None	

Quartz: A framework silicate • Figure 2.14

Quartz, one of the most abundant minerals, is composed of silica ions that are tightly linked in a three-dimensional lattice. It has no cleavage planes and is very resistant to wear. For the purposes of illustration, the oxygen atoms (blue) are shown as being smaller than the silicon atoms (red).

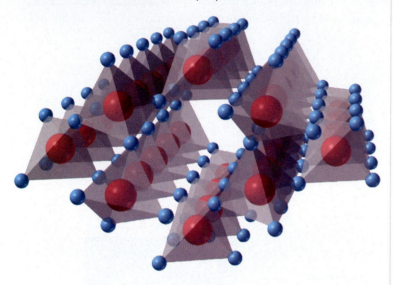

polymerized); (2) which cations are present; and (3) how the cations are distributed throughout the crystal structure. For example, as shown in Figure 2.13, polymerization in mica produces a sheet, and the pronounced cleavage of mica is parallel to the sheets. In quartz the three-dimensional framework, shown in **Figure 2.14**, is equally strong in all directions, so quartz is a hard, tough mineral that lacks cleavage. The two most common mineral families in the continental crust both have three-dimensional frameworks of polymerized silicate tetrahedra. The two families are the feldspars (most abundant) and quartz.

Other important mineral families

Silicates and oxides are the two main rock-forming mineral families, but other mineral groups are also important as constituents of the crust, as mineral resources, and as life-supporting materials. These groups are based on ions and ionic groupings that are different from those in the silicates and oxides.

For example, **carbonate** family minerals are based on the complex $(CO_3)^{2-}$ anion. The carbonate minerals calcite and dolomite are the main constituents of limestone and marble, as well as the shells of small marine organisms such as mollusks. They are also the principal building blocks of coral reefs. Recall from our discussion of mineral identification that calcite typically dissolves in weak acids; as it turns out, ordinary groundwater is a weakly acidic solution. Thus, most terrains that are underlain by carbonate-bearing rocks have complex systems of underground caves that form in areas where carbonate minerals have been dissolved and carried away by circulating groundwater.

Two other important mineral families are the **sulfates** and **sulfides**, which both contain the element sulfur (or S, which is distinct from Si, silicon). Sulfates are based on the complex $(SO_4)^{2-}$ anion, whereas sulfides are based on the simple S^{2-} anion. One example of an important sulfate mineral is gypsum, used to make plaster, which forms as a salt that precipitates as a result of evaporation of seawater. Sulfide minerals typically consist of one or more metal ions (e.g., copper, nickel, or iron) bonded with the S^{2-} anion. Thus, sulfides tend to be important ore minerals from which a wide variety of metals are extracted; galena (PbS) is an example of a sulfide mineral.

Phosphates, another important mineral family, are based on the complex $(PO_4)^{3-}$ anion. Phosphorus is a life-supporting element and, as mentioned earlier, the phosphate mineral apatite is a central constituent of bones and teeth.

Mineral Resources in Modern Society

Can you imagine a world without machines? Our modern world, now with more than 7 billion inhabitants, could not operate without them. We use machines to produce our food, make our clothes, transport us, and communicate. The metals that are needed to build machines all come from minerals dug from Earth. So great has our dependence on minerals become that today we have industrial uses for almost all the naturally occurring chemical elements, and more than 200 kinds of minerals are mined and used.

Ore deposits

Recall from Figure 2.11 that only 12 chemical elements make up a little more than 99.2% of the mass of Earth's crust. Of the 12, 6 are useful metals—silicon, aluminum, iron, magnesium, titanium, and manganese. All the other useful metals are present in Earth's crust in tiny amounts. In order to mine such scarce materials, it is necessary to find a place where some natural process has produced a localized concentration of a scarce element; such a localized concentration is called an **ore deposit**.

> **ore deposit** A localized concentration in the crust from which one or more minerals can be profitably extracted.

Impacts of unregulated mining • Figure 2.15

This photo shows small-scale gold mining in Sulawesi, Indonesia. First, the large boulders and gravel are removed and discarded. Then the bottom layers of sediment, where the gold is concentrated, are pumped out through the white hoses and treated in the devices visible in the upper-right side of the image. When mining is completed, the miners typically walk away and leave the mess.

Ore deposits are, by definition, limited in abundance and distinctly localized in the crust. As a result, no nation is self-sufficient in mineral supplies, and supplies must be searched out around the globe. The quantity of a given mineral resource in any one country is rarely known with accuracy, and the likelihood that new deposits will be found is hard to assess. A country that is an exporter of a mineral resource today may be an importer tomorrow. Unlike plants and animals, which are renewable resources because they can be harvested seasonally and replenished by growth, ore deposits are nonrenewable resources because they are depleted by mining and may eventually be exhausted. This disadvantage can be offset only by finding new deposits or by using the same material repeatedly—that is, by recycling. We will discuss the nature of nonrenewable resources in general, and mineral resources in particular, in greater depth in Chapter 16.

Mining An additional problem with the extraction of mineral resources is that mining disturbs Earth's surface in many ways. The search for ore deposits has increasingly taken miners to places of topographic and climatic extremes, and the result has been destruction of some environmentally sensitive areas. Small-time and poorly regulated miners, in particular, tend to mine an area and leave it, without any effort at reclamation (**Figure 2.15**). In many countries mining is under increasingly strict control; for example, to open a mine in North America today, a mining company must demonstrate that both the plans and the financial re-

sources exist to manage the mine throughout its lifetime, the closure of the mine, and the reclamation and restoration of both the environment and the surrounding communities. Major damage to the environment is nevertheless widespread. For example, the United Nations estimates that as much as 1000 tons of mercury is released to the environment each year by small-scale miners who utilize the mercury in a variety of poorly controlled mining and refining procedures. Mining can also be extremely hazardous to the health of these small-scale miners, many of whom lack the technical training to handle the dangerous materials, such as mercury and cyanide, used to extract metals from rocks.

CONCEPT CHECK　STOP

1. **Which** two chemical elements make up a little more than 70% of Earth's crust by weight?
2. **Which** two mineral families are the most abundant in Earth's crust?
3. **Describe** how polymerization occurs in silicate minerals and how it can affect the properties of the minerals.
4. **What** are the common characteristics of ore deposits?
5. **Describe** some of the ways that mining can create environmental disturbances.

Rock: A First Look

LEARNING OBJECTIVES

1. **Explain** the difference between a rock and a mineral.

2. **Identify** the three major families of rock.

3. **Explain** what holds rocks together and why some rocks are more cohesive than others.

If **rocks** are the words that tell the story of Earth's long history, minerals are the letters that form the words. An important distinction between minerals and rocks is that rocks are aggregates; this means that rocks are collections of minerals (and sometimes other types of particles) stuck together or intergrown. Rocks usually consist of several different types of minerals, but sometimes they are made of just one type of mineral. In any case, a rock will contain many grains of the constituent mineral or minerals.

> **rock** A naturally formed, coherent aggregate of minerals and possibly other nonmineral matter.

Igneous, Sedimentary, and Metamorphic Rock

Rocks are grouped into three large families, according to the processes that form them (**Figure 2.16**). Within each of these families, called **igneous**, **sedimentary**, and **metamorphic**, is a range of possible mineral assemblages—the types and relative proportions of minerals that constitute the rock. Mineral assemblages help geologists identify and classify rocks, and they reveal much about the geologic environment in which a particular rock formed.

Igneous rock Igneous rock (named from the Latin *ignis*, meaning "fire") is formed by the cooling and solidification of **magma**. Magma, or molten rock, often contains mineral grains and dissolved gases. Depending on its environment, magma may cool slowly or very rapidly. When it reaches a temperature that is sufficiently low, mineral grains begin to crystallize from the magma, just as ice crystals form in water as it cools. The physical properties of

> **igneous rock** Rock that forms by cooling and solidification of molten rock.
>
> **magma** Molten rock that may include fragments of rock, volcanic glass and ash, or gas.

the rock will differ, depending on whether the cooling process is slow, giving the crystals a lot of time to grow, or fast. We will have much more to say about these differences in Chapter 6.

Sedimentary rock The next major rock family forms when mineral and rock particles are transported by water, wind, or ice and then deposited. This fragmented, transported, and deposited material is called **sediment**. In certain geologic circumstances, under conditions of low pressure and low temperature near Earth's surface, sediment may be transformed into **sedimentary rock**. "Low temperature" and "low pressure" here are by geologic standards; they may still be higher than the pressure and temperature where we live. Sediment- and soil-forming processes and transportation processes are discussed in detail in Chapter 7, and sedimentary rocks and rock-forming processes are examined in Chapter 8.

> **sediment** Rock that has been fragmented, transported, and deposited.
>
> **sedimentary rock** Rock that forms from sediment under conditions of low pressure and low temperature near the surface.

Metamorphic rock The third major rock family is **metamorphic rock**, whose original sedimentary or igneous form and mineral assemblage have been changed as a result of exposure to high temperature, high pressure, or both. The term **metamorphic** comes from the Greek *meta*, meaning "change," and *morphe*, meaning "form"—hence, "change of form." Metamorphism occurs in a variety of tectonic and geologic environments. It can occur as an end result of the processes of burial and compaction through which sediments become transformed into rock. It can also occur when rock is subjected to the extreme pressures and temperatures associated with mountain building along convergent plate boundaries. Temperature-induced metamorphism can occur when rock is heated—essentially baked—by nearby magma. How and why metamorphism occurs, the types of metamorphic rock, and their characteristic mineral assemblages are discussed in detail in Chapter 10.

> **metamorphic rock** Rock that has been altered by exposure to high temperature, high pressure, or both.

Igneous rock

This rounded granite boulder on Mt. Desert Island in Acadia National Park, Maine, was transported to its current location by a glacier. The boulder is sitting on a thick platform of granite that was once part of an enormous magma chamber underlying a volcano. The volcano is now gone, and its top has eroded away, leaving the solidified remnants of the magma chamber exposed. The weathered surfaces of these granites make them appear buff-colored. Inset is a close-up photo of granite.

Sedimentary rock

This remarkable landscape is part of Bryce Canyon National Park in Utah. The horizontal rock layers are mainly sandstones and limestones, which are both sedimentary rocks. The sandstone layers were formed from grains of sand that were deposited and cemented together when the climate was hot and dry, as it is today. Limestone layers were formed when this region was covered by a shallow sea. Different kinds of sedimentary rock therefore tell us a great deal about climate change in the past. The inset photo shows a close-up of sandstone, a sedimentary rock.

Metamorphic rock

The beautiful Lauterbrunnen Valley in Switzerland lies within a great mountain range—the Alps—where the rocks have been uplifted and chemically and physically altered by enormous tectonic forces. Metamorphic rocks are often found at the site of great mountain ranges—even ranges that may have eroded completely away. In this photo Lauterbrunnen Falls cascades down a steep cliff of metamorphic rock. The inset close-up shows gneiss, a metamorphic rock. This rock may have started as either a sedimentary or igneous rock, but it has been altered by heat and pressure, creating new minerals and giving the specimen its distinctive banded appearance.

What Holds Rock Together?

The minerals in some types of rock are held together with great tenacity, whereas in others they are easily broken apart. While the forces that hold minerals together are chemical, the forces that hold rock together are partly mechanical and partly chemical. Igneous and metamorphic rock are the most cohesive because both types contain intricately interlocked minerals. During the formation of igneous and metamorphic rock, the growing minerals crowd against each other, filling all spaces and forming an intricate, three-dimensional jigsaw puzzle that will not pull apart very easily. A similar interlocking of grains holds together steel, ceramics, and bricks.

The forces that hold together the grains of sedimentary rock are less obvious. Sediment is a loose aggregate of particles, and it must be transformed into sedimentary rock by one or more rock-forming processes, discussed in detail in Chapter 7. These processes tend to cause the individual grains of the sediment to become interlocking. Sedimentary rock can form by compaction, during which the mineral grains in the sediment are squeezed and compressed by the weight of overlying sediment, becoming compacted into an interlocking network of grains. Second, water may deposit new minerals such as calcite, quartz, or iron oxide into the open spaces that act as a **cement**. A third way that sedimentary rock holds together is by **recrystallization**. As sediment becomes deeply buried and the temperature rises, mineral grains begin to recrystallize. The growing grains interlock and form strong aggregates. The process is the same as that which occurs when ice crystals in a snow pile recrystallize into a compact mass of ice (**Figure 2.17**).

CONCEPT CHECK

1. **What** is a mineral assemblage?
2. **What** are the differences between igneous, sedimentary, and metamorphic rocks?
3. **How** can a loose collection of sediment become a cohesive rock?

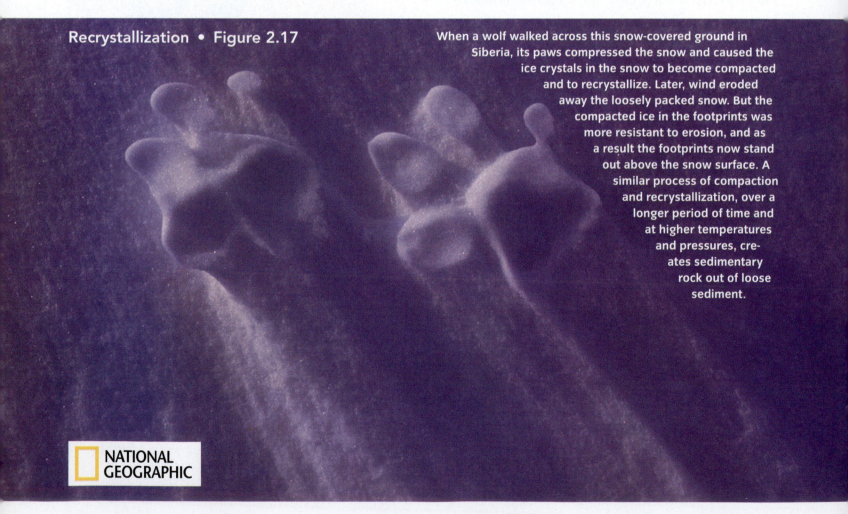

Recrystallization • Figure 2.17

When a wolf walked across this snow-covered ground in Siberia, its paws compressed the snow and caused the ice crystals in the snow to become compacted and to recrystallize. Later, wind eroded away the loosely packed snow. But the compacted ice in the footprints was more resistant to erosion, and as a result the footprints now stand out above the snow surface. A similar process of compaction and recrystallization, over a longer period of time and at higher temperatures and pressures, creates sedimentary rock out of loose sediment.

NATIONAL GEOGRAPHIC

Summary

1 Elements and Compounds 36

- **Elements** are the most fundamental of all naturally occurring substances because they cannot be separated into chemically distinct materials. The minute particles that make up all matter, including elements, are **atoms**. All atoms of a particular element are the same.

- Atoms are made of protons, neutrons, and electrons, which have no independent chemical properties as shown in the diagram. Protons are electrically positive, and electrons are electrically negative. Protons and neutrons reside in the atom's nucleus. The number of protons, the atomic number, identifies the element. The sum of protons and neutrons is the mass number. Most elements have several different **isotopes**, which differ in the number of neutrons. Ordinarily an atom has equal numbers of protons and electrons, but it may gain or lose electrons, in which case it becomes electrically charged and is called an ion.

Inside an atom • Figure 2.1

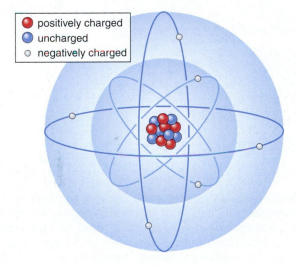

- positively charged
- uncharged
- negatively charged

- **Compounds** consist of multiple elements and therefore multiple types of atoms. The smallest unit that has the properties of a given compound is a **molecule**. Molecules in a compound are held together by **bonds**. The most common form of bonding is ionic bonding, caused by the electrostatic attraction of two oppositely charged ions. Other forms of bonding are covalent, metallic, and Van der Waals bonding.

2 What Is a Mineral? 40

- **Minerals** are naturally occurring solids, each with a unique **crystal structure**. Minerals are formed by inorganic processes and have a fixed chemical composition. One exception to the last rule is atomic substitution, in which other atoms of like size and ionic charge may be substituted for specific atoms in a mineral.

- The visible crystal form of a mineral is a direct consequence of its atomic lattice. In some cases, a mineral will not have enough room to grow identifiable crystals, but the underlying geometric lattice remains the same. Crystal faces bound a crystal.

- Two minerals can have the same chemical formula but different crystal structures. These different forms are polymorphs. Other minerals occur in nature as *native* elements that are uncombined with other elements.

- Several properties can be used to tell minerals apart as shown in the photo. Measuring the angles between faces may be useful in identifying a mineral, for although the size and overall shape of crystals may vary, Steno's Law states that the angle between corresponding faces remains constant. Other factors that can aid in mineral identification include the mineral's **luster**, **habit**, **hardness**, cleavage, and density, or specific gravity. Color is often misleading; however, when a mineral is rubbed on a streak plate, it produces a thin layer of powdered mineral, or **streak**, which is more reliable than color for identification.

Mystery mineral

Identity of the mystery mineral: The mineral is galena (PbS), the main ore mineral of lead.

- Less frequently, such properties as birefringence, magnetism, and luminescence (or fluorescence) may also be of use in identifying minerals.

3 Mineral Families   50

- The distribution of elements in Earth's crust is far from even as shown in the diagram. Only 12 elements are present at a level of more than 1 part in 1000 by mass, and of these oxygen and silicon dominate. As a result, the number of naturally occurring minerals is relatively small, and the number of important rock-forming minerals is even smaller—only about 30 or so.

Composition of earth's crust • Figure 2.11

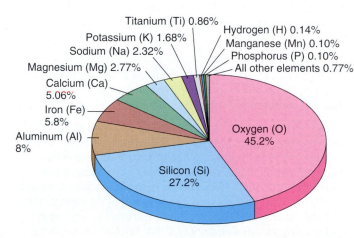

Titanium (Ti) 0.86%
Hydrogen (H) 0.14%
Potassium (K) 1.68%
Manganese (Mn) 0.10%
Sodium (Na) 2.32%
Phosphorus (P) 0.10%
Magnesium (Mg) 2.77%
All other elements 0.77%
Calcium (Ca) 5.06%
Iron (Fe) 5.8%
Aluminum (Al) 8%
Oxygen (O) 45.2%
Silicon (Si) 27.2%

- Silicate minerals, based on the $(SiO_4)^{42}$ anion, are the most abundant family of rock-forming minerals. Oxide minerals are the next most abundant family. Other important mineral families include sulfides and sulfates, carbonates, and phosphates. Less abundant accessory minerals usually do not affect the properties of the rock they are found in, but they may still be of economic importance.

- Silicates can assemble in large, complex structures, such as chains, sheets, and three-dimensional lattices, by **polymerization**, a process in which two silica anions share an oxygen atom. Three factors affect the identity of a silicate mineral: whether the silicate tetrahedra are single or polymerized, which cations are present, and how the cations are distributed.

- Nearly every naturally occurring element has some industrial or commercial use in our society, but the great majority of elements are rare. Thus it is economically feasible to extract elements only where some natural process has concentrated a mineral containing that element, to form an **ore deposit**.

- Ores are nonrenewable resources; thus the supply can be maintained only by finding new deposits or by recycling. Ore mining can cause serious environmental damage if it is not appropriately monitored and regulated.

4 Rock: A First Look   56

- Rock is an aggregate, or complex mineral assemblage, that is sometimes mixed with other materials such as natural glass or organic matter.

- Rock types are grouped into three major families as shown in the photos. **Igneous rock** is formed by the solidification of **magma**, either slowly beneath Earth's surface or rapidly at the surface. **Sedimentary rock** is formed at or near the surface by the deposition of many layers of **sediment**. **Metamorphic rock** starts as either igneous or sedimentary rocks but changes in form and sometimes in mineral assemblage as a result of high temperature, high pressure, or both.

The three rock families • Figure 2.16

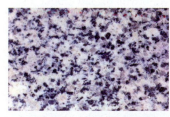

- Igneous and metamorphic rocks are held together by the interlocking of mineral grains. The loose particles of sedimentary rock are held together either by compaction, during which the mineral grains are forced together by the pressure of overlying sediment, cementing of open spaces within a rock, or recrystallization, a process that occurs when growing grains interlock because of increasing pressure and heat.

Key Terms

- atom   36
- bond   38
- compound   37
- crystal structure   42
- element   36
- habit   44
- hardness   44

- igneous rock   56
- isotopes   36
- luster   43
- magma   56
- metamorphic rock   56
- mineral   40
- molecule   38

- ore deposit   55
- polymerization   53
- rock   56
- sediment   56
- sedimentary rock   56
- streak   48

Critical and Creative Thinking Questions

1. When astronauts brought back rock samples from the Moon, the minerals present were mostly the same as those found on Earth. Can you think of reasons this might be so? Would you expect minerals on Mars or Venus to be the same, or at least very similar, to those on Earth?

2. When a volcano erupts, spewing forth a column of hot volcanic ash, the ash particles are tiny fragments of solidified magma that slowly fall to Earth's surface, forming a layer of sediment. Would a rock formed from cemented particles of volcanic ash be igneous or sedimentary? Can you think of other circumstances that might form rocks that are intermediate between two of the major rock families?

3. Identify which of the following materials are minerals and why: water, beach sand, diamond, wood, vitamin pill, gold nugget, fishbone, and emerald. Should a synthetic diamond be considered a true mineral?

4. The minerals calcite and aragonite have the same chemical formula ($CaCO_3$) but different crystal structures. Are they polymorphs? The materials halite (NaCl) and galena (PbS) have the same geometric patterns in their crystal structures but different compositions. Are they polymorphs?

5. Do some research to find out whether any valuable ores are mined in your area. What are the rocks and minerals involved?

What is happening in this picture?

This geologist is part of a team studying the feasibility of establishing a platinum mine near Stillwater, Montana. He is looking at a core sample through a 10× hand lens, one of the simplest and most useful tools in any geologist's backpack.

Think Critically

What features might he be looking for?

NATIONAL GEOGRAPHIC

Self-Test

(Check your answers in Appendix D.)

1. A(n) _____ is an atom that has gained or lost one or more electrons and has a net electric charge.

   a. molecule
   b. isotope
   c. element
   d. ion
   e. compound

2. On this diagram, locate and label the following parts of the atom:

   proton

   nucleus

   electron

   first energy-level electron shell

   neutron

   second energy-level electron shell

   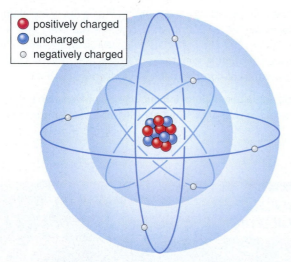

   ● positively charged
   ● uncharged
   ○ negatively charged

3. In _____, electrons from different atoms are shared, and the force of this sharing forms the strongest of chemical bonds.

   a. ionic bonding
   b. covalent bonding
   c. metallic bonding
   d. Van der Waals bonding

4. In _____, the mobility of electrons in the outermost shell allows materials with these types of bonds to act as good conductors of electricity and heat.

   a. ionic bonding
   b. covalent bonding
   c. metallic bonding
   d. Van der Waals bonding

5. To be considered a mineral, a substance must _____.

   a. have a specific chemical composition
   b. be formed by inorganic processes
   c. be a naturally formed solid
   d. have a characteristic crystal structure
   e. have all of the characteristics listed above

6. Volcanic glass is not considered a mineral because it _____.

   a. is amorphous (i.e., it lacks a crystal structure)
   b. is not naturally occurring
   c. does not have enough silicon or oxygen in its chemical composition
   d. All of the above answers are correct.

7. The photograph shows natural samples of the mineral corundum. What best explains the striking difference in color between the red (ruby) and the blue (sapphire) samples?

   a. polymorphism
   b. small variations in composition
   c. differences in crystalline structure
   d. polymerization

8. Which of the following minerals is the hardest?

   a. quartz
   b. calcite
   c. muscovite
   d. feldspar

9. Which of the following physical properties is the least useful in identifying many varieties of common minerals?

   a. cleavage or fracture
   b. hardness
   c. color
   d. density

10. _____ is the most abundant element, by weight, in Earth's continental crust.

   a. Silicon
   b. Iron
   c. Calcium
   d. Aluminum
   e. Oxygen

11. Why is Earth's crust mostly composed of a small number of rock-forming minerals?

    a. Polymorphs are common.

    b. There is an overwhelming abundance of oxygen and silicon.

    c. Igneous rocks lack carbon.

    d. All of the above statements are correct.

12. Gold is an example of a(n) _____, quartz is a _____, and galena is a _____.

    a. native element; silicate mineral; sulfide mineral

    b. oxide mineral; sulfide mineral; phosphate mineral

    c. phosphate mineral; sulfide mineral; carbonate mineral

    d. oxide mineral; silicate mineral; phosphate mineral

13. On the diagram, label each silicate structure with its proper name. For each structure give an example of a mineral (name and chemical formula) and indicate the prominent cleavage.

Structure 1

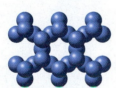

Structure 2

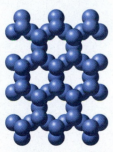

Structure 3

14. The rocks shown in the photographs are examples of _____.

    a. volcanic rock   c. sedimentary rock

    b. plutonic rock   d. metamorphic rock

15. In the rocks depicted in the photographs in question 14, what is most likely holding the mineral grains together?

    a. interlocking crystals formed from cooling magma

    b. compaction and cementation between the mineral grains

    c. interlocking crystals formed from the alteration of preexisting rocks by heat and pressure

    d. Van der Waals bonding

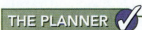

THE PLANNER ✓

Review the Chapter Planner on the chapter opener and check off your completed work.

How Old Is Old? The Rock Record and Geologic Time

The White Cliffs of Dover in England are white because they are made of chalk, a rock formed from the shells of innumerable one-celled sea creatures. There are similar white cliffs in Northern Ireland, France, Germany, and Denmark. The chalk deposits have given their name to the Cretaceous Period, which lasted from about 145 million years ago to 65 million years ago. The name "Cretaceous" comes from the Latin word *creta*, meaning "chalk." In the nineteenth century, geologists realized that the layer of chalk continued from Dover, beneath the English Channel, and into the Champagne province of France, where its excellent water retention helps wine grapes flourish. When geologists realized the extent of chalk deposits of this age, they applied the name Cretaceous to the period.

The Cretaceous Period was the climax of the age of dinosaurs, when such mighty creatures as *Tyrannosaurus rex* roamed the land. Yet, ironically, it draws its name from a sea creature so small that it can barely be seen under a microscope (inset). The discovery that similar deposits of the same age could be found at many different locations was a key to unlocking the secrets of geologic time.

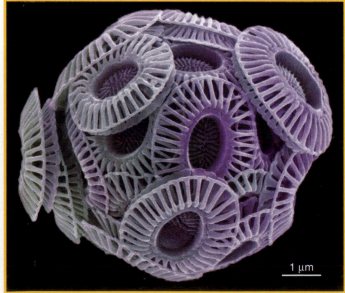

1 μm

CHAPTER PLANNER ✓

- ❏ Study the picture and read the opening story.
- ❏ Scan the Learning Objectives in each section:
  p. 66 ❏   p. 73 ❏   p. 77 ❏   p. 85 ❏
- ❏ Read the text and study all visuals.
  Answer any questions.

Analyze key features

- ❏ Geology InSight, p. 67 ❏   p. 74 ❏
- ❏ Process Diagram, p. 69 ❏
- ❏ What a Geologist Sees, p. 70 ❏
- ❏ Case Study, p. 84
- ❏ Amazing Places, p. 87 ❏
- ❏ Stop: Answer the Concept Checks before you go on:
  p. 72 ❏   p. 77 ❏   p. 84 ❏   p. 86 ❏

End of chapter

- ❏ Review the Summary and Key Terms.
- ❏ Answer the Critical and Creative Thinking Questions.
- ❏ Answer What is happening in this picture?
- ❏ Complete the Self-Test and check your answers.

Relative Age

LEARNING OBJECTIVES

1. **Define** relative age and numerical age.
2. **Define** stratigraphy and identify the four main principles of stratigraphy.
3. **Explain** why gaps are common in the rock record.
4. **Describe** how fossils make it possible for geologists to correlate strata in different places.

The first scientific attempts to determine the numerical extent of geologic time were made almost three centuries ago. Geologists speculated that they might be able to estimate the time needed to erode away a mountain range by measuring the amount of sediment transported by streams. Their attempts were imprecise, but the inescapable conclusion was that Earth must be millions of years old because of the great thickness of sedimentary rocks. One of the founders of geology, James Hutton, whose work led to the **Principle of Uniformitarianism** (Chapter 1), was so impressed by the evidence that in 1788, he wrote that for Earth, there is "no vestige of a beginning, no prospect of an end."

Geologists who followed Hutton agreed with his conclusion that Earth must be very ancient, but they lacked a precise way to determine exactly how long ago a particular event occurred. The only thing they could do was figure out the sequence of past events. They could thus establish the **relative ages** of rock formations or other geologic features, which means that they could determine whether a particular formation or feature was older or younger than another formation or feature. By doing so, they took the essential first steps toward unraveling Earth's geologic history. Relative ages are derived from three basic principles of **stratigraphy**, generally attributed to Hutton and Steno (the same man we met in Chapter 2 in the context of crystal faces), and the principle of cross-cutting relationships, first stated by Charles Lyell, a Scot like Hutton.

> **relative age** The age of a rock, fossil, or other geologic feature relative to another feature.
>
> **stratigraphy** The science of rock layers and the process by which strata are formed.

Stratigraphy

In places where you can find large exposures of sedimentary rocks, such as the Badlands of South Dakota and the American Southwest, you will often see that the rocks have a banded appearance (**Figure 3.1**). These bands are called **strata** (an individual band is called a stratum), from the Latin word for "layer." The bands are often horizontal, but it is not at all unusual to see them tilted or bent.

All of the rocks in a typical stratified formation are sedimentary, deposited over the eons by water. This observation leads to the first of three key principles of stratigraphy: the **Principle of Original Horizontality**, which states that water-laid sediment is deposited in horizontal strata. You can test this principle yourself: Shake up a bottle of muddy water so that all of the particles are suspended. Let the bottle stand and then examine the result—the mud will be deposited at the bottom of the bottle in a horizontal layer (see Figure 3.1a and c). Therefore, whenever we observe water-laid strata that are bent, twisted, or tilted so that they are no longer horizontal, we can infer that some tectonic force must have disturbed the strata after they were deposited (see Figure 3.1d).

A second key to stratigraphy, which, like the first, is based on common sense, is the **Principle of Stratigraphic Superposition**. This principle states that in any undisturbed sequence of strata, each stratum is younger than the stratum below it and older than the stratum above it (see Figure 3.1b and c).

The third key to stratigraphy is the **Principle of Lateral Continuity** of sedimentary strata. This principle is based on the observation that sediment is deposited in continuous layers, and a layer of sediment will extend horizontally as far as it was carried by the water that deposited it. Layers of sediment do not terminate abruptly; they get thinner and ultimately pinch out altogether at their farthest edges. The same is true of the sedimentary rock strata that form from sediment layers; they may thin or pinch out laterally, but they usually do not terminate abruptly unless they are cut by a fracture. This important principle allows the geologist to correlate between outcrops of strata when erosion has removed some of the strata (see Figure 3.1e).

It is quite common for once-horizontal strata to be disrupted by later geologic events. This observation led Lyell

c. Demonstration of the Principles of Original Horizontality and Stratigraphic Superposition
Sediment poured into a bucket of water settles into a stack of horizontal layers. The oldest layer is on the bottom, the youngest on top.

a. The Principle of Original Horizontality
This horizontal layer of sediment was deposited in a lake. The now-dry lake bed is near Death Valley, California.

d. The Principle of Cross-Cutting Relationships
These folded strata in the Hamersley Gorge, Western Australia, were originally horizontal; they were crumpled and contorted by plate tectonic movement. The strata themselves are older than the tectonic disruption that caused the folding of the layers.

b. The Principle of Stratigraphy
Horizontal strata like these, in Badlands State Park, South Dakota, can extend for many kilometers laterally, in all directions. The oldest strata are on the bottom, the youngest on the top. These strata resulted from the deposition of sediment in water.

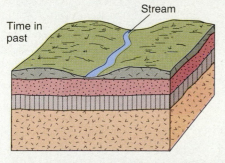

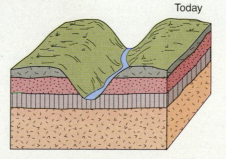

Time in past

Stream

Today

e. The Principle of Lateral Continuity
Even though erosion has removed some of the strata, it is clear that the strata on each side of the valley were originally continuous.

to formulate a fourth principle that can help to determine relative ages: the **Principle of Cross-cutting Relationships**. This principle states that a stratum must always be older than any feature that cuts or disrupts it. If a stratum is cut by a fracture, for example, the stratum itself is older than the fracture that cuts across it, as shown in **Figure 3.2**. When magma fills a fracture, the result is a vein of rock that cuts across the strata. In this case, too, the sedimentary rocks must be older than the cross-cutting vein. Similarly, a "foreign rock" (called a **xenolith** or an **inclusion**) that is encased within another rock unit must predate the rock that surrounds it.

Gaps in the Record

In the 19th century, geologists tried to estimate the numerical ages of rocks—the number of years that have elapsed since a given stratum was deposited. For example, observations might suggest that it would take 1 year for 10 centimeters of sediment to accumulate. If a rock lies 3 meters (300 cm) below the ground surface under younger sedimentary strata, then its

numerical age
The age of a rock or geologic feature in years before the present.

numerical age could be estimated to be about 3000 years (300 cm × 10 cm/year). Three assumptions must be true for this approach to work. First, the rate of sedimentation must have been constant while the layers were deposited. Second, the thickness of sediment must have been the same as the thickness of the sedimentary rock that eventually formed from it. Finally, all strata must be **conformable**. This means that each layer was deposited on the one below it without any interruptions—in other words, there are no depositional gaps in the stratigraphic record.

There are problems with each of these assumptions. For example, rates of sedimentation vary greatly. Sediment is often compressed while it is turning into rock, resulting in a much thinner stratum than was originally deposited. The conformity assumption also fails: Gaps in the stratigraphic record, called **unconformities**, are common and occur for a variety of reasons (**Figure 3.3**). *What a Geologist Sees*, illustrates these principles. In sum, the numerical ages determined by early geologists on the basis of the thickness of stratigraphic sequences were not very accurate.

unconformity
A substantial gap in a stratigraphic sequence that marks the absence of part of the rock record.

Principle of cross-cutting relationships • Figure 3.2

a. These fractures are younger than the strata they cut; they demonstrate the principle of cross-cutting relationships. The fractures are in a sequence of sandstone strata in Merseyside, United Kingdom

Fractures

Dike

Folded strata of metamorphic rock

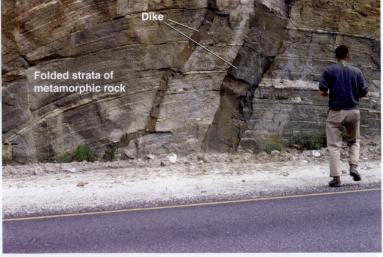

b. At Three Valley Gap in Alberta, Canada, layers of metamorphic rock are sliced by two darker, cross-cutting igneous rock formations called dikes. A geologist sees four stages: deposition of sediment, formation of sedimentary rocks, metamorphism into a rock called *gneiss*, and intrusion of the dikes. By the principle of cross-cutting relationships, the dikes must be the youngest feature.

Unconformities: Gaps in the rock record • Figure 3.3

✓ THE PLANNER

Any boundary that represents a gap in the sedimentary record is called an unconformity. This diagram illustrates the formation of three common types of unconformities: **1** nonconformity, **2** angular unconformity, and **3** disconformity. The contact between deformed rock below and horizontal strata above in part c of *What a Geologist Sees*, on the next page, is an angular unconformity.

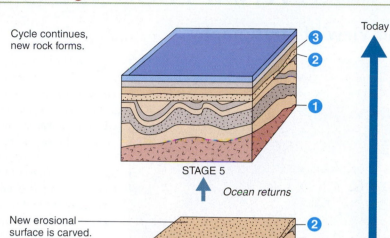

Cycle continues, new rock forms.

3
2
1

STAGE 5

↑ *Ocean returns*

New erosional surface is carved.

2
1

Uplift leads to erosion.

STAGE 4

↑ *Ocean recedes, new surface exposed*

Type of Unconformity	Description/Cause
1 Nonconformity	A surface of erosion that separates younger sedimentary strata above from older igneous or metamorphic rocks below.
2 Angular unconformity	A surface of erosion between two groups of sedimentary rocks in which the orientation of older strata, below, is at an angle to younger strata, above.
3 Disconformity	A surface of erosion in which the orientation of older strata, below, are parallel to younger strata, above.

New sedimentary strata are deposited atop old eroded surface.

2
1

STAGE 3

↑ *Ocean returns*

Tectonic forces distort strata; erosion carves the surface. New unconformity in the making.

1

← Tectonic forces

STAGE 2

↑ *Ocean recedes, exposing new surface*

GEOLOGIC TIME

Think Critically

1. Would a surface between adjacent parallel layers of sediment be a disconformity if erosion had not occurred?
2. What is a common circumstance where such surfaces might be found?

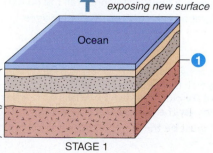

Layers of sediment deposited on an erosional surface become young sedimentary rocks.

Ancient igneous and metamorphic rocks are underlying.

Ocean

1

STAGE 1

Today

Millions of years ago

WHAT A GEOLOGIST SEES

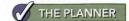

A Grand View of Stratigraphy

a. The Grand Canyon

The Grand Canyon is one of the natural wonders of the world. Whether the view is from a boat on the Colorado River (as seen here), or from the rim looking down, the geologist sees a beautiful example of the principles of stratigraphy. The strata are horizontal; the youngest strata are at the top; and the same strata can be seen on both sides of the Canyon, even though a large piece has been removed by erosion. As a geology student, you owe it to yourself not to stop at the Canyon rim. Make a point of descending to the river level to get a close-up look at 2 billion years of Earth's history from a geologists' perspective.

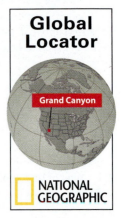

Global Locator

Grand Canyon

NATIONAL GEOGRAPHIC

b. View from the Toroweap Overlook

The view here shows a nearly 2-kilometer thickness of horizontal sedimentary strata lying on top of older strata that were tilted and tectonically deformed before the horizontal strata were deposited. Geologists have named each stratum after its predominant rock type and the location in the Grand Canyon where the best outcroppings occur (such as the Toroweap Overlook). Seeing that some strata stand out boldly, making vertical cliffs, while others have gentler slopes, a geologist would realize that the vertical cliff layers resist erosion more strongly than the others. This means they must differ in the composition or strength of the cement that is holding the grains together.

c. What a geologist sees

Noting the tilted strata at the bottom of the Canyon (the Grand Canyon Supergroup), the geologist would see that they were tilted and deformed before the strata above were laid down, so they must be much older than the horizontal strata. From the view alone, a geologist cannot tell the numerical age of the rocks; stratigraphy can only tell us relative ages.

Think Critically

Using an average rate of deposition of sediment of 1 centimeter per year, estimate the time needed for 2 kilometers of sediment to be deposited above the older, tilted rocks. What might be the errors in estimating the age of a rock sequence by this method?

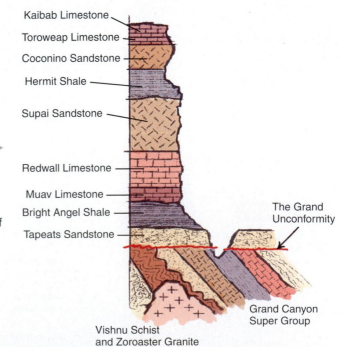

Kaibab Limestone
Toroweap Limestone
Coconino Sandstone
Hermit Shale
Supai Sandstone
Redwall Limestone
Muav Limestone
Bright Angel Shale
Tapeats Sandstone
The Grand Unconformity
Grand Canyon Super Group
Vishnu Schist and Zoroaster Granite

Varieties of fossils • Figure 3.4

Most fossils begin as hard plant or animal parts, such as the bones of this 23-centimeter-long Pachypleurosaurus, a marine reptile that lived about 230 million years ago and was preserved in rocks now found in Switzerland.

These fossil tracks in the Tapeats Sandstone were created when trilobites extended their legs, scooped up some mud, and picked over the mud for food.

▲ Other fossils record the imprint of an animal's body on mud that later solidified. This fossil, found on Snow Hill Island in Antarctica, shows the imprint of a snail-like gastropod, made during the Cretaceous Period.

Fossils and Correlation

Many strata contain remains of plants and animals that were incorporated into the sediment as it accumulated. **Fossils** usually consist of "hard parts" such as shells, bones, or wood whose forms have been preserved in sedimentary rock (**Figure 3.4**). In some cases the imprints of soft animal tissues, such as skin or feathers or the leaves and flowers of plants have been preserved. Even the preserved tracks and footprints of animals are considered to be fossils. Many fossils found in geologically young strata look similar to plants and animals living today (**Figure 3.5**). The farther down in the stratigraphic sequence we go, the more likely we are to find fossils of extinct plants or animals and the less familiar they seem. Nicolaus Steno, whom we introduced in Chapter 2 in connection with crystals, was also one of the first scientists to conclude that fossils were the remains of ancient life. He published his ideas in a landmark paper in 1669, in which he also stated the principles of stratigraphic superposition and original horizontality. His conclusions were ridiculed at the time, but by the next century, the idea of the plant and animal origin of fossils was widely accepted.

Ancient and modern • Figure 3.5

a.

b.

A fossilized imprint of a *Lebachia* (**a**), a conifer that is found in rocks about 250 million years old, strongly resembles a modern Norfolk Island pine (**b**).

NATIONAL GEOGRAPHIC

Fossils and correlation • Figure 3.6

Geologists can use fossils to correlate strata at localities (1, 2, and 3) that are many kilometers apart. Strata B, C, D, and E have fossil assemblages that are different from one another but consistent among the three sites.

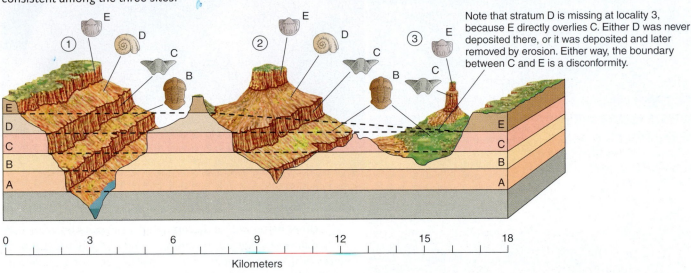

Note that stratum D is missing at locality 3, because E directly overlies C. Either D was never deposited there, or it was deposited and later removed by erosion. Either way, the boundary between C and E is a disconformity.

Kilometers

paleontology
The study of fossils and the record of ancient life on Earth; the use of fossils for the determination of relative ages.

The study of fossils, called **paleontology**, is hugely important for understanding the history of life on Earth, and we will have much more to say about it in Chapter 15. Aside from their biologic interest, fossils also have great practical value to geologists. About the same time that James Hutton was working in Scotland, a young surveyor named William Smith was laying out canal routes in southern England. As the canals were excavated, Smith noticed that each group of strata contained a specific assemblage of fossils. In time, he could look at a specimen of rock from any sedimentary layer in southern England and name the stratum and its position in the sequence of strata. This skill enabled him to predict what kind of rock the canal excavators would encounter and how long it would take to dig through it. It also earned him the nickname "Strata."

The stratigraphic ordering of fossil assemblages is known as the **Principle of faunal and floral succession. Fauna** means animals, **flora** means plants; **succession** means that new species succeed earlier ones as they evolve. William Smith's practical discovery turned out to be of great scientific importance. Floral and faunal successions can be employed in determining the sequence and therefore the relative age of strata. Geologists soon demonstrated that the faunal succession in northern France is the same as that found by Smith in southern England. By the middle of the 19th century, it had become clear that faunal succession is essentially the same everywhere. Thus Smith's practical observations led to a means of worldwide **correlation**, which is the most important way of filling the gaps in the geologic record (**Figure 3.6**).

correlation A method of equating the ages of strata that come from two or more different places.

CONCEPT CHECK 🛑 STOP

1. **How** does the relative age of a stratum differ from its numerical age?

2. **What** are the three basic principles of stratigraphy that were first stated by Hutton and Steno?

3. **Describe** the three main kinds of unconformities

4. **How** are fossils used in stratigraphic correlation?

The Geologic Column

LEARNING OBJECTIVES

1. **Explain** how worldwide observations of strata and the fossils they contain led to a single sequence of relative ages called the geologic column.

2. **Distinguish** four units of geologic time: eons, eras, periods, and epochs.

3. **Explain** how different eras and periods correspond to different fossil assemblages.

Worldwide stratigraphic correlation (**Figure 3.7**) was one of the greatest successes of 19th-century science. It meant that, by using fossils and stratigraphic correlation, a gap in the stratigraphic record in one place could be filled using evidence from somewhere else. Through worldwide correlation, geologists assembled the **geologic column**, or **stratigraphic time scale**, a composite diagram showing the succession of all known strata, fitted together in chronological order, on the basis of their fossils and other evidence of relative age (**Figure 3.8**).

> **geologic column**
> The succession of all known strata, fitted together in relative chronological order.

Eons and Eras

Even the most cursory inspection of the geologic column reveals how closely our understanding of strata is intertwined with the history of life. The vast majority of Earth's history is divided into three **eons** in which fossils are extremely rare or nonexistent. The earliest eon is the time between Earth's formation and the age of the oldest rocks so far discovered. It is sometimes informally called the "Hadean Eon," but this is not a formal stratigraphic term because there is no corresponding rock layer. The most ancient rocks preserved on Earth mark the base of the **Archean** ("ancient") Eon, and the Archean is followed by the **Proterozoic** ("early life") Eon. The Archean is roughly the period when single-celled life developed, and the Proterozoic is when multicelled, soft-bodied organisms first emerged. We now know that each eon spanned several hundred million years of time, as shown in Figure 3.8; however, the 19th-century geologists who first worked out the geologic column had no way of determining the length of the eons.

At the beginning of the fourth and current eon, called the **Phanerozoic** ("visible life"), the fossil record suddenly becomes much more detailed, thanks to the appearance of the first animals with hard shells and skeletons. Thus, paleontologists of the 19th century divided the Phanerozoic Eon into three shorter units called **eras**: the **Paleozoic** ("ancient life"), **Mesozoic** ("middle life"), and **Cenozoic** ("recent life"). These eras are marked by vastly different faunal and floral assemblages. One reason for the dramatic differences between eras is that they were all separated by major extinction events when more than 70% of the species on Earth perished.

Markers of geologic time • Figure 3.7

These ammonite fossils in limestone are from the Jura Mountains in Switzerland. They are typical of those used to correlate strata from place to place. Rock anywhere in the world that contains this species of ammonite can be reliably dated to the Jurassic Period. Dinosaur and other vertebrate fossils attract more attention, but marine invertebrate fossils have been more helpful in allowing scientists to reconstruct the history of this planet, because they are more common and more widely distributed.

NATIONAL GEOGRAPHIC

Geology InSight
The geologic column in words and pictures • Figure 3.8

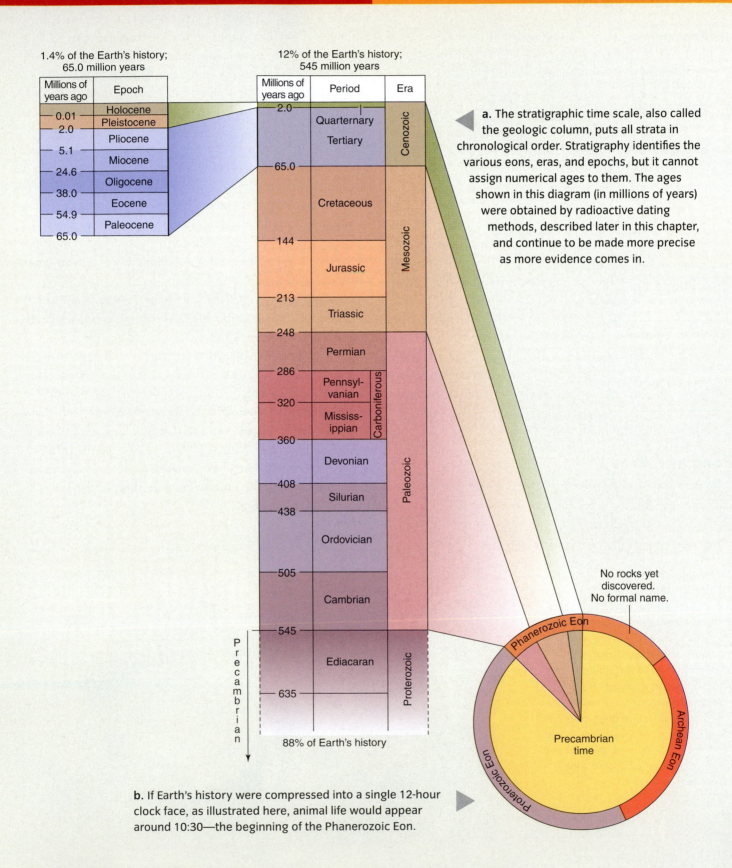

1.4% of the Earth's history; 65.0 million years

Millions of years ago	Epoch
0.01	Holocene
2.0	Pleistocene
5.1	Pliocene
24.6	Miocene
38.0	Oligocene
54.9	Eocene
65.0	Paleocene

12% of the Earth's history; 545 million years

Millions of years ago	Period	Era
2.0	Quaternary	Cenozoic
	Tertiary	Cenozoic
65.0	Cretaceous	Mesozoic
144	Jurassic	Mesozoic
213	Triassic	Mesozoic
248	Permian	Paleozoic
286	Pennsylvanian (Carboniferous)	Paleozoic
320	Mississippian (Carboniferous)	Paleozoic
360	Devonian	Paleozoic
408	Silurian	Paleozoic
438	Ordovician	Paleozoic
505	Cambrian	Paleozoic
545	Ediacaran	Proterozoic
635		Proterozoic

Precambrian

88% of Earth's history

a. The stratigraphic time scale, also called the geologic column, puts all strata in chronological order. Stratigraphy identifies the various eons, eras, and epochs, but it cannot assign numerical ages to them. The ages shown in this diagram (in millions of years) were obtained by radioactive dating methods, described later in this chapter, and continue to be made more precise as more evidence comes in.

No rocks yet discovered. No formal name.

Phanerozoic Eon

Archean Eon

Precambrian time

Proterozoic Eon

b. If Earth's history were compressed into a single 12-hour clock face, as illustrated here, animal life would appear around 10:30—the beginning of the Phanerozoic Eon.

c. The geologic column in pictures

Bats

Messelobunodon

Tiny horses

Jewel beetle

Termites and ants

Primitive woodpecker

Pangolin

Anteater

Cenozoic Era

In the Cenozoic Era, birds and mammals have flourished. In this scene from 50 million years ago, in the Paleogene Period, we can see some possible ancestors of primates.

Bald cypress trees

Anchiceratops

Ground beetle

Magnolia

Plant beetle

Edmontosaurus

Cockroach

Early mammal

Mesozoic Era

The Mesozoic Era saw the rise of dinosaurs, which were the dominant vertebrates (animals with backbones) on land for many millions of years. If you look closely in the corners, you can see two harbingers of the future that first appeared in the Mesozoic Era: the first magnolia-like flowering plants (lower left) and the first shrew-like mammals (lower right).

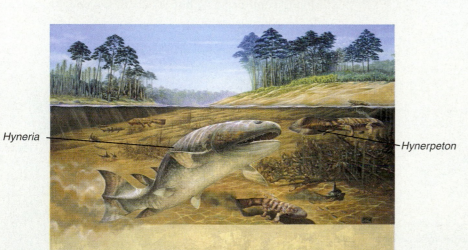

Hyneria

Hynerpeton

Paleozoic Era

During the Paleozoic Era, the evolution of life progressed from marine invertebrates (animals without backbones) to fish, amphibians, and reptiles. In this scene from 350 million years ago during the Carboniferous Period, a lobe-finned fish (foreground) swims alongside some early amphibians whose fins have evolved into legs but who still have fishlike tails.

The Geologic Column **75**

Periods and Epochs

Eras are divided into still shorter units called **periods**. The three eras of the Phanerozoic Eon were the first to be divided into periods, and the divisions were made on the basis of fossils. Geologists named the periods in a somewhat haphazard manner. Some are named for geographic locations (e.g., Jurassic, from the Jura Mountains in Switzerland where this layer was first studied) and others are named for the characteristics of their strata (e.g., Cretaceous, from the Latin word for "chalk").

The earliest period of the Paleozoic Era, the Cambrian Period, is especially noteworthy. This is when animals with hard shells first appeared in the geologic record. It was a time of unprecedented diversification of life, called the **Cambrian explosion** (**Figure 3.9**). Rocks that formed before the Cambrian Period, in the Proterozoic Eon or earlier, with one exception, generally cannot be differentiated on the basis of the fossils they contain. With rare exceptions, the fossils are either nonexistent or microscopic. Thus geologists often lump the time that precedes the Cambrian Period, and the rocks formed then, into a single category, **Precambrian**. The one exception is the Ediacaran Period, which is characterized by remarkable fossils of soft-bodied creatures (**Figure 3.10**). The Ediacaran is the last period of the Proterozoic Eon (and therefore that last of the Precambrian), and it is the only period of the Proterozoic defined by macroscopic fossils. All the periods of the Phanerozoic Eon were defined by geologists in the 19th century. The fossils that define the Ediacaran were first discovered in Australia in the 1940s, then in Canada and Namibia and other places around the world; the Ediacaran Period was finally defined in 2004, the first new addition to the geologic column in 120 years.

Periods, in turn, are split into smaller divisions called **epochs**. The most recent epochs, the Holocene and the Pleistocene, are familiar because humans and their ancestors emerged during these epochs, and their names sometimes appear in the popular press. These recent epochs are not defined by extinction events but by the percentage of their fossils that are still-living species. Many plant and animal fossils found in Pliocene strata, for example, have still-living counterparts, but fossils in Eocene strata have few living counterparts.

The Cambrian explosion • Figure 3.9

Only the sketchiest remains are left of any life forms from Earth's earliest eons. But in the Cambrian Period, life suddenly exploded in a profusion of bizarre and now extinct forms. Trilobites (*below*) ranged from 1 millimeter in size to the largest known specimen, 72 centimeters.

Anomalocaris was the most fearsome predator of the Cambrian seas, a swimming creature up to 1 meter long.

Opabinia, which among other oddities had five eyes, was about 7 centimeters long and would have been a tasty morsel for *Anomalocaris*.

Most ancient macroscopic fossils • Figure 3.10

This curious fossil is *Dickinsonia costata*, a soft-bodied creature that lived about 600 million years ago. Lacking hard parts, *Dickinsonia* is only preserved as an imprint in the sediment in which it was buried.

CONCEPT CHECK   STOP

1. **What** is the geologic column, and how were fossils used in its development?
2. **What** are the major subdivisions of the geologic time scale?
3. **What** major biologic event distinguishes the Phanerozoic Eon from the previous (Precambrian) eons?

Numerical Age

LEARNING OBJECTIVES

1. **Recount** how scientists arrived at a way of quantifying geologic time.
2. **Outline** the process of radioactive decay.
3. **Explain** how and why radioactive decay can be used to date igneous rocks.
4. **Describe** how magnetic polarity dating can be used to date both igneous and sedimentary rocks.

The scientists who worked out the geologic column were tantalized by the challenge of numerical time. They wanted to know Earth's age, how fast mountain ranges rise, how long the Paleozoic Era lasted, and, most challenging of all, when life first appeared and how long humans have inhabited Earth.

Early Attempts

Several methods for solving the problem of numerical time were proposed during the 19th century. All of them were unsuccessful, but the reasons for their failure are nonetheless quite illuminating—and they help explain where our currently accepted estimates came from (**Figure 3.11**). As previously mentioned, early approaches involved estimating

Time line: The debate over numerical age • Figure 3.11

This time line chronicles some of the major episodes in the long debate over how old Earth is. Even the unsuccessful attempts, such as Kelvin's or Joly's, were useful because they pointed out the incorrect assumptions that these scientists and others had made.

1953: Clair Patterson estimates age of Earth = 4.56 billion years, based on radiometric dating of meteorite that is the most ancient fragment of rock ever found. His estimate is still considered correct.

1905: First radiometric dating of rocks (>500 million years) by Rutherford and Boltwood

1913: Arthur Holmes estimates age of Earth >1.6 billion years, by radioactive dating.

1898: Marie Curie discovers that radioactivity releases heat, revealing the flaw in Kelvin's (1862) estimate of Earth's age.

1900

1859: *On the Origin of Species* published. Darwin estimates age of Earth >300 million years. Omits estimate in later editions.

1899: John Joly carries out Halley's suggestion; age of Earth 90 million years.

1788: After viewing a sequence of tilted sedimentary rocks, James Hutton writes that Earth has "no vestige of a beginning."

1862: Lord Kelvin rebuts Darwin; says Earth's age is between 20 and 100 million years (based on cooling time).

(assumes Earth has no internal heat source)

1800

1831–1836: Charles Darwin's voyage on H.M.S. *Beagle*.

1715: Halley suggests using concentration of sea salts to estimate Earth's age.

1700

rates of sedimentation and multiplying by the thickness of stratigraphic sections. Unfortunately, the resulting estimates for Earth's age varied too widely to be useful, from 3 million to 1.5 billion years.

Early scientists recognized that rivers carry dissolved substances to the sea from the erosion of rocks on land. In 1715, Edmund Halley, for whom Halley's Comet is named, suggested that one could measure the rate at which salts are added to the sea by river input and calculate the time needed to transport all the salts now present in the sea. Halley did not carry out his own suggestion, and it was not until 1899 that John Joly made the necessary measurements and calculations. Using this approach, Joly estimated that the ocean and, therefore, Earth, were 90 million years old. Unfortunately, neither Halley nor Joly realized that the ocean, like other parts of the Earth system, is an open system. Salt is removed by reactions between seawater and volcanic rocks on the seafloor, as well as the evaporation of seawater in isolated basins. The addition and removal of salts have balanced each other for hundreds of millions (even billions) of years.

The 19th century also saw the publication of Charles Darwin's *On the Origin of Species* in 1859. The book intensified the ongoing debate over numerical ages. Darwin understood that the evolution of new species by natural selection must be a very slow process that required vast amounts of time. The first edition of his book contained a rough estimate of Earth's age, based on the erosion rate of a mountain range, of at least 300 million years.

William Thomson (Lord Kelvin), a leading physicist and contemporary of Darwin, emphatically rejected Darwin's estimate. Kelvin used the laws of thermodynamics to calculate how long Earth has been a solid body. Kelvin made two assumptions: (1) Earth was once completely molten, and (2) no heat was added to it after it formed. Once it had cooled enough to form a solid outer layer, Kelvin hypothesized that heat would escape only by conduction (the same way that heat moves through the wall of a coffee cup). He measured the current-day rate of heat loss and extrapolated backward to figure out the age of Earth's solid crust, arriving at a figure of 20 million years. Even Darwin conceded, "Thomson's views on the recent age of the world

have been for some time one of my sorest troubles" and removed his age estimate from later editions of *On the Origin of Species*.

Darwin died before geologists vindicated his views on the vast length of geologic time. Kelvin's mistake was his assumption that no heat had been added to Earth's interior since its formation. He did not know about **radioactivity**, a natural physical process that releases heat directly into Earth's interior. Though Earth is a closed system (as we explained in Chapter 1), it does have inputs and outputs of energy, and Kelvin had missed a key input.

> **radioactivity**
> A process in which an element spontaneously transforms into another isotope of the same element or into a different element.

What was needed to solve the problem of numerical geologic time was a way to measure events by some process that runs continuously, is not reversible, is not influenced by such factors as chemical reactions and high temperatures, and leaves a continuous record without any gaps. The discovery of radioactivity not only proved that Kelvin's assumptions were wrong but, by fortunate chance, also provided the breakthrough needed to measure absolute time.

Radioactivity and Numerical Ages

To explain how radioactivity allows us to determine the ages of rocks, we need to go back to the fundamentals of chemistry. Chapter 2 stated that most chemical elements have two or more **isotopes** that have the same number of protons per atom but a different number of neutrons per atom. To put it another way, each isotope of an element has the same atomic number but a different mass number.

Most naturally occurring isotopes have stable nuclei. However, a number of isotopes—such as carbon-14 and potassium-40—are unstable. They spontaneously release particles from their nuclei. In the process, they change their mass number, their atomic number, or both (**Figure 3.12**). Any isotope that spontaneously undergoes such nuclear change is said to be **radioactive**, and the process of change is referred to as **radioactive decay**.

Radioactive decay • Figure 3.12

Radioactive decay involves the nucleus, so it is a nuclear process rather than a chemical process. It releases far more energy than any chemical reaction, which explains why it is still keeping Earth's interior hot after more than 4 billion years. The decay of radioactive isotopes is not influenced by any chemical process, or by heat or high pressure. Thus radioactive decay is a perfect built-in geologic clock. Each radioactive isotope has its own rate of decay, so rock that contains several different isotopes has numerous built-in clocks that can be checked against each other.

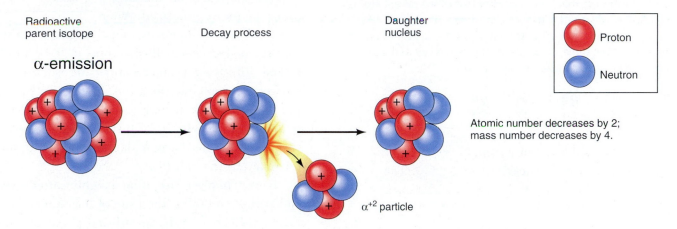

α-emission

Radioactive parent isotope — Decay process — Daughter nucleus

Atomic number decreases by 2; mass number decreases by 4.

α^{+2} particle

A radioactive nucleus releases an alpha particle, which consists of two protons and two neutrons.

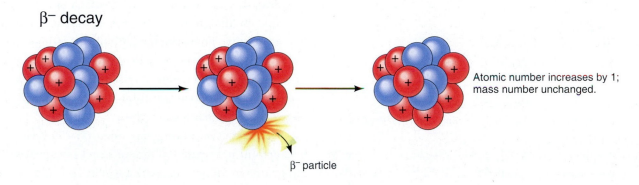

β$^-$ decay

Atomic number increases by 1; mass number unchanged.

β$^-$ particle

A radioactive nucleus releases a beta particle, and one of its neutrons turns into a proton.

Proton
Neutron

Rates of decay In any radioactive decay system, the number of original radioactive parent atoms continuously decreases, while the number of nonradioactive daughter atoms produced by the radioactive decay continuously increases. For this reason, many of the radioactive isotopes that were present when Earth was formed have decayed away because they were short-lived. However, radioactive isotopes that decay very slowly are still present.

All decay rates follow the same basic law: The proportion of parent atoms that decay during each unit of time is

Radioactivity and time • Figure 3.13

This graph illustrates the basic decay law of radioactivity. Suppose that a given isotope has a half-life of 1 hour. If we started with a sample consisting of 100% radioactive parent atoms, after an hour only 50% of the parent atoms would remain and an equal number of daughter atoms would have formed. At the end of the second hour, another half of the parent atoms would be gone, so there would be 25% parent atoms and 75% daughter atoms. After the third hour, another half of the parent atoms would have decayed, leaving 12.5% parent atoms and 87.5% daughter atoms. Note that the total number of atoms, parent plus daughter, remains constant. This is the key to the use of radioactivity to measure geologic time.

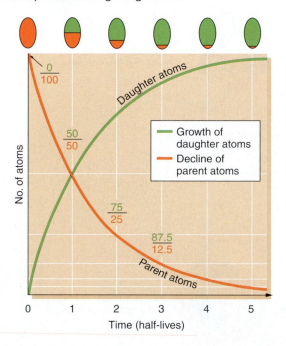

always the same (**Figure 3.13**). The rate of radioactive decay is determined by the **half-life**.

Radiometric dating: Not one clock but many

Following the discovery of radioactivity by Henri Becquerel in 1896, and further work by Marie and Pierre Curie, the first estimates of the ages of rocks using radioactive decay were made in 1905. The long-hoped-for "rock clock" was finally available. The results were, and continue to be, remarkable. **Radiometric dating** has revolutionized the way we think about Earth and its long history.

A radioactive rock clock usually measures the amount of time that has elapsed since the minerals in the rock crystallized. When a new mineral grain forms—for example, a grain of feldspar in cooling lava—all the atoms in the grain become locked into the crystal structure and isolated from the environment outside the grain. In a sense, the atoms in the mineral grain, including any radioactive atoms, are sealed in an atomic time capsule. The trapped radioactive parent atoms decay to daughter atoms at a rate determined by the half-life.

In the simplest case, if no daughter atoms were present in the mineral at the time of formation, we can use Figure 3.13 to work backward and determine how long ago the time capsule was sealed. If the mineral crystal contained some daughter atoms at the time of formation, the process is more difficult. Geologists have developed several ways to estimate the initial contamination of a sample by daughter atoms. Once that is done, and provided that we know the half-life of the radioactive parent, it is a simple matter to calculate how long ago the mineral crystallized. Geologists use different isotopic systems to study rocks, fossils, and biologic materials of different ages and compositions (**Figure 3.14**).

Radiometric dating has been particularly useful for determining the ages of igneous rocks because the mineral grains in an igneous rock form at the same time as the rock that contains them. On the other hand, most sedimentary rocks consist largely of mineral grains that were formed long before the strata that contain them were deposited. Radiometric dating will tell how old the grains are but not when the strata were deposited. This makes it difficult to directly date most sedimentary rocks and to infer the numerical age of ancient life-forms fossilized in the sediments.

Isotopes used in radiometric dating • Figure 3.14

Long-lived isotopes such as uranium-238 are especially useful for dating ancient rocks. Short-lived isotopes such as carbon-14 are useless for dating samples that are more than about 70,000 years old. However, carbon-14 is very useful for dating geologically recent items of biologic origin, such as human remains and artifacts. Geologists have also used it to study the latest ice age, by estimating the ages of wood samples taken from trees killed by the advancing ice sheet.

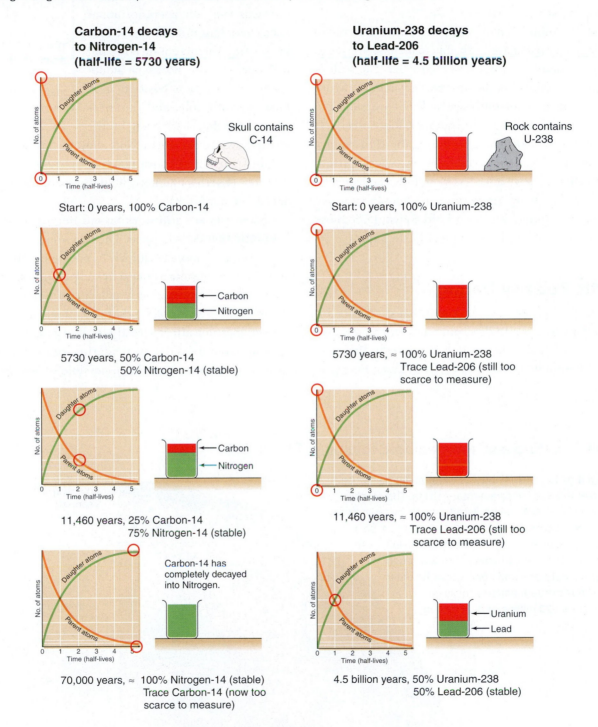

**Carbon-14 decays
to Nitrogen-14
(half-life = 5730 years)**

Skull contains
C-14

Start: 0 years, 100% Carbon-14

← Carbon
← Nitrogen

5730 years, 50% Carbon-14
50% Nitrogen-14 (stable)

← Carbon
← Nitrogen

11,460 years, 25% Carbon-14
75% Nitrogen-14 (stable)

Carbon-14 has
completely decayed
into Nitrogen.

70,000 years, ≈ 100% Nitrogen-14 (stable)
Trace Carbon-14 (now too
scarce to measure)

**Uranium-238 decays
to Lead-206
(half-life = 4.5 billion years)**

Rock contains
U-238

Start: 0 years, 100% Uranium-238

5730 years, ≈ 100% Uranium-238
Trace Lead-206 (still too
scarce to measure)

11,460 years, ≈ 100% Uranium-238
Trace Lead-206 (still too
scarce to measure)

← Uranium
← Lead

4.5 billion years, 50% Uranium-238
50% Lead-206 (stable)

Numerical time and the geologic column As geologists worked out the geologic column, they found many locations where layers of solidified lava and volcanic ash are interspersed with sedimentary strata. Through radiometric dating, they could determine the numerical ages of the lavas and volcanic ash and thereby bracket the ages of the sedimentary strata (**Figure 3.15**).

Through a combination of geologic relations and radiometric dating, scientists have been able to fill in all the dates in the geologic column, as shown in Figure 3.8. The scale is continually being refined, so the numbers given in the figure are considered the best currently available but are subject to change. Further work will make the numbers more precise. It is a tribute to the work of geologists during the 19th century that the geologic column they established by ordering strata according to their relative ages has been fully confirmed by radiometric dating. At the same time, it is humbling to see how one wrong assumption caused Lord Kelvin to be more than 4 billion years off in his estimate of Earth's age.

Magnetic Polarity Dating

Time is so central to the study of Earth that geologists are always seeking new ways to determine numerical ages. An exciting newer method of dating, developed in the 1960s, involves **paleomagnetism**, the study of Earth's past magnetic field (**Figure 3.16**). Both igneous and sedimentary rocks "lock in" information about the magnetic field at their time of formation.

Earth's magnetic field reverses its polarity at irregular intervals but, on average, about once every half million years. This means that the magnetic pole that had been in the northern hemisphere moves to a position near Earth's south pole, and the magnetic pole that had been in the south moves to the north. Note, however, that Earth's geographic North and South poles of rotation do not change position. (Earth's magnetic field is explained in additional detail in Chapter 4.)

Scientists are still working out details of how or why **magnetic reversals** happen. The two important points are that a reversal happens quickly by geologic time standards, and any iron-bearing mineral in an igneous rock retains, or "remembers," the magnetic polarity of Earth at the time that the rock was formed—that is, a change in the magnetic field does not affect already formed minerals. Through a combination of radiometric dating and magnetic polarity measurements, it has been possible to establish a time

> **paleomagnetism**
> The study of rock magnetism in order to determine the intensity and direction of Earth's magnetic field in the geologic past.
>
> **magnetic reversal**
> A period of time in which Earth's magnetic polarity reverses itself.

Radiometric dating and the geologic column • Figure 3.15

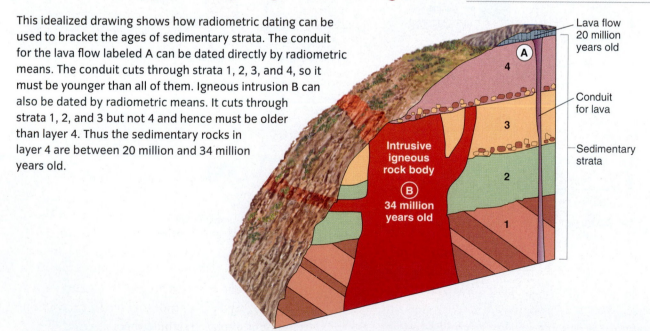

This idealized drawing shows how radiometric dating can be used to bracket the ages of sedimentary strata. The conduit for the lava flow labeled A can be dated directly by radiometric means. The conduit cuts through strata 1, 2, 3, and 4, so it must be younger than all of them. Igneous intrusion B can also be dated by radiometric means. It cuts through strata 1, 2, and 3 but not 4 and hence must be older than layer 4. Thus the sedimentary rocks in layer 4 are between 20 million and 34 million years old.

Lava flow 20 million years old

Conduit for lava

Sedimentary strata

Intrusive igneous rock body

B 34 million years old

Magnetic reversal time scale
• Figure 3.16

Periods of normal polarity, as today, and periods of reversed polarity have been identified and dated, using radiometric dating of lavas, back to the beginning of the Jurassic Period, about 200 million years ago. This figure shows the most recent 7 million years.

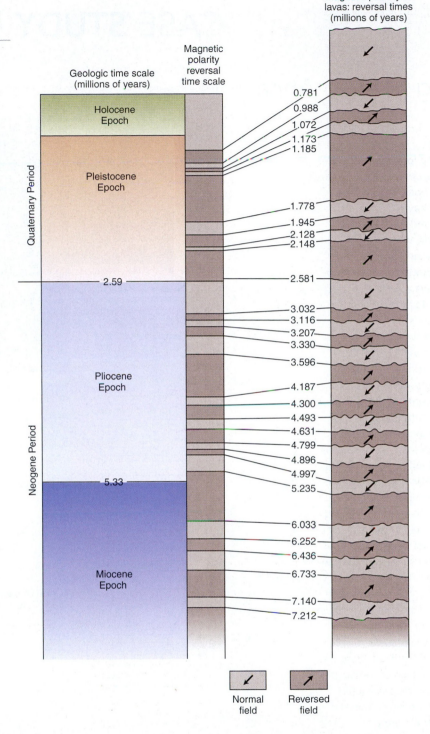

scale of magnetic polarity reversals dating back to the Jurassic Period (see Figure 3.16). Still earlier reversals are a subject of ongoing research.

Correlation on the basis of magnetic reversals differs from other geologic correlation methods. One magnetic reversal looks just like any other in the rock record. When evidence of a magnetic reversal is found in a sequence of rocks, the problem lies in knowing which of the many reversals it actually represents. When a continuous record of reversals can be found, starting with the present, it is simply a matter of counting backward. But if there is not a continuous record of reversals, counting backward must be combined with stratigraphic and radioactive dating techniques. Magnetic polarity studies have been used to date strata that contain the fossils of human ancestor species (see Case Study on the next page). Chapter 4 discusses how magnetic polarity studies played a crucial role in the development of plate tectonic theory.

CASE STUDY

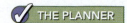 THE PLANNER

Dating Human Ancestors

The Hadar region of northern Ethiopia (photo) has been a fertile site for finding fossils of ancient human ancestors, part of a larger group called *hominids*. Many of the fossils have been found in strata derived from ancient stream gravels, where they were tumbled and battered by long-ago floodwaters. The fossil-bearing gravels are interlayered with sediments that give good magnetic signals.

The problem is to know where in the magnetic polarity time scale the Hadar sediment falls. Geologists answered this question by using the Potassium-Argon (K-Ar) method to establish the ages of a lava flow that lay below some of the hominid fossils and a layer of volcanic ash that lay above them. The two radiometric dates determined the magnetic reversal ages unambiguously and indicated that early hominids lived in the region between 3 and 4 million years ago—that is, during the Pliocene Epoch. Through many studies like this, it is slowly becoming clear that the hominid genus *Homo* (which includes our own species, *Homo sapiens*) evolved a little more than 2 million years ago from an older genus called *Australopithecus*. (Two reconstructed *Australopithecus* skulls from the Hadar are shown here.)

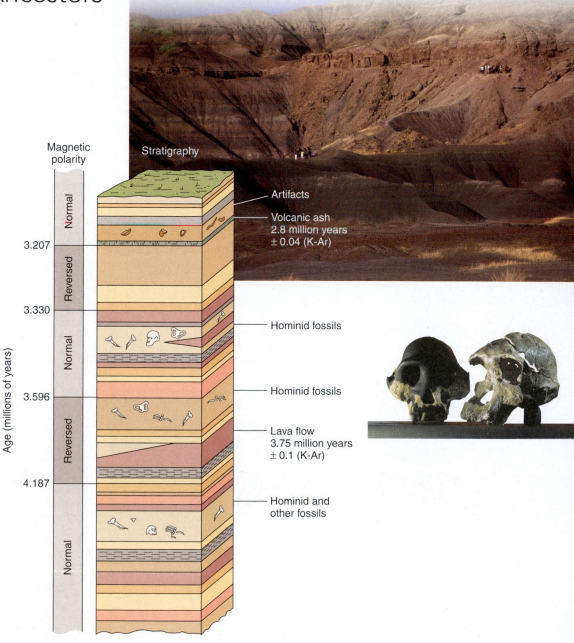

CONCEPT CHECK

1. **What** were some of the early attempts to calculate the age of Earth, and why were they inaccurate?

2. **How** does the process of radioactive decay work?

3. **Why** and how is radioactive decay useful as a "geologic clock"?

4. **How** can magnetic polarity reversals contribute useful information about rock and fossil ages?

The Age of Earth

LEARNING OBJECTIVES

1. **Explain** why the oldest rocks are not necessarily the same age as the planet.

2. **Explain** why scientists currently believe Earth is about 4.56 billion years old.

Throughout this book, we mention examples of actual rates of geologic processes. This would not be possible without the numerical dates obtained through radiometric dating and other numerical age methods. In fact, more than any other contribution by geologists, the ability to determine numerical dates has changed the way humans think about the world and the immensity of geologic time. To learn more about the challenges geologists face and the tools they use to quantify geologic time, see *Where Geologists Click*.

Now that we know how to determine the numerical ages of rocks, can we determine Earth's age? It's not as easy as you might think. The continual recycling of Earth's surface by erosion and plate tectonics means that very few remnants of Earth's original crust, if any, remain. Of the many radiometric dates obtained from Precambrian rocks, the oldest is about 4.0 billion years (**Figure 3.17**). Although no older rocks than this have been found, an individual mineral grain from sedimentary rock in Australia has been dated to 4.4 billion years, so igneous rocks older than 4.0 billion years may someday be located.

However, geologists believe that there is still a gap between the age of these oldest mineral grains and the age of Earth. There is strong geochemical evidence that Earth formed at the same time as the Moon, the other planets, and meteorites. Through radiometric dating, it has been possible to determine the ages of meteorites and of "moon dust" brought back by astronauts. The *Apollo* astronauts found rocks and individual grains of Moon dust that are believed to be pieces of the Moon's original crust. Such rocks should be abundant because the Moon has been geologically much less active than Earth.

Where Geologists CLICK

The U.S. Geological Survey

http://pubs.usgs.gov/gip/geotime/contents.html

To learn more about the age of Earth and how it is determined, visit the U.S. Geological Survey Web site. This section of the USGS Web site also has a discussion of radiometric dating, and an explanation of relative time and how index fossils are used to establish relative time.

Earth's oldest rock • Figure 3.17

The Acasta gneiss in northern Canada, shown here, was formed 4 billion years ago. It is the most ancient body of rock so far discovered on Earth.

Meteorite ages are especially valuable because some meteorites have remained virtually unaltered since the formation of the solar system. Melting and other types of geologic alteration reset radiometric clocks. However, some meteorites, such as the Allende meteorite, which fell to Earth in the Mexican state of Chihuahua on February 8, 1969, belong to a rare category called **carbonaceous chondrites**, which are believed to contain unaltered material of the kind that accreted to form Earth from the solar nebula (see Chapter 1). It is "carbonaceous" because it contains tiny amounts of carbon, some of which is in chemical compounds called amino acids—organic components that are essential for life.

A portion of the Allende meteorite is shown in **Figure 3.18**. The dark, fine-grained part is mostly olivine, with a few flecks of metallic iron and some carbon. The clumps of white material are oxides of calcium and aluminum and are thought to be among the first matter to condense from the gas cloud from which the solar system formed. The white clumps are older than Earth itself. The ages of carbonaceous chondrites, as a group, cluster closely around 4.56 billion years; planetary scientists therefore conclude that Earth, and indeed the entire planetary system, formed at that time.

Today, more than two centuries after Hutton's Insightful observations at the Siccar Point unconformity that now bears his name (see *Amazing Places*), it is widely agreed that Earth's age is approximately 4.56 billion years. So it is no longer correct to say that Earth has "no vestige of a beginning." When will it cease to exist? Astronomers tell us that billions of years in the future, the Sun will become a red giant, at which point it will expand and engulf Earth. However, Hutton is still correct in one sense: Earth's history is profound, and geologic (as opposed to astronomical) evidence shows no prospect of an end.

CONCEPT CHECK

1. **What** is the oldest age that has been obtained from material found on Earth? Does this match the presumed age of the planet as a whole? Why or why not?

2. **How** have meteorites and rock samples from the Moon helped geologists determine the age of Earth as 4.56 billion years?

A cosmic interloper • Figure 3.18

The Allende meteorite, which fell to Earth in Mexico, is one of the most famous meteorites in history. Note the white spots on the meteorite. Some of these inclusions, which are slightly older than the black carbonaceous material around them, are more than 4.6 billion years old, making them the oldest objects of any kind ever found on Earth.

NATIONAL GEOGRAPHIC

AMAZING PLACES

Famous Unconformities

In 1788 James Hutton, whom we first met in Chapter 1, sailed along the eastern coast of Scotland, not far from Edinburgh. He was accompanied by John Playfair and James Hall, each of whom would later play important roles in the development of the science of geology. Hutton was struck by the scene he saw; near-vertical strata that had been truncated and overlain by slightly tilted sandy strata. Hutton was looking at an unconformity. ▼

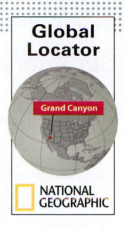

a. Hutton's Unconformity

We now know that the vertical strata of Hutton's Unconformity at Siccar Point in Scotland are sandstones of Silurian age, about 425 million years old. The slightly tilted strata are Devonian-aged Old Red Sandstone, about 345 million years old. The surface between the two sets of strata is the unconformity, and it represents an 80-million-year time gap for which there is no rock record. ▼

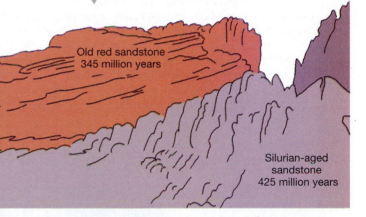

Old red sandstone
345 million years

Silurian-aged sandstone
425 million years

b. The Great Unconformity

◀ In 1869, just 81 years after Hutton's voyage along the Scottish coast, John Wesley Powell, an eminent American geologist, led the first recorded trip by boat through the Grand Canyon. Powell was so struck by the unconformity below the Tapeats Sandstone that he named it the Great Unconformity. It represents a time gap of several hundred million years. In this panoramic view, you can see the Great Unconformity between the horizontally layered sedimentary strata at the top and the older, tilted metamorphic and igneous rock units underneath.

Tapeats Sandstone

Great Unconformity

Grand Canyon Supergroup

Summary

1 Relative Age 66

- Geologists study the chronologic sequence of geologic events—that is, their **relative age**. Relative age is derived from **stratigraphy**, the study of rock layers and how those layers are formed.

- There are three basic principles of stratigraphy. (1) Strata, or sedimentary rock layers, are horizontal when they are deposited as water-laid sediment (Principle of Original Horizontality). (2) Strata accumulate in sequence, from the oldest on the bottom to the youngest on top (Principle of Stratigraphic Superposition). (3) Strata extend outward horizontally; they may thin or pinch out at their farthest edges, but they generally do not terminate abruptly unless cut by a younger rock unit (Principle of Lateral Continuity). In addition, the Principle of Cross-cutting Relationships states that a rock stratum is always older than any geologic feature, such as a fracture, that cuts across it.

- **Numerical age**, the exact number of years of a geologic feature, is more difficult to determine than relative age. The sequence of strata in any particular location is not necessarily continuous in time. An **unconformity** is a break or gap in the normal stratigraphic sequence. It usually marks a period during which sedimentation ceased and erosion removed some of the previously laid strata. The three common types of unconformities are nonconformities, angular unconformities, and disconformities as shown in the diagram.

Unconformities • Figure 3.3

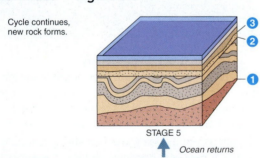

2 The Geologic Column 73

- The **geologic column**, a stratigraphic time scale, is a composite diagram that shows the succession of all known strata, arranged in chronological order of formation, based on fossils and other age criteria as shown in the diagram.

The Geologic column in words and pictures • Figure 3.8

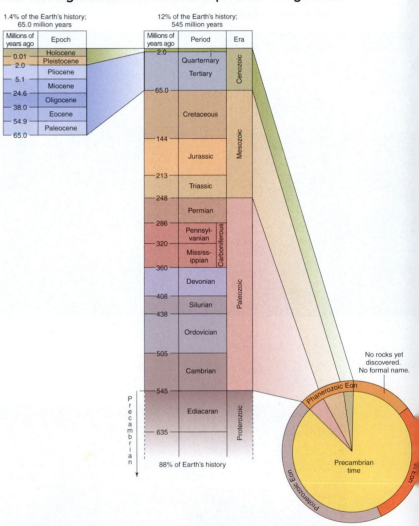

- **Correlation** of strata is the establishment of the time equivalence of strata in different places. Fossil assemblages, usually consisting of hard shells, bones, and wood, have been the primary key to correlation of strata across long distances. The study of fossils and the record of ancient life on Earth is called **paleontology**. The Principle of Faunal and Floral Successions (animals and plants, respectively) says that fossil assemblages will occur in the same sequence, regardless of location.

- The geologic column is divided into several different units of time, called eons, eras, periods, and epochs. The majority of Earth's history is divided into three eons, each spanning several hundred million years. During the first two, the Archean and Proterozoic, fossils are very rare or nonexistent except at the end of the Proterozoic, when curious soft-bodied fossils allow the Ediacaran Period to be defined. The third and most recent eon, the Phanerozoic, is the only eon in which fossils are abundant. Very dramatic changes in fossil assemblages occur between the three eras of the Phanerozoic eon—the Paleozoic, Mesozoic, and Cenozoic—which were separated by major extinction events.

3 Numerical Age   77

- Many scientists in the 19th century and earlier proposed different ways to find the age of Earth, and came up with widely differing estimates. Though many of those approaches were later shown to be imprecise, each was an important step in finding a method of numerical dating.

- **Radioactivity** is the process in which an element spontaneously transforms itself into another isotope of the same element or into a different element through the release of particles and heat energy. The radioactive decay of isotopes of chemical elements provides a basis for radiometric dating, which gives values for the **numerical ages** (age in years) of rock units and thus values for numerical dates of geologic events. Because radioactive decay is not influenced by chemical processes or by heat and high pressure, it is an extremely accurate gauge of numerical age.

- **Radiometric dating** is based on the principle that in any sample containing a radioactive isotope, half of its atoms of that isotope will change to daughter atoms in a specific length of time, called the **half-life** as shown in the diagram. Radioactive isotopes with long half-lives, such as uranium, are most useful for dating rocks. Carbon-14, which has a much shorter half-life, is most useful for dating organic materials of relatively recent origin (less than 70,000 years).

- Though radiometric dating is primarily useful for igneous rocks, a complementary technique called magnetic polarity dating works for sedimentary rocks, too. Magnetic polarity dating involves **paleomagnetism**, and the study of reversals in Earth's magnetic field. The cause of these **magnetic reversals** is still not fully understood.

4 The Age of Earth   85

- Through measures of numerical age, it has become clear that most of Earth's history took place in Precambrian time. The oldest Earth rocks discovered are about 4.0 billion years old as shown in the photo.

Earth's oldest rock • Figure 3.17

- Earth is not a good place to look for the oldest rocks in the solar system. Earth's surface has been subjected to a lot of geologic activity. This has reset some radiometric clocks and destroyed the earliest fragments of the crust. Samples from the Moon and from meteorites indicate that the solar system formed about 4.56 billion years ago, and by inference this is also the age of Earth.

Radioactivity and time • Figure 3.13

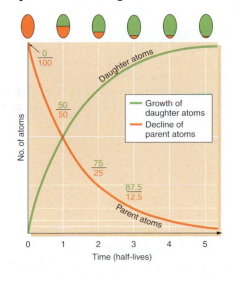

Key Terms

- correlation   72
- geologic column   73
- half-life   80
- magnetic reversal   82

Critical and Creative Thinking Questions

1. Do the same principles of stratigraphy apply on the Moon as on Earth? Bear in mind that the geologic processes on the Moon have been very different from those on Earth. If you had to determine the relative age of features on the Moon, based entirely on satellite photographs, what would you look for, and how might you proceed?

2. Check the area in which you live to see if there is an excavation—perhaps one associated with a new building or road repair. Visit the excavation and note the various layers, the paving (if the excavation is in a road), and the soil below the surface. Is any bedrock exposed beneath the soil?

3. How old are the rock formations in the area where you live and attend college or university? How can you find out the answer to this question?

4. Choose one of the geologic periods or epochs listed in Figure 3.8 and find out all you can about it. How are rock formations from that period identified? What are its most characteristic fossils? Where are the best samples of rock from your chosen period found?

5. Do some research to determine the ages of the oldest-known fossils. What kind of life-forms were they?

What is happening in this picture?

This skier is hauling a sled past a cliff face on Ellesmere Island, Canada.

Think Critically

1. Why do you think the rock strata in the background tilt at such a steep angle?
2. Why are the strata wavy instead of straight?

Self-Test

(Check your answers in Appendix D.)

1. A _____ is the age of one rock unit or geologic feature compared to another.
   a. relative age
   b. numerical age

2. The principle of cross-cutting relationships says that _____.
   a. water-borne sediments are deposited in nearly horizontal layers
   b. a sediment or sedimentary rock layer is younger than the layers below it and older than the layers that lie above
   c. a rock unit is older than a feature that disrupts it, such as a fault or igneous intrusion
   d. a sediment or sedimentary rock layer is older than the layers below it and younger than the layers that lie above

3. The _____ states that water-borne sediments are deposited in nearly horizontal layers.
   a. law of superposition
   b. principle of faunal succession
   c. principle of original horizontality
   d. principle of cross-cutting relationship

4. In a conformable sequence, _____.
   a. each layer must have been deposited on the one below it, without any interruptions
   b. there must not be any depositional gaps in the stratigraphic record
   c. Both a and b are correct.
   d. Neither a nor b is correct.

5. An unconformity represents _____.
   a. a gap in the stratigraphic record
   b. a period of erosion or nondeposition
   c. Both a and b are correct.
   d. Neither a nor b is correct.

6. On this diagram, label each unconformity as one of following:
   nonconformity
   angular unconformity
   disconformity

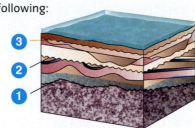

7. Fossils found in strata _____.
   a. are the records of ancient life
   b. allow the correlation of strata separated by many kilometers
   c. have been useful to geologists in creating the geologic column
   d. All of the above answers are correct.

8. The three eras that make up the Phanerozoic Eon are the _____.
   a. Hadean, Archean, and Proterozoic
   b. Paleozoic, Mesozoic, and Cenozoic
   c. Triassic, Jurassic, and Cretaceous
   d. Pliocene, Pleistocene, and Holocene

9. The most distinctive changes in the fossil record occur across the boundaries between _____.
   a. periods   b. eras   c. epochs

10. The dinosaurs were dominant during _____.
   a. the Cenozoic Era   c. the Paleozoic Era
   b. the Mesozoic Era   d. the Precambrian time

11. Label the two decay sequences depicted in this diagram as either alpha emission or beta decay. For each decay sequence, also label the following:
    parent nucleus   daughter nucleus
    alpha particle   (or)   beta particle

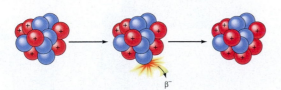

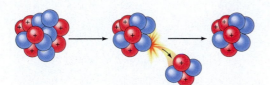

●	Proton
●	Neutron

12. Potassium-40 is a naturally occurring radioisotope that decays to Argon-40 and is common in many rocks of the continental crust. The half-life of Potassium-40 is 1.3 billion years. Assuming no contamination, what would be the age of a sample that contains a 1:3 ratio of Potassium-40 to Argon-40?
    a. 1.3 billion years   c. 2.6 billion years
    b. 650 million years   d. 325 million years

13. What are carbonaceous chondrites? Why are meteorites—and carbonaceous chondrite meteorites, in particular—useful to scientists in determining the age of Earth?
    a. The sample would appear too young.
    b. The sample would appear too old.
    c. Rocks are a "closed system"; there would be no error.

14. List all of the labeled items (1, 2, 3, 4, A, and B) on the cross section in chronological order from youngest (at the top of your list) to oldest (at the bottom of your list), using the rules of stratigraphy to determine their relative ages.

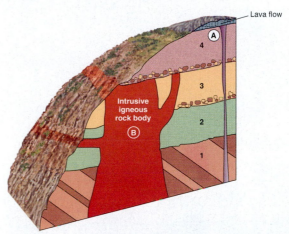

Lava flow

Intrusive igneous rock body

15. Earth is not considered a good place to look for the oldest rocks in the solar system because _____.
    a. contamination from atmospheric tests of nuclear weapons have contaminated the crust of Earth
    b. Earth's magnetic field interferes with the radiometric clocks in most igneous rocks
    c. melting has reset radiometric clocks in the rocks of Earth's crust
    d. the earliest crustal rocks have been destroyed by geologic activity
    e. All of the above answers are correct.
    f. Both c and d are correct.

THE PLANNER

Review the Chapter Planner on the chapter opener and check off your completed work.

Plate Tectonics

4

K2, the mountain rising 8611 meters behind these climbers, is the world's second-tallest peak. Only Mt. Everest (8848 meters) is higher. Both mountains are in the Himalaya, a range formed over the past 40 million years by a collision between India and Asia.

More than 300 people have climbed K2 and more than 4000 have climbed Everest. But only two—a Swiss named Jacques Piccard and a U.S. Navy lieutenant, Donald Walsh—have reached the lowest point on Earth. The Mariana Trench, near Guam, bottoms out at 11,000 meters below sea level. Piccard and Walsh made their historic descent in 1960, in a spherical submersible (inset) capable of withstanding the immense pressures at that depth.

Surprisingly, both the Himalaya and the Mariana Trench were created by the same geologic process—the collision of two plates of lithosphere. In the first case, the colliding plates thrust up the world's highest mountain range. In the second, one plate dove beneath the other, dragging the seafloor down with it.

The slow motion of plates, known as **plate tectonics**, continually reshapes our world. Over millions of years, it has restructured and relocated continents, built mountain ranges, opened and closed oceans. Without plate tectonics the Himalaya and the Mariana Trench would not exist.

CHAPTER PLANNER ✓

- ❏ Study the picture and read the opening story.
- ❏ Scan the Learning Objectives in each section:
  p. 94 ❏   p. 102 ❏
- ❏ Read the text and study all visuals. Answer any questions.

Analyze key features

- ❏ Geology InSight, p. 96 ❏   p. 104 ❏
- ❏ Process Diagram, p. 101 ❏
- ❏ What a Geologist Sees, p. 106 ❏
- ❏ Amazing Places, p. 110 ❏
- ❏ Stop: Answer the Concept Checks before you go on:
  p. 102 ❏   p. 114 ❏

End of chapter

- ❏ Review the Summary and Key Terms.
- ❏ Answer the Critical and Creative Thinking Questions.
- ❏ Answer What is happening in this picture?
- ❏ Complete the Self-Test and check your answers.

A Revolution in Geology

LEARNING OBJECTIVES

1. **Describe** the supercontinent called Pangaea.
2. **Identify** the early arguments for and against Wegener's theory of continental drift.
3. **Explain** how paleomagnetism provided the definitive evidence for continental drift.
4. **Describe** the process of seafloor spreading.
5. **Measure** the rates at which continents move.

Scientific revolutions challenge us to look at the world in a new way. They turn accepted ideas upside down. However, they don't take place overnight. For example, for millennia, people thought that Earth was the center of the universe, with the Sun and all the planets revolving around it. In 1543, Nicolaus Copernicus argued that it is the other way around—that the planets revolve around the Sun. It took decades of debate, new inventions such as the telescope, and finally the persuasive writing of such scientists as Galileo and Newton to convince astronomers that Copernicus was right.

Geology underwent a revolution in the 1960s, when geologists discovered new evidence for a 50-year-old idea called **continental drift**.

> **continental drift** The slow lateral movement of continents across Earth's surface.

Just like Copernicus's theory, the hypothesis that continents move provoked controversy at first. But in recent decades, thanks to supporting evidence from many different sources, including sensitive global positioning systems that measure the actual rates and directions of motion, continental drift has become an integral part of the **Theory of Plate Tectonics**. (See Chapter 1 to review the important distinction between hypotheses and theories.) This theory has encouraged us to think of Earth as a system in constant flux, a dynamic world whose appearance has changed considerably over the eons. Early geologists gathered information about Earth and its processes painstakingly, one piece at a time. Plate tectonics shows us how the pieces fit together.

Wegener's Hypothesis of Continental Drift

In 1912, a German meteorologist named Alfred Wegener began lecturing and writing scientific papers about **continental drift**. Wegener hypothesized that the continents had once been joined together in a single "supercontinent," which he called *Pangaea* (pronounced "Pan-JEE-ah"), meaning "all lands" (**Figure 4.1**). He suggested that Pangaea had split into fragments like

Pangaea • Figure 4.1

In 1915, Alfred Wegener drew a map much like this one, showing the distribution of the continents about 320 million years ago, in the Carboniferous Period. Wegener proposed that the continents at that time were joined in one supercontinent, which he named Pangaea. The apparent close fit of the coastlines of the modern continents, particularly Africa and South America, had been observed by Leonardo da Vinci and Francis Bacon as early as the 1500s. Modern reconstructions of Pangaea differ from Wegener's version in some respects (e.g., Africa is located farther south on this map than on Wegener's original), but his basic idea was correct.

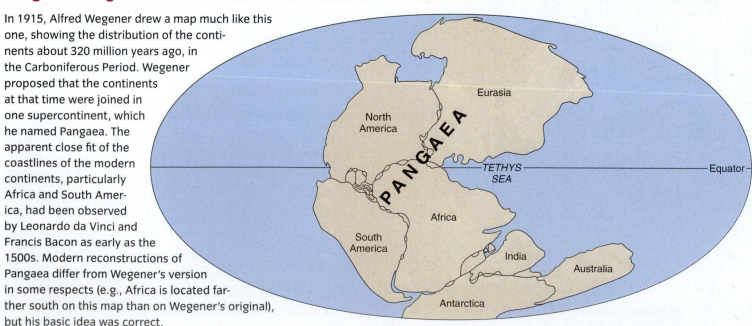

What is the "true" edge of a continent? • Figure 4.2

The edge of a continent usually contains several components. The shallow and gently sloped **continental shelf** lies just offshore. The steeper **continental slope** leads down to the bottom of the ocean. At the base of the continental slope lies the **continental rise**, which is usually composed of a layer of sediment. The main part of the ocean floor is the flat **abyssal plain**. In the center of the ocean, between two continents, we often find a submarine mountain range called a **midocean ridge**. This structure turns out to be key to the mystery of what causes continental drift.

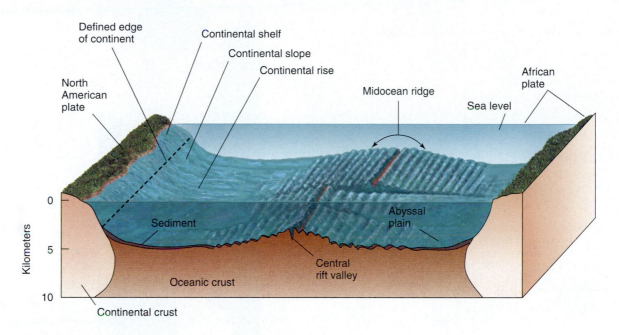

pieces of ice floating on a pond and that the continental fragments had slowly drifted to their present locations.

Wegener marshaled a good deal of evidence in favor of the continental drift hypothesis. Nevertheless, his proposal created a storm of protest in the international scientific community—one that continued for decades. Some of the criticisms were well founded. Contemporary geologists could not envision any reasonable mechanism by which continents could be moved. The terminology suggests continents somehow "adrift" in a sea of water. But remember that the continents are anchored in solid rock, and a drifting continent would have to plow through or across a seafloor also made of solid rock. Even Wegener could not explain how this could happen.

Let's put this problem aside for a moment and look at the evidence as the geologists of the time did. Importantly, no single piece of evidence is conclusive on its own. It took the combined weight of all of these arguments (and more) to convince geologists several decades later that Wegener was right.

The Puzzle-Piece Argument

It's easy to see from a map that the Atlantic coastlines of Africa and South America seem to match, almost like puzzle pieces. But is this an accident, or does it truly support the hypothesis that the continents were once joined together?

To decide whether the continents really do match, we must first note that there is more to a continent than meets the eye. The Atlantic coasts of South America and Africa, like those of other continents, do not terminate abruptly at the shoreline but slope gently seaward. This gently sloping land, part above sea level and part below, is called the **continental shelf**. At a water depth of about 100 meters, there is an abrupt change of slope called the **shelf break**, below which lies the steeper **continental slope** (**Figure 4.2**). The true edge of a continent is the point where the continental crust meets the oceanic crust, which is roughly halfway down the continental slope.

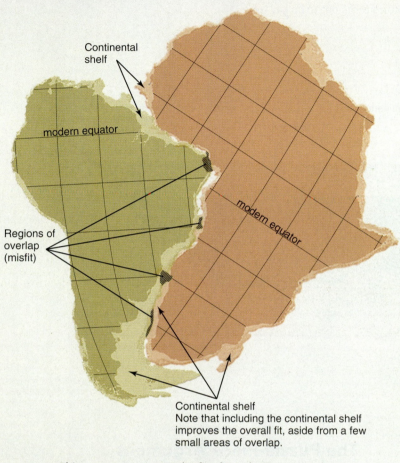

Continental shelf

modern equator

Regions of overlap (misfit)

modern equator

Continental shelf
Note that including the continental shelf improves the overall fit, aside from a few small areas of overlap.

a. This map reconstructs the fit of South America and Africa along the *true* edges of the continents—not along the shoreline, but the submerged edge of the continental crust. The darkly shaded areas show overlap. Note that the inclusion of the shallow off-coastal areas makes the fit better, not worse. Wegener did not have access to this information. Note, too, that Africa and South America have subsequently rotated in different ways, a point we will return to later in the chapter.

	Older than 550 million years
	About 550 million years old
	Younger than 550 million years

Africa

Close match in rock ages

Modern Brazil

South America

b. When the continents are rotated back together, the ages of rock units generally match, particularly in the regions of northeastern Brazil and West Africa.

c. In this reconstruction of the northern part of Pangaea, mountain belts of similar ages match up.

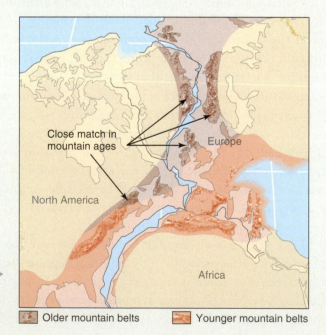

Close match in mountain ages

Europe

North America

Africa

| | Older mountain belts | | Younger mountain belts |

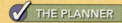

THE PLANNER

d. The tongue-shaped fossil leaves shown here came from a tree of the Carboniferous Period called *Glossopteris* (named after the Greek word for "tongue"). Similar fossils have turned up in Africa, India, Australia, and South America, providing strong evidence that these regions were once contiguous.

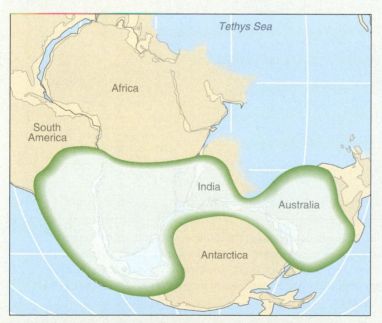

e. This map outlines the locations where *Glossopteris* fossils have been found. The territories match well when the continents are moved back to their probable locations in the Carboniferous Period.

When we fit together South America and Africa along the true edges of the continents (**Figure 4.3a**), we get an even better match than we might expect. In the "best-fit" position, the average gap or overlap between the two continents is only 90 kilometers. In addition, the most significant overlaps consist of wedges of sedimentary or volcanic rock that were added *after* the continents are thought to have split apart.

Matching Geology

The close fit between Africa and South America suggests that they were once joined, like pieces of a jigsaw puzzle. But as you know, it is possible to be fooled by jigsaw puzzle pieces that have nearly matching shapes. To be sure that the pieces fit together, you also need to match the designs on the puzzle pieces.

There is an analogous test in geology: If the continents were once joined together, we should find similar geologic features on both sides of the join. However, matching the geology of rocks on opposite sides of an ocean is more difficult than you might imagine. Erosion and rock formation since the breakup of Pangaea may have destroyed or covered up some of the evidence, so we will be putting together a puzzle with some pieces missing and others defaced.

Matching rocks A starting point in matching up the geologic puzzle pieces is to see if the ages of similar rocks match up across the ocean. In Wegener's time, the technique of radiometric dating was just being developed, so it was not easy to determine the exact age of a rock. But now we know that there is, indeed, some similarity in the ages of rocks and correlation between rock sequences on both sides of the ocean. As shown in **Figure 4.3b**, the match is particularly good between rocks about 550 million years old in northeastern Brazil and West Africa.

We can also check for the continuity of mountain chains. **Figure 4.3c** shows a reconstruction of the northern part of the supercontinent Pangaea. Notice, again, how mountain chains of similar ages seem to line up when the continents are moved back into this position. The oldest portions of the Appalachian Mountains, extending from the northeastern United States through eastern Canada, match up with the Caledonides of Ireland, Britain, Greenland, and Scandinavia. A younger part of the Appalachians lines up with a belt of similar age in Africa and Europe.

A Revolution in Geology **97**

Matching fossils If Africa and South America were really joined at one time, with the same climate and matching geologic features, then we should expect that they would have been inhabited by the same plants and animals. To check this hypothesis, Wegener looked at fossils. He found that some communities of plants and animals apparently evolved together until the time that Pangaea split apart, and after that time they evolved separately.

For example, Wegener found fossils of an ancient tree, *Glossopteris* (**Figure 4.3d**), in matching areas of southern Africa, South America, Australia, India, and Antarctica (**Figure 4.3e**). Could the seeds of this plant have been carried by wind or water from one continent to another? Probably not. The seeds of *Glossopteris* were large and heavy, and they would not have traveled far on the wind or water currents. Not only that, *Glossopteris* flourished in a cold climate; it would not have thrived in the warm present-day regions where its fossil remains are found. This, too, is consistent with the idea that these continents were once joined together in a more southerly location.

Certain animal fossils, too, match up well. The fossil remains of *Mesosaurus*, a small, freshwater reptile from the Permian Period, are found both in southern Brazil and in South Africa. The types of rocks in which the fossils are found are very similar. *Mesosaurus* did swim, but it was too small (about 0.5 meter long) to swim all the way across the ocean. Fossil remains of certain types of earthworms also occur in areas that are now widely separated. Since an earthworm could not have hopped across a wide ocean, the landmasses in which they lived must once have been connected.

Evidence left by glaciers Additional evidence that the southern hemisphere continents were once connected can be found in glacial deposits of the same age (Permian–Carboniferous) in South America, Africa, Australia, and India.

As glacial ice moves, it cuts grooves and scratches in underlying rocks and produces folds and wrinkles in soft sediments (**Figure 4.4a**). These features provide evidence

Evidence from glaciers • Figure 4.4

a. As glaciers advance, the sediment and rocks contained in them grind and abrade the bedrock, producing a polished and grooved surface. This polished surface, at Nooitgedacht in South Africa (see map), was created by the Pangaean ice sheet. The grooves indicate the direction of ice movement.

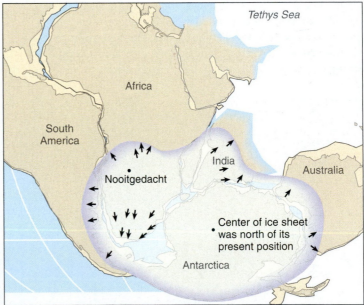

b. This reconstruction of the southern continents of Pangaea shows the rough extent of glacial deposits. Small arrows indicate the direction in which the ice was moving during the glaciation, as deduced from evidence such as grooves and scratches in the bedrock.

not only of the extent of glaciation but also of the direction the ice was moving during the glaciation. When today's southern hemisphere continents are moved back together, the direction of ice movement on them is consistent, radiating outward from the center of the former ice sheet. It's hard to imagine how such similar glacial features could have been created if the continents had not been joined together. The glacial deposits suggest also that Africa, South America, Australia, and India were once closer to the South Pole than they are today (**Figure 4.4b**) and that they had cooler climates.

Apparent Polar Wandering Paths

Wegener had some supporters for his ideas. Two in particular, Arthur Holmes in England and Alexander du Toit in South Africa, were distinguished scientists. He and his supporters gathered more and more evidence in support of continental drift, but most scientists remained unconvinced, mainly because there was no reasonable hypothesis explaining *how* continents could move. Wegener died in 1930 without seeing a resolution to the problem.

In the 1950s, **paleomagnetism** emerged as a new tool for studying Earth's history, and it was destined to break the intellectual logjam. Recall from Chapter 3 that when magma cools and solidifies into rock, grains of iron-bearing minerals become magnetized and take on the prevailing polarity (north–south orientation) of Earth's magnetic field at that time. We explained in Chapter 3 how this fact allows us to detect reversals in Earth's magnetic field. It also allows us to reconstruct the apparent location of Earth's past magnetic poles.

In the 1950s, geophysicists studying paleomagnetic pole positions found evidence suggesting that Earth's magnetic poles had wandered all over the globe over the past several hundred million years. They plotted the pathways of the poles on maps like the one shown in **Figure 4.5** and called the phenomenon **apparent polar wandering**. Geophysicists found this puzzling because they knew that because Earth's magnetism is related to Earth's rotation about its axis, the magnetic poles could not actually wander very far from the axis of rotation. Even more puzzling, geologists found that the path of apparent polar wandering measured from North American rocks differed from that of European rocks. Earth certainly cannot have two different magnetic north poles at the same time! The only solution is that the magnetic pole is fixed, and the continents themselves moved, carrying the rocks with them. In that

case, the apparent motion of the poles would be an illusion, like the apparent motion of trees when you drive by them. In fact, the apparent polar wandering path of a continent, determined from rocks of various ages, provides a historical record of the position of that continent relative to the magnetic poles.

Notice also that the apparent wandering paths for Europe and North America actually look quite similar from 600 million years ago to 200 million years ago. In fact, if you rotated Europe and America so that they were side by side, the two paths would overlap each other exactly. This indicates, once again, that Europe and America were moving as a single continent during that time.

The Missing Clue: Seafloor Spreading

By the early 1960s, many clues had been amassed in support of continental drift. The hypothesis made testable predictions, such as matching geology and fossils, and those predictions had been confirmed. The hypothesis was also consistent with evidence that was not known during Wegener's lifetime—the polar wandering paths. (Remember from

Wandering poles • Figure 4.5

This map traces out the *apparent* path of the magnetic north pole over the past 600 million years. The numbers indicate millions of years before the present. Rocks in North America point to the apparent magnetic poles shown on the red curve, and rocks in Europe point to the apparent poles shown on the black curve. The *actual* magnetic pole probably never wandered far from the geographic North Pole (center). Its apparent motion is created by the motion of the continents, which carried magnetically oriented rocks along with them.

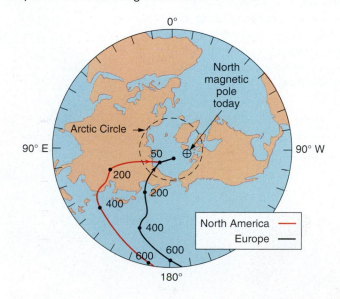

Chapter 1 that testing a hypothesis against observations is one of the most important steps of the scientific process.) Even so, many scientists were still skeptical. To that point, all the evidence was circumstantial. The ripping apart and relocation of an entire continent should have left signs that would still be visible today.

Then, three decades after Wegener's death, geologists found the evidence they needed—at the bottom of the sea. The missing clue turned up when oceanographers applied the then-new technique of magnetic polarity dating to rocks at the bottom of the Atlantic Ocean. They were astounded to find that parts of the seafloor consist of magnetized rocks with alternating bands of normal and reversed polarities. The bands are hundreds of kilometers long. More importantly, they are symmetrical on either side of a centerline that coincides with the ridge running down the middle of the Atlantic Ocean. In other words, if you could fold the seafloor in half along the midocean ridge, the bands on either side would match. Later evidence would show that not only the paleomagnetic polarities but also the numerical ages of the rocks were symmetrically banded, and they were mirrored on either side of the ridge (**Figure 4.6**), with the youngest rocks closest to the ridge itself.

The symmetrical pattern of magnetic bands provided powerful support for a hypothesis, first proposed in 1960, called **seafloor spreading**. According to this hypothesis, the midocean ridge is a place where the seafloor has split apart and the rocks are moving away from one another (**Figure 4.7**). When magma from below wells up into the crack, it solidifies into new volcanic rock on the seafloor. Over time, the expanding seafloor operates as a conveyer belt, carrying the newly magnetized bands of rock away from the ridge in either direction.

The decisive piece of evidence for seafloor spreading is that the ages of seafloor rocks increase with distance from the ridge. The youngest rocks are found along the center of the ridge, where new molten material wells up (as shown in Figure 4.6). This final piece of evidence convinced the great majority of geologists that seafloor spreading is real. The magnetic evidence of the seafloor also allowed geologists to calculate the spreading rate of the seafloor; it turned out to range between 1 and 10 centimeters a year. Apparent polar wandering paths

> **seafloor spreading**
> The processes through which the seafloor splits and moves apart along a midocean ridge and new oceanic crust forms along the ridge.

Banded rocks on the seafloor • Figure 4.6

This map shows the ages of magnetically banded rocks on either side of the Mid-Atlantic Ridge. The numbers indicate the ages of the rocks, in millions of years. The youngest bands lie along the midocean ridge, and the oldest are far away from the ridge, indicating that the seafloor has been spreading over time.

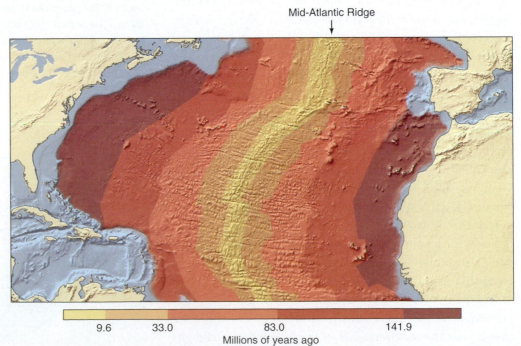

Mid-Atlantic Ridge

9.6   33.0   83.0   141.9
Millions of years ago

Seafloor spreading • Figure 4.7

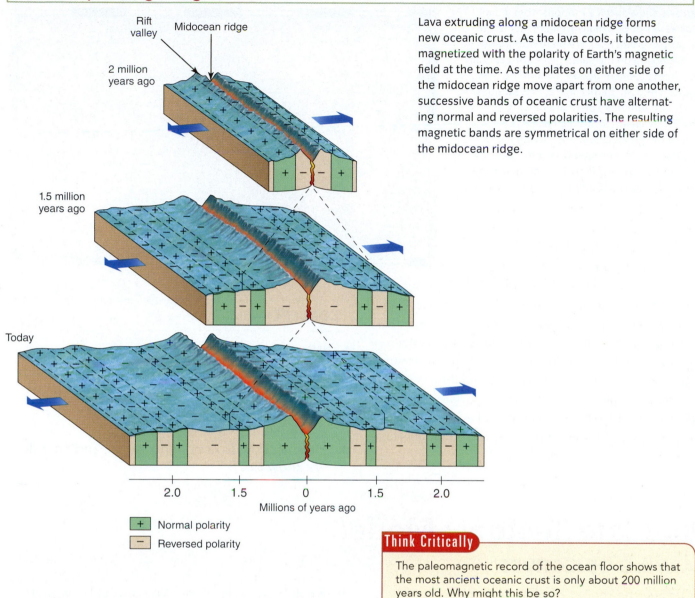

Lava extruding along a midocean ridge forms new oceanic crust. As the lava cools, it becomes magnetized with the polarity of Earth's magnetic field at the time. As the plates on either side of the midocean ridge move apart from one another, successive bands of oceanic crust have alternating normal and reversed polarities. The resulting magnetic bands are symmetrical on either side of the midocean ridge.

Rift valley
Midocean ridge
2 million years ago
1.5 million years ago
Today

2.0   1.5   0   1.5   2.0
Millions of years ago

+ Normal polarity
− Reversed polarity

Think Critically

The paleomagnetic record of the ocean floor shows that the most ancient oceanic crust is only about 200 million years old. Why might this be so?

established that continents move, and seafloor magnetism showed that the seafloor moves, too. Scientists quickly realized that continents are not trundling along on top of a static ocean floor, which would be physically impossible. Instead, they are conveyed in opposite directions by a dynamic ocean floor that is constantly spreading and replenishing itself. Ironically, it was geophysicists—the group that had most vigorously opposed Wegener's ideas—who found the final piece of evidence that proved Wegener correct. The many investigations seeking to test the continental drift hypothesis provide a good example of the scientific method at work.

Recently, an even more direct piece of evidence for continental drift became available. In the 1980s, the U.S. Department of Defense declassified the data from a network of satellites that form the Global Positioning System

The arrows show surface motions from a continuous series of GPS measurements on North America. The green lines are plate boundaries. Note that points on the same plate are generally moving in the same direction, confirming that the plates are rigid. In this figure, GPS stations on the North American plate are moving slowly southwest, while those on the Pacific plate (to the left of the green curve) are moving more rapidly northwest. The speeds are too slow for humans to be aware of (1 to 10 centimeters per year) but are easily detectable by GPS sensors.

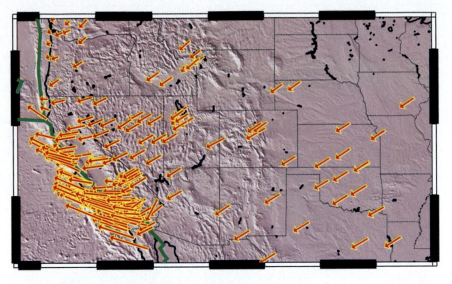

(GPS). GPS receivers help motorists, hikers, boaters, and pilots find their exact location quickly and easily, and they can also be used to track continental drift. Geologists have attached GPS receivers permanently to the ground, so that they do not move unless the ground moves. This allows them to plot maps of Earth's surface velocities (**Figure 4.8**). The GPS data confirm that the continents are moving—and they are moving at the same rates and in the same directions as the seafloor!

CONCEPT CHECK STOP

1. **What** does the name Pangaea mean?

2. **What** were the arguments that Wegener gave in favor of his hypothesis of continental drift?

3. **How** does seafloor spreading create paleomagnetic bands?

4. **How** quickly do continents move, and how do we know this?

The Plate Tectonic Model

LEARNING OBJECTIVES

1. **Outline** the theory of plate tectonics.
2. **Describe** three different types of plate margins.
3. **Explain** why different kinds of plate margins are susceptible to different kinds of earthquakes.
4. **Explain** the role of mantle convection in plate tectonics.
5. **Present** evidence of supercontinents before Pangaea and the present tectonic cycle.

O nce the evidence for continental drift and seafloor spreading had been established, researchers began seeking a mechanism for the motions. A lot of previously puzzling phenomena in geology suddenly began to make sense. For instance, geologists could now explain the location of mountain ranges and deep ocean trenches. They could explain why, in some places (e.g., the Tibetan Plateau), Earth's crust seems to be squeezed together, making high mountains, while in other places (e.g., the East African Rift valleys), it seems to be pulling apart and making deep, steep-sided trenches. Geol-

ogists also realized that the distribution of earthquakes and volcanic activity around the planet, which is far from uniform, must somehow be related to the motions (**Figure 4.9**). Other chapters in this book discuss each of these developments in greater detail. The unifying theory that emerged from all the research is called **plate tectonics**. Note that we call it a *theory* rather than a *hypothesis*. As explained in Chapter 1, a theory is an explanatory model that is strongly supported by scientific evidence.

plate tectonics
The movement and interactions of large fragments of Earth's lithosphere, called plates.

Distribution of earthquake epicenters around the world • Figure 4.9

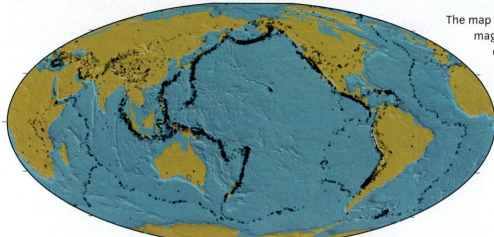

The map shows locations of earthquakes, Richter magnitude 4.0 and above during 2010, as catalogued by the US National Earthquake Information Center. The distribution of earthquakes provides clues that geologists use to locate the edges of tectonic plates. Compare this figure with Figure 4.10, a map showing the plate boundaries as we currently understand them.

Plate Tectonics in a Nutshell

Earth's **lithosphere**, or rocky outer layer, is very thin relative to Earth as a whole, no more than about 100 kilometers thick. The solid rock that makes up the lithosphere is strong, but it lies on top of a vast mantle of hotter, weaker material that is constantly in motion (albeit very slow motion). The layer directly below the lithosphere, called the **asthenosphere**, is especially weak because it is close to the temperature at which melting begins. (The lithosphere and asthenosphere were introduced in Chapter 1 and will be discussed in greater detail in Chapter 5.) The relationship between these two layers is a condition called **isostasy**, which means that the lithosphere is essentially "floating" on the asthenosphere, like a sheet of ice floating on water.

If you place a very thin, cool, hard shell on top of hot, ductile material that is moving around, what do you expect will happen? It's almost a certainty that the shell will crack, and that is exactly what has happened to Earth's lithosphere. It has broken into a set of enormous rocky fragments we call **plates**. Today there are six large plates, extending for thousands of miles, as well as a number of smaller ones (**Figure 4.10**). It is one thing to realize that the lithosphere has fractured into plates; it is another thing altogether to know where the boundaries are. A fracture that separates one piece of moving lithosphere from another is called a **fault**, and geologists have long understood that movement along faults is the main cause of earthquakes. To locate plate edges, geologists turned to evidence from earthquakes (see Figure 4.9).

> **plate** A large fragment of rigid lithosphere bounded on all sides by faults.
>
> **fault** A fracture in Earth's crust along which movement has occurred.

Tectonic plates • Figure 4.10

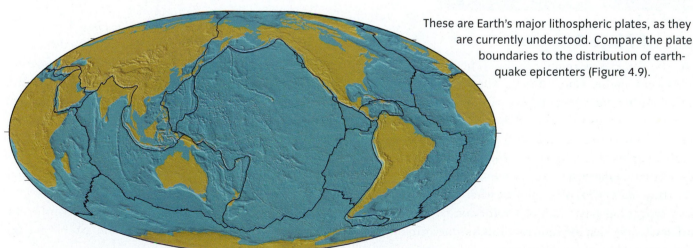

These are Earth's major lithospheric plates, as they are currently understood. Compare the plate boundaries to the distribution of earthquake epicenters (Figure 4.9).

a. Divergent margin

Rift Valley

The only place in the world where a midocean ridge rises to the surface is in Iceland. The rift valley seen here lies directly atop the Mid-Atlantic Ridge, a divergent margin where the North American and Eurasian plates are gradually moving apart. The valley is getting wider at a rate of about 2 centimeters per year, and part of it has filled in with water, creating Lake Thingvallavatn (at right). The rift itself extends in width off the right side of the photo.

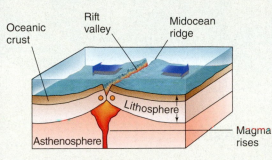

Oceanic crust · Rift valley · Midocean ridge · Lithosphere · Asthenosphere · Magma rises

Divergent Margin
At divergent margins, earthquakes tend to be fairly weak and shallow. Earthquakes can only occur in rock that is cold and brittle enough to break; this means they cannot be very deep along midocean ridges.

b. Convergent margin

Subduction zone

Indonesia is a volcanic island arc that has formed parallel to an ocean–ocean convergent margin, or subduction zone. Merapi, an active volcano on the island of Java that erupted in 2006, is in the foreground, and several other volcanoes can be seen in the background.

Oceanic crust · Oceanic trench · Continental crust · Magma rises · Lithosphere · Melting · Asthenosphere · Benioff zone

● Shallow earthquakes
● Deep earthquakes

Subduction Margin
The deepest and most powerful quakes occur in subduction zones. Here an oceanic plate sinks downward relative to a continental plate. The earthquake foci are shallow near the oceanic trench but become deeper along the descending edge of the subducting plate. These zones of shallow-and-deep focus earthquakes, called **Benioff zones**, first alerted scientists to the phenomenon of subduction.

Types of Plate Margins

There are three fundamentally different ways in which two adjacent plates can interact. They can move away from each other (or diverge), they can move toward each other (or converge), or they can slide past each other more or less horizontally, along a long fracture. We will take a look at examples of each of these types of margins and the kinds of earthquakes that happen along them.

Earthquakes occur along faults, where huge blocks of rock are grinding past each other (discussed in greater detail in Chapter 5). Tectonic motions produce directional pressure, which causes rocks on either side of a large

fracture to move past each other. The movement is rarely smooth; usually the blocks stick because of friction, which slows their movement. Eventually, the friction is overcome and the blocks slip abruptly, releasing pent-up energy with a huge "snap"—an **earthquake**.

The actual location beneath the surface where an earthquake begins is called the **focus**. This should not be confused with the better-known **epicenter**, which is the point on Earth's surface that lies directly over the focus. The depths of foci provide useful information about the characteristics of the plate margin (**Figure 4.11**).

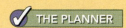

c. Continental collision margin

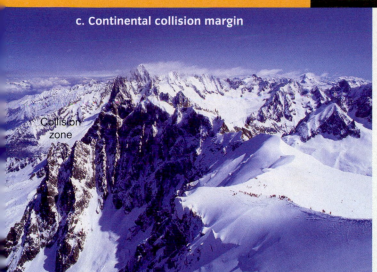

Collision zone

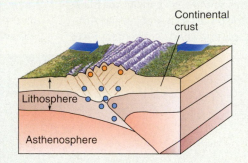

In this photograph, the snow-covered Himalaya Mountains are on the right, the lower-lying Indian peninsula is at the upper left, and the convergent margin lies roughly on the border between them.

Continental crust

Lithosphere

Asthenosphere

Continental Collision Zone
When two continental masses meet at a convergent boundary, a continental collision occurs. Earthquakes in continental collision zones can be deep and also very powerful.

d. Transform fault margin

Transform fault margin

The constant grinding of the Pacific plate and the North American plate has made the area near the San Andreas fault, a transform fault margin, highly unstable. There have been several notable earthquakes related to this fault, including the 1906 earthquake that destroyed much of San Francisco.

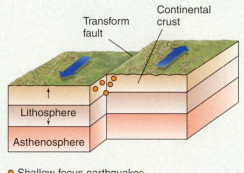

Transform fault

Continental crust

Transform Fault Margin
Transform fault margins have shallow earthquakes, but they can be very powerful.

Lithosphere

Asthenosphere

○ Shallow-focus earthquakes
● Deep-focus earthquakes

Divergent margins, also called **rifting centers** or **spreading centers**, occur where two plates are moving apart. They can occur either in continental or oceanic crust. In East Africa, for example, the African plate is being stretched and torn apart, creating long **rift valleys**. Eventually, a new ocean may form in the widening rift; a modern example of this is the Red Sea (see *What a Geologist Sees* on the next page). Where oceanic crust is splitting apart, the result is a **midocean ridge**, and the one place in the world where a midocean ridge can be seen above

> **divergent margin**
> A boundary along which two plates move apart from one another.

the sea is Iceland (see Figure 4.11a). Earthquakes along divergent margins tend to be weak and have shallow foci.

Convergent margins occur where two plates move toward each other. This leads to different types of features, depending on whether the boundary is between two oceanic plates, two continental plates, or one of each. When one or both plates of lithosphere are capped by oceanic crust, one plate—it is always one capped by oceanic crust—will typically slide beneath the other plate, plunging into the asthenosphere, by a process called **subduction**.

> **convergent margin** A boundary along which two plates come together.

The Plate Tectonic Model **105**

WHAT A GEOLOGIST SEES

The Red Sea and the Gulf of Aden

This spectacular photo, taken by astronauts on the *Gemini 11* mission, shows the southern end of the Red Sea (left top, center) and the Gulf of Aden (right), separating the southern tip of the Arabian Peninsula (right) from the northeastern corner of Africa. A geologist looking at this scene sees parallel coastlines and realizes that two lithospheric plates (the Arabian and African plates; see inset map) are splitting apart. A divergent plate margin runs down the center of the Gulf of Aden and joins another divergent margin that runs down the center of the Red Sea (both underwater). The Red Sea is widening at a rate of about 1 centimeter per year. If the process continues for several million years, the long narrow sea will become a wide gulf. The red dots on the map (inset) show the locations of recent earthquakes and volcanic activity.

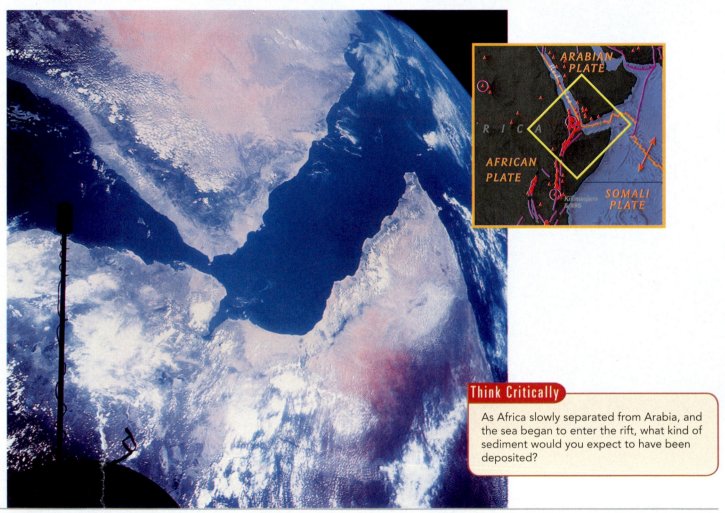

Think Critically

As Africa slowly separated from Arabia, and the sea began to enter the rift, what kind of sediment would you expect to have been deposited?

Subduction zones are marked by very deep **oceanic trenches**—the deepest points in the ocean; the Mariana Trench, mentioned at the beginning of the chapter, is an example. As the plate capped by oceanic lithosphere sinks, it heats up and releases water. An arc of volcanoes typically forms parallel to the edge of the overriding plate, formed by the magma generated by water released from the subducting plate

> **subduction zone** A boundary along which one lithospheric plate descends into the mantle beneath another plate.

(see Figure 4.11b). Earthquakes along subduction zones are commonly strong and often have deep foci.

When one continent meets another continent along a convergent margin, they crumple upward and downward as the lithosphere thickens, in a **continental collision zone**. The Himalaya Mountains (see Figure 4.11c), Earth's highest mountain range, were thrust up in just this way by an ongoing violent collision over 40 million years between the Australian–Indian plate and the Eurasian plate. Some continental collision zones were previously subduction zones, where the process of subduction terminated when the overriding plate encountered continental rock that was too buoyant to be subducted.

Transform fault margins are the most common boundaries on the ocean floor, and hundreds of them run perpendicular to the mid-ocean ridges (see Figure 4.10). This causes the plates to have very complicated, jagged boundaries, in which spreading centers alternate with transform faults.

transform fault margin A fracture in the lithosphere where two plates slide past each other.

Transform faults also occur in continental crust; the San Andreas fault in California is a famous example.

The types of plate interactions and the plate boundaries that result from them are illustrated schematically in **Figure 4.12**. Note, again, that along some plate boundaries new crust is being created, whereas at others crust is being consumed—either by compression and crumpling or by subduction into the mantle. There is no solid evidence to suggest that Earth has either grown or shrunk significantly since its formation; this means that the total area of Earth's crust is constant. For every square kilometer of

Plate margins: A summary • Figure 4.12

Several different types of plate margins are illustrated schematically in this diagram. In reality, these different types of margins would not be so close to one another. At middle right is a divergent margin, in oceanic crust. Offsetting the divergent margin are transform fault margins. At far left and far right are two types of convergent margins. At ocean–continent and ocean–ocean margins, one plate subducts under the other. The subducting oceanic plate, which typically carries ocean floor sediments, has a relatively low melting temperature because of its high water content (Chapter 6). Therefore, it will begin to melt at a relatively shallow depth. The melt rises to the surface, creating a line of volcanoes called a **volcanic arc** along the edge of the overriding plate.

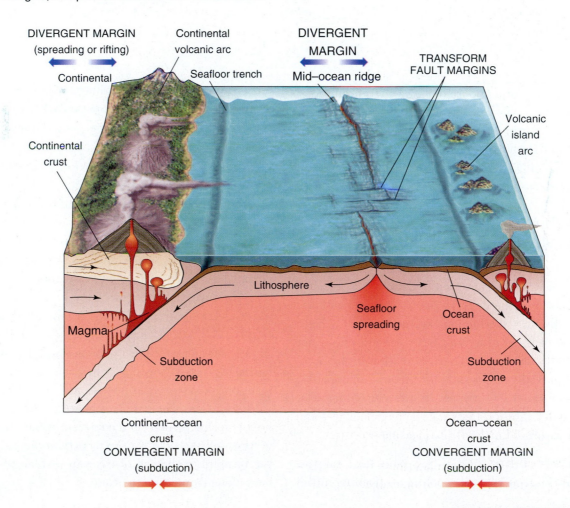

Where Geologists CLICK

Interactive Tectonic Map

http://mineralsciences.si.edu/tdpmap/

The Smithsonian Institution, the U.S. Geological Survey, and the U.S. Naval Research Laboratory offer an extremely detailed interactive world map showing plate boundaries, earthquakes, and volcanoes, which you can explore. Use the map the way a geologist does; enlarge an area you are interested in and see where the volcanoes and earthquake centers are in relation to a nearby plate boundary.

crust that is created at a midocean ridge, therefore, somewhere else a square kilometer is consumed by subduction at a convergent margin. Overall, there is a balance between creation and destruction in the tectonic cycle.

The U.S. Geological Society provides a detailed map of world tectonic activity (see *Where Geologists Click*).

The Search for a Mechanism for Plate Motion

Although virtually all geologists accept the basic theory of plate tectonics, some questions remain. What, exactly, drives plate motion? How does the mantle interact with the crust? What initiates subduction? Scientists have a basic understanding of these processes, but the details have not been completely worked out. Thermal motion in the mantle is at least partially responsible for the motion of plates. This thermal motion in turn results from the release of heat from Earth's interior. Let's take a closer look at some of the complexities of Earth's heat-releasing processes.

Earth's internal heat Earth gives off heat for two main reasons. First, it is slowly cooling off from its initial formation processes, including formation of a molten iron core. Second, heat is constantly being generated by the decay of radioactive elements in the interior, primarily uranium, potassium, and thorium. If Earth did not release heat into outer space, the entire interior would eventually melt.

Some of Earth's heat is released through **conduction**. This is a gentle and slow process similar to what you feel when you hold a cup of hot coffee in your hands. The heat moves through the wall of the cup by conduction, a gradual transfer of energy from atom to atom.

When you boil water in a pot on a stove, you see the water churning around in big circles called *convection cells* (**Figure 4.13a**). A mass of hot water at the bottom is slightly less dense than the cooler water at the top, and hence it rises. When it reaches the surface, it sheds its heat, moves sideways as it cools, and then sinks back down to the bottom, where it is reheated. This mechanism of heat transfer is more efficient than conduction. The convection cells act like couriers, carrying heat directly from the burners to the top of the pan instead of passing the heat along from atom to atom.

Convection as a driving force Even though Earth's mantle is composed mostly of solid rock, it too releases heat through **convection**. Solid rock, if it is hot enough, can behave as a flowing viscous fluid, just as the solid ice in a glacier flows. Rock deep in the mantle heats up and expands, becoming buoyant. Very, very slowly it moves upward, in huge convection cells (**Figure 4.13b**). Just like the water in the boiling pot, the hot rock moves laterally near the surface as it sheds its heat. This lateral movement of rock in the asthenosphere is believed to be one of the causes of the motion of lithospheric plates.

Convection in the mantle is complex and incompletely understood. Many challenging questions remain unan-

> **convection** A form of heat transfer in which hot material circulates from hotter to colder regions, loses its heat, and then repeats the cycle.

swered. For example, does the whole mantle convect as a unit, or does the top part convect separately from the bottom, creating rolls upon rolls? In subduction zones, are lithospheric plates pushed down or pulled down, or do they sink into the asthenosphere under their own weight? (Plates with subduction edges move faster than plates without subduction edges, which suggests that the sinking plate edge is pulling the rest of the plate.) How do mid-ocean ridges start, and what role do they play in plate motion? What are the shapes and distribution of convection cells? We know that hot rocks sometimes do not travel in neatly packaged cells and that they may rise in long, thin blobs called **plumes** (see *Amazing Places* on the next page). Do plumes originate in the middle of the mantle, or even deeper, where the core meets the mantle? Geologists continue to actively research questions such as these.

Mantle convection • Figure 4.13

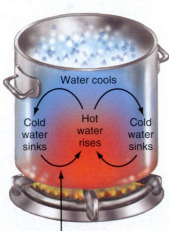

Convection Cell

a. An everyday example of convection can be seen when you boil a pot of water. The water closest to the burners is hotter than the rest of the water. As it heats up, it becomes less dense and rises to the top. At the surface, it cools down and moves sideways to make room for the hot water rising beneath it. As the water at the surface cools, it becomes denser and sinks.

b. The same process of convection happens in Earth's mantle on a much grander scale and over a much longer time. Hot rock rises slowly and plastically from deep inside Earth; then it cools, flows sideways, and sinks. The relationship between convection cells and lithospheric plates is far more complex than what we see in a pot of boiling water.

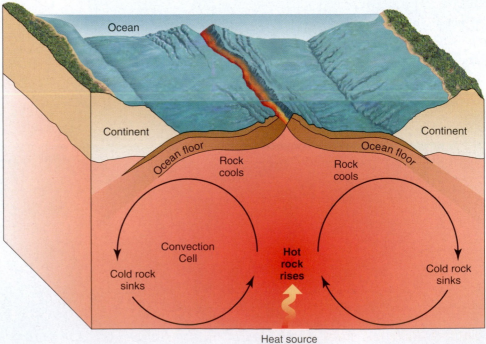

AMAZING PLACES

The Hawaiian Islands: A Growing Island Chain

One of the most striking geologic features of the Hawaiian Islands, first noted by geologist and explorer James Dwight Dana on an expedition to the islands in 1840–1841, is the increasing ages of the islands from southeast to northwest. In **Figure a**, the highly eroded cliffs of the Na Pali Coast of Kauai (the northernmost of the major islands), you can see the effects of more than 3 million years of weathering. By contrast, in some places on the south end of the Big Island of Hawaii, the land is less than 30 years old. **Figure b** shows an ongoing eruption of the Kilauea volcano that has added new land to the island almost ceaselessly from 1983 to the present day. Just 15 kilometers south of Hawaii, the next Hawaiian island is already being born. Loihi seamount (**Figure c**), an underwater volcano that appears as a raised feature in this topographical map of the seafloor, already has a peak that is only 1 kilometer below the ocean surface (shown in red), and it is still growing. Loihi erupted as recently as 1996.

Even though Hawaii is thousands of kilometers from the nearest plate boundary, the reason for the islands' differing ages has a great deal to do with plate tectonics. The Big Island of Hawaii is currently sitting above a long, thin plume of hot material, called a hot spot, rising from deep in the mantle. The plume itself is stationary, but the lithospheric plate above it is moving to the northwest (**Figure d**). The plume supplies magma to the currently active volcanoes of Mauna Loa and Kilauea and the undersea volcano of Loihi. The older islands, such as Kauai, once lay above the hot spot but moved off it as the Pacific plate moved northwest. With no new volcanic additions to their landmass, they will eventually erode into the sea.

a. The oldest Hawaiian Island

b. The youngest Hawaiian island

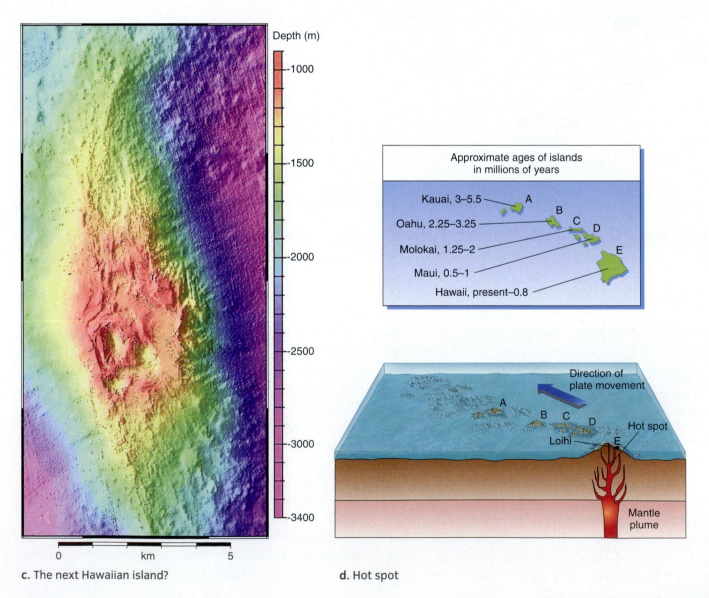

Depth (m)

-1000
-1500
-2000
-2500
-3000
-3400

0   km   5

c. The next Hawaiian island?

Approximate ages of islands
in millions of years

Kauai, 3–5.5 — A
B
Oahu, 2.25–3.25 — C
D
Molokai, 1.25–2 —
E
Maui, 0.5–1 —
Hawaii, present–0.8 —

Direction of
plate movement

A
B   C
D
Loihi   E   Hot spot

Mantle
plume

d. Hot spot

The tectonic cycle • Figure 4.14

This diagram should look familiar—you saw it in Chapter 1. The first of the three rings is called the tectonic cycle, and it includes all of Earth's processes that result directly from plate tectonics. In this chapter you have learned about the processes at the top of the cycle—seafloor spreading and subduction. In the next two chapters, you will learn about earthquakes and volcanoes, which also result from processes in the tectonic cycle. In subsequent chapters you will learn more about the hydrologic cycle and the rock cycle, the two other cycles in this interacting system.

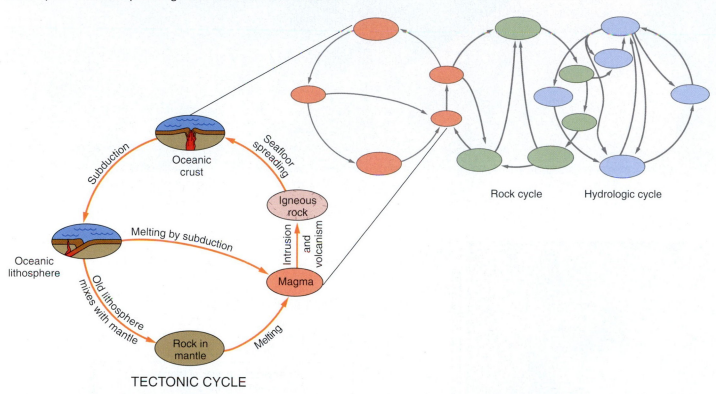

TECTONIC CYCLE

Rock cycle   Hydrologic cycle

The Tectonic Cycle: Past, Present, Future

Recall that in Chapter 1 we introduced the concept of interacting cycles in the Earth system (refer to Figure 1.10). In this chapter we have introduced the **tectonic cycle**, the right-hand side of the diagram in **Figure 4.14**. New material is constantly added to oceanic crust by volcanism along divergent plate boundaries, while subduction consumes an equal amount of oceanic crustal material at convergent margins. Because of these processes, the seafloor renews itself on the order of every 200 million years.

Continental crust lasts much longer than oceanic crust. Because of its lower density (greater buoyancy), it cannot easily be subducted. For that reason, the continental crust preserves a much longer record of the tectonic cycle than

> **tectonic cycle**
> Movements and interactions in the lithosphere by which rocks are cycled from the mantle to the crust and back; this cycle includes earthquakes, volcanism, and plate motion, driven by convection in the mantle.

does the oceanic crust. While oceanic crust provided the clinching evidence of the breakup and dispersal of the ancient continent Pangaea, the continental crust—interpreted in the light of plate tectonics—shows that Pangaea was not the first supercontinent in Earth's history.

Geologists have discovered that Pangaea itself arose out of a collision between fragments of continental crust between 350 million and 450 million years ago. The collision (or collisions) formed mountain ranges, such as the Appalachians in North America and the Urals in Russia, that are highly eroded now but were once as massive as the present-day Himalayas. These remnants remain as evidence of the assembly of Pangaea.

But even older and more highly eroded remnants of ancient mountain ranges, along with paleomagnetic evidence, reveal an even earlier cycle of supercontinent formation and breakup. Roughly 1100 million years ago, during the Proterozoic Eon, another supercontinent existed, which geologists call Rodinia (**Figure 4.15**).

Before Pangaea • Figure 4.15

A proposed arrangement of continental fragments in the 1100-million-year-old supercontinent Rodinia. The dark red strip marks the location of a now deeply eroded mountain chain that formed as a result of collisions as Rodinia was assembled. Pieces of the old mountain range are now dispersed among Earth's present-day continents—scrambled by the breakup of Rodinia and the subsequent assembly and breakup of Pangaea.

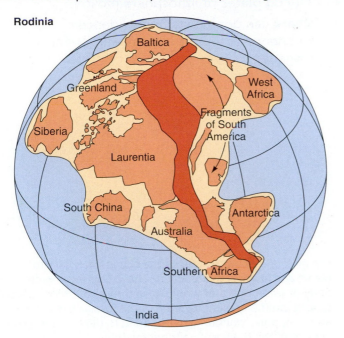

Rodinia

Fragmentary evidence indicates that prior to Rodinia, other ancient supercontinents formed, existed for a few hundred million years, and then broke apart. This repeated process of formation and breakup is called the **supercontinent cycle**, or sometimes the **Wilson cycle**, in honor of J. Tuzo Wilson of Canada, one of the major contributors to the plate tectonic revolution.

What will Earth be like in the future? Refer to Figure 1.19 in Chapter 1; the panel showing Earth 500 million years ago shows the scene after Rodinia broke apart, when the pieces were starting to be assembled into Pangaea. Then in the panel for 100 million years ago, Pangaea is breaking up, and the fragments are moving toward the world we know today. But the plates are still growing at spreading edges and being consumed at convergent margins. In the future new spreading edges and convergent margins will form and slowly new supercontinents will arise. Based on much research, **Figure 4.16** illustrates what the world may look like 150 million years in the future.

Plate tectonics affects all life on Earth, sometimes in subtle ways and sometimes in dramatic ways. It influences climate through the distribution of continents and ocean basins, for example. These changes are slow but profound; they can lead to ice ages or periods of unusual warmth, both of which strongly affect the evolution of species.

Earth of the future • Figure 4.16

The world map 150 million years ahead may well look like this. Note the subduction zones on both sides of the Americas, and the welding together of Africa and Eurasia into a giant continent.

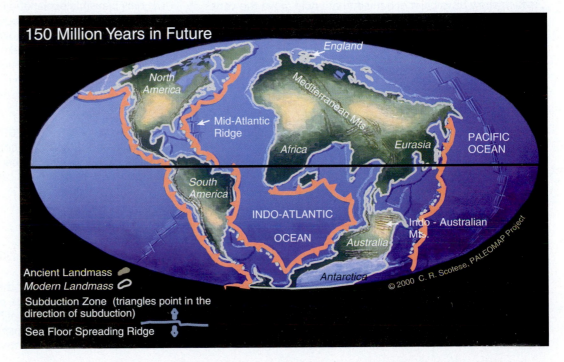

Plate tectonics also controls the distribution of mineral deposits, which we will explore in greater detail in Chapter 16. Other effects are more immediately obvious, such as major earthquakes and volcanic eruptions.

At the beginning of this chapter, we described plate tectonics as a "unifying" theory of how the Earth system works. This is because the plate tectonic model brings together many diverse observations of Earth's geologic features and unifies them into a single, reasonably straightforward "story." As a model, plate tectonics is truly representative of the Earth system science approach because it illustrates how internal Earth processes are integrated with all other parts of the Earth system.

CONCEPT CHECK

1. **Why** has Earth's lithosphere broken into plates?
2. **What** are the principal types of plate margins?
3. **How** have geologists determined the shapes of Earth's plates?
4. **Why** do plates of lithosphere move?
5. **Where** can geologists find evidence of continental assembly and breakup before Pangaea?

 THE PLANNER

Summary

1 A Revolution in Geology 94

- A revolution in geology began almost 100 years ago, with the hypothesis that the continents have not always been in their present positions but moved into their present-day positions after the breakup of a "supercontinent" known as Pangaea, as seen in this reconstruction. This became known as the **continental drift** hypothesis. At first it was quite controversial because scientists could not envision a satisfactory physical mechanism that could move the continents through solid rock.

Pangaea • Figure 4.1

- The evidence for continental drift included the fit between the coastlines of continents. The evidence is stronger when the fit is made along the "true" edge of the continent—the **continental shelf**. In addition, geologic features such as mountain ranges, ancient glacial deposits, and rock types match very closely on both sides of the Atlantic Ocean, which now separates these continents. Matching fossils

can be found on both sides of the ocean. Many are from plants and animals that could not have crossed a wide body of water and that would have thrived at different latitudes than the present location of the fossils. In the 1950s, apparent polar wandering was added to the list of evidence in favor of the theory of continental drift.

- The most significant piece of evidence in support of continental drift was the discovery of bands of magnetized rock on the seafloor with alternating normal and reversed polarities, aligned symmetrically on either side of the midocean ridges. The only plausible explanation for this discovery was **seafloor spreading**. Scientists quickly realized that continents and the adjacent seafloor move together. This evidence was decisive because it finally led to a mechanism that could account for the continents' movement. At midocean ridges, the ocean floor is constantly replenished, pushing older rocks apart. At deep-sea trenches, old ocean floor sinks back into the mantle.

- Global Positioning System (GPS) satellites can now measure the speed and direction of continental drift directly and precisely. They show that the continents are still moving today at the rate of several centimeters per year.

2 The Plate Tectonic Model 102

- According to **plate tectonic** theory, the lithosphere has fragmented into several large **plates**. These plates essentially float on an underlying layer of hot, ductile rock called the asthenosphere. The asthenosphere is in slow but constant motion, and this forces the more rigid lithospheric plates to move around, collide, split apart, or slide past each other. At present there are six major lithospheric plates and many smaller ones.

- Plate tectonics predicts three different kinds of interactions along plate margins, as shown in this diagram. At **divergent margins**, two plates move apart. If the boundary occurs in oceanic crust, a divergent margin coincides with a midocean ridge. When the divergent boundary occurs in continental crust, it produces long and relatively straight rift valleys. Eventually, if the divergence continues for a long enough time, the rift valley will become wide and deep enough to form a sea or an ocean.

Plate margins: A summary • Figure 4.12

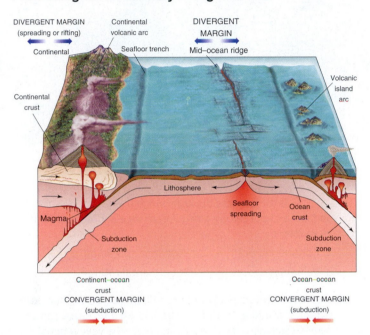

- At **convergent margins**, two plates move toward each other. When an oceanic plate collides with another oceanic plate or with a continental plate, one oceanic plate slides beneath the other plate. This creates a **subduction zone**, which is marked by a deep **oceanic** trench and a great deal of volcanic activity. A second kind of convergent margin occurs where two continental plates collide over an ancient subduction zone, forming a collision zone. Continental crust is generally too buoyant to subduct, so instead the lithosphere crumples and thickens, building up giant mountain ranges.

- At **transform fault** margins, which are fractures in the lithosphere, two plates slide past each other. Transform faults are common on the ocean floor, where they run perpendicular to midocean ridges.

- Many **earthquakes** and active volcanoes are located along plate margins. Studies of the locations and depth of earthquakes enable scientists to determine the shapes of lithospheric plates and the type of margin between the plates. Earthquakes occur as a result of motion of rocks along a **fault**. The **focus**, where the motion starts, is underground; the epicenter is the place on the surface that lies directly above the focus.

- The release of heat from Earth's interior creates huge **convection** cells, which are at least partly responsible for driving plate motion. Convection brings hot rock up from deep in the mantle and recycles cold rock back into the mantle. Many unanswered questions remain about **convection cells**; for example, hot rock also rises in **plumes** that do not seem to be related to the convection cells.

- Plate tectonics unifies continental drift, seafloor spreading, mountain building, faulting, earthquakes, and volcanism into the **tectonic cycle**. The tectonic cycle creates and recycles oceanic crust on a time scale of roughly 200 million years. Continental crust lasts much longer, and it preserves earlier cycles of continental assembly and disassembly.

Key Terms

- continental drift   94
- convection   109
- convergent margin   105
- divergent margin   105

- fault   103
- plate   103
- plate tectonics   102
- seafloor spreading   100

- subduction zone   106
- tectonic cycle   112
- transform fault margin   107

Critical and Creative Thinking Questions

1. Why was the discovery of paleomagnetic bands on the Atlantic Ocean floor such an important turning point in the acceptance of the theory of plate tectonics? Suppose you could rewrite history so that the satellite measurements of continental drift came first. Would the discovery of seafloor spreading still have been important?

2. What are some of the important questions about plate tectonics that remain unanswered today?

3. Why do geologists call plate tectonics a "unifying" theory?

4. We have called plate tectonics a scientific revolution. What other scientific revolutions do you know about? How is the plate tectonics revolution similar to, or different from, other revolutions, such as the Copernican revolution in astronomy?

5. What is the tectonic environment of the place where you live or attend school? Do you live near a plate margin? If so, what type is it? Or do you live in the middle of a plate? If so, which one?

What is happening in this picture?

From September 14 to October 4, 2005, a series of earthquakes and eruptions in the Afar desert in Ethiopia opened up the rift seen in this photograph, which is 60 meters wide at its widest point. The rift is part of a much more extensive depression where two plates, the African and the Somalian plates, are spreading apart. (Older rifts can also be seen in the background.) Compare this photo to the map and satellite image in *What A Geologist Sees* on page 106.

Think Critically

What will happen if the spreading continues?

Self-Test

(Check your answers in Appendix D.)

1. The work of geologists over the years has supported Wegener's contention that the current continental masses were assembled into a single supercontinent, which Wegener called _____.

   a. Pangaea
   b. Transantarctica
   c. Gondwana
   d. Tethys
   e. Laurasia

2. Which of the following lines of evidence supporting continental drift did Wegener not use when he first proposed his hypothesis?

   a. the apparent fit of the continental margins of Africa and South America
   b. ancient glacial deposits of the southern hemisphere
   c. the apparent polar wandering of the magnetic north pole
   d. the close match of ancient geology between West Africa and Brazil
   e. the close match of ancient fossils on continents separated by ocean basins

3. Analysis of apparent polar wandering paths led geophysicists to conclude _____.

   a. that Earth's magnetic poles have wandered all over the globe in the past several hundred million years
   b. that the continents had moved because it is known that the magnetic poles themselves are essentially fixed
   c. that the apparent wandering path of a continent provides a historical record of the position of that continent over time
   d. Both b and c are correct.

4. _____ is the process through which oceanic crust splits and moves apart along a midocean ridge and new oceanic crust forms.

   a. Continental drift
   b. Paleomagnetism
   c. Seafloor spreading
   d. Continental rifting

5. This map shows the age of the seafloor, across the northern extent of the Atlantic Ocean. The Mid-Atlantic Ridge can be seen stretching roughly north–south (in the yellow band) down the middle of the map. Yellow through red colors show rocks of similar age. Number them on the map from 1 (oldest) through 5 (youngest).

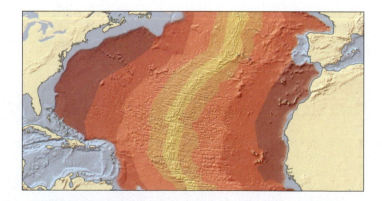

6. _____ technology has allowed scientists to measure the movement of continental crust.

   a. Global Positioning System (GPS)
   b. Seismic recording
   c. Magnetometer
   d. Gravity meter

7. This map shows radiometric ages for the Hawaiian island chain in the middle of the Pacific plate. These islands formed over a hot spot in Earth's mantle. Draw an arrow on the map to show the direction of movement of the Pacific plate over this hot spot, as indicated by the ages of the islands in the Hawaiian chain.

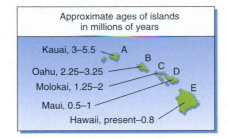

Approximate ages of islands in millions of years

Kauai, 3–5.5 A
B
Oahu, 2.25–3.25 C D
Molokai, 1.25–2
E
Maui, 0.5–1
Hawaii, present–0.8

8. At a _____, two lithospheric plates slide past one another horizontally.

a. divergent boundary

b. transform fault boundary

c. subduction zone boundary

d. continental collision boundary

9. At a _____, oceanic crust is consumed back into the asthenosphere.

a. divergent boundary

b. transform fault boundary

c. subduction zone boundary

d. continental collision boundary

10. At a _____, new oceanic crust forms along midocean ridges.

a. divergent boundary

b. transform fault boundary

c. subduction zone boundary

d. continental collision boundary

11. A _____ is a convergent margin along which subduction is no longer active and high mountain ranges are formed.

a. divergent boundary

b. transform fault boundary

c. subduction zone boundary

d. continental collision boundary

12. _____ is the horizontal movement and mutual interaction of large fragments of Earth's lithosphere.

a. Continental drift   d. Chemosynthesis

b. Polar wandering   e. Plate tectonics

c. Paleomagnetism

13. Heat from the solid mantle is released through a process of _____.

a. polar wandering   c. convection

b. paleomagnetism   d. magnetic reversal

14. These block diagrams depict different types of plate boundaries. For each block diagram, label the appropriate plate boundary from the following list:

Divergent margin   Continental collision margin

Transform fault margin   Subduction zone margin

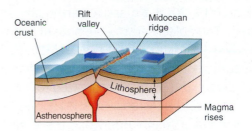

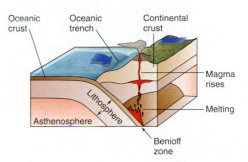

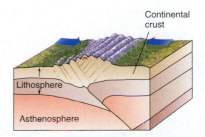

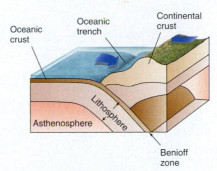

15. On the block diagrams in question 14, indicate the locations of earthquakes for each type of plate boundary. Use a red dot to show the locations of shallow-focus earthquakes and a blue dot to show the locations of deep-focus earthquakes.

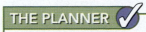

THE PLANNER ✓

Review the Chapter Planner on the chapter opener and check off your completed work.

Earthquakes and Earth's Interior

On March 11, 2011, a massive earthquake shook Japan. The epicenter was offshore of Sendai, a city in northeast Honshu, where the Pacific plate is being subducted beneath the Eurasian plate. Slippage on the fault between the plates is estimated to have been 30 to 40 meters over a distance of 300 kilometers.

The earthquake, magnitude 9.0, is the largest ever recorded in Japan; damage was massive. But worse was to follow. As the plates moved, the sea floor was suddenly pushed upward, and a great tsunami was generated. On the nearby shore water rose to a height approaching 40 meters above sea level, and sweeping inland, entire towns were washed away. Carried ashore by the tsunami, this ship was left high and dry amidst the wreckage of a fishing village.

Japan has suffered many earthquakes and tsunamis, and its citizens, from kindergarten onward, attend regular survival drills. They also place ancient stone monuments to record maximum water rise from tsunamis of the past (see inset). For safety you live uphill from the monuments. In the March 11 event, water rose far beyond the ancient markers.

Disciplined survival training saved many lives, but even so, on May 1 2011, the Red Cross reported a count of more than 15,000 deaths, over 7000 missing, and over 110,000 displaced persons.

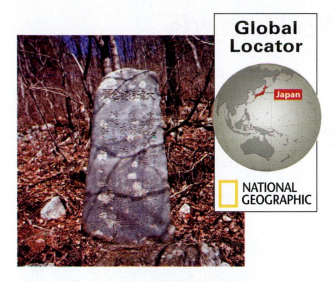

Global Locator

Japan

NATIONAL GEOGRAPHIC

CHAPTER PLANNER ✓

- ❏ Study the picture and read the opening story.
- ❏ Scan the Learning Objectives in each section:
  p. 120 ❏   p. 129 ❏   p. 135 ❏   p. 140 ❏
- ❏ Read the text and study all visuals. Answer any questions.

Analyze key features

- ❏ Case Study, p. 122
- ❏ Geology InSight, p. 124
- ❏ What a Geologist Sees, p. 134
- ❏ Process Diagram, p. 136
- ❏ Amazing Places, p. 142
- ❏ Stop: Answer the Concept Checks before you go on:
  p. 128 ❏   p. 133 ❏   p. 140 ❏   p. 143 ❏

End of chapter

- ❏ Review the Summary and Key Terms.
- ❏ Answer the Critical and Creative Thinking Questions.
- ❏ Answer What is happening in this picture?
- ❏ Complete the Self-Test and check your answers.

Earthquakes and Earthquake Hazards

LEARNING OBJECTIVES

1. **Explain** the connections between giant earthquakes and plate tectonics.
2. **Outline** the theory of elastic rebound.
3. **Identify** several earthquake-related hazards.
4. **Compare** short-term prediction and long-term forecasting of earthquakes.

The Tohoku tsunami that hit the Japanese island of Honshu, pictured on the previous two pages, occurred as a result of an earthquake in a subduction zone off the northeast coast of Japan, where oceanic lithosphere of the Pacific plate is being subducted beneath continental lithosphere on the leading edge of the Eurasian plate. It was the sixteenth giant earthquake, magnitude 8.5 or higher, since 1900. Five of the quakes were magnitude 9.0 or above; besides Tohoku in 2011, the other giants were in Kamchatka, Russia, in 1952; southern Chile in 1960; Prince William Sound, Alaska, in 1964; and off the northwest coast of Sumatra, Indonesia, in 2004 (**Figure 5.1**).

The association between subduction zones and large earthquakes suggests that the convergent motion of two plates must be responsible for the very largest earthquakes. But plate motion is very gradual, typically on the order of a few centimeters per year. Why, then, should earthquakes be so sudden, the big ones so catastrophic? And why, after long time intervals, do they recur in the same places? Through the science of **seismology** we seek to answer these and related questions.

> **seismology** The scientific study of earthquakes and seismic waves.

Earthquakes and Plate Motion

Most earthquakes are caused by the sudden movement of stressed blocks of Earth's crust along a fault. If the rocks could slide past one another smoothly, like the parts of a well-oiled engine, big quakes such as the Tohoku quake would not happen. In the real world, smooth sliding is rare; friction between the huge blocks of rock causes them to seize up, bringing the motion along the locked part of the fault to a temporary stop. While the fault remains locked

Megathrust earthquakes • Figure 5.1

Major earthquakes in subduction zones are sometimes called "megathrust" quakes. The largest earthquakes since 1900, including five with magnitudes of at least 9.0, are shown here; almost all were megathrust earthquakes. The one exception is the Assam-Tibet earthquake of 1950, which occurred in a former subduction zone, now a continental collision zone. Compare this map with Figures 4.9 and 4.10.

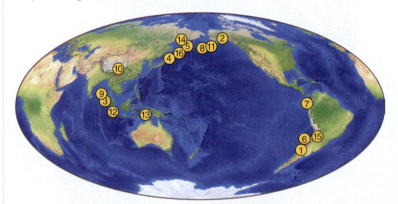

	Location	Date	Magnitude
1.	Chile	1960 05 22	9.5
2.	Prince William Sound, Alaska	1964 03 28	9.2
3.	Off the west coast of northern Sumatra	2004 12 26	9.1
4.	Near the east coast of Honshu, Japan	2011 03 11	9.0
5.	Kamchatka	1952 11 04	9.0
6.	Offshore Maule, Chile	2010 02 27	8.8
7.	Off the coast of Ecuador	1906 01 31	8.8
8.	Rat Islands, Alaska	1965 02 04	8.7
9.	Northern Sumatra, Indonesia	2005 03 28	8.6
10.	Assam, Tibet	1950 08 15	8.6
11.	Andreanof Islands, Alaska	1957 03 09	8.6
12.	Southern Sumatra, Indonesia	2007 09 12	8.5
13.	Banda Sea, Indonesia	1938 02 01	8.5
14.	Kamchatka	1923 02 03	8.5
15.	Chile-Argentina Border	1922 11 11	8.5
16.	Kuril Islands	1963 10 13	8.5

by friction, energy continues to build up as a result of the plate motion, causing rocks adjacent to the jammed section to bend and buckle. Finally, the stress becomes great enough to overcome the friction along the fault. All at once, the blocks slip, and the pent-up energy in the rocks is released as the violent tremors of an earthquake. This cycle of slow buildup of energy followed by abrupt movement along a fault repeats itself many times.

Although movement along a large fault may eventually total many kilometers, this distance is the sum of numerous smaller slips happening over many millennia. In some places these small slips (collectively called seismic creep) are frequent, though they are imperceptible to humans. Nevertheless, over time they can create visible distortion of surface features, as shown in **Figure 5.2a**. Earthquakes can also cause vertical dislocations of the ground surface (**Figure 5.2b**). The largest abrupt vertical displacement on land in recent history occurred in 1899 at Yakutat Bay, Alaska, when a long stretch of the Alaskan shore was suddenly lifted 15 meters above sea level during a major earthquake.

The elastic rebound theory The initial vertical or horizontal motion of plates, dramatic as it may appear, is often not what does the most damage during an earthquake. It is the sustained shaking of the ground that destroys buildings, bridges, and cities, sometimes many kilometers away from the location of the quake. In 1910, Harry Fielding Reid,

> **elastic rebound theory** The theory that continuing stress along a fault results in a buildup of elastic energy in the rocks, which is abruptly released when an earthquake occurs.

a member of a commission appointed to investigate the infamous 1906 San Francisco earthquake that destroyed much of the city, proposed the most widely accepted explanation for the shaking.

Reid's **elastic rebound theory** says that rocks, like all other solids, are elastic (within limits). This means that rocks will stretch or bend when subjected to stress, and they snap back when the stress is removed, which happens when the two blocks on either side of a fault manage to overcome friction and slip past one another. But the strained rocks on either side of the fault don't just snap back and stop. Like a guitar string after it is plucked, they continue vibrating. These vibrations are called **seismic waves**. Like sound waves from a guitar string, they can travel a long distance from their place of origin. (We will investigate the scientific concepts of *stress* and *strain* in greater depth in Chapter 9.)

> **seismic wave** An elastic shock wave that travels outward in all directions from an earthquake's source.

Observations show that earthquakes occur in the same places repeatedly. The elastic rebound theory tells us that this is just what should be expected. Along a portion of a fault that is locked, stress builds up over time. When the frictional lock is broken, the stress is suddenly and (sometimes) violently relieved, and an earthquake occurs. Tectonic movement of the plates continues, and locked portions of a fault will once again begin accumulating stress over time; a sudden release of energy occurs when the blocks slip past one another. The cycle repeats, and earthquakes happen repeatedly along the same faults (see *Case Study* on the next page).

Evidence of lateral and vertical fault motion • Figure 5.2

a. When these orange trees were planted on land that lies over the San Andreas fault in southern California, the rows were straight. In 1938, earthquake motion along the fault displaced the trees significantly. Arrows show the direction of movement of the plates.

b. The second most powerful earthquake on record, the Alaska "Good Friday" earthquake, struck the Anchorage area on March 27, 1964. The vertical motion along the fault amounted to several meters in some places. Here, the two plates moved apart, causing the ground between them to subside by about 2 meters.

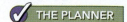
Proving the Elastic Rebound Theory

The first evidence to support Reid's elastic rebound model came from studies of the San Andreas fault, a large, complex fault in California that generated both the 1906 quake and the 1989 "World Series" quake near Oakland. Beginning in 1874, scientists from the U.S. Coast and Geodetic Survey had been measuring the precise positions of many points both adjacent to and distant from the fault. As time passed, movement of the points revealed that at some places, the two sides of the fault were smoothly slipping in opposite directions. Near San Francisco, however, the fault appeared to be locked by friction and did not reveal any slip. Then, on April 18, 1906, the two sides of this locked section of fault shifted abruptly. The elastically stored energy in the rocks was released as the crust snapped to its new position, creating a violent earthquake. Reid's measurements after the quake revealed that the bending, or strain, stored in the crust had disappeared.

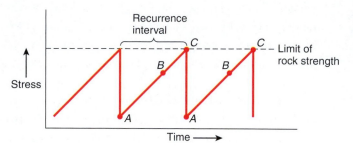

The elastic rebound theory says that stress builds up until the rock strength at the fault lock is exceeded, and then the lock breaks and an earthquake occurs. **Figure a** corresponds to the unstressed state; **Figure b** corresponds to the period during which stress is building up; and **Figure c** is the moment when the frictional lock is broken, the blocks slip past one another, and the fault returns to an unstressed state.

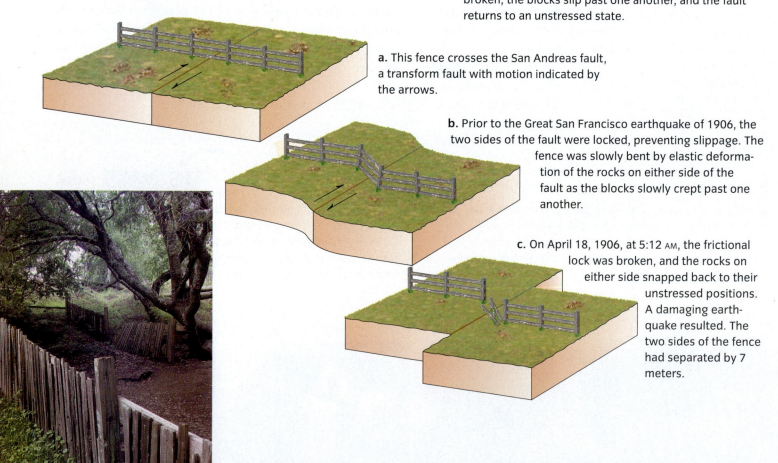

a. This fence crosses the San Andreas fault, a transform fault with motion indicated by the arrows.

b. Prior to the Great San Francisco earthquake of 1906, the two sides of the fault were locked, preventing slippage. The fence was slowly bent by elastic deformation of the rocks on either side of the fault as the blocks slowly crept past one another.

c. On April 18, 1906, at 5:12 AM, the frictional lock was broken, and the rocks on either side snapped back to their unstressed positions. A damaging earthquake resulted. The two sides of the fence had separated by 7 meters.

Along the San Andreas fault near Point Reyes, California, sections of a fence show the dramatic displacement of land that occurred in the 1906 quake. This fence was built in the 1970s along the track of an old fence that was wrenched apart by the quake.

Earthquake Hazards and Readiness

Each year several million earthquakes occur around the world. Fortunately, only a few are large enough, or close enough to major population centers, to cause much damage or loss of life. A great deal of research focuses on earthquake prediction and hazard assessment. Geologists are working hard to improve their forecasting ability to the point where effective and accurate early warnings can be issued. Let's look briefly at the hazards associated with earthquakes and at efforts to predict them.

Earthquake hazards Ground motion, with the resulting collapse of buildings, bridges, and other structures, is usually the most significant **primary hazard** to cause damage during an earthquake. In the most intense quakes, the surface of the ground can be observed moving in waves. In cases where the earthquake's focus is shallow (close to the surface), shaking can be surprisingly intense. In 2011, nine people were killed and thousands of buildings were damaged or destroyed by an earthquake of only moderate intensity (magnitude 5.1); ground shaking was intense because the earthquake had a particularly shallow focus (about 1 kilometer depth).

Where a fault breaks the ground surface, buildings can be split, roads disrupted, and anything that lies on or across the fault broken apart. Sometimes large cracks and fissures open in the ground. To make matters worse, movement on one part of a fault can cause stress along another part of the fault, which in turn slips, generating another earthquake, called an aftershock. Aftershocks triggered by large earthquakes tend to be on the same fault system as the original quake, though they may be quite far from the original location, causing the damage to be spread more widely as time passes. Some of the hundreds of aftershocks that followed the 2011 Tohoku earthquake had magnitudes greater than 7; these **aftershocks**, triggered by the initial great quake, are major earthquakes in their own right.

Ground motion is not the only source of damage in an earthquake. Sometimes the after-effects, or **secondary hazards** related to an earthquake, can cause even more damage than the original quake (**Figure 5.3**). Examples of secondary hazards that can be initiated by earthquakes include landslides, fires, ground liquefaction, and tsunamis (described in **Figure 5.4**).

Earthquake-related secondary hazards • Figure 5.3

a. Landslide, Huascaran, Peru

b. Fissure, Santa Cruz, California

c. Fire, San Francisco, California

d. Ground liquefaction, Niigata, Japan

Tsunamis are less common than other secondary hazards associated with earthquakes, but they can be even more deadly, as demonstrated by the great Sumatra–Andaman tsunami in the final week of 2004 and the Tohoku tsunami of March, 2011. In those two events, the world witnessed the devastating potential of tsunamis as it never had before. Tsunamis can also be generated by submarine landslides and volcanism, but about 85 percent of all tsunamis are generated by earthquakes in subduction zones.

On December 26, 2004, a 9.1-magnitude earthquake shook the Indian Ocean floor 160 kilometers off the island of Sumatra, Indonesia. The quake caused the most deadly tsunami in history. The quake began when the Pacific plate, which is being subducted beneath the Eurasian plate, suddenly moved downward 15 meters (**Figure a**). Along a distance of 1200 kilometers, the rebounding motion pushed the seafloor up as much as 5 meters on the Eurasian side. On the surface of the ocean, killer waves generated by the sudden movement of the seafloor swept toward Indonesia, Thailand, Sri Lanka, and India. When they reached the shore, the waves were amplified to 20 to 30 meters in height and swept far inland. Although earthquakes and tsunamis are common in the Indian Ocean, there was no warning system in place when the disaster occurred. The resulting devastation caused hundreds of billions of dollars in damage and at least 283,000 deaths.

A series of computer-simulated maps (**Figure b**) shows the progress of the tsunami across the Indian Ocean. The red color indicates wave crests, the blue the wave troughs. Satellite images show a coast in Sri Lanka before and during the tsunami. The height of the 2004 tsunami wave approaching the shore can be judged from the photo of a boat mooring in Thailand (**Figure c**). The mooring and all the boats were destroyed.

a. How the Tohoku tsunami was unleashed.

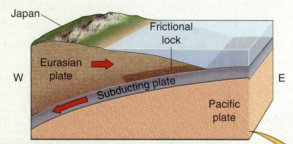

1. The subducting Pacific plate sticks to the overriding Eurasian plate as a result of friction.

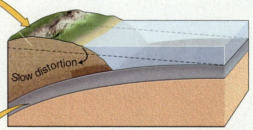

2. Stress builds and, in accord with elastic rebound theory, the seafloor gradually deforms as the Eurasian plate is bent downward.

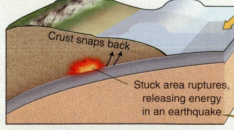

3. A sudden rupture frees the stuck section of the Eurasian plate. It rebounds elastically and releases energy in the form of seismic waves. The seafloor along the fault is lifted, causing the displacement of a very large volume of water.

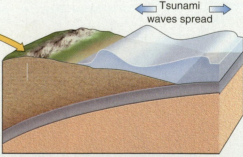

4. The tsunami waves spread out in two directions from the uplifted region, traveling mainly to the east and west at right angles to the subduction zone. Tsunami waves from the Tohoku earthquake of 2011 traveled across the Pacific Ocean and reached the western shores of North and South America. Fortunately the size of the waves decreased as they traveled, so damage in those coastal regions was slight.

b. Progress of the Sumatra-Anderman tsunami

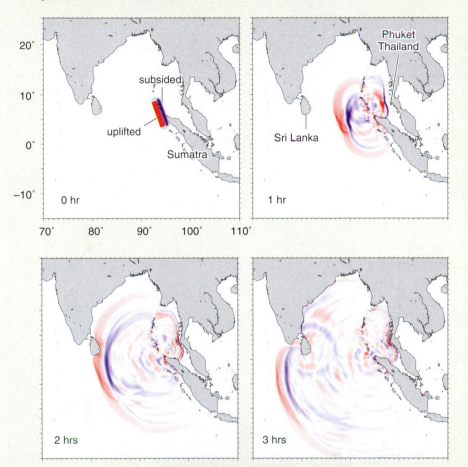

Before

During

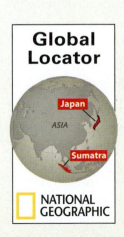

Global Locator

Japan

ASIA

Sumatra

NATIONAL GEOGRAPHIC

c.

Think Critically

What kind of wave would you expect to travel faster: a seismic wave or a tsunami wave?

Earthquake readiness Earthquakes can cause total devastation in a matter of seconds. The most disastrous quake in history occurred in Shaanxi Province, China, in 1556, killing an estimated 830,000 people. The earthquake caused the caves in which most of the population lived to collapse. In all, 19 earthquakes in history have caused 50,000 or more deaths apiece (**Table 5.1**).

The most powerful quakes are not necessarily the deadliest. The death toll depends to a great extent on the population of the affected region and how well prepared they are for a major quake. The Great Sichuan earthquake of May 12, 2008, in China, and the Haiti earthquake of 2010 were much greater disasters than they need have been because many buildings, including schools, had not been constructed to withstand earthquakes, even though both regions are known to be prone to quakes (**Figure 5.5a**).

Every earthquake hazard, from fires to tsunamis, can be reduced in severity (though not eliminated) with proper design and preparedness. For instance, skyscrapers can be built with reinforced concrete and large counterweights to help them resist shaking. **Figure 5.5b** illustrates the inadequacy of unreinforced concrete during an earthquake. Other aspects of earthquake-resistant design include bolting wood-frame buildings to their foundations; isolating buildings from vibrations by supporting them on steel and rubber pads; enclosing utilities such as water pipes and electric cables in common ducts; and building tunnels with flexible joints. It is also important for people living in earthquake-prone areas to prepare for them—for example, by securing heavy furniture to the walls and by preparing emergency supplies. During an earthquake, your best shelter is underneath a sturdy desk or table (**Figure 5.5c**).

Earthquake Prediction

Charles Richter, inventor of the Richter scale for quantifying the severity of earthquakes, once said, "Only fools, charlatans, and liars predict earthquakes." Today, unfortunately, this is still more or less correct: No one can predict the exact magnitude and time of occurrence of an earthquake, although many reputable people are trying hard. Scientists' understanding about seismic mechanisms and the tectonic settings in which earthquakes occur has improved greatly since Richter's time, and advances in modern seismology may yet prove him wrong.

There are two aspects to the problem of earthquake prediction. Short-term prediction would identify the

Earthquakes during the past 800 years that have caused 50,000 or more deaths   Table 5.1

Place	Year	Estimated number of deaths
Silicia, Turkey	1268	60,000
Chihli, China	1290	100,000
Naples, Italy	1456	60,000
Shaanxi, China	1556	830,000
Shemakha, Russia	1667	80,000
Catania, Italy	1693	60,000
Beijing, China	1731	100,000
Calcutta, India	1737	300,000
Lisbon, Portugal	1755	60,000
Calabria, Italy	1783	50,000
Messina, Italy	1908	160,000
Gansu, China	1920	180,000
Tokyo and Yokohama, Japan	1923	143,000
Gansu, China	1932	70,000
Quetta, Pakistan	1935	60,000
T'ang-shan, China	1976	240,000*
Sumatra–Andaman, Indian Ocean	2004	283,000**
Sichuan, China	2008	69,000
Port-au-Prince, Haiti	2010	220,000

*Unofficial reports list up to 779,000 deaths
**Most of the deaths due to tsunami

precise time, magnitude, and location of an earthquake in advance of the actual event, providing an opportunity for authorities to issue an early warning. Long-term forecasting involves the prediction of a large earthquake years or even decades in advance of its occurrence.

Short-term prediction and early warning Unfortunately, the short-term prediction of earthquakes has not been very successful to date. Attempts at short-term prediction are based on observations of anomalous **precursor phenomena**—that is, unusual activity preceding and leading up to the occurrence of an earthquake. For example, the magnetic or electrical properties of rock could change, the level of well water could drop, or the amount of radon gas in groundwater could rise in advance of an earthquake, any of which

a. Poorly constructed buildings in Port au prince, Haiti, collapsed as a result of the earthquake, 22 January, 2010.

b. A laboratory experiment in California shows how unreinforced concrete crumbles during earthquake-like conditions.

c. Preparedness can save lives during an earthquake. Earthquake drill in a Japanese school. This four-year-old boy, wearing a padded protective hood, has taken cover under a table. Drills like these helped to save many lives during the Tohoku earthquake, but the toll of dead and injured was still staggering.

may indicate unusual activity in the underlying rock. Strange animal behavior, glowing auras, and unusual radio waves have also been reported as precursors near the sites of large earthquakes; there are plausible scientific explanations for these. Small cracks and fractures can develop in severely strained rock and cause swarms of tiny earthquakes—**foreshocks**—that may presage a big quake. In retrospect, scientists now realize that the Tohoku earthquake was preceded by several large foreshocks in the days immediately before the event.

The most famous successful short-term earthquake prediction, made by Chinese scientists in 1975, was based on combined observations of slow tilting of the land surface, changes in groundwater, fluctuations in the magnetic field, and numerous foreshocks that preceded a large quake that struck the town of Haicheng. Half of the city was destroyed, but because authorities had evacuated more than a million people beforehand, only a few hundred were killed. However, less than two years after the successful prediction of the Haicheng earthquake, the devastating 1976 T'ang-shan earthquake struck, with no apparent precursory activity. With an official death toll of 240,000 (and unofficial reports suggesting many more), it was the second-most-disastrous earthquake in history. In 1976, and still today, short-term prediction and early warning of earthquakes remain elusive goals for seismologists.

Long-term forecasting Long-term earthquake forecasting is based mainly on our understanding of the elastic rebound theory, the tectonic cycle, and the geologic settings in which earthquakes occur. In places where earthquakes are known to occur repeatedly, seismologists have detected patterns in the **recurrence intervals** of large quakes—that is, the timing between earthquakes occurring at the same location on a fault. Because historical records seldom go back as far as seismologists would like, scientists also use

> **paleoseismology**
> The study of prehistoric earthquakes.

the information provided by **paleoseismology** to investigate earthquake recurrence intervals.

Prehistoric quakes left evidence in the stratigraphic record, such as vertical displacement of sedimentary layers, indications of liquefaction, or horizontal offset of geologic features (**Figure 5.6**). If the pattern of recurrence suggests regular intervals of, say, a century between major quakes, it may be possible to predict within a decade or two when a large quake is due to happen next in that location.

By studying ancient earthquakes, scientists have identified a number of **seismic gaps** around the Pacific Rim. A seismic gap is a place along a fault where a large earthquake has not occurred for a long time, even though tectonic movement is still active and stress is building. Some geophysicists consider seismic gaps to be the places most likely to experience large earthquakes.

Evidence of ancient quakes • Figure 5.6

Dark, carbon-rich layers of sediment, offset by ancient earthquakes, have yielded carbon-14 dates, which help scientists pinpoint the ages of ancient quakes.

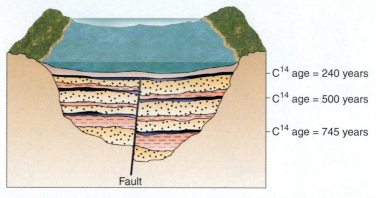

C^{14} age = 240 years

C^{14} age = 500 years

C^{14} age = 745 years

Fault

- ⋯ Sand layer
- Clay layer
- Carbon-containing layer

Long-term forecasting has met with reasonable success. Seismologists know where most (but not all) of the hazardous areas are located. They can calculate the probability that a large earthquake will occur in a particular area within a given period. They have a theory of earthquake generation that successfully unites their predictions and observations in the context of plate tectonic theory. Forecasting helps people who live in seismically active areas to plan and prepare well in advance of a major event. If short-term prediction could advance as much as long-term forecasting has done, many lives could be saved.

CONCEPT CHECK 🛑 STOP

1. **Why** are earthquakes more common along plate edges than in plate interiors?

2. **How** does the elastic rebound theory explain the violent tremors that occur during earthquakes?

3. **What** are some of the primary and secondary hazards associated with earthquakes?

4. **How** do scientists predict earthquakes?

The Science of Seismology

LEARNING OBJECTIVES

1. **Explain** how a seismograph works.
2. **Define** body waves and surface waves.
3. **Identify** the two kinds of body waves and explain how they differ.
4. **Explain** how seismologists locate the epicenter of an earthquake.
5. **Describe** the Richter and moment magnitude scales and compare them to the Modified Mercalli Intensity scale.

Seismologists can quickly locate an earthquake anywhere on Earth and tell how strong it is. They are also very good at telling the difference between earthquakes and other seismic disturbances, such as explosions and landslides.

Seismographs

The earliest known **seismographs** (also called seismometers) were invented in China in the 2nd century. The first seismographs in Europe were invented much later, in the 19th century. Modern seismographs provide a printed or digital record of seismic waves, called a **seismogram** (**Figure 5.7**).

> **seismograph** An instrument that detects and measures vibrations of Earth's surface.
>
> **seismogram** The record made by a seismograph.

The most advanced seismographs measure the ground's motion optically and amplify the signal electronically. Vibrations as tiny as one-hundred-millionth (10^{-8}) of a centimeter can be detected. Indeed, many instruments are so sensitive that they can sense vibrations caused by a moving automobile many blocks away.

Ancient and modern seismographs • Figure 5.7

a. Ancient Chinese seismograph. Tremors would cause the central rod to tilt, releasing a ball from the mouth of the dragon and alerting people to the earthquake danger.

b. Modern seismographs use the principle of inertia—the resistance of a heavy mass to motion. In this schematic diagram, seismic waves cause the support post and the roll of paper to vibrate back and forth. However, the large mass attached to the pendulum and the pen attached to it barely move at all. It looks to an observer as if the pen is moving, but in reality it is the paper that moves underneath the pen to create the seismogram.

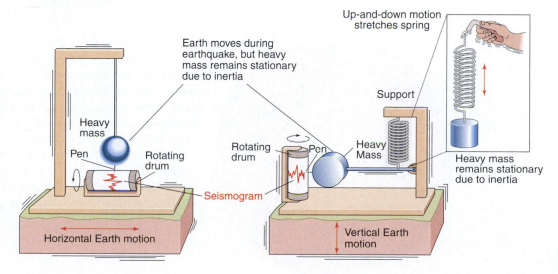

Seismic Waves

The energy released by an earthquake is transmitted to other parts of Earth in the form of seismic waves. The waves elastically deform the rocks they pass through; they leave no record behind them once they have passed, so they must be detected while they pass. The waves, which include **body waves** and **surface waves**, travel outward in all directions from the earthquake's **focus** (**Figure 5.8**).

Body waves Body waves can be further subdivided into two types. **Compressional waves** (**Figure 5.9a**) can pass

through solids, liquids, and gases. They have the highest velocity of all seismic waves—typically 6 km/s in the uppermost portion of the crust—and thus they are the first waves to arrive and be detected by a seismograph after an earthquake. For this reason they are called **P waves** (for **primary waves**).

Shear waves, the other type of body wave, travel through materials by generating an undulating motion in the material (**Figure 5.9b**). Solids tend to resist a shear force and bounce back to their original shape afterward, whereas liquids and gases do not. Without this elastic rebound, there can be no wave. Therefore, shear waves cannot be transmitted through liquids or gases. This has important consequences for the interpretation of seismic waves, as you will see. Shear waves travel more slowly than compressional waves, at about 3.5 km/s. Because they arrive at a seismograph after the P waves from the same earthquake, they are called **S waves** (for **secondary waves**).

Surface waves Surface waves travel along or near Earth's surface, like waves along the surface of the ocean. They travel more slowly than P and S waves, and they pass around Earth rather than through it. Thus, surface waves are the last to be detected by a seismograph. **Figure 5.10a** shows a typical seismogram, in which the P waves arrive first, followed by the S waves, and finally the surface waves. Surface waves are responsible for much ground shaking and structural damage during major earthquakes.

Locating Earthquakes

The **epicenter**, or surface location of an earthquake, can be determined through simple calculations, provided that at least three seismographs have recorded the quake. The first step is to find out how far each seismo-

Travel path of seismic body waves • Figure 5.8

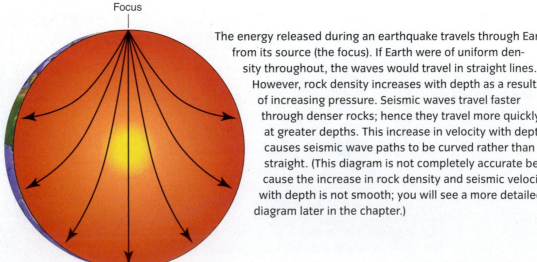

Focus

The energy released during an earthquake travels through Earth from its source (the focus). If Earth were of uniform density throughout, the waves would travel in straight lines. However, rock density increases with depth as a result of increasing pressure. Seismic waves travel faster through denser rocks; hence they travel more quickly at greater depths. This increase in velocity with depth causes seismic wave paths to be curved rather than straight. (This diagram is not completely accurate because the increase in rock density and seismic velocity with depth is not smooth; you will see a more detailed diagram later in the chapter.)

Seismic body waves • Figure 5.9

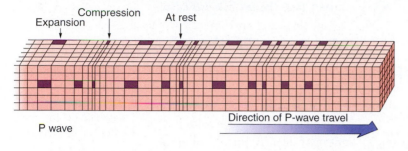

P wave

Expansion
Compression
At rest

Direction of P-wave travel

a. A compressional wave alternately squeezes and stretches the rock as it passes through. The grid is intended to help you visualize how the rock responds. All the divisions in the grid start out square, but the wave alternately squeezes them down into narrow rectangles and stretches them into long rectangles.

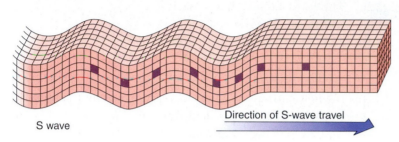

S wave

Direction of S-wave travel

b. A shear wave causes the rock to vibrate up and down, like a rope whose end is being shaken. In this case, the squares do not expand or contract but do get distorted, changing shape alternately from squares to parallelograms and back to squares again.

Using seismograms to locate an earthquake • Figure 5.10

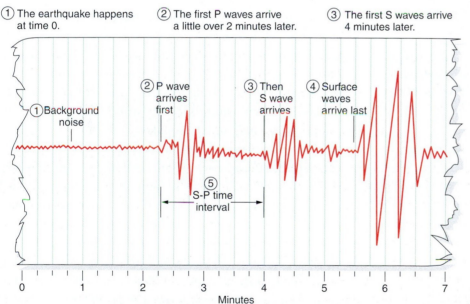

① The earthquake happens at time 0.

② The first P waves arrive a little over 2 minutes later.

③ The first S waves arrive 4 minutes later.

a. Seismogram from a typical earthquake

① Background noise

② P wave arrives first

③ Then S wave arrives

④ Surface waves arrive last

⑤ S-P time interval

Minutes

④ The surface waves, which travel the long way around Earth's surface, arrive last.

⑤ The S-P interval, here slightly less than 2 minutes, tells the seismologist how far away the earthquake was.

b. The method of **triangulation**. Three seismic stations—shown here in Stockholm, Honolulu, and Manila—record an earthquake. Each one independently determines its own distance from the focus of the quake, generating a circle on which the epicenter of the earthquake must lie. Each black arrow indicates the radius of a circle, which is equal to the calculated distances from the epicenter. The three circles have a unique intersection point, which is the location of the epicenter—in this case Kobe, Japan (the site of a major earthquake in 1995).

Stockholm

Honolulu

Kobe (epicenter)

Manila

Where Geologists CLICK

Earthquakes at Your Fingertips

http://earthquake.usgs.gov/earthquakes/

To learn more about earthquakes, report one that you have felt, or monitor the occurrence of earthquakes as they happen around the world, visit the U.S. Geological Survey Earthquake Hazard Program Web site, at http://earthquake.usgs.gov/earthquakes/. The site is extremely comprehensive; be sure to explore it thoroughly!

graph is from the source of the earthquake. The greater the distance traveled by the seismic waves, the more the S waves will lag behind the P waves. Thus, the lag time between the P and S waves on a seismogram (**Figure 5.10b**) provides geologists the necessary distance information.

After determining the distance from each seismograph to the source of the earthquake, a seismologist draws a circle on a map, with the seismic station at the center of the circle. The radius of the circle is the distance from the seismograph to the focus. It is a circle because the seismologist knows only the distance travelled by the seismic waves, not the direction from which they came. When this information is calculated and plotted for three or more seismographs, the unique point on the map where the three circles intersect is the location of the epicenter (see Figure 5.10b). This process is called **triangulation**. Geologists and others can report information about earthquakes and share information about where and when earthquakes have occurred,

through the Web site maintained by the U.S. Geological Survey (see *Where Geologists Click*).

Measuring Earthquakes

Geologists use several different scales to quantify the strength, or **magnitude**, of an earthquake, by which we mean the amount of energy released during the quake. The most familiar of these is the **Richter magnitude scale**.

> **Richter magnitude scale** A scale of earthquake intensity based on the recorded heights, or amplitudes, of the seismic waves recorded on a seismograph.

The Richter magnitude scale Charles Richter developed his famous magnitude scale in 1935. Though it was not the first earthquake intensity scale, it was an important advance because it used data from seismographs rather than subjective estimates of damage. Also, it compensated for the distance between the seismograph and the focus. This means that each seismic station will (in principle) calculate the same magnitude for a given earthquake, no matter how far from the epicenter it is located.

The Richter scale is **logarithmic**, which means that each unit increase on the scale corresponds to a 10-fold increase in the amplitude of the wave signal. Thus, a magnitude 6 earthquake has a wave amplitude 10 times larger than that of a magnitude 5 quake. A magnitude 7 earthquake has a wave amplitude 100 times larger (10×10) than that of a magnitude 5 quake.

However, even this comparison understates the difference because the amount of damage done by an earthquake is more closely related to the amount of energy released in the quake. Each step in the Richter scale corresponds roughly to a 32-fold increase in energy (see *What a Geologist Sees*). The actual amount of damage done by a quake also depends, of course, on local conditions—how densely populated the area is, how the buildings are constructed, how deep the focus is, and how severe the secondary effects are.

Moment magnitude scale

Seismologists today determine magnitudes using both the Richter scale and the **moment magnitude scale**, which are calculated using different starting assumptions. Richter scale calculations are based on the assumption that an earthquake focus is a point. Therefore, the Richter scale is best suited for earthquakes

> **moment magnitude scale** A measure of earthquake strength that is based on the rupture size, rock properties, and amount of displacement on the fault surface.

Earthquake magnitudes, frequencies, and effects   Table 5.2

Richter and Moment Magnitude*	Number per Year	Modified Mercalli Intensity Scale*	Characteristic Effects in Populated Areas
<3.4	800,000	I	Recorded only by seismographs
3.5–4.2	30,000	II–III	Felt by some people who are indoors
4.3–4.8	4,800	IV	Felt by many people; windows rattle
4.9–5.4	1,400	V	Felt by everyone; dishes break, doors swing
5.5–6.1	500	VI–VII	Slight building damage; plaster cracks, bricks fall
6.2–6.9	100	VIII–IX	Much building damage; chimneys fall; houses move on foundations
7.0–7.3	15	X	Serious damage, bridges twisted, walls fractured; many masonry buildings collapse
7.4–7.9	4	XI	Great damage; most buildings collapse
>8.0	<1	XII	Total damage; waves seen on ground surface, objects thrown in the air

*The correspondence between Richter and moment magnitudes and the Mercalli Intensity is not exact because they are calculated on the basis of very different parameters.

in which energy is released from a relatively small area of a locked fault. In contrast, the calculation of seismic moment takes account of the fact that energy may be released over a large area; this is particularly true of bigger earthquakes. A classic example is the Sumatra–Andaman earthquake of 2004, during which a 1200-kilometer length of fault moved. Though their methods of calculation differ and the results often differ somewhat, the scales both measure the same thing—the amount of energy released. In either system, magnitude 9 is catastrophic, and magnitude 2.5 is imperceptible to humans.

Modified Mercalli Intensity scale
Scientists measure the magnitude of an earthquake because they are concerned with the amount of energy released. What is important for engineers designing buildings and authorities charged with civilian safety is the amount of damage done during an earthquake. Earthquake damage is measured by using the **Modified Mercalli Intensity (MMI) scale**, which was developed by an Italian scientist in 1902 and later modified. The Modified Mercalli scale is based on descriptions of vibrations that people felt, saw, and heard, as well as on the extent of damage to buildings. The scale ranges from I (not felt, except under unusual circumstances) to XII (visible waves on ground, practically all buildings destroyed). The MMI of an earthquake varies

with distance from the epicenter; a quake could have an intensity of X near the epicenter, whereas 100 kilometers away the intensity might be only II. The MMI of the Tohoku earthquake in March 2011 reached IX in some areas, even though the epicenter was located some distance offshore.

The correspondence between Mercalli Intensity and Richter and moment magnitudes is not exact because they are calculated on the basis of very different parameters. In **Table 5.2** and *What a Geologist Sees*, on the next page, we compare the Richter magnitudes, extent of damage, and Mercalli intensities for earthquakes of various magnitudes.

CONCEPT CHECK

1. **What** physical property does a modern seismograph depend on?

2. **What** evidence do body waves leave in rocks as they pass through Earth?

3. **How** does a P wave differ from an S wave?

4. **How** can travel times of body waves be used to reveal the location of an earthquake?

5. **How** is the Richter magnitude of an earthquake determined, and how does it differ from the determination of a Modified Mercalli estimate?

WHAT A GEOLOGIST SEES

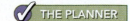

Richter Magnitude and Modified Mercalli Intensity Scales

The energy released in an earthquake increases exponentially with its magnitude. A magnitude 6 earthquake releases as much energy as the atomic bomb dropped on Hiroshima, the largest nuclear weapon ever used in combat. A magnitude 7 quake is equivalent in energy to about 32 Hiroshima bombs (but without the radioactivity associated with those bombs), and a magnitude 8 quake would be equivalent to 32×32, or about 1000, of them. A magnitude 9 quake, such as the Tohoku quake, is equivalent to $32 \times 32 \times 32$, or about 32,000 bombs. Note that the Mercalli scale, which measures damage, and the Richter scale are not directly comparable; one major difference is that the Mercalli magnitude of an earthquake differs with distance from the epicenter, while the Richter magnitude does not.

Parkfield, CA, 2004

Richter magnitude 6
Damage on surface close to the epicenter: A geologist would notice that small objects have been broken, that sleepers have wakened, and buildings are still standing but bricks have fallen from the walls. The geologist would estimate a Modified Mercalli Intensity of ≈ VII.
Energy released: about the same as 1 atomic bomb

Richter magnitude 7
Damage on surface close to the epicenter: A geologist would see that some buildings have moved, some have collapsed, and damage has been severe. The presence of police suggests panic and the need for control. The geologist would estimate a Modified Mercalli Intensity of ≈ IX or X.
Energy released: about the same as 32 atomic bombs

Kobe, Japan, 1995

Richter magnitude 8
Damage on surface close to the epicenter: A geologist would observe widespread destruction. With many buildings collapsed, thousands would be dead or injured in densely populated areas. The geologist would estimate a Modified Mercalli Intensity of ≈ XI.
Energy released: about the same as 1000 atomic bombs

San Francisco, CA, 1906

Think Critically

Might there be an upper limit to the possible magnitude of an earthquake? If so, what might cause this?

Studying Earth's Interior

LEARNING OBJECTIVES

1. **Explain** how the materials in Earth's interior affect seismic waves.

2. **Explain** why seismic data point to the existence of a liquid core.

3. **Discuss** the limitations of direct sampling of Earth's interior.

4. **Identify** several ways by which scientists can study Earth's interior indirectly or remotely.

 arthquakes are important because of the damage they can cause, but they also have benefits from a scientific perspective. They provide us with some of our most detailed information about Earth's interior—including parts that we can never hope to observe directly.

When scientists cannot study something by direct sampling, a second method comes to the forefront: indirect study, or **remote sensing**. Some familiar objects—including the human eye—are actually remote-sensing devices. A camera, for instance, is a remote-sensing instrument that collects information about how an object reflects light. Medical techniques such as X-rays allow doctors to study the inside of the body remotely without opening it up surgically.

The seismic waves from an earthquake are much like X-rays in the sense that they enter Earth near the surface, travel all the way through it, and emerge on the other side. They travel along different paths, depending on the different kinds of materials they encounter. We will first discuss how earthquakes reveal Earth's structure and then describe other sources of information about Earth's interior.

How Geologists Look into Earth's Interior: Seismic Methods

Before 1906, scientists' understanding of seismic waves was limited. But that year, British geologist Richard Dixon Oldham first identified the difference between P waves and S waves and then suggested an explanation for the complicated patterns recorded after an earthquake: Underneath thousands of kilometers of solid rock, he postulated, Earth has a liquid core. Oldham's conclusion is now universally accepted by geologists.

Seismic discontinuities and the liquid core

When a major earthquake strikes, the tremors are measured at seismic stations around the world, and the arrival times of P waves and S waves can be used to determine what kinds of rock the seismic waves passed through. Even a crude diagram of Earth's structure shows that the pattern of arrival times is quite complex. Because Earth's interior is not homogeneous, seismic waves must pass through the boundaries between different materials as they travel through Earth's interior. Seismic waves behave differently depending on the properties of the materials they pass through, including the density of the material and whether it is a liquid or a solid.

Three distinct things can happen to seismic waves when they meet such boundaries:

1. They can be **refracted**, or bent, as they pass from one material into another. This is the same thing that happens to light waves when they pass from air to water.

2. They can be **reflected**, which means that all or part of the wave energy bounces back, like light from a mirror.

3. They can be absorbed, which means that all or part of the wave energy is blocked.

The abrupt changes in velocity that result when seismic waves are refracted or reflected across boundaries or absorbed by materials as they travel through Earth's interior are called **seismic discontinuities**.

> **refraction** The bending of a wave as it passes from one material into another material, through which it travels at a different speed.
>
> **reflection** The bouncing back of a wave from an interface between two different materials.
>
> **seismic discontinuity** A boundary inside Earth where the velocities of seismic waves change abruptly.

Seismic waves in Earth's interior • Figure 5.11

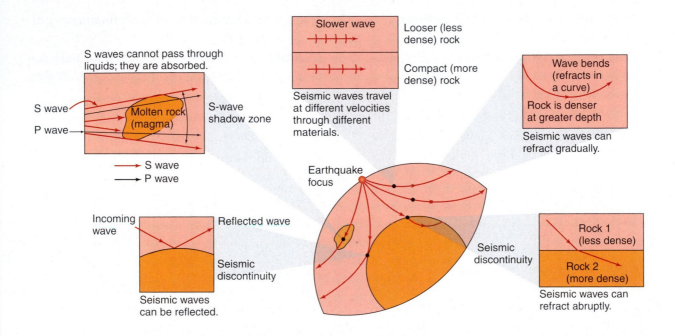

S waves cannot pass through liquids; they are absorbed.

S wave
P wave

Molten rock (magma)

S-wave shadow zone

→ S wave
→ P wave

Slower wave

Looser (less dense) rock

Compact (more dense) rock

Seismic waves travel at different velocities through different materials.

Wave bends (refracts in a curve)

Rock is denser at greater depth

Seismic waves can refract gradually.

Earthquake focus

Incoming wave

Reflected wave

Seismic discontinuity

Seismic waves can be reflected.

Seismic discontinuity

Rock 1 (less dense)

Rock 2 (more dense)

Seismic waves can refract abruptly.

The combination of the various behaviors (above) creates a complex pattern of arrival times of seismic waves at distant locations following an earthquake. Note especially the **shadow zones**, where P waves and S waves are not observed.

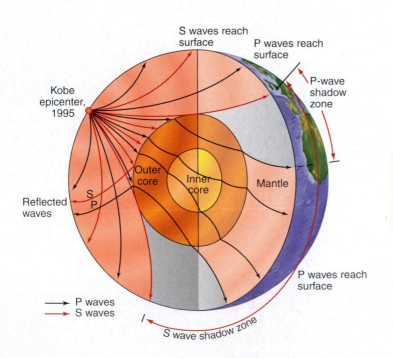

S waves reach surface

P waves reach surface

P-wave shadow zone

Kobe epicenter, 1995

Outer core

Inner core

Mantle

S
P

Reflected waves

P waves reach surface

→ P waves
→ S waves

S wave shadow zone

Think Critically

If you were on a ship in the ocean, would you be able to feel an earthquake that occurred below you, on the ocean floor?

Refraction, reflection, and absorption all play a role in Oldham's model of Earth's core. P waves are refracted dramatically, and some are reflected when they encounter the boundary between the mantle and the outer core (as shown in **Figure 5.11**). This refraction and reflection creates a ring-shaped **P-wave shadow zone** on the opposite side of Earth from the earthquake. S waves, on the other hand, are blocked completely by the outer core because shear waves cannot pass through liquid. This creates an even larger **S-wave shadow zone** and also provides firm evidence that Earth has a liquid core.

Seismic tomography The boundary between the liquid outer core and the mantle was the first seismic discontinuity to be identified and explained. Since then, as scientists have developed more sophisticated seismic equipment and more detailed observations, they have discovered many other boundaries and layers within Earth. Today seismologists use seismic waves to probe Earth's interior in much the same way that a doctor uses X-rays and CAT scans to probe the interior of a human body. In CAT (computerized axial tomography) scanning, a series of X-rays along successive planes are used to create a three-dimensional picture of the inside of the body. Similarly, **seismic tomography** allows seismologists to superim-

pose many two-dimensional seismic "snapshots" to create a three-dimensional image of the inside of Earth. These techniques have also helped us to understand more about how plate tectonics works. They also allow scientists to map the locations of seismic discontinuities, the distribution of hot and cold masses, and the distribution of dense and less dense materials inside Earth (**Figure 5.12**).

How Geologists Look into Earth's Interior: Other Methods

Earthquakes have provided a great deal of information about Earth's interior, but geologists can make use of many other tools and techniques to study the deepest parts of our planet. Some of these tools, such as the use of seismic information, involve indirect or remote observation of materials and processes deep within the planet. Others are more direct and give geologists access to actual samples from deep in the crust, and even from the mantle.

Beneath the surface • Figure 5.12

Geologists don't always have to wait for earthquakes to happen in order to study Earth's interior. For shallow, near-surface studies, they can use explosive charges (or large "thumper trucks") as the source for seismic waves and then map out the underground rock layers with seismic tomography. Here, two oil-prospecting geologists in Texas are walking through a virtual three-dimensional image of the underground rocks at a potential drilling site in Alaska.

Diamonds: Messengers from the deep
• Figure 5.13

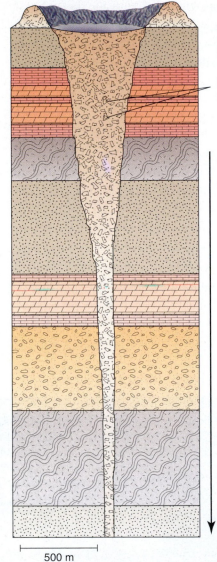

Magma vent is circular when viewed from above.

Xenoliths of mantle rock

Pipe extends 150–200 km down into mantle

a. Diamonds form only at extremely high pressures, found at depths of 100 to 300 kilometers below the surface. The different colors in this diamond—nicknamed the "Picasso" diamond—show various zones of growth. The diamond is approximately 1 millimeter across, and the colors are revealed by a special type of photography that highlights small variations in composition. Would a synthetic diamond, grown in the laboratory, record a similar complex growth history?

b. The beautiful uncut diamond below is the 253.7-carat Oppenheimer Diamond from the Smithsonian Institution; it was discovered in South Africa in 1964.

500 m

c. To reach the surface from their great depths, diamonds must be carried by an eruption of unusual ferocity. These eruptions leave behind a long, cone-shaped tube of solidified magma, called a **kimberlite pipe**. Although most people treasure diamonds for their beauty and luster, geologists treasure them also as "messengers"—samples from an otherwise inaccessible region of Earth's interior.

Direct observation: Drilling and xenoliths Perhaps the most obvious tool for the retrieval and study of samples from Earth's interior is drilling. To date, Earth's deepest mine (in South Africa) is 3.6 kilometers deep, and the deepest hole ever drilled (in the Kola Peninsula of Russia) reached a depth of just over 12 kilometers. Recall from Chapter 1 that Earth's crust varies from an average thickness of 8 kilometers for oceanic crust to an average of 45 kilometers for continental crust. Therefore, a 12-kilometer hole sounds just about right for sampling the top part of the mantle—or does it?

The problem with drilling is that areas where the crust is thin tend to have high heat flow. In other words, if you try to drill a hole through thin oceanic crust all the way to the mantle, you will quickly encounter temperatures that could destroy your drilling equipment. Another problem with oceanic crust is that it's deep underwater, which makes drilling especially difficult. The only place where the rocks are both accessible and cool enough for drilling to great depth is precisely where the crust is very thick—on the continents. The hole in the Kola Peninsula went through more than 12 kilometers of thick continental crust and never even came close to reaching the mantle. Thus, although drilling has yielded much interesting and useful information about the composition and properties of the crust, it hasn't even penetrated to the outermost seismic discontinuity of the mantle (the crust–mantle boundary).

If we can't obtain samples from deep within Earth by reaching in to retrieve them, perhaps we can wait for them to come to us. This does happen—in two different ways. Molten rock, or **magma**, is formed in the upper portions of Earth's mantle, in areas where the temperature is high enough. By studying magma that originates at depth and erupts to the surface, scientists learn about the temperature, pressure, and composition of the mantle in the region where the magma formed. Furthermore, as magma rises toward the surface, it often breaks off and carries with it fragments of the unmelted surrounding rock. We call these fragments **xenoliths**, from the Greek words *xenos* ("foreigner") and *lithos* ("stone"). A xenolith that reaches Earth's surface is a sample of the deep crust or mantle, accessible for direct scientific study (**Figure 5.13**).

Indirect observation: Magnetism and planetary characteristics
The characteristics of Earth as a planet, including its orbital characteristics and its magnetic field, provide another powerful indirect source of information to enhance our understanding of Earth's interior. Some of these techniques are also useful in the study of other planets in the solar system.

As we have pointed out in Chapters 3 and 4, Earth's magnetic field has provided important tools for dating rocks and reconstructing the past motion of lithospheric plates. It turns out that magnetism also gives us

information about Earth's deep interior. **Magnetism** is a force created either by permanent magnets (ferromagnets) or by moving electrical charges. We can try thinking of Earth as having a huge dipole bar magnet with north and south poles at its center, offset slightly from the geographic North and South poles. The problem is that solids, including bar magnets, lose their magnetism at temperatures above a critical transition temperature, called the **Curie point**, which is specific to each material. The Curie point for iron (the material of bar magnets) is about 770°C, but we know that the temperature deep inside Earth is *much* higher than this—at least 5000°C. This means that the "bar magnet" analogy for Earth's magnetic field must be incorrect, and a different explanation is needed.

Physicists have shown that the movement of an electrically conducting liquid inside a planet could generate a self-sustaining magnetic field, much like a rotating coil of wire in an electric motor (**Figure 5.14**). This is consistent with the observation from seismology that at least the outer part of Earth's core is liquid. However, molten rock is not a good enough electrical conductor to generate

Earth's magnetic field • Figure 5.14

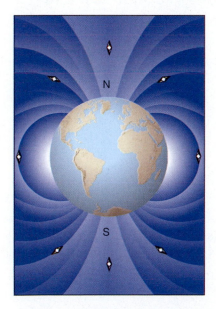

a. Earth is surrounded by a magnetic field, which causes a compass needle to point north. More precisely, the needle is aligned along the field lines that lead to the magnetic north and south poles, which are almost—but not exactly—aligned with Earth's geographic North and South poles.

b. This is a photograph of the aurora borealis, or northern lights, as seen from Fairbanks, Alaska. This phenomenon is caused by charged particles from the Sun entering Earth's atmosphere at high latitudes along magnetic field lines.

a magnetic field in this manner; for this and other reasons, geologists believe that the liquid outer core is made of molten iron and nickel rather than molten rock. This is consistent with evidence from meteorites, as discussed later in this chapter.

We can gain a certain amount of information about any planet's interior—including Earth's—from astronomical observations that reveal the overall characteristics of the planet. The first step is to determine the planet's *mass*. This can be deduced from the planet's gravitational influence on other planets and satellites. Second, we need to know the diameter of the planet. Knowing the dimensions of the planet and its shape (in the case of Earth, a very slightly flattened sphere), it is a simple matter to figure out its volume and average density (mass divided by volume).

What do these kinds of measurements reveal about Earth's interior? For one thing, we can determine whether material is distributed evenly throughout the planet. The rocks at Earth's surface are very light (low-density) compared to the planet as a whole. Surface rocks have an average density of about 2.8 g/cm^3, whereas Earth's overall density is 5.5 g/cm^3. (For comparison, water has a density of 1 g/cm^3 at 4°C.) For the planet as a whole to have such a high density, with such low-density rocks at the surface, there must be a concentration of denser material somewhere inside the planet. This evidence is consistent with a model where Earth's core is primarily a mixture of two metals, iron and nickel, with a density of about 10 g/cm^3.

A final way to study Earth's interior—last in our summary, but definitely not least in importance—is to analyze the building blocks that formed it. Planetary scientists have discovered that many (though not all) meteorites were formed at about the same time and in the same part of the solar system as Earth. Some of these meteorites are **primitive**—that is, they have remained unaffected by melting and other geologic processes since the beginning of the solar system. These meteorites give scientists an idea of the overall composition of the solar system and its constituent bodies (as discussed in Chapter 1). Other meteorites—the **irons**, **stony-irons**, and some kinds of **stony meteorites**—show signs of melting and differentiation and may be more representative of Earth's core and mantle. It is highly significant that a core with the composition of a typical iron meteorite (mostly iron and nickel) would bring Earth's overall density up to the observed value of 5.5 g/cm^3.

CONCEPT CHECK STOP

1. **Why** do seismic waves undergo refraction as they pass through Earth?

2. **Describe** three pieces of evidence that indicate that Earth has a molten, iron-rich outer core.

3. **Why** can't geologists drill a hole down to the mantle?

4. **What** does Earth's magnetic field tell us about Earth's interior?

A Multilayered Planet

LEARNING OBJECTIVES

1. **Define** crust, mantle, and core.
2. **Summarize** the composition of the crust, mantle, and core.
3. **Describe** the layering within the mantle and core.

By piecing together information from various sources, geologists have arrived at a very detailed understanding of Earth's interior. Let's take a brief tour, starting at the top and working down to the innermost layers. As we proceed, keep in mind that some boundaries inside Earth mark the transition between layers with differing composition, whereas others separate layers with the same composition but different physical properties.

The Crust

Earth's outermost compositional layer is the **crust**. The thickness of the crust varies greatly, from an average of 8 kilometers for oceanic crust to an average of 45 kilometers for continental crust (**Figure 5.15**). Even at

crust The outermost compositional layer of the solid Earth; part of the lithosphere.

its thickest spots, the crust is extremely thin compared with Earth as a whole. It's like a thin, brittle eggshell, or about the same relative thickness as the glass of a lightbulb.

Like its thickness, the composition of the crust varies from place to place. About 95% of the crust is igneous rock or metamorphic rock derived from igneous rock. In general, the rocks of the crust are lighter (less dense) than the material of Earth's interior because the crust is composed of material that "floated" to the top during planetary differentiation. The very outermost layer of the crust, both on land and on the ocean floor, is quite different from the crust as a whole. This surface layer—the ground surface we see around us every day—consists of about 75% sediment and sedimentary rock,

formed by the constant action of erosion, weathering, and deposition.

Moving downward through the crust, we encounter the boundary that separates the crust from the mantle. This is a major seismic discontinuity, the next to be discovered after the discovery of the seismic discontinuity that marks the outer liquid core. It was named the **Mohorovičić discontinuity** after the seismologist who discovered it in 1909, but it is usually called the **Moho** for short. Mantle rocks, being denser and compositionally different from crustal rocks, transmit P waves much more quickly. The Moho is thus an example of a boundary between two layers of rock (the crust and the upper mantle) that have different compositions and densities but similar physical characteristics (rigidity).

Inside view of Earth • Figure 5.15

This diagram shows Earth's internal structure and summarizes what we have learned from seismic studies and other direct and indirect observations. The upper part of the cutaway shows compositional layers, and the lower part shows layers with different rock properties. Note that the boundaries between zones that differ in strength, such as the rigid lithosphere and the more plastic asthenosphere, do not always coincide with compositional boundaries.

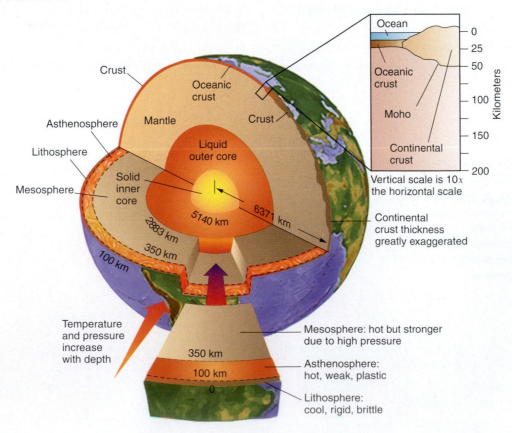

The Mantle

The **mantle** extends from the Moho to the core (see Figure 5.15). About 80% of Earth's volume is contained in the mantle. Geologists believe, on the strength of evidence from xenoliths, meteorites, and seismic analyses, that the mantle consists mainly of iron and magnesium silicate minerals. The upper part of the mantle has a composition similar to that of **peridotite**, an igneous rock not typically found in the crust, which consists mainly of the minerals olivine and pyroxene.

> **mantle** The middle compositional layer of Earth, between the core and the crust.

Seismic studies have revealed boundaries within the mantle, but not all of them are compositional boundaries. Extending from about 100 to 350 kilometers below the surface is a layer called the **asthenosphere**, from the Greek words meaning "weak sphere." In this zone, some of the rocks are very near the temperatures at which rock melting begins, so they have the consistency of butter or warm tar. The composition of the asthenosphere appears to be the same as that of the mantle just above and below it. Thus, the asthenosphere is a layer whose distinctiveness is based on its physical properties—reduced rigidity—rather than its composition.

> **asthenosphere** A layer of weak, ductile rock in the mantle that is close to melting but not actually molten.

The outermost 100 kilometers of Earth, which includes the crust and the part of the uppermost mantle just above the asthenosphere, is called the **lithosphere** (see Figure 5.15). The rocks of the lithosphere are cooler, more rigid, and much stronger than the rocks of the asthenosphere. In plate tectonic theory, it is the entire lithosphere—not just the crust—that forms the plates. These plates can move around because they rest on the underlying weaker rocks of the asthenosphere, which slowly deform and flow like an extremely thick, viscous liquid. The asthenosphere deforms by ductile flow, while the lithosphere deforms mainly by fracturing (faulting), as illustrated by the San Andreas fault at Point Reyes, discussed in *Amazing Places*.

> **lithosphere** Earth's rocky, outermost layer, comprising the crust and the uppermost part of the mantle.

The rest of the mantle, from the bottom of the asthenosphere (at about 350 kilometers depth) down to the core–mantle boundary, is the **mesosphere**. Although temperatures in the mesosphere are very high, the rocks are a bit stronger than in the asthenosphere because they are highly compressed. Additional seismic discontinuities exist within the mesosphere, with transitions at about 400 kilometers and at about 670 kilometers below the surface (**Figure 5.16**). These discontinuities are not well understood; they do not seem to be compositional boundaries but rather to result from changes in physical properties. For example, when the mineral olivine is squeezed at a pressure equal to that found at a depth of 400 kilometers, the atoms rearrange themselves into a more compact

Seismic discontinuities in the mantle • Figure 5.16

Earth's mantle is not uniform but has several seismic boundaries within it. We know the boundaries exist because P waves and S waves slow down or speed up abruptly and are refracted or reflected at these boundaries. However, the exact nature of the boundaries is not completely understood.

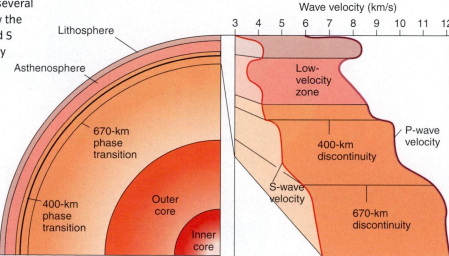

AMAZING PLACES

Point Reyes, California

Point Reyes peninsula, just north of San Francisco, is a spectacular place to view the history of the San Andreas fault, one of the world's most famous faults. The fence dislocated by the 1906 earthquake in *Case Study: Proving the Elastic Rebound Theory* is part of a walking tour at Point Reyes State Park.

Perhaps the most visible nonhuman evidence of the fault in this location is Tomales Bay, a narrow, straight inlet 20 kilometers long and only 1.4 kilometers wide on the northeast side of the peninsula. The San Andreas fault runs directly down the center of the bay, which is seen from a satellite in **Figure a** and from an airplane in **Figure b**. Over the millennia, earthquakes have ground the rocks on both sides of the fault together and weakened them. Erosion of these weakened rocks formed a linear valley, which has been filled in by water from the Pacific Ocean.

The peninsula itself has been on an amazing journey for the past 100 million years, as it has slipped northward along the San Andreas fault. The ridge on the west side of Tomales Bay (seen on the left background, in Figure b) contains outcrops of granite and diorite that must have originated in a continental mountain range—possibly the Tehachapi Mountains of southern California 300 kilometers to the south. As it moved northward, the peninsula scraped up souvenirs of the places it passed. The scenic Point Reyes lighthouse (**Figure c**) is built on a rock formation called the Point Reyes Conglomerate. An identical rock formation is found at Point Lobos, 180 kilometers to the south. It is very likely that Point Reyes and Point Lobos were adjacent to each other 60 million years ago, when the conglomerate layer formed.

Where will Point Reyes go next?

Tomales Bay

Global Locator

Point Reyes, California

NATIONAL GEOGRAPHIC

a.

Point Reyes lighthouse

b. Aerial view of Tomales Bay, looking southwesterly towards Pt. Reyes.

c.

structure, or polymorph, of olivine (see Chapter 2). Perhaps this change to a more compact form is responsible for causing the 400-kilometer seismic discontinuity.

It is important to remember that the mantle is mostly solid rock, with the exception of small pockets of melt in the asthenosphere. We know that the mantle must be solid because both P waves and S waves can travel through it. Nevertheless, pressures and temperatures deep within Earth are so high that even solid rock can flow in very, very slow convection currents, as described in Chapter 4. Seismologists have recently detected hot regions in the mantle that may coincide with the rising limbs of convection cells that help drive plate motion.

The Core

At a depth of 2883 kilometers, there is a huge decrease in the velocity of P waves, and the velocity of S waves drops to zero. This is the core–mantle boundary (see Figure 5.15), which represents a change in both the composition and

> **core** Earth's innermost compositional layer, where the magnetic field is generated and much geothermal energy resides.

the physical properties of the materials. The **core**, the innermost of Earth's compositional layers, is the densest part of Earth. It consists of material that "sank" to the center during the process of planetary differentiation. As discussed earlier, geologists believe that Earth's core is composed primarily of iron–nickel metal.

The S-wave shadow zone (and other evidence) tells us that the **outer core**, from 2883 kilometers to 5140 kilometers depth, must be liquid. For three decades geologists thought the core was a homogeneous liquid, but in 1936 the Danish seismologist Inge Lehmann showed that the core, too, has layers. She detected faint seismic waves within the P-wave shadow zone that had reflected off the **inner core**. The pressure in the inner core is so great that iron must be solid there, in spite of the very high temperature. We know this from high-pressure experiments on iron. The main difference between the inner core and outer core is thus a difference in the physical state rather than the composition. As heat escapes from the core and works its way to the surface, the core is gradually crystallizing. Thus the solid inner core must be growing larger, although very slowly.

It should be clear by now that what happens deep in the interior of Earth profoundly affects the surface. The release of heat from the interior is an important driving force for plate tectonics, which in turn is the major uplifting force in shaping Earth's varied landscapes and topographies.

CONCEPT CHECK STOP

1. **What** is the nature of the boundary between the crust and mantle?

2. **What** are the major layers of the mantle, and what are their distinguishing characteristics?

3. **How** and why does the inner core differ from the outer core?

✓ THE PLANNER

Summary

1 Earthquakes and Earthquake Hazards 120

- **Seismology** relates earthquakes to the processes of plate tectonics. Although the motion of tectonic plates is very gradual, friction causes the rocks in the crust to jam together for long periods and then to break suddenly and lurch forward, causing an earthquake to occur. Earthquakes can cause large vertical or horizontal displacements of the ground, but much of the damage they cause results from the violent shaking that accompanies the displacement.

- The shaking motion experienced during an earthquake can be explained by the **elastic rebound theory**, which says that the energy stored in bent and deformed rocks is

released as **seismic waves**. After an earthquake, the rocks return to their undeformed state.

- In many cases the destructiveness of earthquakes is magnified by secondary hazards, such as fires, landslides, soil liquefaction (see photo), and tsunamis. Proper building design and earthquake preparedness can greatly reduce the loss of life from earthquakes and secondary hazards.

- Short-term forecasting of earthquakes is still very unreliable. Scientists have concentrated their efforts on finding precursor phenomena, such as foreshocks, but with limited success. However, long-term forecasting can provide a good idea of which regions are at risk. One of the main tools of long-term forecasting is **paleoseismology**, which reveals when past earthquakes occurred in a given region, as well as the periodicity and magnitudes of past earthquakes.

Secondary hazards: soil liquefaction • Figure 5.3d

2 The Science of Seismology 129

- **Seismographs** produce recordings of seismic waves that are called **seismograms**. In a basic seismograph (see figure), a pen is attached to a heavy suspended mass. Seismic waves cause the paper to shake while the pen stays still and traces a wavy line on the vibrating paper.

- Earthquakes produce three main types of seismic waves: **compressional waves**, or **P waves** (primary waves); **shear waves**, or **S waves** (secondary waves); and a variety of **surface waves**. Compressional and shear waves are called **body waves** because they travel through Earth's interior.

- Compressional waves travel faster than shear waves and hence arrive at seismographs first. The difference in arrival times between the P and S waves allows seismologists to compute the distance, but not the direction, to the **focus** of an earthquake. To determine the precise location of the

epicenter, seismologists need measurements from three separate seismic stations. They can then determine the location by triangulation.

- The **Richter** and **moment magnitude scales** are measures of earthquake intensity that can be determined regardless of the distance to the earthquake or the amount of damage done. Both are logarithmic scales, in which each unit of magnitude corresponds roughly to a 10-fold increase in the amplitudes of seismic waves, but a 32-fold increase in the amount of energy released by the earthquake. The Modified Mercalli Intensity scale is a descriptive scale based on the extent of earthquake damage. On the Mercalli scale, the intensity is highest near the epicenter.

Seismograph • Figure 5.7

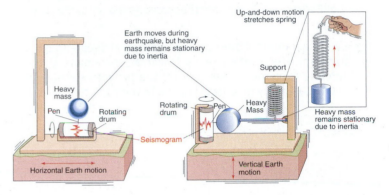

3 Studying Earth's Interior 135

- After an earthquake, seismic waves travel downward into Earth's interior as well as upward and along the surface (see figure). Seismic waves travel at different velocity through different materials, and they change velocity and direction when they pass from one material to another material with different physical and/or compositional properties. This understanding of seismic waves has allowed seismologists to identify many **seismic discontinuities** in Earth's interior.

Seismic waves in Earth's interior • Figure 5.11

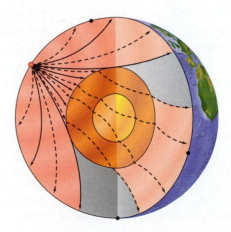

- Seismic discontinuities can result from either a change in composition or a change in physical properties of the material. They may **refract**, **reflect**, or even block seismic waves. P waves are strongly refracted, or bent, when they pass from the **mantle** to the **core**. S waves, on the other hand, are completely blocked by the core. This discovery provides evidence that Earth has a liquid outer core.

- Geologists use a variety of remote or indirect techniques to understand Earth's interior, in addition to the study of naturally occurring earthquakes. They can use seismic tomography, analogous to medical tomography, to detect seismic discontinuities underground.

- Other sources of information about Earth's interior include drilling, the magnetic field, the mass and density of Earth, and *meteorite* studies. Evidence points to the likelihood that Earth has a dense core that consists mainly of iron and nickel.

- So far, geologists have been unable to drill deep enough to sample the mantle directly. However, some mineral samples from the mantle come to the surface as *xenoliths*, carried along by magma that rises to the surface.

- The crust consists of solid rock that is mostly igneous, with a thin veneer of sediment and sedimentary rock at the surface. It varies in average thickness from 8 kilometers (for oceanic crust) to 45 kilometers (for continental crust).

- The mantle contains several physical boundaries (see figure). The most important are the boundaries between the **lithosphere**, the **asthenosphere**, and the **mesosphere**. The lithosphere is about 100 kilometers thick; the asthenosphere begins about 100 kilometers beneath the surface and ends at a depth of about 350 kilometers; and the mesosphere extends from a depth of 350 to 2883 kilometers.

- The core has two layers, a liquid outer core and an inner core that is solid—despite its high temperature—because of the extremely high pressure. The boundary between them is physical, not compositional. The outer core begins at a depth of 2883 kilometers and extends to a depth of 5140 kilometers. The inner core, which was first discovered by its reflection of P waves, is almost certainly growing very gradually as Earth cools down from its formation.

Seismic discontinuities in the mantle • Figure 5.16

4 **A Multilayered Planet   140**

- Earth's three main compositional layers are the **crust**, the **mantle**, and the **core**. Each of these layers creates a seismic discontinuity, and this is how geologists can determine their thickness.

Key Terms

- asthenosphere 142
- body wave 130
- compressional wave 130
- core 144
- crust 140
- elastic rebound theory 121
- epicenter 130
- focus 130

- lithosphere 142
- mantle 142
- moment magnitude scale 132
- paleoseismology 128
- reflection 135
- refraction 135
- Richter magnitude scale 132
- seismic discontinuity 135

- seismic wave 121
- seismogram 129
- seismograph 129
- seismology 120
- shear wave 130
- surface wave 130

Critical and Creative Thinking Questions

1. Use the elastic rebound theory to describe what happens to rocks at the focus just before, during, and after an earthquake.

2. Why is short-term prediction of earthquakes so much less successful than long-term prediction? Why do you think seismologists are extremely cautious about making predictions? Do you think it will ever be possible to predict earthquakes accurately and issue effective early warnings? Research your answer.

3. If you were asked to determine the exact shape and size of Earth, how would you go about it? What would you do differently if you were not allowed to use Space-age technology such as satellite photographs and orbital data?

4. Some of the boundaries inside Earth represent transitions between layers with differing compositions, whereas others represent transitions between layers with different physical states. Find out more about these different layers and draw a detailed diagram to show the layering.

5. Which of the techniques used to study Earth's interior could also be used to study other planets? Which ones cannot, and why? Scientists know more about the surface of the sun than about the interior of our own planet; why do you think this is so?

What is happening in this picture?

Scotland's Loch Ness has formed along a former fault, now inactive, where the rocks were ground together and weakened by earthquakes for tens of millions of years.

Think Critically

1. Why do you think the lake is so straight and narrow?
2. Could it be similar to Tomales Bay, discussed in *Amazing Places* on page 143?

Self-Test

(Check your answers in Appendix D.)

1. The largest recorded earthquakes have occurred at
   _____.
   a. divergent boundaries
   b. transform fault boundaries
   c. subduction zone boundaries
   d. continental collision boundaries

2. According to the elastic rebound model, earthquakes are
   caused by the _____
   a. slow release of gases from the asthenosphere
   b. sudden release of elastic energy stored in rocks
   c. sudden movement of otherwise stable tectonic plates
   d. rapid release of gases from the asthenosphere

3. _____ and the resulting collapse of buildings, bridges,
   and other structures are usually the most significant primary
   hazards to cause damage during an earthquake.
   a. Fire   c. Ground liquefaction
   b. Tsunami   d. Ground shaking

4. _____ can provide a good idea of which regions are at risk
   for severe earthquakes.
   a. Short-term forecasting
   b. Long-term forecasting
   c. Unusual animal behavior studies
   d. Studies of groundwater levels

5. Which of the following are true of body waves?
   a. They move through Earth's interior.
   b. They cannot penetrate Earth's liquid outer core.
   c. They move along Earth's surface, causing great
      destruction.
   d. Both b and c are correct.

6. Using seismograms from three different seismic recording stations A, B, and C, you determine the epicenter of an
   earthquake. Stations A and B both had an S–P interval of 3
   seconds and C had an S–P interval of 11 seconds. Which of
   the following statements most accurately depicts the location of the epicenter?
   a. The epicenter is closest to station A and equally far from
      B and C.
   b. The epicenter is closest to station B and equally far from
      A and C.
   c. The epicenter is closest to station C and equally far from
      A and B.
   d. The epicenter is equally close to A and B and farthest
      from station C.

7. This diagram shows a seismogram of a hypothetical earthquake. On the seismogram, label the following:
   S–P interval
   first arrival of P wave
   first arrival of S wave
   background noise
   first arrival of surface waves

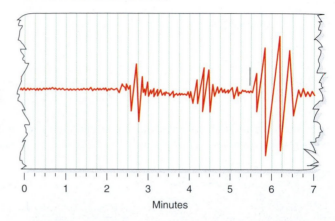

Minutes

8. Diagrams a and b depict two different types of seismic
   waves. Which of the following statements can be made
   about these two seismic waves?

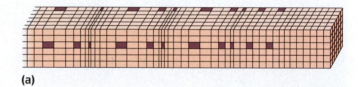

(a)

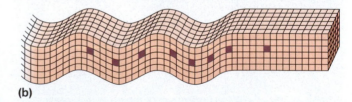

(b)

   a. The wave depicted in a is a P wave and has a greater
      velocity through Earth's crust than other types of seismic
      waves.
   b. The wave depicted in a is an S wave and has a greater
      velocity through Earth's crust than other types of seismic
      waves.
   c. The wave depicted in b is a P wave and has a greater
      velocity through Earth's crust than other types of seismic
      waves.
   d. The wave depicted in b is an S wave and has a greater
      velocity through Earth's crust than other types of seismic
      waves.

9. For the earthquake mentioned above, which seismic recording station would have recorded the P wave first?

    a. station A

    b. station B

    c. station C

    d. Both stations A and B would have recorded the P wave before station C.

10. A magnitude 8 earthquake releases approximately _____ times more energy than a magnitude 7 event.

    a. 2   b. 10   c. 20   d. 21.5   e. 32

11. How does moment magnitude differ from Richter magnitude?

    a. Richter magnitude assumes that earthquakes are generated at a point source, whereas moment magnitude takes into account that earthquakes can be generated over a large area of rupture.

    b. Moment magnitude assumes that earthquakes are generated at a point source, whereas Richter magnitude takes into account that earthquakes can be generated over a large area of rupture.

    c. Moment magnitude uses Roman numerals to designate strength of an earthquake.

    d. Richter magnitude uses Roman numerals to designate strength of an earthquake.

12. Which of the following is true of seismic waves reaching a discontinuity inside Earth's interior?

    a. They can be refracted, or bent, as they pass from the first material into the second.

    b. They can be reflected, which means that all or part of the wave energy bounces back.

    c. They can be absorbed, which means that all or part of the wave energy is blocked by the second material.

    d. All of the above answers are correct.

13. Earth's mantle is composed of _____, which surrounds a(n) _____ core.

    a. rock that contains iron- and magnesium–silicate minerals; iron–nickel metallic

    b. iron–nickel alloy; rocky iron- and magnesium–silicate

    c. rock that contains iron- and magnesium–silicate minerals; molten iron–nickel

    d. molten rock; solid iron–nickel metallic

14. On this diagram, label Earth's internal structure using the following terms:

    | mantle | lithosphere | oceanic crust |
    | asthenosphere | outer core | Moho |
    | inner core | continental crust | mesosphere |

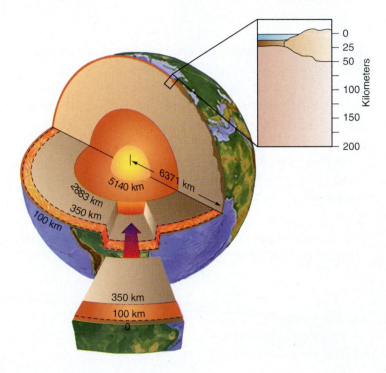

15. The asthenosphere is a layer whose distinctiveness from the rest of the mantle is based on its _____.

    a. differences in composition

    b. reduced rigidity

    c. increased rigidity

    d. relatively low temperature

THE PLANNER ✓

Review your Chapter Planner on the chapter opener and check off your completed work.

Volcanoes and Igneous Rocks

Iceland is home to many active volcanoes, and some are covered by ice caps. The consequences of an eruption beneath a glacier can be dangerous. Eyjafjallajökull, a glacier-covered volcano, erupted in 2010. The glacier—really an ice cap—covers 100 square kilometers, and the volcano is 1651 meters high.

Eyjafjallajökull began to erupt on April 14, 2010; part of the ice cap melted, causing a *jökulhlaup*—a flood of glacial melt water. Some of the water flowed into the volcanic vent, causing the eruption to become so explosive that plumes of gas carried volcanic ash high into the atmosphere. Upper atmosphere winds spread the ash cloud across northern Europe and the Atlantic Ocean, shutting down air travel because ash in a jet engine can cause a crash. Hundreds of thousands of passengers were stranded, some for weeks. By May 23 the eruption had settled down and the worst peace-time travel disruption from natural causes came to an end.

The scenario was repeated in May 2011, with the eruption of Grimsvötn (see inset photo), another subglacial volcano. Air travel was not as severely disrupted because lessons from Eyjafjallajökull allowed air traffic controllers to refine their decision making process about volcanic eruptions and air travel.

Global Locator

Iceland

NATIONAL GEOGRAPHIC

CHAPTER PLANNER ✓

- ☐ Study the picture and read the opening story.
- ☐ Scan the Learning Objectives in each section:
  p. 152 ☐   p. 166 ☐   p. 174 ☐   p. 179 ☐
- ☐ Read the text and study all visuals.
  Answer any questions.

Analyze key features

- ☐ Process Diagram, p. 165 ☐   p. 169 ☐
- ☐ Geology InSight, p. 154 ☐   p. 172 ☐
- ☐ Case Study, p. 159 ☐
- ☐ Amazing Places, p. 164 ☐
- ☐ What a Geologist Sees, p. 175 ☐
- ☐ Stop: Answer the Concept Checks before you go on:
  p. 166 ☐   p. 171 ☐   p. 179 ☐   p. 182 ☐

End of chapter

- ☐ Review the Summary and Key Terms.
- ☐ Answer the Critical and Creative Thinking Questions.
- ☐ Answer What is happening in this picture?
- ☐ Complete the Self-Test and check your answers.

Volcanoes and Volcanic Hazards

LEARNING OBJECTIVES

1. **Identify** several different categories of volcanic eruptions.

2. **Explain** why stratovolcanoes tend to erupt explosively, whereas shield volcanoes tend to erupt nonexplosively.

3. **Describe** how volcanic features such as calderas, geysers, and fumaroles arise.

4. **Identify** the hazards of volcanoes and the beneficial effects they can have.

5. **Describe** how scientists monitor volcanic activity.

F or many people, the thought of a **volcano** conjures up visions of fountains of **lava** spurting up into the air and pouring out over the landscape (**Figure 6.1a**). Although it's true that most volcanoes produce at least some liquid lava, many other types of materials can emerge from volcanoes as well, such as fragments of rock, and glassy volcanic ash (**Figure 6.1b, c,** and **d**). A fragment of rock ejected during a volcanic eruption is called a **pyroclast** (from the Greek words meaning "fire broken"). Collectively, all the ejecta from a volcano are known as **tephra**, and they can range from car-sized rocks to ultrafine **volcanic ash** whose individual particles can be seen only under a microscope. To see what's going on at an active volcano, see *Where Geologists Click*.

Gases are also released during eruptions, and stored-up gases can cause a volcano to explode, covering the surrounding area with a catastrophic shower or flow of tephra. Or gases can seep out silently and poison a whole town overnight, as happened in Cameroon, a country in central Africa, in 1984. The different kinds of eruptions and the

> **volcano** A vent through which lava, solid rock debris, volcanic ash, and gases erupt from Earth's crust to its surface.
>
> **lava** Molten rock that reaches Earth's surface.

Different eruption styles, different hazards • Figure 6.1

a. Hawaii's volcanic eruptions, such as this eruption of Kilauea that began in 1983, pose minimal danger to humans, although they can cause extensive property damage if lava flows into populated areas.

b. Violent release of gas ejects volcanic bombs, fist-sized and larger, from Mt. Etna in Sicily.

Volcanic bombs

c. Intermediate-sized tephra called **lapilli** were erupted by Kileaua volcano and cover the Kau Desert in Hawaii.

volcanoes they build have much to do with the physical properties of the **magma** that lies at their source.

magma Molten rock, which may include gas and fragments of rock, volcanic glass, and ash.

We will begin our discussion by taking a look at some of the different kinds of volcanoes.

Eruptions, Landforms, and Materials

All active volcanoes are dangerous on some level, but they have a range of eruption styles. We can distinguish several types of eruptions on the basis of their explosiveness and the materials they produce. These differences are also reflected in the kind of terrain they build. **Figure 6.2** (on the next page) shows several types of volcanoes, arranged in order from most explosive to least explosive. The eruptive style of volcanoes can change from year to year, month to month, or even hour to hour; there is no such thing as a completely safe volcano or a completely predictable one.

Volcanoes and eruptions **Strombolian** eruptions (**Figure 6.2d**) are explosive. The volcano may eject showers of lava or pyroclasts hundreds of feet into the air.

d. Volcanic ash, the smallest tephra, blankets a farm in Oregon after the violent eruption of Mt. St. Helens in 1980. Though we call it ash, it is different from what you find in your fireplace because it consists of microscopic pieces of volcanic glass.

Where Geologists CLICK

The Hawaiian Volcano Observatory

http://hvo.wr.usgs.gov/kilauea/update/images.html

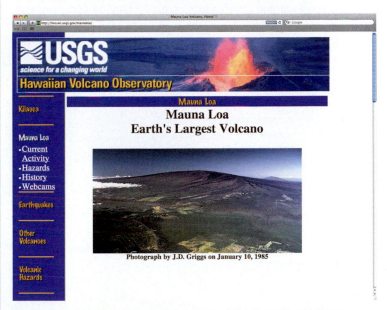

This interesting site records the events at the Hawaiian Volcano Observatory, on top of one of the world's most active volcanoes. As magma rises it causes swarms of small earthquakes; geologists go to this site to see the seismic record. Something is almost always happening here; webcams have been installed around the area of activity, so this site is the place to go to see the action.

This type of eruption creates cones of loose volcanic rock called **spatter cones**, **tephra cones**, or **cinder cones** (**Figure 6.2e**). Loose pyroclasts are often welded together during the eruption or cemented together afterward, forming pyroclastic rock. The rock is called **agglomerate** when the tephra particles are large and **tuff** when the particles are small. Spatter cones and tephra cones are made of exactly this kind of rock.

Vulcanian eruptions (**Figure 6.2b**) are much more explosive than Strombolian eruptions. They propel billowing clouds of ash to a height of 10 km or so. They also produce **pyroclastic flows** of hot cinders and ash that sweep down the mountainside like an avalanche. These flows travel much faster than flowing lava, and they are the most dangerous consequence of a volcanic eruption.

pyroclastic flow Hot volcanic fragments (tephra) that flow very rapidly, buoyed by heat and volcanic gases.

a. The classic volcano profile is that of a stratovolcano, which builds up over time from pyroclastic flows released in explosive eruptions and from less frequent lava flows. This creates steep-sided volcanoes like Mt. St. Helens.

1 Magma is very viscous, preventing the escape of gas bubbles. The trapped gas and upward movement of magma create increasing pressure inside the volcano. Thick deposits of pyroclastic material, like these, are often a sign of a past violent explosion—and a sign of possible future eruptions.

2 Gas continues to build and magma rises until the pressure causes an explosion. Small bits of lava and rock (called tephra) are ejected in all directions.

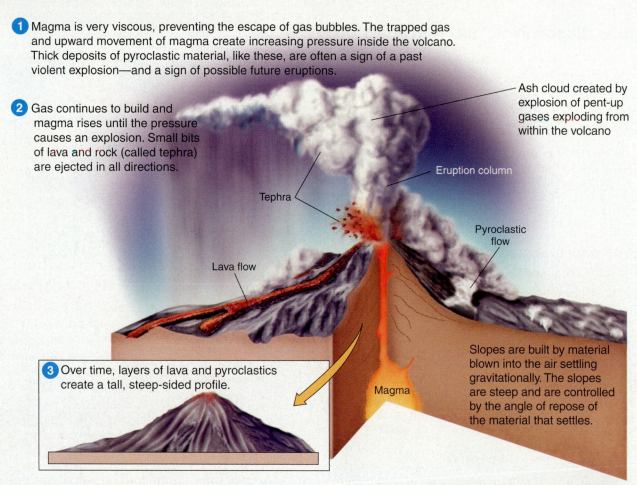

Ash cloud created by explosion of pent-up gases exploding from within the volcano

Eruption column

Tephra

Pyroclastic flow

Lava flow

3 Over time, layers of lava and pyroclastics create a tall, steep-sided profile.

Magma

Slopes are built by material blown into the air settling gravitationally. The slopes are steep and are controlled by the angle of repose of the material that settles.

b. The 1993 Vulcanian eruption of Mt. Mayon, a stratovolcano in the Philippines included pyroclastic flows.

c. The Plinian eruption of stratovolcano Mt. St. Helens in May 1980 produced an ash column and released destructive pyroclastic flows and debris avalanches down the steeply sloping sides of the volcano.

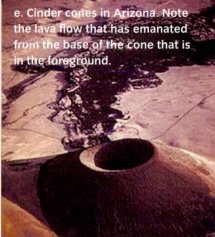

d. In 2002, Mt. Etna in Sicily experienced a Strombolian eruption.

e. Cinder cones in Arizona. Note the lava flow that has emanated from the base of the cone that is in the foreground.

f. Mauna Kea and Mauna Loa are shield volcanoes formed by Hawaiian eruptions.

g. Fissure eruptions, like this one in Hawaii, issue from long cracks in the ground.

h. These flood basalts are on the Snake River Plains, Idaho.

Mauna Loa
Mauna Kea

1 Magma is very hot when erupted and has low viscosity, so it flows easily. The slope of the volcano is low.

2 As magma flows, it cools, becomes more viscous, and flows slowly. The slope of the volcano may steepen at the edges, creating a gently curved profile.

i. Shield volcanoes have a unique profile, due to their different kind of lava and eruption style. They have very shallow, broad slopes that in some cases are reminiscent of a warrior's shield laid horizontally on the ground.

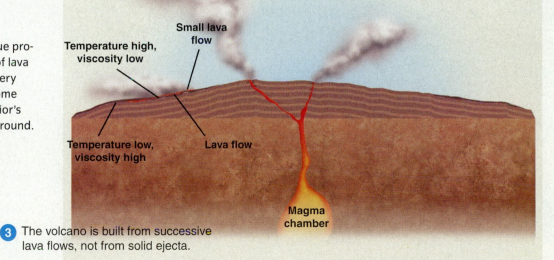

Small lava flow

Temperature high, viscosity low

Temperature low, viscosity high

Lava flow

Magma chamber

3 The volcano is built from successive lava flows, not from solid ejecta.

Volcanoes and Volcanic Hazards **155**

stratovolcano A volcano composed of solidified lava flows interlayered with pyroclastic material. Such volcanoes usually have steep sides that curve upward.

The most violent and famous eruptions in history are **Plinian**, named after Pliny the Elder, a Roman scholar who died during the eruption of Mt. Vesuvius in 79 CE (**Figure 6.2c**). They produce ash columns, driven by violent streams of magmatic gas, that reach into the stratosphere 20 km or more and create pyroclastic flows. Both Vulcanian and Plinian eruptions tend to build steep-sided volcanoes, called **stratovolcanoes** (**Figure 6.2a**). Stratovolcanoes have a somewhat complicated structure, with alternating layers of pyroclastic material and solidified lava flows. Their height stems from the layers of lava, which act as cement that holds the pyroclastic layers together.

What causes the diversity of eruption types? The answer lies mostly in the kind of magma that provides the source for the volcano. Two factors are important: the **viscosity** of the magma and the amount of gas dissolved in it. If gas is present in the magma, it must escape somehow. If magma has low viscosity (i.e., if it is runny), the dissolved gas can escape relatively easily. The lava may bubble and fountain dramatically, especially at the beginning of an eruption, but the volcano will not explode. However, if the magma is relatively viscous (i.e., thick), it is harder for gas bubbles to form and escape. When the gas finally does escape, it usu-

viscosity The degree to which a substance resists flow; a less viscous liquid is runny, whereas a more viscous liquid is thick.

ally vents explosively. Viscous magmas of the kind that build stratovolcanoes contain 60 to 70% SiO_2 by weight.

Hawaiian eruptions consist of very runny lava that flows easily from a volcanic vent and contains about 50% SiO_2 by weight. These flows gradually build up to form broad, flat volcanoes with very gently sloping sides, called **shield volcanoes** (**Figure 6.2i**). Sometimes growing to enormous size, they resemble a warrior's shield lying flat. Mauna Kea and Mauna Loa (**Figure 6.2f**), on the Big Island of Hawaii, rise more than 10 km from their bases (6 km below sea level) to their peaks, 4 km above sea level. That makes them the tallest mountains on Earth, measured from base to peak.

shield volcano A broad, flat volcano with gently sloping sides, built of successive lava flows.

Sometimes lava rises to the surface through long fissures rather than central craters (**Figure 6.2g**). These fissures can produce vast, flat lava plains called flood basalts or basalt plateaus (**Figure 6.2h**). Shield volcanoes, like Mauna Loa, often display some fissure activity as well.

Other volcanic features Near the summit of most volcanoes is a **crater**, a funnel-shaped depression from which gas, tephra, and lava are ejected. Some volcanoes have a much larger depression known as a caldera, a roughly circular, steep-walled basin that may be several kilometers in diameter. Calderas form when the chamber of magma underlying a volcano partially empties due to eruption and the unsupported roof of the chamber collapses under its own weight. Crater Lake in Oregon (**Figure 6.3**) occupies a caldera 8 km in diameter that formed after an immense

Crater Lake • Figure 6.3

Beautiful Crater Lake, Oregon, the deepest lake in the United States, is all that remains of a once-lofty stratovolcano that geologists have named Mt. Mazama. Wizard Island, a small tephra cone in the middle of the lake, formed by resurgent activity after the collapse that created the caldera.

A resurgent dome • Figure 6.4

A resurgent dome of sticky lava forms a small peak in the crater of Mt. St. Helens, Washington, in May 1982. The plume rising above the dome is steam.

eruption about 6600 years ago. Tephra deposits from that eruption can still be seen in Crater Lake National Park and over a vast area of the northwestern United States and southwestern Canada.

Volcanoes do not necessarily become inactive after a major eruption. If magma begins to enter the chamber again, it may lift the floor of the caldera or crater and form a resurgent dome. The caldera of Mt. St. Helens contains a dome that has been growing since the eruption of 1980 (**Figure 6.4**).

When volcanism finally ceases, the magma chamber still contains hot (though not necessarily molten) rock for hundreds of thousands of years. When groundwater comes into contact with this hot rock, it heats up and may create a **thermal spring**. Many such springs have been turned into famous health spas. Some thermal springs have a natural system of plumbing that allows intermittent eruptions of water and steam. These are called **geysers**, a name that comes from the Icelandic word *geysir*, meaning "to gush" (**Figure 6.5**). Finally, some volcanic vents emit only gas—usually water vapor that's sometimes mixed with foul-smelling sulfur compounds. These features are known as **fumaroles**.

The Great Geysir • Figure 6.5

Stokkur geyser, Haukadalur, Iceland, is close to the now dormant Great Geysir from which all geysers take their name. Haukadalur is a literal hotbed of geothermal activity.

Lava flow • Figure 6.6

This house in Kalapana, Hawaii, is about to succumb to the slow but unstoppable advance of a lava flow (June 1989). The grass of the lawn burns on contact with the molten rock.

Volcanic Hazards

Like other natural hazards, such as earthquakes, a volcanic eruption has **primary effects**, which are directly caused by the eruption itself; **secondary effects**, which are indirectly triggered by the eruption; and **tertiary effects**, which are long-lasting or even permanent changes brought about by the eruption. Many effects of volcanism are harmful, but some, such as volcanic ash supplying nutrients to the soil, are beneficial.

Primary effects Most volcanoes produce at least some lava flows. Because people are usually able to outrun them, lava flows typically cause more property damage than injuries. In Hawaii, where Kilauea has erupted almost continuously for more than two decades, homes, cars, roads, and forests have been buried by lava, but not a single life has been lost (**Figure 6.6**). It is sometimes possible to control a lava flow, at least partially, with retaining walls or a water spray, but otherwise nothing can be done to stop an eruption from occurring.

The greatest threats to human life during volcanic eruptions do not come from lava but from pyroclastic flows and volcanic gases. Unlike slowly moving lava, pyroclastic flows move extremely fast and can easily outrace a running (or even a driving) human. The most destructive pyroclastic flow in the 20th century (in terms of lives lost) occurred on the island of Martinique in 1902, when an avalanche of searing ash descended Mt. Pelée at a speed of more than 160 km/hr and killed 29,000 people. (There were 2 survivors.) In 79 CE, the Italian towns of Pompeii and Herculaneum were buried under hot pyroclastic material, entombing the bodies and buildings in a natural time capsule (**Figure 6.7**). However, most of these people

were dead already, due to another hazard of volcanic eruptions: poisonous gases. More recently, at least 1700 people and 3000 cattle lost their lives when poisonous gas erupted from a volcano at Lake Nyos in Cameroon. In 1783 the Laki eruption in Iceland released so much acidic gas that nearly one-third of the people and half of the domestic animals in the country perished.

Secondary effects Secondary effects related to volcanic activity but not a direct result of it include fires (which are often caused by lava flows) and flooding (which may happen if a river channel is blocked or a crater lake bursts). The famous eruption at Krakatau, Indonesia, in 1883 claimed most of its victims due to a tsunami, the same kind of ocean wave that can also be caused by earthquakes. Volcanoes can also produce **volcanic tremors**, a type of seismic activity that helps scientists predict eruptions but rarely poses a threat itself.

Mudslides have often been a major cause of volcano-related deaths. A deadly mudflow called a **lahar** can result from volcanic ash mixing with snow at the volcano's summit or rain falling on recently deposited volcanic ash. Figure b in the *Case Study* shows a lahar from the eruption of Mt. Pinatubo. Lahars can occur months after the eruption. A related phenomenon is a volcanic **debris avalanche**, in which many different types of materials, such as mud, pyroclastic material, and downed trees, are mixed together. A devastating debris avalanche caused much of the damage from the 1980 eruption of Mt. St. Helens.

Victim of Mt. Vesuvius • Figure 6.7

This is not an actual body but a plaster cast of a citizen of Pompeii, Italy, who was killed during the eruption of Mt. Vesuvius in 79 CE. Death was caused by poisonous gases, and then the body was encased in pyroclastic material. Over the centuries, the body decayed, but a mold of its shape remained in the tephra. The cast was made when the natural mold was discovered during modern archaeological excavations.

Plinian Eruption in the Philippines

Mt. Pinatubo is 90 km northwest of Manila, in the Philippines. In 1990 Pinatubo was a vegetation-covered mountain that had not erupted for 500 years. Then, in July 1990, the region was rocked by a 7.8-magnitude earthquake, suggesting that Pinatubo might be waking up. By March 1991 villages around the volcano were feeling quakes, and on April 2 a small eruption of volcanic ash occurred.

Geologists first began monitoring events out of interest and then out of concern. They knew that the plains around the volcano were underlain by layers of fertile volcanic ash, all of them thousands of years old. Pinatubo had once been a dangerous volcano; the 500-years-ago eruption was big (about the same size as the 1980 eruption of Mt. St. Helens), but could a cataclysmic eruption be on the way? On June 7 a lava dome began to form; on June 9 geologists made a gutsy call: they recommended evacuation for everyone in a 20-km radius; 25,000 people departed. The following day, Clark Airfield, the main U.S. Air Base in the Philippines, was evacuated; 18,000 personnel and families left. Concerns rose, and on June 13, the danger radius was extended to 30 km; the number of evacuees rose to 58,000.

The eruption of Mt. Pinatubo commenced on June 15, 1991. It was the second-largest volcanic eruption of the 20th century. (The largest was Novarupta, Alaska, in 1912.) The top of the mountain exploded (**Figure a**), blasting a hole 2.5 km in diameter and propelling volcanic ash and sulfurous gases more than 30 km into the atmosphere. The cloud lingered in the stratosphere and lowered worldwide temperatures for the next year by half a degree.

By unfortunate coincidence, Tropical Storm Yunya was bearing down on the island at the time of the eruption. The rain-soaked ash caused roofs to collapse, and the unstable mud continued to flow downhill for months, burying towns, wiping out bridges, and causing more damage than the eruption itself. Figures b and c show the town of Bamban, 30 km away from Mt. Pinatubo, one month (**Figure b**) and three months after the eruption (**Figure c**).

Thanks to early warnings from geologists, most of the area around the volcano had been evacuated so the eruption killed relatively few people. Although 847 people died (mainly due to mudslides), scientists estimate that up to 20,000 lives were saved by the timely evacuation.

a.

b.

c.

Global Locator

Philippines

NATIONAL GEOGRAPHIC

Think Critically

1. What do you think might have happened to local residents if the geologists had issued their early warning too late?
2. What if they had issued the warning too early, and the eruption had not occurred as predicted?

Volcanic hazards • Figure 6.8

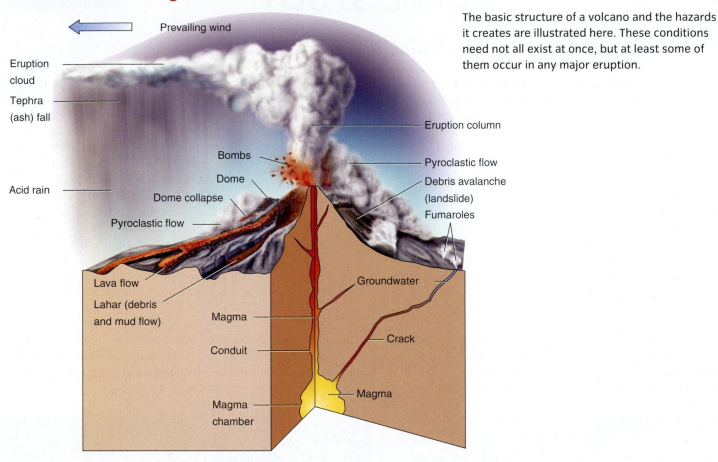

Prevailing wind

Eruption cloud

Tephra (ash) fall

Acid rain

Bombs

Dome

Dome collapse

Pyroclastic flow

Lava flow

Lahar (debris and mud flow)

Magma

Conduit

Magma chamber

Eruption column

Pyroclastic flow

Debris avalanche (landslide)

Fumaroles

Groundwater

Crack

Magma

The basic structure of a volcano and the hazards it creates are illustrated here. These conditions need not all exist at once, but at least some of them occur in any major eruption.

Figure 6.8 summarizes the various primary and secondary hazards from an eruption, and **Figure 6.9** shows where the most deadly eruptions of the past two centuries occurred. Note that in many cases, the secondary effects are responsible for the greatest loss of life.

Deadly eruptions • Figure 6.9

Since 1800, there have been 19 volcanic eruptions in which 1000 or more people have died from eruption-related causes. These eruptions are marked on the map. One major event that doesn't appear on this map is the 1991 eruption of Mt. Pinatubo in the Philippines because scientific monitoring and timely evacuations saved thousands of lives (see *Case Study: Plinian Eruption in the Philippines*).

El Chichón, 1985
1700 eruption

Santa Maria, 1902
6000 eruption

Nevado del Ruíz, 1985
23,000 mudflow

Cotopaxi, 1877
1000 mudflow

Mt. Pelée, 1902
29,000 eruption

Soufrière, 1902
1565 eruption

Lake Nyos, 1986
1700 lethal gas

Krakatau, 1883
36,417 tsunami

Galunggung, 1822
1500 eruption
4000 mudflow

Merapi, 1930
1300 eruption

Kelud, 1919
5110 mudflow

Agung, 1963
1900 eruption

Mayon, 1814
1200 eruption

Mayon, 1825
1500 mudflow

Awu, 1826
3000 mudflow

Awu, 1892
1532 mudflow

Lamington, 1951
2942 eruption

Tambora, 1815
12,000 eruption
80,000 famine

Tertiary effects Volcanic activity can change a landscape. Eruptions can block river channels and divert the flow of water. They can dramatically alter a mountain's appearance, as in the 1980 eruption of Mt. St. Helens. They can form new land, such as the black sand beaches of Hawaii, which are made of dark pyroclastic fragments, or the volcanic island of Surtsey, which emerged from the ocean near Iceland in 1963 and is composed of both lava flows and pyroclastic cones.

Volcanoes can also affect the climate on regional and global scales (see **Figure 6.10**). Major eruptions can cause toxic rain and acid rain, spectacular sunsets, or extended periods of darkness. Sulfur dioxide, a common gaseous emission of volcanoes, forms small droplets, or **aerosols**. If they get into the stratosphere, these aerosols spread around the world, absorb sunlight, and cool Earth's surface. An example from recent times is the 1815 eruption of Mt. Tambora in Indonesia, which caused three days of near darkness as far away as Australia. The following year was so cool in Europe and North America that it was called "the year without a summer." Farther back in time, the eruption of flood basalts, such as the Deccan Traps in India and the Siberian Traps in Russia, may have caused or contributed to several of the mass extinctions that divide geologic periods.

Beneficial effects Not all the effects of volcanic eruptions are negative, and it is no accident that many people choose to live near active volcanoes. Periodic volcanic eruptions renew the mineral content of soils and replenish their fertility; some of the most fertile soils of the world are adjacent to active volcanoes (**Figure 6.11**) (on the next page). Volcanism also provides geothermal energy and some types of mineral deposits. One rare kind of volcanism brings up diamond-bearing magma from deep in the mantle. All natural gem-quality diamonds on Earth reach the surface through volcanism.

Predicting Eruptions

It isn't possible to stop volcanic eruptions, but it is sometimes possible to predict them. Prediction is based on a combination of understanding the geologic history of a volcano and monitoring present activity for any changes or anomalies.

Volcanoes and climate • Figure 6.10

Fissure

Crater

a. The fissure eruption of Laki, a volcano in Iceland, lasted from 1783 to 1784 and was the largest flow of lava in recorded history.

b. In the winter after Laki's eruption, the average temperature in the northern hemisphere was about 1°C below normal. In the eastern United States, the decrease was closer to 2.5°C. At the same time, ice cores from Greenland recorded a dramatic spike in acidity due to acid precipitation.

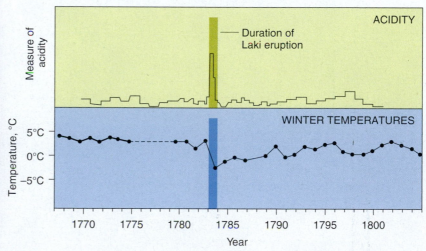

Fertile but dangerous • Figure 6.11

Farming villages speckle the slopes of Mt. Merapi, an active volcano on the island of Java, Indonesia. Merapi has a long history of dangerous, often fatal, eruptions, but the fertility of its soils lures farmers to its hazardous slopes. Although volcanic soils cover just 1% of Earth's land surface, they support 10% of the world's population.

AMAZING PLACES

Mt. St. Helens

Mt. St. Helens is a young and active volcano in the Cascade Range of the Pacific Northwest. It has erupted many times in its 37,000-year history. Even so, geologists and nearby residents alike were stunned by the violence of its eruption in 1980. The mountain blew off its top and most of its northern flank. The initial blast rattled windows all over the state of Washington, flattened trees as far as 27 km away, and killed 57 people.

If you visit the Mt. St. Helens National Volcanic Monument, you will see many signs of revival amid vistas of incredible desolation.

a. This photo, showing widespread devastation from the 1980 eruption, was taken in 2002 by an astronaut from the International Space Station.

b. Weeks after the eruption, not a living thing could be found in the vicinity of the volcano. Here, a forester inspects a forest that was flattened instantaneously by the blast wave, as if the trees were just blades of grass.

c. Here you see part of the largest debris avalanche in recorded history. Immediately after the eruption, this entire landscape was a barren gray. This photo is a downstream view of the North Fork Toutle River valley, north and west of St. Helens, with part of the 2.3 cubic km of debris avalanche material. The avalanche traveled approximately 24 km downstream at a velocity exceeding 240 km/hr. It left behind a hummocky deposit with an average thickness of 45 m.

d. Today the volcano is quiescent, though not dormant. You may be able to spot steam rising from the crater, as shown in this photo taken in 2004. This is a reminder that Mt. St. Helens may one day grow back to its former height and violently erupt again.

Pumice plain

Terminus of debris avalanche

Establishing a volcano's history The first step in prediction is to identify a volcano as active, dormant, or extinct. An **active** volcano has erupted within recorded history; a **dormant** one has not erupted in recent history. A volcano is called **extinct** when it shows no signs of activity and is deeply eroded. Mt. Pinatubo in the Philippines had been dormant for about 500 years prior to its awakening in 1991.

Another important step in prediction is identifying the volcano's past eruptive style. For example, Mt. Pinatubo is surrounded by thick deposits of pyroclastic material, a sign that the volcano has erupted violently in the past. Subduction zone volcanoes such as Mt Pinatubo and Mt. St. Helens are more likely to erupt explosively than are shield volcanoes and fissure eruptions, so understanding the tectonic setting of the volcano is also important. The type of rock that has solidified from past eruptions, either silica rich or silica poor, also indicates the volcano's style of eruption. Mt. St. Helens is discussed in *Amazing Places*.

Monitoring changes and anomalies When a volcano begins to show signs of increasing activity, scientists begin to monitor it more closely (**Figure 6.12**). They watch for changes—sudden as well as gradual, cumulative ones—and anomalies—things that don't fit the pattern of activity that the volcano had previously established.

As magma intrudes upward from an underlying magma chamber, the crust is stressed and cracked by the intrusion, causing small earthquakes. Therefore, the first indication of a forthcoming eruption is usually the onset of swarms of small earthquakes. As an eruption approaches, the volcanic edifice swells in response to the intrusion of magma. Warning signs include changes in shape or elevation, such as bulging or tilting, or the formation of a dome. Swelling can be detected using tiltmeters, lasers, and other instruments that measure the shape and orientation of the ground surface. The presence of these features indicates that the underground reservoir of magma is growing.

Rising magma releases volcanic gases, and careful monitoring of changes in the rate of emission (volume per unit of time) and composition of gases is often helpful in predicting how soon an eruption will occur. In particular, an increase in the ratio of sulfur to chlorine in volcanic gas

Volcano monitoring from the ground • Figure 6.12

THE PLANNER

Volcanic eruptions are almost always preceded by a host of physical changes that geologists can monitor, such as tremors, releases of gas, and changes in the slopes of the volcano.

WILEY PLUS | NATIONAL GEOGRAPHIC Video

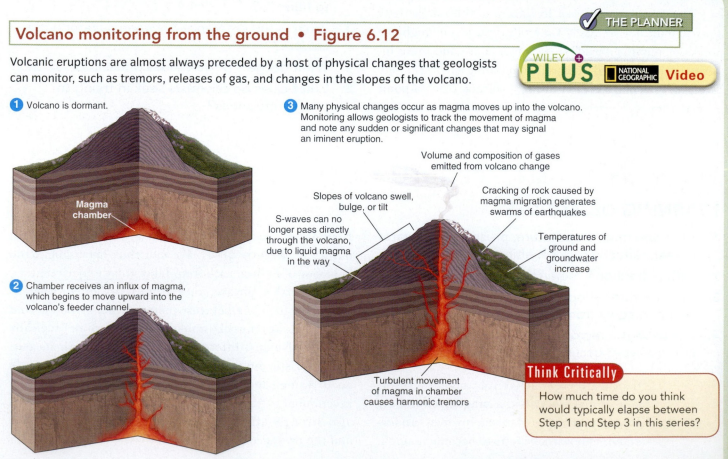

1 Volcano is dormant.

Magma chamber

2 Chamber receives an influx of magma, which begins to move upward into the volcano's feeder channel.

3 Many physical changes occur as magma moves up into the volcano. Monitoring allows geologists to track the movement of magma and note any sudden or significant changes that may signal an iminent eruption.

Volume and composition of gases emitted from volcano change

Slopes of volcano swell, bulge, or tilt

S-waves can no longer pass directly through the volcano, due to liquid magma in the way

Cracking of rock caused by magma migration generates swarms of earthquakes

Temperatures of ground and groundwater increase

Turbulent movement of magma in chamber causes harmonic tremors

Think Critically

How much time do you think would typically elapse between Step 1 and Step 3 in this series?

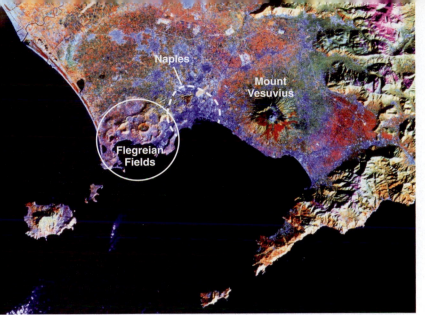

Naples

Mount Vesuvius

Flegreian Fields

Volcano monitoring from orbit • Figure 6.13

This false-color satellite image shows the area around Mt. Vesuvius (center right) and the Bay of Naples, Italy, a densely populated region. Recent lava flows show up bright red in this image, which records infrared radiation (indicating heat). Older lavas and volcanic ash show up in shades of yellow and orange. The dark blue and purple region at the head of the bay is the city of Naples. West of Naples lies a cluster of smaller volcanoes called the Flegreian Fields. By comparing successive satellite images, geologists can detect changes in ground temperature.

emissions is often an indication of an imminent eruption. Changes in ground temperature and changes in the temperature or composition of water in crater lakes, wells, or hot springs can also be warning signs.

Many of these changes can be detected and monitored from space, via remote sensing. Ground temperature changes can be detected using infrared imaging, for example (**Figure 6.13**). Even changes in the shape of the land can be detected remotely: Scientists obtain two images of the area taken days or weeks apart and compare them digitally, highlighting even tiny changes in slope or orientation of the ground surface.

The first indications of forthcoming volcanic activity do not indicate the likely time of eruption—it might be weeks or months ahead. However, as more and more information is obtained, the predicted timing can be refined. Eventually, geologists monitoring a volcano reach a point where they can recommend to civil authorities that

nearby settlements be evacuated until after the eruption. As a consequence, volcanic eruptions—even very large ones—tend to cause fewer fatalities than earthquakes because earthquakes are much more difficult to predict.

CONCEPT CHECK STOP

1. **How** does the eruption of a stratovolcano differ from that of a shield volcano?

2. **Which** kinds of eruption pose the greatest risk to humans?

3. **What** is the difference between a crater and a caldera?

4. **What** dangers do volcanoes present for humans?

5. **What** clues do scientists seek in trying to predict eruptions?

How, Why, and Where Rock Melts

LEARNING OBJECTIVES

1. **Describe** how temperature, pressure, and water content affect a rock's melting point.

2. **Define** fractional melting.

3. **Identify** three properties that distinguish one lava or magma from another.

4. **Describe** the tectonic settings in which major magma types occur.

nderneath every active volcano lies a reservoir of magma, called a **magma chamber**. Understanding volcanism involves understanding how rock melts to become magma.

Fortunately, rock can be melted artificially as well as naturally (**Figure 6.14**). We can thus learn about the behavior of molten rock from laboratory experiments.

At Earth's surface, rock begins to liquefy when it is heated to a temperature between about 800°C and 1000°C. However, rock (unlike ice, for example) typically consists of many different minerals, each with its own characteristic melting temperature. Thus we cannot talk about a single melting point for a rock. Complete melting is commonly attained by about 1200°C. Two other factors also strongly affect the melting temperature: pressure and the presence of water in the rock.

Molten rock: Artificial vs. natural • Figure 6.14

a. In a steel mill, workers heat metal ores to the melting point in order to separate the metal from the surrounding rock.

b. A geologist in a protective suit measures the temperature of lava erupting from Mauna Loa, Hawaii. Bright orange, yellow, and white lava is hotter, whereas dull red, brown, and black colors indicate cooler lava.

Heat and Pressure Inside Earth

If you descend into a mine, it becomes apparent that the farther down you go, the hotter it gets. The rate at which temperature increases with depth, called the **geothermal gradient**, is quite different underneath continental surfaces than it is under the seafloor. In the continental crust, temperature rises initially at about 30°C per kilometer; at depth temperature rises increasingly slowly, for an average of about 6.7°C per kilometer, reaching 1000°C at a depth of 150 km. Underneath the ocean floor, the rate of increase is about twice as rapid. The temperature increases by 13°C per kilometer, reaching 1000°C at a comparatively shallow depth of 80 km (**Figure 6.15**). Below the asthenosphere–lithosphere boundary, the geothermal gradient becomes more gradual (0.5°C/km) and the temperature difference between suboceanic and subcontinental rock disappears.

Geothermal gradient • Figure 6.15

a. Temperature increases with depth. The dashed lines are *isotherms*, lines of equal temperature. Notice how the lines "sag" underneath the continental crust because the rate of increase of temperature is slower there.

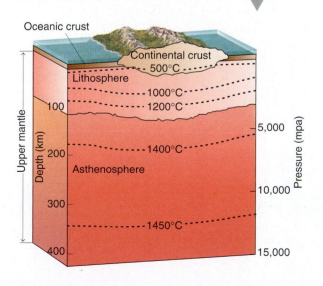

- - - Temperature rise with depth in oceanic lithosphere
——— Temperature rise with depth in continental lithosphere

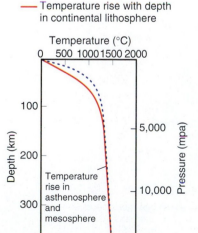

b. This graph represents the same information as shown in a. Earth's surface is at the top, so depth (and pressure) increase as you move down. The dashed curve shows the geothermal gradient under oceans, and the solid curve shows the gradient under continental crust. Note that the two curves merge (and the isotherms become level) below 200 km.

As you can see in Figure 6.15, the temperature in the upper mantle is higher than the temperature at which most rock types melt at Earth's surface. Yet the upper mantle is mostly solid. How is this possible? The answer is that the pressure also rises very dramatically with increasing depth, and increasing pressure causes rock to resist melting (**Figure 6.16**). For example, albite, a common rock-forming mineral (a feldspar), melts at 1104°C at the surface. At a depth of 100 km, the pressure is 35,000 times greater than it is at sea level. At that pressure, the melting temperature of albite rises to 1440°C, which still slightly exceeds the normal temperature at that depth. Thus albite remains solid when it is beneath the surface.

The presence of water (or water vapor) in rock dramatically reduces its melting temperature (Figure 6.16). By analogy, as anyone who lives in a cold climate knows, salt can melt the ice on an icy road because a mixture of salt and ice has a lower melting temperature than pure ice. Similarly, a mineral-and-water mixture has a lower melting temperature than the dry mineral alone. The effect of water on the melting of rock becomes particularly important in subduction zones, where water is carried down into the mantle by oceanic crust, as described in Chapter 4.

Fractional melting Most rock is composed of many different minerals, and each mineral melts at a different temperature, so rock typically melts over a temperature range of 200 degrees or more. This means that the boundary between solid rock and melt (molten rock) is not crisp, as in the melting of an ice cube, but blurry, as in **Figure 6.17**. When the temperature rises enough for part of the materials in the rock to melt and part to remain solid, it becomes a **fractional melt**. Only if the temperature continues to increase or the pressure decreases will the rock melt completely. **Fractionation**, an important process that can lead to the development of a diversity of rock types, is caused by fractional melting.

> **fractional melt** A mixture of molten and solid rock.
>
> **fractionation** Separation of melted materials from the remaining solid material in the course of melting.

Effects of temperature and pressure on melting • Figure 6.16

a. The melting temperature of a dry mineral (albite, in this case) increases at high pressures. A mineral at depth (shown by the small square) can melt in two different ways: either by an increase in temperature (red arrows) or by a decrease in pressure (blue arrows). The latter effect is called *decompression melting*, and it is an important reason many magmas stay molten all the way to the surface.

b. The melting temperature of a mineral in the presence of water typically decreases as pressure increases. This is exactly the opposite of what happens to dry minerals. Magmas that contain dissolved water typically solidify before they reach the surface.

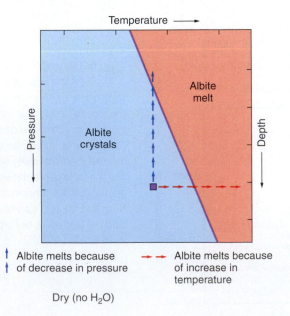

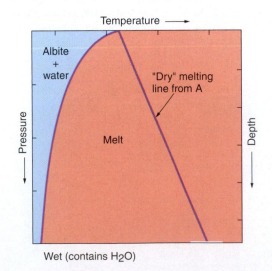

Effects of temperature and pressure on rock • Figure 6.17

Because most rock types contain a mixture of materials, they do not melt all at once, at a single temperature; instead, there is a range of temperatures and pressures in which they contain a mixture of melted and unmelted crystals. The dots on this diagram show the stages in the melting process, which are illustrated further below.

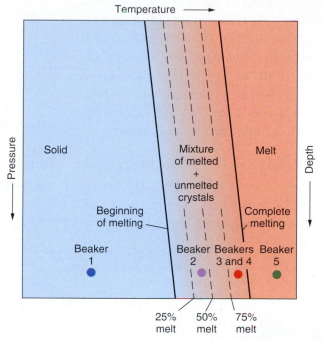

The process of fractional melting can lead to another effect, called fractionation (illustrated for convenience in a laboratory beaker rather than buried in Earth's mantle).

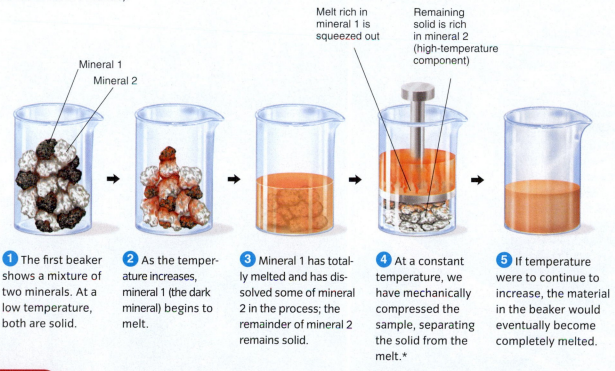

Melt rich in mineral 1 is squeezed out

Remaining solid is rich in mineral 2 (high-temperature component)

Mineral 1
Mineral 2

1 The first beaker shows a mixture of two minerals. At a low temperature, both are solid.

2 As the temperature increases, mineral 1 (the dark mineral) begins to melt.

3 Mineral 1 has totally melted and has dissolved some of mineral 2 in the process; the remainder of mineral 2 remains solid.

4 At a constant temperature, we have mechanically compressed the sample, separating the solid from the melt.*

5 If temperature were to continue to increase, the material in the beaker would eventually become completely melted.

Think Critically

If you were to heat up a glass beaker full of crushed rock, the beaker would melt before you could finish studying the rock-melting process. How do you think geologists study rock melting?

*This kind of mechanical separation of melt from solid residue can occur in the lithosphere as a result of tectonic forces.

How, Why, and Where Rock Melts **169**

Magma and Lava

As mentioned earlier, molten rock below ground is called magma. When magma reaches the surface, it is called lava. A lot of magma never reaches the surface but instead remains underground, trapped in a magma chamber, until it crystallizes and hardens to igneous rock. We cannot study magma underground in its natural setting, but we can study lava, and we can experiment with synthetic magma. From our direct observations of lava, we know that magmas differ in composition, temperature, and viscosity.

Composition Most magma is dominated by silicon, aluminum, iron, calcium, magnesium, sodium, potassium, hydrogen, and oxygen—Earth's most abundant elements. Oxygen combines with the other elements to form oxides, such as SiO_2, Al_2O_3, CaO, and H_2O. Silica, or SiO_2, usually accounts for 45 to 75% of the magma, by weight. In addition, a small amount of dissolved gas (between 0.2 and 3% of the magma, by weight) is usually present, primarily water vapor (H_2O) and carbon dioxide (CO_2). Despite their low abundance, these gases strongly influence the properties of magma. The proportion of silica (SiO_2) also has a strong effect on magma's appearance and properties.

Temperature We know from direct measurements at erupting volcanoes that lavas vary in temperature from

Viscosity of lava • Figure 6.18

Geologists refer to the two kinds of lava commonly associated with shield volcanoes with the Hawaiian words *pahoehoe* (the lava shown in **a**) and *aa* (pronounced "ah-ah," for the lava shown in **b**).

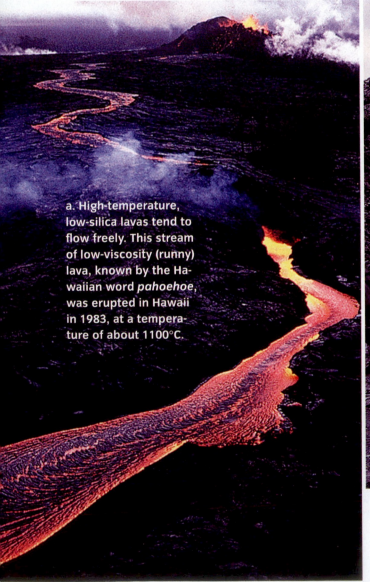

a. High-temperature, low-silica lavas tend to flow freely. This stream of low-viscosity (runny) lava, known by the Hawaiian word *pahoehoe*, was erupted in Hawaii in 1983, at a temperature of about 1100°C.

b. Two strikingly different lava flows are visible here at Kilauea, Hawaii. The smooth, ropy rock on which the geologist is standing formed from a low-viscosity **pahoehoe** lava. The rough, chunky rock that the geologist is sampling is known by the Hawaiian word **aa**; it came from a more viscous, slow-moving flow that erupted years later.

about 750°C to 1200°C. From laboratory experiments with synthetic magma, geologists know that magma temperatures in the mantle must rise as high as 1400°C. They also know that magmas with high H_2O contents tend to melt at lower temperatures.

Viscosity All magma is liquid and has the ability to flow, but magmas differ to a marked extent in how readily they flow. This is certainly true for lavas. As shown in **Figure 6.18**, some lava is very fluid, almost like a stream of water. But other lava creeps along slowly and steadily, like molasses. Two properties in particular control viscosity: temperature and silica content. The higher the temperature, the lower the viscosity of lava. In **Figure 6.18b** the two lavas have the same composition, but the runny pahoehoe flow was erupted at higher temperature than the sticky aa flow. Lavas with high silica contents tend to flow slowly because of the tendency of silica molecules to polymerize, or form long chains (see Chapter 2). Thick, slow-moving lavas have high viscosity, and these lavas have the greatest tendency to erupt explosively.

The three main kinds of magma Fractionation leads to a wide diversity in magma compositions, but observations indicate that three compositions are predominant. Basaltic magma contains between 45 and 50% SiO_2 by weight and a little dissolved gas. Formed in the mantle by decompression melting, basaltic magma has low viscosity, flows readily, and forms shield volcanoes.

In a subduction zone, oceanic crust on top of sinking lithosphere contains water and hydrous minerals; at a depth of about 100 km, water is released and wet fractional melting in the mantle produces andesitic magma. Andesitic magma contains about 60% SiO_2 by weight and 2 to 4 percent by weight of dissolved gases—mainly H_2O. Such magmas are viscous and tend to be erupted explosively with a lot of pyroclasts, forming tephra cones and stratovolcanoes.

When wet fractional melting occurs within the continental crust, the rhyolitic magma so formed contains between 70 and 75% SiO_2 by weight plus 3 to 8% by weight of dissolved H_2O. Rhyolitic magma is extremely viscous; it erupts violently and forms masses of volcanic ash. When most of the gas has escaped from a rhyolitic magma chamber, the remaining magma may be extruded as a sticky, slow-moving lava.

Tectonic setting and volcanism The location of a volcano has a great deal to do with the type of lava or magma that is found there and, hence, with the type of volcanic rock that is formed. As discussed in Chapter 4, volcanoes are mostly found in two tectonic settings: near plate margins and above "hot spots" in the mantle. **Figure 6.19** (on the next page) illustrates the types of lava associated with these different locations.

At oceanic divergent margins such as along midocean ridges (**Figure 6.19a**), the oceanic crust is quite thin, and the geothermal gradient is steep. This setting favors the eruption of hot, low-viscosity basaltic lavas. Basalt is generated by fractional melting of the underlying mantle. The magma rises through crustal fissures along the midocean range, creating new oceanic crust.

At ocean–ocean or ocean–continent subduction zones (**Figures 6.19e** and **c**), the subducting rock has a high water content and therefore melts at a lower temperature, as shown in Figure 6.15. Such magma will often solidify before reaching the surface, but it may stay molten for a variety of reasons. For example, fractional melting may separate the minerals with a higher melting point from those with a lower melting point and allow the latter to erupt to the surface. The resulting andesitic lava is cooler and more viscous than basaltic lava. Andesitic lava is more likely to erupt explosively, producing pyroclastic deposits and building stratovolcanoes.

The lava generated at mantle hot spots (**Figure 6.19d**) tends to be hot and basaltic, and it builds giant shield volcanoes by layering one fluid lava flow on top of another. Finally, the lava formed in continental crust as a result of heat supplied from below by a hot spot (**Figure 6.19b**) tends to be rhyolitic and especially high in silica. Eruption of such material produces abundant pyroclasts, particularly in the form of volcanic ash.

CONCEPT CHECK STOP

1. **What** effect does H_2O have on the melting properties of rocks?

2. **How** does fractional melting separate different minerals from one another?

3. **What** properties distinguish one kind of lava from another?

4. **Why** do different tectonic settings lead to the formation of different kinds of lava?

Geology InSight

Different lava types are generated in different tectonic settings. Yellow dots on the map indicate hot spots, and red triangles indicate volcanoes.

a. Midocean ridge: submarine basaltic pillow lavas on the East Pacific Rise.

b. Melting in continental crust by heat from a hot spot: rhyolitic lavas in Yellowstone National Park.

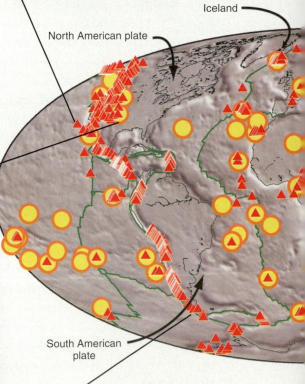

Iceland

North American plate

South American plate

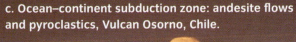

c. Ocean–continent subduction zone: andesite flows and pyroclastics, Vulcan Osorno, Chile.

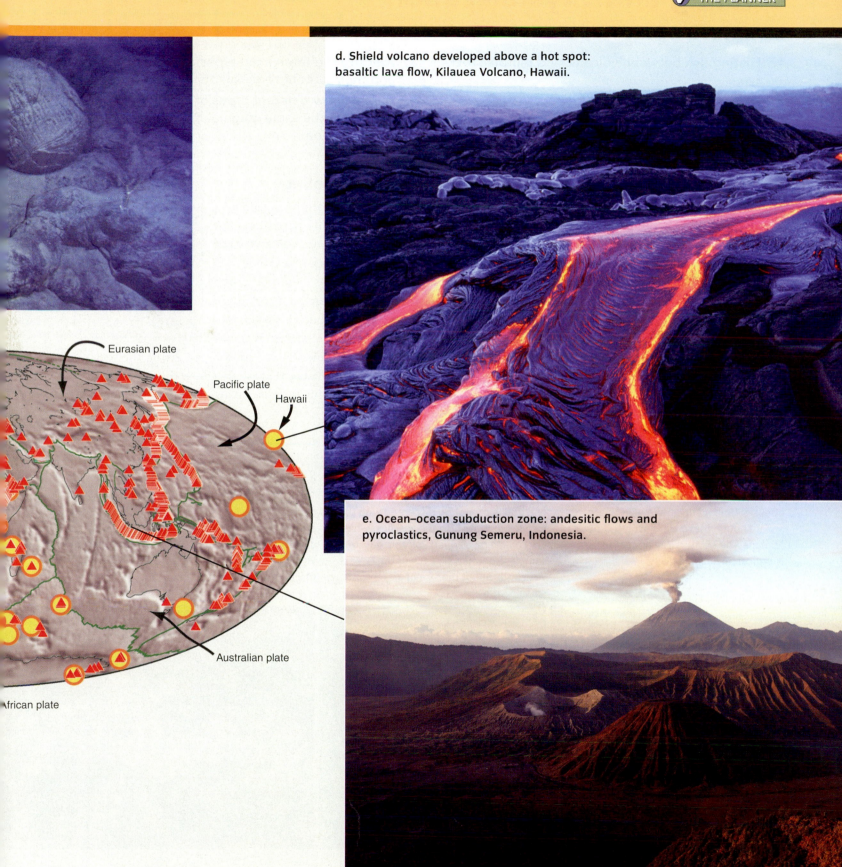

d. Shield volcano developed above a hot spot: basaltic lava flow, Kilauea Volcano, Hawaii.

Eurasian plate

Pacific plate

Hawaii

Australian plate

African plate

e. Ocean–ocean subduction zone: andesitic flows and pyroclastics, Gunung Semeru, Indonesia.

How, Why, and Where Rock Melts 173

Cooling and Crystallization

LEARNING OBJECTIVES

1. **Distinguish** between volcanic and plutonic rock types.

2. **Identify** different textures characteristic of igneous rock and explain the processes that produce them.

3. **Identify** the three main types of volcanic rock and their plutonic equivalents.

4. **Explain** how different types of igneous rock can form from the same magma, through fractional crystallization.

Whereas melting influences the properties of magma, cooling and **crystallization** influence the properties of igneous rock. For example, the rate of cooling determines how large the individual mineral grains in the rock will grow. Grain size affects the appearance or texture of the rock. The composition of the magma determines the final mineral assemblage in the solidified rock. Let's examine each of these factors more closely.

> **crystallization**
> The process whereby mineral grains form and grow in a cooling magma (or lava).

Rate of Cooling

Even a quick study of igneous rock types reveals that there are two large families. **Volcanic rock** forms from magma extruded at Earth's surface. **Plutonic rock** forms when magma crystallizes deep underground. This process is much slower and therefore gives the mineral grains time to grow larger, giving the rock an easily distinguishable texture. Because the rate at which magma cools influences the sizes of mineral grains that

> **volcanic rock**
> Igneous rock that solidifies on or near the surface, from lava.
>
> **plutonic rock**
> Igneous rock that solidifies underground, from magma.

Volcanic rock textures • Figure 6.20

a. Glassy texture
This photograph shows obsidian obelisks from a Mayan grave in Guatemala. Though these were sculpted by humans, the shiny, curvy appearance is typical of volcanic glass.

b. Aphanitic texture
In this fine-grained rock, individual mineral grains cannot be discerned with the naked eye. Note also the vesicles, which are caused by trapped volcanic gas. In this sample, from Hawaii, each of the largest vesicles is about the size of a small pea.

c. Porphyritic texture
This volcanic rock sample from Nevada contains large mineral grains, or **phenocrysts**, suspended in an aphanitic material called the **groundmass**. The largest grains visible in the photo are approximately 6 millimeters in length.

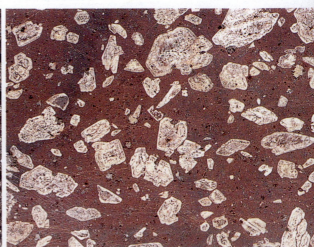

WHAT A GEOLOGIST SEES

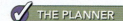

Putting Rocks Under a Microscope

Faced with identifying the minerals present in an aphanitic igneous rock, a geologist turns to a microscope for help. The geologist first polishes a flat surface on a small fragment of the rock sample and then glues the polished surface to a glass plate. Careful grinding of the free surface of the sample to a thickness of just 0.03 mm makes the fragment thin enough for light to pass through.

a. The specimen of aphanitic volcanic rock from Figure 6.20b has been cut and polished into a wafer that is thin enough to allow light to pass through. Geologists call these rock slices **thin sections**. This magnified thin section is about 3.5 mm across.

b. Next, our geologist puts the thin section on the stage of a microscope and examines it under polarized light. Different minerals appear differently in polarized light. In this specimen, the geologist has focused on an area about 0.05 mm across and can see white, needle-shaped plagioclase feldspar grains and smaller, brightly colored pyroxene gains. The dark background is volcanic glass. The geologist identifies this sample as aphanitic, vesicular basalt.

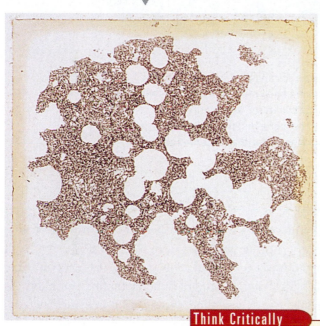

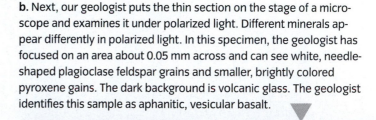

Think Critically

Notice that the individual mineral grains in the igneous rocks shown in the photographs in this chapter have irregular, interlocking boundaries.

1. Why would this be so?
2. Would a metamorphic or sedimentary rock look different?

---

form, volcanic rock tends to have smaller mineral grains than more slowly cooled plutonic rock.

Rapid cooling: Aphanitic igneous rock

Sometimes magma cools so rapidly that mineral grains do not have a chance to form at all. The resulting rock, called obsidian, is not crystalline but glassy (**Figure 6.20a**). More often, mineral grains do form in rapidly solidifying magma, but they are extremely small and can be seen only with magnification (see *What a Geologist Sees*). Rock with this very fine-grained texture is said to be **aphanitic** (**Figure 6.20b**). Volcanic rock cools rapidly, so the texture is commonly aphanitic.

> **aphanitic** An igneous rock texture with mineral grains so small they can be observed only under a magnifying lens.

granite

granite

pegmatite

Plutonic rock textures Figure • 6.21

Granite is a plutonic igneous rock. Two distinct phaneritic textures can be seen in this granite specimen from California's Sierra Nevada. The two outside layers have phaneritic texture, with small but visible grains of plagioclase, potassium feldspar, quartz (white), and biotite (black). Sandwiched between them is a vein of pegmatite, which contains the same minerals in much larger grains. **Pegmatite** is a phaneritic igneous rock with mineral grains larger than 2 centimeters.

Slow cooling: Phaneritic igneous rock Unlike aphanitic rock with its tiny mineral grains, **phaneritic** rock usually has time to form mineral grains that can be readily seen by the unaided eye (**Figure 6.21**). Exceptionally large mineral grains (sometimes up to several meters!) typically form in the last stage of crystallization of a plutonic rock body, when gases build up in the remaining magma. The vapor facilitates the growth of large crystals because chemicals can migrate quickly to the growing crystal faces. Phaneritic texture is common in plutonic rock.

> **phaneritic** An igneous rock texture with mineral grains large enough to be seen by the unaided eye.

Two cooling rates Some igneous rocks have a **porphyritic** texture (**Figure 6.20c**), which consists of large mineral grains embedded in an aphanitic matrix. This happens when the magma starts to cool slowly at depth and crystallizes large grains. When the magma is partly crystallized, it is erupted, and the remaining liquid lava cools rapidly to form the aphanitic matrix in which are embedded the large mineral grains that formed earlier.

Dissolved gases, too, can affect the texture of volcanic rock. An erupting lava may froth and bubble; if the froth is blasted into the air and cools quickly, it forms **pumice**, which is a glassy mass full of bubbles (called **vesicles**). As lava cools, the viscosity increases, and it becomes increasingly difficult for gas bubbles to escape. When the lava finally solidifies into rock, the last bubbles to form may become trapped. In basaltic lava, this process can create volcanic rock with lots of bubble holes that resembles Swiss cheese. Such a rock is said to be **vesicular**.

Chemical Composition

Geologists subdivide the most common igneous rock types into three broad compositional categories, based on their silica contents. Rock that contains a large amount of silica (about 70% SiO_2 by weight) is usually light colored. Geologists call such rock **felsic** (a word formed from "feldspar" and "silica") because feldspar is typically the most abundant mineral. At the other end of the scale is **mafic** rock (a word formed from "magnesium" and "ferric," or iron rich), with a large amount

Volcanic and plutonic equivalents   Table 6.1

Silica Content of Magma	Resulting Volcanic Rocks	Grain Size →	Resulting Plutonic Rocks	
High (= 70%–75%)	**Rhyolite** lies at the felsic, high-silica end of the scale, and consists largely of quartz and feldspars. It is usually pale, ranging from nearly white to shades of gray, yellow, red, or lavender.		**Granite**, the plutonic equivalent of rhyolite, is common because felsic magmas usually crystallize before they reach the surface. It is found most often in continental crust, especially in the cores of mountain ranges.	
Intermediate (= 60%)	**Andesite** is an intermediate silica rock, with lots of feldspar mixed with darker mafic minerals such as amphibole or Pyroxene. It is usually light to dark gray, purple, or green.		**Diorite** is the plutonic equivalent of andesite, an intermediate silica rock.	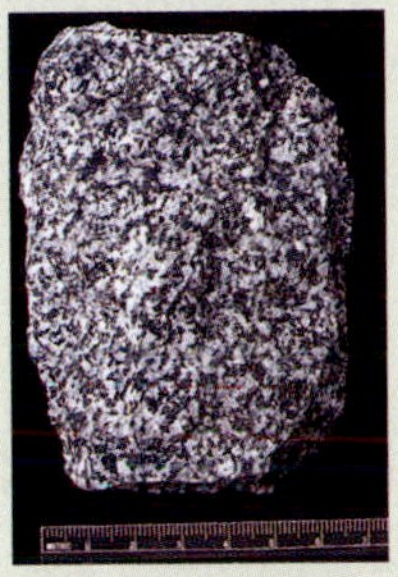
Low (= 45%–50%)	**Basalt**, a mafic rock, is dominant in oceanic crust, and the most common igneous rock on earth. Large-volume, low-viscosity lava flows from shield volcanoes and fissures are usually basaltic. Dark-colored pyroxene and olivine give it a dark gray, dark green, or black color.		**Gabbro** is the plutonic equivalent of basalt, a low-silica rock.	

Silica Content (axis, vertical)

of dark-colored minerals rich in magnesium and iron. Mafic rock is usually lower in silica content (about 50% SiO_2 by weight). Finally, igneous rock with about 60% SiO_2 by weight is said to be **intermediate**, in both composition and color.

Geologists thus organize igneous rock types in two ways: by grain size and by silica content. The results are shown in **Table 6.1**. The rock samples on the left are volcanic (generally aphanitic), and the ones on the right are plutonic (almost always phaneritic). The ones on the top contain the most silica and are lightest in color, and the ones on the bottom contain the least silica and are darkest in color.

The composition of any igneous rock is closely related to the composition of its parent magma. Basaltic magma flows freely, tends to contain less trapped gas, and erupts less explosively. Rhyolitic magma, because of its high silica content, flows less freely, traps more gas, and thus tends to erupt explosively. Therefore, the rock content of a volcano gives an important clue to its history, showing how it has erupted in the past and how it is likely to erupt in the future.

Fractional Crystallization

Although the six igneous rock types summarized in Table 6.1 are the most common, there are literally hundreds of other kinds of igneous rock on Earth. The reason for this diversity is that a single magma source can differentiate into several kinds of igneous rock through **fractional crystallization**, a sort of reversal of fractional melting (**Figure 6.22**).

> **fractional crystallization**
> Separation of crystals from liquids during crystallization.

Crystallization occurs over the same range of temperatures as melting, and the last minerals to melt are the first to crystallize. As shown in Figure 6.22, the newly formed crystals can become separated from the remaining magma in several different ways. The result is a rock and magma with different compositions, both of them different from the original magma.

This process was first investigated experimentally by Norman Bowen in the early 1900s. Bowen melted powdered rock samples and observed the sequence of minerals that crystallized from the melts. He hypothesized that rock types with widely varying compositions could form by differentiation from a single, homogeneous starting melt of basaltic composition. The sequence of mineral assemblages and resulting rock compositions, from early-crystallizing mafic to late-crystallizing felsic assemblages, is called **Bowen's reaction series**. Real crystallization is more complex than Bowen's model suggests; however, these processes are certainly very important in generating the wide range of igneous rock types on Earth.

Separating crystals from melt • Figure 6.22

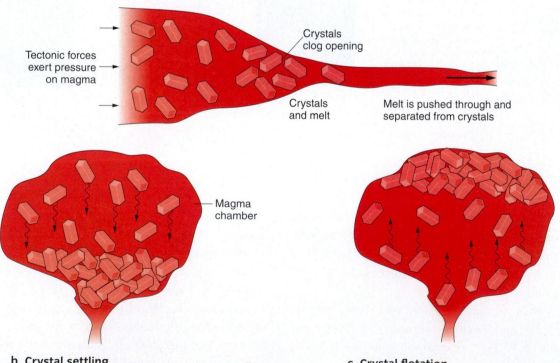

a. Filter pressing
Magma is squeezed through a small opening by tectonic forces. Only the liquid gets through, and the newly formed crystals are left behind.

Crystals clog opening

Tectonic forces exert pressure on magma

Crystals and melt

Melt is pushed through and separated from crystals

Magma chamber

b. Crystal settling
The first minerals to crystallize are denser than the melt and may sink to the bottom.

c. Crystal flotation
The first crystals are lighter than the liquid and may rise to the top.

1. **How** does volcanic rock differ from plutonic igneous rock?

2. **What** are the main textural differences observed in igneous rock?

3. **What** are the plutonic equivalents of basalt, andesite, and rhyolite?

4. **What** is the difference between fractional crystallization and fractional melting?

Plutons and Plutonism

LEARNING OBJECTIVES

1. **Describe** the most common plutonic formations.

2. **Explain** why volcanoes also create plutonic rock and plutons.

A lthough perhaps less familiar than volcanic rock, plutonic rock gives rise to some dramatic geologic formations, known as **plutons** (**Figure 6.23**). Plutons are always intrusive

> **pluton** Any body of intrusive igneous rock, regardless of size or shape.

bodies, different from the rock that surrounds them. They originate as magma underground, solidify underground, and are eventually exposed as plutons at the surface by erosion.

Most stocks and batholiths are granitic or between granite and diorite in composition. The magma that forms batholiths probably results from extensive fractional melting of the lower continental crust. Despite their huge size, the

Plutons • Figure 6.23

This diagram shows the origin of various forms taken by plutons. Note the vertical volcanic necks; sills parallel to the layering in the surrounding rocks; and dikes, which cut across the surrounding rock layers. In every case, the magma intrudes into previously existing rock.

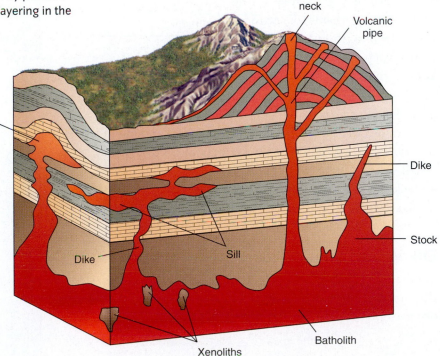

How magma rises • Figure 6.24

Magma forces its way upward by three different mechanisms:

a. By wedging open preexisting cracks
b. By wedging off fragments of rock, called **xenoliths**
c. By melting some of the invaded rock

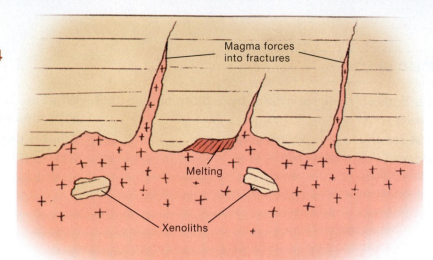

Magma forces into fractures

Melting

Xenoliths

magma bodies that form batholiths migrate upward, squeezing into preexisting fractures and pushing overlying rock out of the way (**Figure 6.24**).

Batholiths and Stocks

Plutons are named according to their shapes and sizes. The largest type of pluton is a **batholith** (from the Greek words meaning "deep rock"). Some batholiths exceed 1000 km in length and 250 km in width (**Figure 6.25**).

Where they are visible at the surface as a result of erosion, the walls of batholiths tend to be nearly vertical. This early observation led geologists to believe that batholiths extend downward to the base of Earth's crust. However, geophysical measurements suggest that this perception is incorrect. Most batholiths seem to be only 20 to 30 km thick. A smaller version of a batholith, only 10 km or so in its maximum dimension, is called a **stock**. In some cases, as shown in Figure 6.23, a stock may be associated with a batholith that lies underneath it.

> **batholith** A large, irregularly shaped pluton that cuts across the layering of the rock into which it intrudes.

Dikes and Sills

Smaller plutons tend to take advantage of fractures in the surrounding rock. Two of the most obvious indicators of past igneous activity are dikes and sills **Figure 6.26**.

Batholiths • Figure 6.25

Because batholiths are so immense, we cannot show one in a single photo; however, we can illustrate them on a map. The Coast Range batholith of southern Alaska, British Columbia, and Washington dwarfs the largest batholiths of Idaho and California. Many of the individual stocks and batholiths shown on this map are just the exposed tops of much larger intrusive bodies that lie underground.

Pacific Ocean

Coast Range batholith

Idaho batholith

Sierra Nevada batholith

Southern California batholith

0   200 mi.

a. A dike of gabbro cuts across horizontal sedimentary rock strata in Grand Canyon National Park, Arizona.

b. The sill is the middle piece of this rock "sandwich," a layer of dark brown gabbro intruded between sedimentary rock strata above and below, in Big Bend National Park, Texas.

c. This volcanic neck called Devil's Tower, in Wyoming, is all that remains of an ancient eroded volcano. You might remember this location for the role it played in the movie *Close Encounters of the Third Kind*.

A **dike** forms when magma squeezes into a cross-cutting fracture and then solidifies. If the magma intrudes between two layers and is parallel to them, it forms a **sill**. Sometimes this intrusion causes the overlying rock to bulge upward, forming a mushroom-shaped pluton called a laccolith. As shown in Figure 6.23, all of these intrusive forms may occur as part of a network of plutonic bodies.

Dikes and sills can be very large. For instance, there is a large and well-known sill-like mass made of gabbro that is visible in the Palisades, the cliffs that line the Hudson River opposite New York City. The Palisades Intrusive Sheet is about 300 m thick. It formed from multiple charges of magma intruded between layers of sedimentary rock about 200 million years ago. The sheet is visible today because tectonic forces raised that portion of the crust upward, and then the covering sedimentary rock strata were largely removed by erosion.

As Figure 6.23 shows, plutons can also be connected to volcanoes. Beneath every volcano lies a complex network of channels and chambers through which magma reaches the surface. When a volcano becomes extinct, the magma in the channels solidifies into various kinds of plutons. A **volcanic pipe** is the remnant of a channel that originally fed magma to the volcanic vent; when exposed by erosion, it is called a **volcanic neck** (Figure 6.26c).

CONCEPT CHECK

1. **What** are the names of two commonly observed tabular plutons?

2. **How** do plutons become visible at Earth's surface, if they are formed underground?

Summary

1 Volcanoes and Volcanic Hazards 152

- **Volcanoes** eject a wide variety of materials, including **lava** (molten rock, or **magma**, that has reached Earth's surface), gases, volcanic ash, larger pebbles, and rocks. The solid fragmental ejecta are collectively called tephra. Large tephra particles are called volcanic bombs, intermediate-sized particles are called lapilli, and the smallest particles are called volcanic ash. When tephra is consolidated into a rock, it is called agglomerate if the particles are large and tuff if they are small.

- The diversity of volcanic eruption types is mostly due to two factors: the **viscosity** of the magma and the amount of gas present in it. The viscosity of the magma in turn depends on its chemical composition, temperature, and gas content.

- **Shield volcanoes** and fissure eruptions tend to be quiet and nonexplosive. They gradually build up the volcano

through a series of lava flows. **Stratovolcanoes** tend to erupt explosively. They are built up from a series of layers of lava and pyroclastic material.

- Funnel-shaped craters and much larger depressions known as calderas are common features at the summits of volcanoes. New magma may push up the floor of a crater or caldera and form a resurgent dome. If groundwater in the vicinity of an active magma chamber becomes heated, it may form thermal springs or geysers.

- Direct, or primary, volcanic hazards of volcanoes are hazards directly caused by the volcanic eruption. They include **pyroclastic flows**, lava flows, and poisonous gases (see figure). Secondary hazards are those triggered by the eruption, such as volcanic tremors or lahars. Some extremely large eruptions produce tertiary effects, long-lasting and even permanent changes brought about by the eruption. One example is a worldwide drop in temperature because of aerosols in the upper atmosphere. Tertiary effects can also be beneficial—for example, the creation of new land or rich volcanic soil.

Volcanic hazards • Figure 6.8

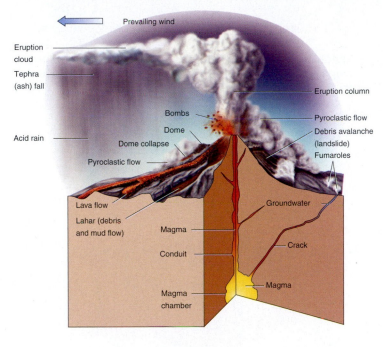

- For purposes of prediction, volcanoes are categorized as active, dormant, or extinct. The past eruption style of a volcano is a guide to its future behavior. Volcanologists can often tell an eruption is coming because of changes in seismic activity, gas emissions, ground and water temperatures, and the slope of a volcano's sides. They cannot yet predict exact eruption times, but by monitoring a volcano and watching for signs of anomalies, they can often recommend evacuation of the area days or weeks before an eruption.

2 How, Why, and Where Rock Melts 166

- Both temperature and pressure increase with depth. The increase of temperature with depth is called the **geothermal gradient** (see figure); the rate of increase with depth is less

Geothermal gradient • Figure 6.15

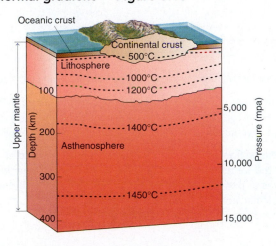

through and under the continental crust than it is through and under the oceanic crust.

- Minerals can melt in two ways: through an increase in temperature or a decrease in pressure (known as decompression melting). The presence of water in a rock typically lowers its melting point.

- Rock-forming minerals melt at different temperatures, so rocks do not have a single melting point; rather, they melt over a range of temperatures. A **fractional melt** is a body of rock in which some materials have melted and others have not. Fractional melting often separates minerals through a process called **fractionation.** When rock begins to melt, only a small volume of melt—a partial melt—forms at first. In some circumstances, the melt may become segregated from the remaining solid rock (e.g., by filter pressing), which may then continue to melt on its own. The separated material will have a different mineral composition from the original rock.

- Magmas can be distinguished from one another by composition, temperature, and viscosity. Magma that has a high silica content usually has a low melting temperature and high viscosity. Medium- to high-silica magma typically forms at convergent margins, where one plate subducts under another and where water is involved in the melting process. Very high-silica magma may have passed through continental crust on its way to the surface. Low-silica basaltic magma commonly forms at divergent margins, such as midocean ridges, and at hot spots.

3 Cooling and Crystallization 174

- **Crystallization,** the process whereby mineral grains form and grow in cooling magma or lava, influences some properties of igneous rock, such as texture and grain size. The crystallization process and the final mineral assemblage in turn depend on factors such as the magma composition and rate of cooling. Magma that cools from the molten state more rapidly will have smaller grains because the grains do not have as much time to crystallize.

- Igneous rock can be classified as **volcanic** or **plutonic.** Volcanic rock crystallizes from lava; plutonic rock crystallizes underground, from magma. Because volcanic rock cools rapidly, the crystals are usually small, microscopic, or even absent, with a texture known as **aphanitic.** Plutonic rock is typically **phaneritic,** with easily visible crystal grains.

- Felsic rock is typically light in color, with high silica content; mafic rock is typically darker, with lower silica and higher iron contents. The most common volcanic rock is basalt, a dark, low-silica (mafic) rock that is the main constituent of oceanic crust. Granite, a light-colored, high-silica (felsic) plutonic rock, is a dominant constituent of continental crust.

- **Fractional crystallization** occurs when mineral grains become separated from the melt from which they are crystallizing. This separation can happen if the crystals sink to the bottom (crystal settling) or float to the top of the magma (crystal flotation). If magma flows through an opening that is too constricted to allow the crystals to pass through, separation by filter pressing may occur. The combination of different magma compositions, rates of cooling (see photo), and fractional crystallization leads to the great diversity of igneous rock types on Earth.

Plutonic rock textures Figure • 6.21

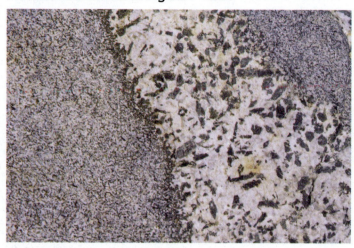

4 Plutons and Plutonism   179

- Plutonic rock bodies, or **plutons**, form underground when magma intrudes into preexisting rock strata, either cutting through them or flowing between them (see figure).

Plutons • Figure 6.23

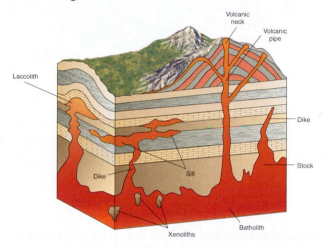

- Plutons, which occur in a variety of sizes and shapes, can be exposed at the surface when surrounding rock is stripped away by weathering and erosion. The largest type of pluton is a **batholith**, an irregularly shaped igneous body more than 50 km in diameter, which cuts across the rock it intrudes on. A stock is a smaller version of batholith. Two smaller types of plutons, dikes and sills, are indicators of past magmatic activity.

Key Terms

- aphanitic   175
- batholith   180
- crystallization   174
- fractional crystallization   178
- fractional melt   168
- fractionation   168

- lava   152
- magma   153
- phaneritic   176
- pluton   179
- plutonic rock   174
- pyroclastic flow   153

- shield volcano   156
- stratovolcano   156
- viscosity   156
- volcanic rock   174
- volcano   152

Critical and Creative Thinking Questions

1. For many years, scientists debated the reasons for the existence of Earth's geothermal gradient. What explanations can you think of for the hotter temperatures toward the center of Earth? Why is the geothermal gradient steeper under the oceans than elsewhere?

2. Volcanic rock is sometimes called **extrusive**, and plutonic rock is sometimes called **intrusive**. Why do you think geologists describe them this way? (You might want to look up these words in a dictionary.)

3. What factors might prevent magma from reaching Earth's surface?

4. The slopes of active volcanoes tend to be populated. What reasons can you think of for living near a volcano? Do you think the advantages outweigh the disadvantages? Why or why not?

5. Several flood basalt eruptions apparently occurred at roughly the same time as mass extinctions that divide geologic eras or periods. But geologists are not sure yet whether volcanic activity actually causes mass extinctions. What are some arguments for and against this theory?

What is happening in this picture?

Pumice, a volcanic rock, is so light that this person has no trouble holding a large armful.

Think Critically

1. How can rock be so light?
2. What does this rock sample tell us about the magma from which it came?

Self-Test

(Check your answers in Appendix D.)

1. _____ are explosive eruptions characterized by pyroclastic flows and ash plumes that extend into the stratosphere.
   a. Vulcanian eruptions
   b. Hawaiian eruptions
   c. Plinian eruptions
   d. Strombolian eruptions

2. _____ consist of low-viscosity lava that flows easily from a volcanic vent.
   a. Vulcanian eruptions
   b. Hawaiian eruptions
   c. Plinian eruptions
   d. Strombolian eruptions

3. These two photographs show the Mauna Loa, Mauna Kea, and Mt. St. Helens volcanoes. Of these volcanoes, which has the greatest potential for an explosive eruption?

a. Mauna Loa

b. Mt. St. Helens

c. Mauna Kea

d. They all have equal potential for an explosive volcanic eruption.

4. Which of the volcanoes shown in the photos in question 3 is being fed by magma with the highest viscosity?

a. Mauna Loa

b. Mt. St. Helens

c. Mauna Kea

d. The magma composition is probably identical for all of these volcanoes.

5. _____ form when the chamber of magma underlying a volcano empties due to eruption and the unsupported roof of the chamber collapses under its own weight.

a. Fumaroles

b. Geysers

c. Calderas

d. Craters

6. The greatest threats to human life during volcanic eruptions do not come from lava but from _____ and _____.

a. pyroclastic flows; volcanic gases

b. pyroclastic flows; ash fall

c. ash fall; mudflows

d. ash fall; volcanic gases

7. One way geologists monitor volcanic activity is by studying changes in the shape of volcanic features by using _____.

a. temperature gauges

b. seismographs

c. geologic studies of past eruptions

d. tiltmeters

8. Melting of rock can occur because of _____ and can be facilitated by the presence of water.

a. increasing temperature or increasing pressure

b. increasing temperature or decreasing pressure

c. decreasing temperature or decreasing pressure

d. decreasing temperature or increasing pressure

9. These two photographs are close-up views of two igneous rock samples. Label each appropriately with the following terms:

porphyritic texture   volcanic rock

phaneritic texture   plutonic rock

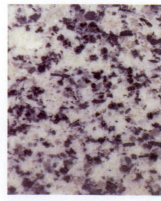
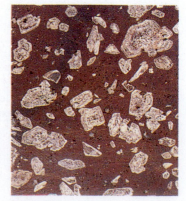

a.   b.

10. Which of the samples depicted in the photographs in question 9 cooled more slowly?

   a. The sample labeled A.

   b. The sample labeled B.

   c. Both rocks cooled quickly.

   d. Both rocks cooled slowly.

11. Which of the samples depicted in the photographs from question 9 records two distinct phases of cooling—slow followed by rapid?

   a. The sample labeled A.

   b. The sample labeled B.

   c. Neither. Both rocks cooled at the same rate.

12. These six photographs show volcanic rock samples paired with their plutonic equivalents. Identify each and label the photograph with the proper rock name from the following list:

   basalt   andesite

   granite   gabbro

   rhyolite   diorite

   1. _____   2. _____

   3. _____   4. _____

   5. _____   6. _____

13. In the process of _____, crystals that have already formed in magma become separated from the remaining melt.

   a. magmatic redistribution

   b. distributed crystallization

   c. fractional crystallization

   d. fractional melting

14. Label this block diagram, depicting various plutonic bodies, using the following terms:

   dike   volcanic neck

   sill   stock

   batholith   xenoliths

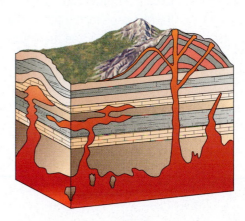

15. The plutonic features depicted in the block diagram will become exposed at Earth's surface _____.

   a. through continued volcanic eruptions

   b. through uplift and erosion

   c. through melting of the overlying rock

   d. only after an explosive volcanic eruption

THE PLANNER ✓

Review your Chapter Planner on the chapter opener and check off your completed work.

Weathering and Erosion

Utah's Bryce Canyon National Park is a fantastic landscape of multicolored spires called *hoodoos*. The Paiute Indians believed the shapes were people turned to stone as punishment for evil deeds.

Hoodoos are a testament to the relentless geological processes of weathering and erosion. They form from the narrow rock walls, or fins, that can be seen in the photograph. Water seeps into fractures and by repeatedly freezing and thawing pries the cracks open.

Boulders come loose and break away from the fin, creating a window or an arch (inset photo, at Arches National Park in Utah). Eventually the roof of the arch collapses, leaving a spire on either side.

Many of the hoodoos at Bryce Canyon have names, such as "Indian Princess" or "Three Wise Men." But they are among the most short-lived of all geologic features—each hoodoo loses about 1 meter of height per century. In a few centuries, the hoodoos of today will be weathered away.

No matter how large or magnif-icent they are, rock formations are constantly broken down by weath-ering and erosion. Without these processes, we would not have soils or beaches. At the same time, new rock is constantly forming in Earth's crust. The never-ending creation and destruction is called the rock cycle.

Global Locator

North America

Utah

NATIONAL GEOGRAPHIC

CHAPTER OUTLINE

CHAPTER PLANNER ✓

- ❏ Study the picture and read the opening story.
- ❏ Scan the Learning Objectives in each section:
  p. 196 ❏   p. 201 ❏   p. 206 ❏
- ❏ Read the text and study all visuals. Answer any questions.

Analyze key features

- ❏ What a Geologist Sees, p. 192 ❏
- ❏ Amazing Places, p. 194 ❏
- ❏ Process Diagram, p. 197 ❏
- ❏ Geology InSight, p. 198 ❏
- ❏ Case Study, p. 205 ❏
- ❏ Stop: Answer the Concept Checks before you go on:
  p. 200 ❏   p. 205 ❏   p. 212 ❏

End of chapter

- ❏ Review the Summary and Key Terms.
- ❏ Answer the Critical and Creative Thinking Questions.
- ❏ Answer What is happening in this picture?
- ❏ Complete the Self-Test and check your answers.

Weathering—The First Step in the Rock Cycle

LEARNING OBJECTIVES

1. **Introduce** the rock cycle.

2. **Name** the two main types of weathering.

3. **Describe** several processes that contribute to mechanical weathering.

4. **Explain** the three types of chemical reactions involved in chemical weathering.

5. **Summarize** the conditions of climate, topography, and rock composition that are conducive to different types of rock weathering.

Earth's surface is a meeting place. It is where the activities of Earth's internally driven processes—plate motion, seismicity, rock deformation, and volcanism—confront the quicker-paced activity of Earth's surface layers: the atmosphere, hydrosphere, and biosphere. The external forces of wind, water, and ice constantly modify the surface, cutting away material here, depositing material there, and sculpting the landscapes that surround us. But the internal forces of the tectonic cycle have an important role, too, as they recycle sedimentary rocks deposited at the surface and thrust them up again. Sometimes, in the process, the rocks change form; that is, they undergo metamorphism, or melt and form magma.

All of this activity comprises the **rock cycle**, one of the three great cycles that drive the Earth system, as shown in **Figure 7.1**. (This figure also appears in Chapters 1, 4, and 11; compare the four versions of the diagram.) The Earth system has three interconnected parts: the hydrologic cycle, the rock cycle, and the tectonic cycle. The rock cycle is in the middle because it is affected by the other two. The tectonic cycle is the subject of Chapters 4 through 6, and the rock cycle is the subject of this chapter and Chapters 8 through 10.

The rock cycle has no beginning or end; it is an endless process, powered by Earth's internal heat energy and by incoming energy from the Sun. Nevertheless, we have to jump in somewhere, so we will begin our discussion with **weathering**, the process that breaks bedrock into smaller rock and mineral fragments, and eventually into soils.

> **weathering** The chemical and physical breakdown of rock exposed to air, moisture, and living organisms.

The rock cycle • Figure 7.1

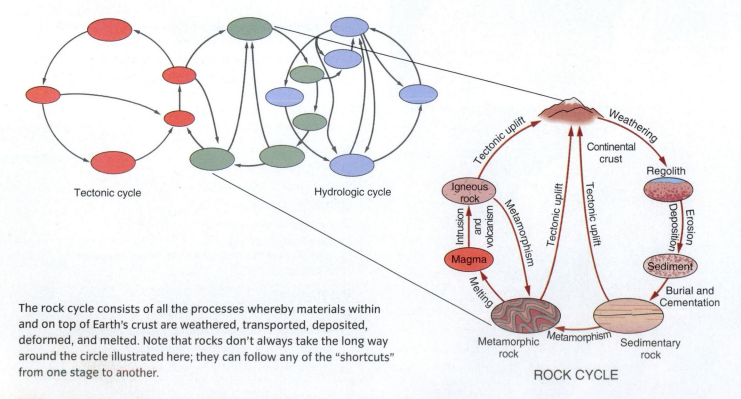

The rock cycle consists of all the processes whereby materials within and on top of Earth's crust are weathered, transported, deposited, deformed, and melted. Note that rocks don't always take the long way around the circle illustrated here; they can follow any of the "shortcuts" from one stage to another.

How rocks disintegrate • Figure 7.2

Weathering causes the progressive breakdown of rock into smaller units. Each time a cube **(a)** is subdivided, the available surface area doubles **(b)**. This renders the rock more susceptible to attack by agents of weathering. Little by little, the corners and edges become rounded, and the particles become smaller as the rock disintegrates.

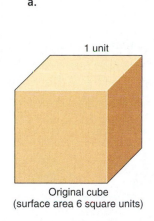
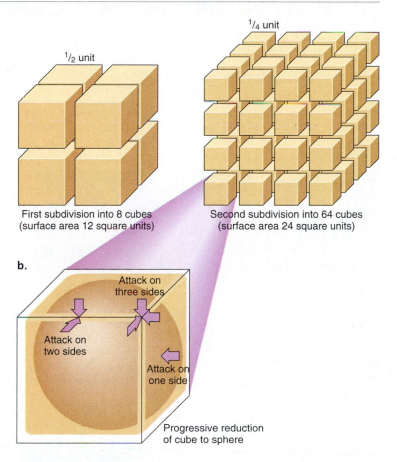

a.

1 unit

Original cube
(surface area 6 square units)

½ unit

First subdivision into 8 cubes
(surface area 12 square units)

¼ unit

Second subdivision into 64 cubes
(surface area 24 square units)

b.

Attack on three sides

Attack on two sides

Attack on one side

Progressive reduction of cube to sphere

How Rocks Disintegrate

Weathering takes place throughout the zone in which materials of the geosphere, hydrosphere, atmosphere, and biosphere can mix. This zone extends downward, below Earth's surface as far as air, water, and microscopic organisms can readily penetrate, and it ranges from 1 m to hundreds of meters in depth. Rock in the weathering zone usually contains numerous **fractures** (cracks) and **pores** (small spaces between mineral grains) through which water, air, and organisms can enter. Given enough time, they produce major changes in the rock (**Figure 7.2**).

The product of weathering is called **regolith**, from the Greek words meaning "blanket" and "stone." Fragments in the regolith range in size from microscopic to many meters across, but all of them have formed by chemical and physical breakdown of bedrock. As the particles get smaller and smaller, plants become capable of growing roots into the regolith and extracting mineral nutrients from it. At this point, the regolith becomes **soil** (**Figure 7.3**).

regolith A loose layer of broken rock and mineral fragments that covers most of Earth's surface.

soil The uppermost layer of regolith, which can support rooted plants.

From rock to soil • Figure 7.3

In this photograph taken in South Africa, soil has formed through disintegration of sedimentary rock. Water and air penetrate through fractures in the rock and react with the minerals. Rock near the surface is heavily weathered because it has been exposed to more water and air. At the surface, the rock has completely turned to soil. Weathering does not have to be accompanied by erosion. Here, the products of weathering have stayed where they formed.

WHAT A GEOLOGIST SEES

Joint Formation

a. Looking at these rocks in Joshua Tree National Monument in California, a geologist would realize from the uniform texture and lack of layering that the rocks are igneous. The geologist would look at the fractures, which he or she would understand were joints. The geologist would see that erosion has widened some of the joints to the point of detaching stones from the main rock body.

c. Now the rocks are at the surface, so a geologist would wonder about the sequence of events after the fracturing happened. The geologist would reason that as the rock rose to the surface as a consequence of the overlying rock being eroded, the pressure on the rock mass was decreased, causing the rock to expand and crack. Later, weathering rounded off and widened the joints even further.

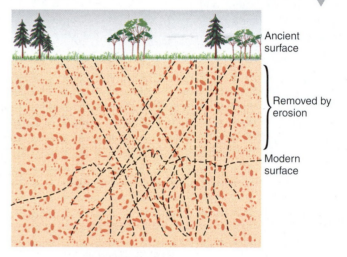

Ancient surface

Removed by erosion

Modern surface

b. A geologist would reason that the joints probably originally formed underground, where the rock mass was subject to great pressure from the overlying and surrounding rock. Squeezing and twisting by tectonic forces caused the rock to fracture and form joints.

Think Critically

1. Why are some granite bodies extensively jointed, while others are essentially joint free?
2. How might the presence of joints affect the rate of weathering?

---

mechanical weathering The breakdown of rock into solid fragments by physical processes that do not change the rock's chemical composition.

chemical weathering The decomposition of rocks and minerals by chemical and biochemical reactions.

The breakdown processes involved in weathering fall into two general categories. In **mechanical weathering**, the rock physically breaks down into pieces, but there is no change in its mineral content. **Chemical weathering** involves dissolving of minerals or chemical reactions that replace the original minerals with new minerals that are stable at Earth's surface. Although mechanical weathering is distinct from chemical weathering, the two processes almost always occur together, and their effects are sometimes difficult to separate.

Mechanical Weathering

Rocks in the upper half of the crust are brittle, and like any other brittle material, they break when they are twisted, squeezed, or stretched by tectonic forces. Though we cannot always determine the timing and origin of tectonic forces, we can see the results in the form of **joints**. *What a Geologist Sees* shows how joints commonly form.

joint A fracture in a rock, along which no appreciable movement has occurred.

Joints are the main passageways through which rainwater, air, and small organisms enter the rock and lead to mechanical and chemical weathering, as you saw in Figure 7.2. They differ from faults (Chapter 5) in that there has not been any noticeable slippage along the fracture.

Joints are not always straight. In a process called **sheet jointing** or **exfoliation** (**Figure 7.4a**), large, curved slabs of rock peel off from the surface of a uniformly textured igneous rock. As with other types of joints, sheet jointing may be due to pressure release or a combination of forces that contribute to mechanical weathering.

Mechanical weathering • Figure 7.4

a. Sheet jointing
Sheet jointing results in curved domes, like the famous Half Dome in Yosemite National Park, California.

b. Frost wedging
This granite boulder in the San Andres Mountains of New Mexico has been split apart by repeated freezing and thawing of water that penetrated along the joints.

c. A geology student takes notes on a steep scree slope in Victoria Land, Antarctica. All of the loose rubble in this photograph was separated from the bedrock by frost wedging.

d. Root Wedging
This tree began growing in a crack in the hard limestone of the Niagara Escarpment in Southern Ontario. The tree's roots have widened the crack over time.

Mechanical weathering takes place in four main ways—through freezing of water, formation of salt crystals, penetration by plant roots, and abrasion. By far the most widespread type of mechanical weathering involves the freezing of water.

Water is an unusual substance. Most liquids contract when they freeze, and the volume of the resulting solid is smaller than the volume of the liquid. However, when water freezes, it expands, increasing in volume by about 9%. If you put a full, capped bottle of water in the freezer, the bottle will burst when the water freezes because it cannot contain the larger volume of ice. Wherever temperatures fluctuate around the freezing point for part of the year, water in the ground will alternately freeze and thaw. If the water gets inside a joint in the rock, the freeze–thaw cycles act like a lever, prying the rock apart, and eventually the rock shatters. This process is known as **frost wedging** (see **Figure 7.4b** and **c**).

The formation of salt crystals can also cause mechanical weathering. Water moving slowly through rock fractures will dissolve soluble material, which may later precipitate to form salt crystals. The force exerted by growing crystals within rock cavities or along grain boundaries can cause rock to fall apart. This process occurs mostly in desert

AMAZING PLACES

Monadnock—and Monadnocks

Mt. Monadnock (**Figure a**), a 1156-m peak in New Hampshire, is one of the world's most frequently climbed mountains—no doubt because it is easy to climb yet rewards the climber with a beautiful view of all six New England states. Its name, which came from an Algonquin Indian phrase meaning "mountain standing alone," has actually become a generic term for a mountain that rises out of a surrounding plain. (A synonym used more often by geologists is **inselberg**.) Many of the world's most scenic and best-loved peaks are of this form.

Monadnocks (or inselbergs) are isolated by erosion, either because they are unjointed or because they were made of more resistant material than the surrounding landmass. They are often domed because of exfoliation, like Stone Mountain in Georgia (**Figure b**). They can be made of any of the three rock types. Mt. Monadnock is made of schist (a metamorphic rock), and Stone Mountain is made of granite. Another famous example, Uluru (also known as Ayers Rock) in Australia (**Figure c**), is made of **arkose**, a sedimentary rock.

b.

a.

c.

regions, where calcium carbonate and calcium sulfate salts are precipitated from groundwater as a result of evaporation.

Another type of mechanical weathering is caused by penetration by plant roots. Trees are very resourceful and can grow where there seems to be hardly any soil. A tree may become rooted in a crack in the bedrock, eventually widening the crack and wedging apart the bedrock, as shown in **Figure 7.4d**. Large trees swaying in the wind can also cause fractures to widen. When trees are blown over, they can cause additional fracturing. Although it is difficult to measure, the total amount of rock breakage caused by plants must be very large.

The final method of mechanical weathering is **abrasion.** This is the gradual wearing away of bedrock by the constant battering of loose particles transported by water, wind, or ice.

Some of the world's most scenic and best-loved peaks are joint-free rock masses that resist mechanical weathering because water cannot find an entry. **Monadnock,** an Algonquin Indian term meaning "mountain standing alone," has actually become a generic term for a mountain that rises out of the surrounding plain as a result of resistance to erosion. Several important monadnocks are described in *Amazing Places*.

This series of diagrams illustrates the formation of Uluru, seen here (d) in an aerial view. Around 550 million years ago, the strata of the area were folded and buckled during a tectonic event called the Petermann Ranges Orogeny. The high mountain ranges that resulted underwent rapid erosion (**Figure e**). The sediment that washed down the mountain slopes created **alluvial fans** (Chapter 8), and a remnant of one of these fans eventually became Uluru. By about 500 million years ago, the area was covered by a shallow sea (**Figure f**). The sediment in the alluvial fan was covered by sand and marine mud, and it eventually became compacted and cemented into a sedimentary rock called arkose. Between 400 and 300 million years ago, the region again underwent folding and fracturing, during the Alice Springs Orogeny (**Figure g**). The rocks that eventually became Uluru were tilted to an almost vertical position (as shown in Figures **d** and **g**), and the whole area was raised above sea level. The surrounding sandstones eroded rapidly, leaving the more resistant arkose of Uluru as the visible tip of a much larger rock stratum that extends far beneath the ground (**Figure h**).

d.

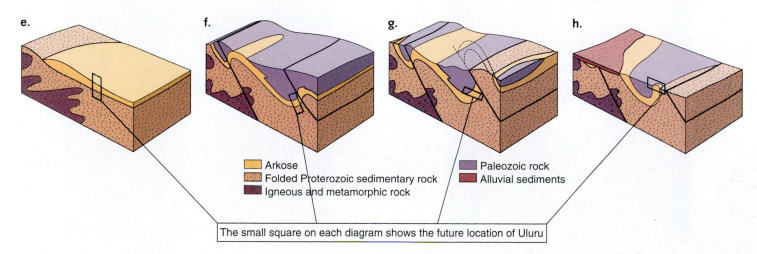

e.   f.   g.   h.

	Arkose
	Folded Proterozoic sedimentary rock
	Igneous and metamorphic rock
	Paleozoic rock
	Alluvial sediments

The small square on each diagram shows the future location of Uluru

Chemical Weathering

Chemical weathering is primarily caused by slightly acidic water. As raindrops fall through the air, they dissolve atmospheric carbon dioxide:

$$H_2O + CO_2 \rightarrow H_2CO_3$$

Rainwater thus is a weak solution of carbonic acid (H_2CO_3)—like Perrier water! When weakly acidified rainwater becomes soil water or groundwater, it may dissolve additional carbon dioxide from decaying organic matter, becoming more strongly acidified. Another way that rainwater can become acidified is by interacting with **anthropogenic** (human-generated) sulfur and nitrogen compounds released into the atmosphere. This produces a phenomenon called **acid rain.** Human-caused acid rain is stronger than natural acid rain and causes accelerated weathering.

> **dissolution** The separation of a material into ions in solution by a solvent, such as water or acid.

Through **dissolution**, minerals can be completely removed without leaving a residue. Some common rock-forming minerals, such as calcite (calcium carbonate) and dolomite (calcium magnesium carbonate), dissolve in slightly acidified water (**Figure 7.5**).

Water can also alter the mineral content of rock without dissolving all of it. One reaction of special importance in chemical weathering is **ion exchange**. Ions, which have a positive or negative electric charge, exist both in solution and in minerals. They form when atoms give up or accept electrons (see Chapter 2). Ions in minerals are tightly bonded and fixed in a crystal lattice, but ions in solutions can move about randomly and cause chemical reactions. Ion exchange is important in a common chemical weathering process known as **hydrolysis**, in which hydrogen ions (H^+) released from acidic water enter and alter a mineral by displacing larger positively charged ions such as potassium (K^+), sodium (Na^+), and magnesium (Mg^{2+}) (**Figure 7.6**). This type of weathering alters the composition of both the minerals and the water solutions that fill the pore spaces and fractures in the rock.

Where do the potassium and other ions go after they are replaced by hydrogen ions? Some remain in the

Dissolution of calcite • Figure 7.5

▼ **a.** This marble tombstone, which has stood in a New England cemetery since the early 19th century, has gradually been dissolved by rainwater. Over the years, the once sharply chiseled inscriptions have become barely legible. Marble, which contains the soluble mineral calcite, is a rock type that is vulnerable to dissolution.

▼ **b.** In the same cemetery, only a few meters away from the tombstone in **(a)**, a granite tombstone has weathered much less. Note how sharp the letters are and how fresh and unaltered the rock appears. Feldspar and quartz, the main minerals in granite, react with rainwater much more slowly than calcite does.

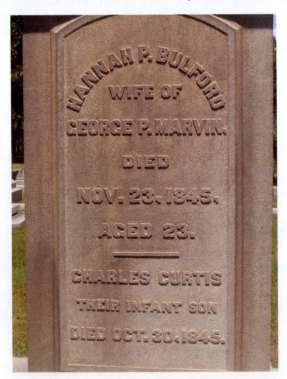

Ion exchange • Figure 7.6

This photo taken down a microscope shows where a feldspar grain has been altered by ion exchange. The clay residue has been removed to make the pattern of alteration more visible.

Clay residue forms along cleavage planes

Unaltered feldspar

1 Acidified water containing hydrogen ions (H⁺) enters feldspar crystal along existing fractures.

Potassium ion (K⁺) leaves in solution.

Unaltered feldspar

Alteration products (clay)

2 Where potassium has washed away, an insoluble residue of clay remains.

Think Critically

On Earth, clay minerals are the most common products of weathering. Samples from the surface of the Moon brought back by astronauts do not contain any clay minerals. Why?

---

groundwater, accounting for the taste of "mineral water" that some people find pleasant and others do not, and some flow out to sea and form part of the ocean's reserve of dissolved salts. On the other hand, if the water evaporates, the dissolved materials can precipitate out again as solid evaporites, such as halite and gypsum.

Another very important process of chemical weathering is **oxidation**, a reaction between minerals and oxygen dissolved in water. Oxidation and hydrolysis often operate together (**Figure 7.7**). Iron and manganese, in particular, are present in many rock-forming minerals; when such minerals undergo chemical weathering, the iron and manganese are released into the solution and are immediately combined with oxygen.

Hydrolysis and oxidation • Figure 7.7

The photograph shows two types of igneous rock, a pale granite and a darker gabbro. The granite is resistant to chemical weathering, but groundwater has seeped into joints in the gabbro and chemically attacked the rock. This chemical alteration of the gabbro has happened in two ways: first, by hydrolysis, in which original minerals are gradually altered to produce clay and soluble salts; and second, by oxidation of iron in the minerals, to produce the brown iron oxide alteration seen in the photo.

Oxidized iron commonly forms an insoluble yellowish hydrous material called **limonite**, and manganese forms an insoluble black mineral called **pyrolusite**.

Geology InSight

Tectonic setting
Young, rising mountain ranges, such as those of the Himalaya, weather very rapidly. This view of the Indus River in Pakistan shows many boulders that have separated from the bedrock because of mechanical weathering.

Rock structure
The pace of weathering is also strongly affected by the closeness of joints. Sugarloaf Mountain in Rio de Janeiro, Brazil, is a large, unjointed mass of granite, and it stands out against an otherwise deeply eroded landscape.

Unjointed rock weathers slowly.

Steep slope weathers quickly.

Topography
Weathering proceeds more quickly on a steep slope than on a gentle one. This rockslide on the island of Madeira in the Atlantic Ocean will expose new bedrock to weathering.

Factors Affecting Weathering

Many factors influence the susceptibility of a rock to chemical and mechanical weathering, as you can see in

Figure 7.8. The most important factors are tectonic setting, composition of the rock, rock structure (the abundance of openings such as joints), topography, amount of

Bacterial activity can promote chemical weathering.

Deforestation accelerates weathering and erosion

Biologic activity
Animals—even microorganisms—contribute significantly to the breakdown of rock. Here, a student lifts a mat of algae. The algae feed on a bacterium, **Thiobacillus ferrooxidans**, that thrives in acidic runoff from abandoned sulfur- and iron-bearing mines.

Resistant quartzite knob weathers slowly

Composition
Different minerals weather at different rates. Calcite weathers quickly by dissolution, and feldspar weathers at an intermediate rate by ion exchange. On the other hand, quartz is very resistant to weathering because it dissolves very slowly and is not affected by ion exchange. The knob atop Pilot Mountain, in North Carolina, is made of quartzite that has weathered much more slowly than the surrounding sedimentary rock.

Vegetation
Plants contribute to both mechanical and chemical weathering. They tend to hold a deeper regolith in place, which promotes weathering because it retains water. However, the removal of plants through slash-and-burn agriculture, clear-cut logging, or natural landslides can also accelerate weathering, as seen here in a photo of Madagascar.

vegetation and biologic activity, and climate (especially temperature and rainfall).

From the perspective of a human lifespan, the chemical weathering of rocks happens very slowly. For example, granite and other hard bedrock surfaces in New England, Canada, and Scandinavia still display polish and fine grooves made by the scraping of glaciers of the last Ice Age, which ended more than

▼ **a.** Different climates cause rock to weather at different rates and by different types of weathering.

▼ **b.** This map of North and South America illustrates some locations where the climate corresponds to the different zones of weathering shown in **a**.

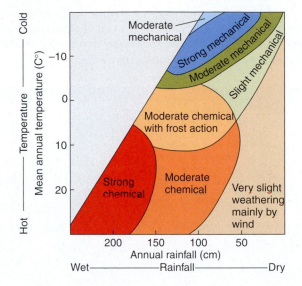

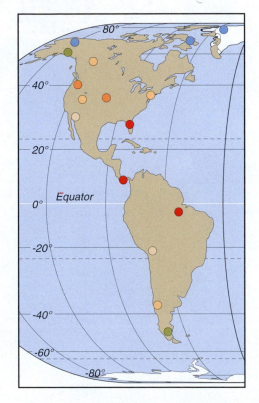

10,000 years ago. In such regions, which have plentiful rainfall but cool climates, it takes hundreds of thousands of years for a regolith to develop. In warmer regions, chemical weathering occurs more quickly at the surface and extends to depths of many tens of meters. Through the use of various dating techniques, it has been estimated that deep tropical weathering of 500 m or more requires many millions and possibly tens of millions of years.

Climate affects the rate of weathering in two different ways. As noted earlier, chemical weathering is more intense and extends to greater depths in a warm, wet, tropical climate than in a cold, dry, arctic climate. In cold, dry climates, such as in Greenland or Antarctica, chemical weathering proceeds very slowly. On the other hand, mechanical weathering is fairly rapid in these harsh environments. The only environment where both kinds of weathering proceed very slowly is a hot, dry climate (**Figure 7.9**).

CONCEPT CHECK 🛑 STOP

1. **What** are the main steps of the rock cycle?

2. **What** are the main processes that cause rocks to disintegrate?

3. **How** do joints contribute to mechanical weathering?

4. **What** are three processes that contribute to chemical weathering?

5. **Why** is mechanical weathering prevalent in Arctic regions?

Products of Weathering

LEARNING OBJECTIVES

1. **Identify** the end products of weathering.
2. **Describe** the soil horizons found in most soils.
3. **Explain** why soils from different climates have different profiles.
4. **Discuss** the links between human activity and soil erosion.

Have you ever wondered where a boulder in a mountain stream or the sand on a beach or the mud in a swamp actually comes from? The answer is the same in each case: from the mechanical and chemical weathering of rock exposed at Earth's surface. Chunks of rock freshly broken loose from bedrock are usually similar to the original material. However, with prolonged exposure to chemical weathering,

minerals that are stable at higher temperatures and pressures begin to decompose, while new minerals, stable at the conditions of Earth's surface, are formed. For example, the feldspar crystals in granite break down to form **clay** minerals (**Figure 7.10**). Although tiny (typically less than 0.004 mm), clay minerals are some of the most stable minerals on Earth's surface. They are a major component of mud, both on land and in the sea.

> **clay** A family of hydrous aluminosilicate minerals; also, tiny mineral particles of any kind that have physical properties like those of clay minerals.

On the other hand, some minerals, such as quartz, are more resistant to weathering and hence do not break down into grains as small as those found in clay. Instead,

Kaolin—From rock to fine china • Figure 7.10

The white clay found near Kao Ling, China, has for centuries been used to produce some of the finest porcelain in the world. The same kind of clay has now been found in many other places, but it is still called kaolin, in honor of the place that made it famous.

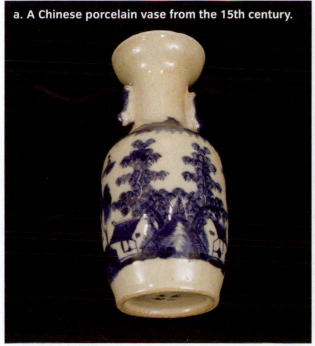

a. A Chinese porcelain vase from the 15th century.

b. A worker in Jingdezhan, China, fashions cups out of kaolin clay.

Earth and lunar "soil"—Not the same! • Figure 7.11

These microscopic views of Earth soil (**a**) and lunar regolith (**b**) show some significant differences. Most importantly, Earth soil contains organic material and hydrous minerals such as clay, while lunar regolith contains none of either.

a.   **b.**

Think Critically
Why is Earth the only planet (that we know of) where true soils occur?

---

sand A sediment made of relatively coarse mineral grains.

quartz usually ends up as **sand**, which has grain sizes as coarse as 1–2 mm—roughly 100 to 1000 times larger than the grains in clay. Sediment with grain sizes between those of sand and clay is called **silt**. The grain sizes of particles weathered from bedrock decrease with the distance traveled by the sediment from its source. Some of the finest sand is found at the beach because it has had an especially long journey from the mountains.

Soil

Perhaps the most complex product of weathering is also the most familiar: **soil**. It usually contains a mixture of minerals with different grain sizes, all the way from fine clay to coarse sand, along with some material of biologic origin. If you look at soil under a magnifying glass, you will find fragments of **humus**, and possibly some tiny insects and worms. With a very strong microscope, you might even observe bacteria living on the humus.

Soil is a complex medium in which all parts interact and play important roles. Humus retains some of the chemical nutrients released by decaying organisms and by the chemical weathering of minerals. Humus is critical to soil fertility, which is the ability of a soil to provide nutrients such as phosphorus, nitrogen, and potassium needed by growing plants. All of the processes that involve living organisms and other soil constituents produce a continuous cycling of plant nutrients between the regolith and the biosphere. Earth is the only planet in the solar system that has true soil. The other rocky bodies in the solar system have blankets of loose rocky material (**regolith**) that have sometimes been pulverized to a very fine texture, but their regoliths lack humus (**Figure 7.11**).

humus Partially decayed organic matter in soil.

Soil Profiles

Soil evolves gradually, from the top down. As erosion removes the top layer, weathering of the underlying material continually creates new soil.

soil horizon One of a succession of zones or layers within a soil profile, each with distinct physical, chemical, and biologic characteristics.

soil profile The sequence of soil horizons from the surface down to the underlying bedrock.

When fully developed, soil consists of **soil horizons**, each of which has distinct physical, chemical, and biologic characteristics. All of the soil horizons in a particular location, from the surface down to the bedrock, comprise the **soil profile**. Soil profiles (**Figure 7.12**) vary considerably, being influenced by such factors as climate, topography, and rock type. However, certain kinds of horizons are common to many profiles.

The uppermost horizon in many soil profiles, the **O horizon**, is an accumulation of organic matter. Below it lies the **A horizon**, which is typically dark in color because of the humus present. An **E horizon**, which is sometimes present below A, is typically grayish in color because it contains little humus, and the mineral grains do not have dark coatings of iron and manganese hydroxides. Both the A and E horizons have had the soluble minerals leached out of them. E horizons are most common in the acidic soils of evergreen forests.

The **B horizon** underlies the A horizon (or E, if one is present). B horizons are brownish or reddish in color because of the presence of iron hydroxides that have been transported downward from the horizons above. The B horizon is a zone of accumulation, where materials that were leached from the A horizon are redeposited. Clays are usually abundant in the B horizon. The **C horizon** (commonly known as the **subsoil**) is deepest, consisting of parent rock material in various stages of weathering. Oxidation of iron in the parent rock gives the C horizon a yellowish or rusty color.

Beneath the C horizon lies the unweathered bedrock. Geologists studying the rock of a region search for samples of "fresh," unweathered bedrock in order to form a true picture of a rock's original properties. Such samples can be found in natural outcrops along stream banks, in steep sides of mountains, in cliff faces where the soils are thin, and at artificial exposures such as highway road cuts, quarries, or mines.

As we have emphasized repeatedly in this chapter, all of Earth's systems interact at the surface. You might not think that the climate above the ground would have anything to do with the soil profile underground, but in

Soil profile • Figure 7.12

This is a typical sequence of soil horizons that would commonly develop in moist, temperate climates. The A horizon, which lies within reach of plant roots, is commonly called the **topsoil**.

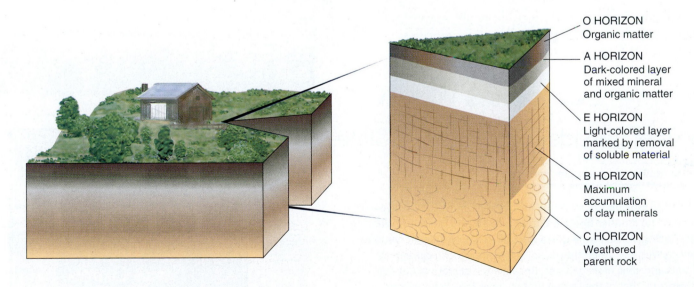

O HORIZON
Organic matter

A HORIZON
Dark-colored layer of mixed mineral and organic matter

E HORIZON
Light-colored layer marked by removal of soluble material

B HORIZON
Maximum accumulation of clay minerals

C HORIZON
Weathered parent rock

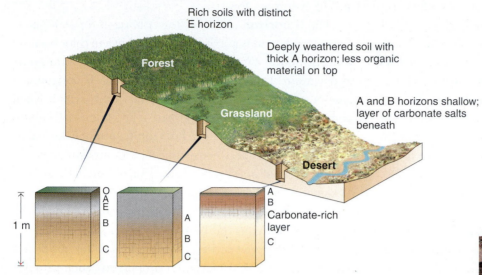

Rich soils with distinct E horizon

Deeply weathered soil with thick A horizon; less organic material on top

Forest

Grassland

A and B horizons shallow; layer of carbonate salts beneath

Desert

1 m

O A E B C

A B C

A B Carbonate-rich layer C

Soil horizons are strongly influenced by the climatic zone in which they form. For example, a layer of carbonate minerals forms in many desert soils because the hot, dry conditions cause groundwater to evaporate and precipitate the dissolved minerals. The precipitates form a hard layer, sometimes called **hardpan** or **caliche**. If the hard layer is near the surface, plant roots may not be able to grow to their normal depth.

The photographs show two similar-looking soils from different locations. The "Steedman profile," left, from Kansas, has a dark A horizon whose bottom is indicated by the white arrow. The lighter B horizon extends to the bottom of the picture, and the C horizon (not seen here) is gray and starts just below this photo. The "Windsor profile," right, from Connecticut, also has a dark A horizon and a brown-colored B horizon. The yellowish-brown C horizon is the sandy parent material, which was deposited by glacial meltwaters.

fact they are closely related. As **Figure 7.13** shows, the soil profile in an arid region is quite different from the soil profile in a region that gets plenty of rainfall.

Go to *Where Geologists Click* to see how geologists compare and contrast soil management practices in adjacent countries with different climates.

Because it is part of the never-ending rock cycle, soil is not static. Soil can be formed, and it can be depleted by natural processes. But unlike some other parts of the rock cycle, it is very strongly affected by human activities, and this makes proper soil management a vital issue (see *Case Study*).

Where Geologists CLICK

Soils of Canada and the United States

http://soils.usda.gov/gallery/photos/profiles/

Be sure to look up the state soil for the state in which you live—you will find it under Photo Gallery/State Soils in the USDA Web site. Geologists go to the USDA Natural Resources Conservation Service Web site when they wish to look into technical aspects of soils and soils management. Because the climate of Canada differs from that of the United States, a geologist compares Canadian soils practices by going to the Canadian Soil Information Service, maintained by Agriculture and Agrifood Canada.

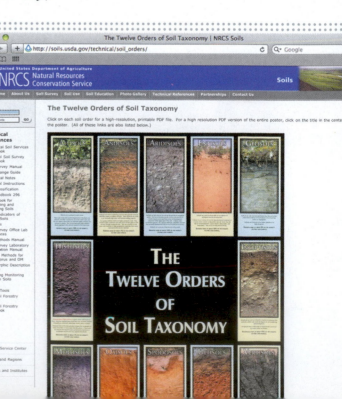

CASE STUDY

Bad and Good Soil Management

Providence Canyon in Georgia is a gorgeous example of deeply weathered soil, but it is also a dreadful example of poor soil management. In **Figure a**, in the canyon wall you can readily spot the dark brown A horizon, the bright red B horizon that is full of clay, and the paler E horizon. This is a good, productive soil. Some people have called Providence Canyon the "Little Grand Canyon," but in reality, the two canyons are very different. While the layered appearance of the Grand Canyon is due to the strata of bedrock, the appearance of Providence Canyon is caused by soil alone.

Another difference is age. Providence Canyon, believe it or not, is less than 200 years old. There was no canyon here when settlers from Europe began farming in the early 1800s. But the farmers plowed straight up and down the hills, and the furrows rapidly developed into gullies. By 1850, the gullies were 1 to 2 m deep. The farmers had to abandon their fields, but by then, erosion in the gullies was running amok. The canyon is now more

a.

Providence Canyon

NATIONAL GEOGRAPHIC

A horizon

E horizon

B horizon

b.

than 50 m deep. Unfortunately, there are many such locations in North America.

In the early 20th century, scientists involved with erosion studies pointed out that water flowing in plowed land needed to be controlled. To fight erosion, farmers should use contour plowing (**Figure b**). Instead of going in straight lines, the furrows follow the contour of the land, slowing runoff and inhibiting the formation of gullies. Crop rotation can also significantly help prevent erosion.

Despite measures such as contour plowing, the erosion of farmland soil is a massive worldwide problem. In the United States, the amount of agricultural soil eroded each year exceeds the amount of replenished soil by about 1 billion tons. For every kilogram of food we eat, the land loses 6 kilograms of soil. Although there is a small "sustainable farming" movement, we are very far from consuming only as much as we can put back. A lot more critical thinking and action are needed in order to control soil erosion.

CONCEPT CHECK 🛑 STOP

1. **What** are the primary differences between clay, sand, and soil?

2. **What** is the difference between a soil horizon and a soil profile?

3. **What** are the soil horizons normally observed in temperate climates?

4. **How** do good and bad soil management practices affect the landscape?

Erosion and Mass Wasting

LEARNING OBJECTIVES

1. **Explain** the distinction between weathering, erosion, and mass wasting.

2. **Define** turbulent and laminar flow.

3. **Describe** how ice, water, and air transport regolith across Earth's surface.

4. **Define** and give examples of mass wasting by slope failure and by sediment flow.

s distinct from **weathering**, **erosion** is a term that describes the transport of regolith from one place to another (weathering happens in place). Both processes can, of course, happen at the same time; for instance, a rock can be abraded (a weathering process), and the particles that break off the bedrock will be transported elsewhere by the wind (an erosion process).

> **erosion** The wearing-away of bedrock and transport of loosened particles by a fluid, such as water.

Erosion requires a natural fluid to pick up and transport the eroded material. That requirement differentiates erosion from mass wasting, when regolith moves downslope under the pull of gravity, with no transporting medium required. (We will discuss mass wasting later in this section.)

The fluids that cause the most erosion on Earth are water, wind, and ice. (Even though we usually think of ice as a solid, it does flow when it forms a glacier or an ice sheet.) As discussed in Chapter 6, different fluids display different resistance to flow, or **viscosity**. Ice, which flows so slowly that its motion cannot be seen by the human eye, behaves as an extremely viscous fluid. Water, which flows freely, is much less viscous, and air is the least viscous. A fluid's viscosity partly determines whether its flow is **laminar** or **turbulent**. In laminar flow, all fluid particles travel in parallel layers. Turbulent flow is erratic and complex, full of swirls and eddies. Turbulent flow is more effective at picking up particles off the ground. Air flow is almost always turbulent; water flow is usually turbulent, except when the velocity of flow is very low.

Erosion by Water

Erosion by water begins even before a distinct stream has formed on a slope. This happens in two ways: by impact, which occurs when raindrops hit the ground and dislodge small particles of soil; and by overland flow, which occurs during heavy rains. Overland flow involves water moving as sheets over the ground, not in channels. When the water starts flowing in a channel, particles are moved in several ways: The largest particles, which form the **bed load** (boulders, cobbles, and pebbles), roll or slide along the stream bed due to the force of the flowing water. Smaller (sand-sized) particles move along the stream bed by **saltation** (**Figure 7.14**).

The particles in the **suspended load**—silt and clay—are small. Although they do not actually float, they do not sink to the bottom, as long as the water is flowing. If you took a sample of water from a muddy stream and let it stand, the silt and clay would settle to the bottom as mud. But in the turbulent environment of a stream, the upward-moving currents keep them from sinking. Thus mud deposits form only where velocity decreases and turbulence ceases, as in a lake, the sea, or a reservoir.

> **bed load** Sediment that is moved along the bottom of a stream.
>
> **saltation** A mechanism of sediment transport in which particles move forward in a series of short jumps along arc-shaped paths.
>
> **suspended load** Sediment that is carried in suspension by a flowing stream of water or wind.

Streams also carry a **dissolved load** of soluble materials released by chemical weathering. The dissolved load may also contain organic matter, which accounts for the infamous "black water" found in rivers that drain swamps.

Erosion by Wind

Because the density of air is about 800 times less than that of water, air cannot move the large particles that water flowing at the same velocity can move. In exceptional cases, such as hurricanes and tornadoes, winds can reach speeds of 300 km/h and sweep up coarse rock particles several centimeters in diameter. In most regions, however, wind speeds rarely exceed 50 km/h, a velocity that is

Bed load and suspended load • Figure 7.14

a. A stream's bed load consists of the particles that are too heavy to stay suspended in the water. Pebbles creep or roll along the bottom, while sand-sized particles move in small jumps called **saltation.** Very fine silt and clay particles form the suspended load and give the water a muddy appearance.

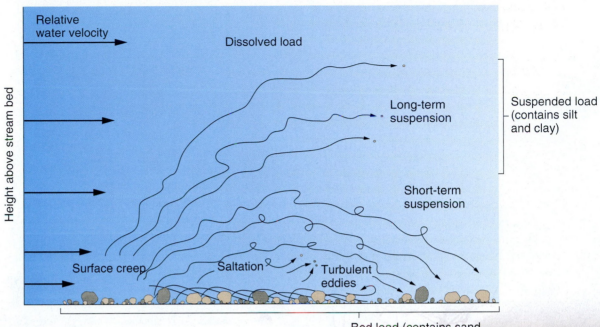

Relative water velocity

Dissolved load

Height above stream bed

Long-term suspension

Suspended load (contains silt and clay)

Short-term suspension

Surface creep Saltation Turbulent eddies

Bed load (contains sand, pebbles, boulders)

b. The turbulent flow of this river in Gabon, Africa, enables it to carry a large suspended load.

Massive dust storm • Figure 7.15

In this photo a wall of dust approaches Lubbock, Texas, the result of a massive dust storm on the afternoon of October 17, 2011.

described as a strong wind. As a result, most (at least 75%) of the sediment transported by wind occurs through saltation of sand grains. Only the finest particles, the dust, remain aloft long enough to be moved by suspension. Even so, enough suspended particles are moved during a dust storm that visibility is reduced to a few feet (**Figure 7.15**).

Erosion by Ice

Ice is a solid. However, it flows under the influence of gravity, albeit very slowly, in parts of the world where there

> **glacier** A semi-permanent or perennially frozen body of ice, consisting largely of recrystallized snow, that moves under the pull of gravity.

is enough year-round ice to form a **glacier**. Compared with water and air, ice is extremely viscous. Glacial ice therefore moves only by laminar flow.

A glacier plays a three-part role in erosion and transport: It acts as a plow, a file, and a sled. As a plow, a glacier scrapes up weathered rock and soil and plucks out blocks of bedrock (**Figure 7.16a**). As a file, the load of sediment rasps away and polishes the bedrock (**Figure 7.16b**). As a sled, a glacier carries away the load of sediment ac-

quired by plowing and filing, as well as additional debris that falls onto it from adjacent slopes (**Figure 7.16c**).

Gravity and Mass Wasting

Landscapes may seem fixed and unchanging, but if you made a time-lapse movie of almost any hillside for a few years, you would see that the slope changes constantly as a result of **mass wasting**. There is no such thing as a static hillside—a lesson learned all too often by people who live at the top, the bottom, or on a steep slope. Exactly how movement happens and how fast it happens are controlled by the composition and texture of the regolith and bedrock, the amount of air and water in the regolith, and the steepness of the slope. For convenience, we divide mass wasting into two categories: **slope failures** and **flows**.

> **mass wasting** The downslope movement of regolith and/or bedrock masses due to the pull of gravity.

> **slope failure** The falling, slumping, or sliding of relatively coherent masses of rock.

> **flow** Any mass-wasting process that involves a flowing motion of regolith containing water and/or air within its pores.

Glacial erosion • Figure 7.16

a. This rocky debris was deposited at the edge of Matunuska Glacier, Alaska. Note the extreme range in size of the rock fragments, from boulders much larger than the person in the photo down to tiny pebbles.

b. This polished and grooved rock surface was produced by the Findelen Glacier in the Swiss Alps. The glacier has retreated in modern times, exposing the weathered rock beneath. A famous mountain, the Matterhorn, is in the background.

c. Two ice streams, bearing debris that has fallen from adjacent mountain slopes, merge to form the Kaskawulsh Glacier in the Yukon, Canada. The smooth, parallel streams are a hallmark of laminar flow.

Slope Failures

Slope failures occur as one of three basic types. A **fall** (**Figure 7.17**) is a sudden vertical, or nearly vertical, drop of rock fragments or debris. Rockfalls and debris falls are sudden and usually very dangerous. **Slides** (see **Figure 7.17b**) involve rapid displacement of a mass of rock or sediment in a straight line down a steep or slippery slope. A **slump** (**Figure 7.17c**) involves **rotational** movement of rock and regolith—that is, downward and outward movement along a curved surface. Slumps often result from poor engineering practices, such as after slopes have been oversteepened for construction of buildings or roads.

Flows

Flowing regolith can be either wet or dry (**Figure 7.18**). **Slurry flows** occur when the regolith is saturated with water; they can occur either rapidly or slowly. Rapid slurry flows can move at speeds up to 160 km/h and are very dangerous. Slow slurry flow, a process known as **solifluction**, is common in areas with high rainfall, where soil is thin over bedrock, and in areas where the ground is frozen at depth. Flowing regolith that is not water saturated is called a **granular flow** (**Figure 7.18b**). Like slurry flows, granular flows can be either slow or fast. The most common kind of granular flow (and the most common kind of mass wasting)

Slope failures • Figure 7.17

a. Kaibito Canyon in Arizona has been the site of repeated rockfalls, as you can see both from the debris at the base of the cliff and from the scars on the cliff face where rocks have detached themselves in the past.

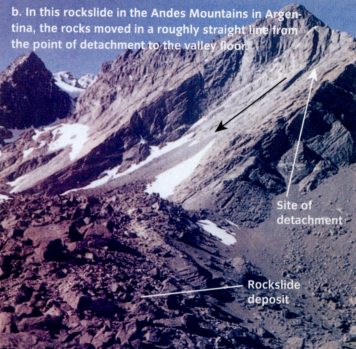

b. In this rockslide in the Andes Mountains in Argentina, the rocks moved in a roughly straight line from the point of detachment to the valley floor.

Site of detachment

Rockslide deposit

c. A slump is a slower kind of failure in which the debris moves rotationally (as shown by the curved arrow). This slumping failure occurred in central California.

Direction of motion

Downslope movement of regolith is controlled by the amount of water present and particle size.

a. Slurry (wet) flows

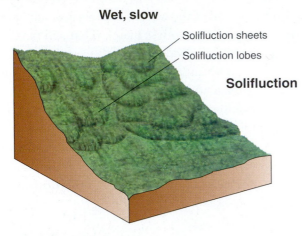

Wet, slow

Solifluction sheets
Solifluction lobes

Solifluction

Wet, fast

Mudflow

b. Granular (dry) flows

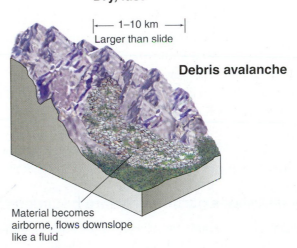

Dry, slow

Leaning or
curved tree trunk

Creep

Slope failure
under road

Leaning
gravestones

Dry, fast

|← 1–10 km →|
Larger than slide

Debris avalanche

Material becomes
airborne, flows downslope
like a fluid

---

creep The imperceptibly slow downslope granular flow of regolith.

is called **creep**. One type of rapid granular flow is **debris avalanches**, which are rare, spectacular, and extremely dangerous. A debris avalanche is likely to start as a rockfall or slide but gains speed when the material pulverizes and begins to flow downslope like a fluid. Debris avalanches can also be triggered by earthquakes or volcanic eruptions.

Factor of Safety and Landslide Prediction

It is not often possible to stop a large landslide, but with sufficient warning, it may be possible to move people and sometimes property out of harm's way. As with other natural hazards, landslide prediction relies on close monitoring of potentially unstable slopes, in an effort to identify any

sudden changes or anomalies that could indicate an upcoming failure.

For landslides, the central concern is the balance between **destabilizing** forces (pushing the material downhill) and **stabilizing**, or **resisting**, forces (holding the material in place).

The downslope forces cause **shear stress** on the slope materials. The resisting forces result from the **shear strength** of the slope materials. The ratio of shear strength to shear stress for a given slope is the **factor of safety (FS)**. If FS > 1, shear strength is greater than shear stress, and the slope is stable. If FS < 1, the slope is unstable because shear stress is greater than shear strength.

factor of safety (FS) The balance between destabilizing forces (shear stress) and stabilizing forces (shear strength) on a slope.

The factors that contribute to shear stress and shear strength are numerous and complex. The **cohesive strength** of the soil or regolith itself is the main resisting factor; this in turn depends on rock or soil properties such as the mineral composition, water content, pore spaces, and presence or absence of fractures. Downslope drivers include the weight of the soil or regolith itself, the steepness of the slope, the load on top of the slope (e.g., a building), water saturation, and many other factors. The role of a factor can sometimes vary. For example, plant roots typically bind soils and contribute to slope stability. However, vegetation can become a destabilizing influence, as when a large tree at the top of a slope is caught and pulled by the wind. The likelihood that triggering events, such as earthquakes, will occur also contributes to the potential for slope instability.

The world's major historic and prehistoric landslides tend to cluster along belts that lie close to the boundaries between converging lithospheric plates. They do so for two main reasons.

First, the world's highest mountain chains lie at or near plate boundaries, and the rocks of many mountain ranges consist of well-jointed strata that were strongly fractured and deformed as they were uplifted. Both the joint planes and the bedding surfaces are potential zones of failure. Stratovolcanoes, which are found in the same regions, also tend to have steep slopes conducive to landslides.

Second, most large earthquakes occur along the boundaries between plates, where plate margins slide past or over one another. Earthquakes often trigger landslides in areas where the regolith is unstable. Several historic landslides were directly related to major earthquakes (**Figure 7.19**).

It may seem as if landslides and mass wasting should ultimately level all of the world's mountains and leave the continents as flat, featureless plains. That will never happen because uplift is always taking place at the same time. For example, at Nanga Parbat, a Himalayan mountain that lies at the boundary between the Indian and Eurasian plates, uplift rates are as high as 5 mm/yr. At such rates, the mountain should increase in altitude by 5000 m every million years. However, high mountains also mean high erosion rates, and erosion is tearing down the mountain almost as rapidly as it is rising. Much of this destruction is a result of mass wasting.

CONCEPT CHECK   STOP

1. **How** are weathering, erosion, and mass wasting related, and in what ways are they distinct?

2. **What** are the differences between a stream's bed load, its suspended load, and its dissolved load?

3. **Which** three fluids are responsible for most of the erosion on Earth?

4. **Which** kind of mass wasting is most common, and which kinds are most dangerous?

Why landslides occur near plate boundaries • Figure 7.19

Earthquakes, particularly those close to plate edges, often trigger landslides. The subduction-related quake of March 27, 1964, the Great Alaska Earthquake, caused many slides. Here, a landslide triggered by the earthquake has covered part of the Sherman Glacier.

Think Critically

Why is mass wasting most rapid in regions of active tectonic uplift?

Summary

1 Weathering—The First Step in the Rock Cycle 190

- The rock cycle is the continuous cycle of processes by which rock is formed, modified, transported, decomposed, and reformed. Most of Earth's surface is covered by a blanket of weathered rock, which we call **regolith**. Regolith fragments can range in size from many meters to microscopic (see figure). When small regolith particles are altered by biologic processes, the result is **soil**, a material from which rooted plants can extract nutrients.

How rocks disintegrate • Figure 7.2

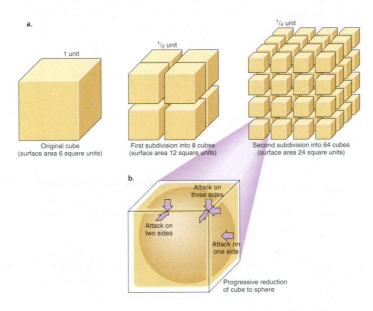

a.

1 unit

1/2 unit

1/4 unit

Original cube
(surface area 6 square units)

First subdivision into 8 cubes
(surface area 12 square units)

Second subdivision into 64 cubes
(surface area 24 square units)

b.

Attack on three sides

Attack on two sides

Attack on one side

Progressive reduction of cube to sphere

- When rocks are exposed at Earth's surface, they are constantly subjected to **weathering**, the process by which air, water, and microbes break down bedrock into smaller rock and mineral fragments. Weathering extends as far down as air, water, and living organisms can readily penetrate Earth's crust. Weathering may be **mechanical**, or **chemical**. The agents of weathering enter bedrock along **joints** and via pores.

- Joints are fractures along which no appreciable movement has occurred. Mechanical weathering takes place in four main ways: by frost wedging, or the freezing of water; by the growth of salt crystals in confined spaces; by the prying action of roots; and by abrasion.

- Chemical weathering involves the removal of some minerals in solution and the transformation of others into new minerals that are stable at Earth's surface. This type of weathering is caused primarily by water that is slightly acidic. Acid rain can be naturally occurring or human generated. Human-generated acid rain, which is created when rainwater interacts with anthropogenic sulfur and nitrogen compounds, is stronger than natural acid rain and causes accelerated weathering.

- There are several important processes by which chemical weathering occurs, including **dissolution**, in which minerals are completely dissolved and carried away in solution; ion exchange and hydrolysis, in which hydrogen ions from acidic water enter and alter a mineral by displacing larger, positively charged ions; and oxidation, a reaction between minerals and oxygen dissolved in water.

- From a human perspective, chemical and mechanical weathering occur very slowly, over many thousands of years. The effectiveness of weathering depends on the type and structure of rock, the tectonic setting, the steepness of the slope, the local climate, and the amount of biological activity. Chemical weathering is most active in moist, warm climates, whereas mechanical weathering is most active in cold, dry climates.

2 Products of Weathering 201

- If given enough time, weathering eventually breaks down rock into very fine particles known as **clay**, silt, or **sand**. Clay refers to a family of hydrous aluminosilicate minerals, but it is more commonly used to describe any tiny mineral particles that have physical properties similar to those of clay minerals. Silt particles are larger than clay, and sand is larger than silt.

- Soil is regolith that can support rooted plants. Earth's soil is different from the "soil" of other planets because it contains **humus**, or partially decayed organic matter, as well as many small living organisms. Humus retains some chemical nutrients released by decaying organisms and the chemical weathering of minerals. These nutrients, such as phosphorus, nitrogen, and potassium, are critical for soil fertility.

- Soils weather from top to bottom and develop distinctive **soil horizons**, whose properties are a function of the duration, intensity, and nature of the weathering process. In a typical **soil profile** (see figure), the O horizon is the topmost layer of accumulated organic matter. The A horizon is rich in humus; the A and E horizons are layers from which soluble material, especially iron and aluminum, has been lost through leaching. Farther down, clay minerals accumulate in the B horizon. Deeper still, the C horizon consists of slightly weathered parent rock. Soil profiles can vary greatly from location to location, as they are strongly affected by climate.

Soil profile • Figure 7.12

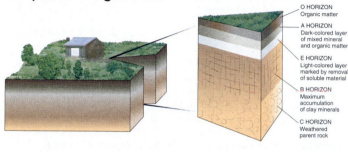

O HORIZON
Organic matter

A HORIZON
Dark-colored layer of mixed mineral and organic matter

E HORIZON
Light-colored layer marked by removal of soluble material

B HORIZON
Maximum accumulation of clay minerals

C HORIZON
Weathered parent rock

3 Erosion and Mass Wasting 206

- **Erosion** involves the removal and transport of regolith through the combined actions of ice, water, wind, and gravity. This is different from weathering, which happens in place, though both processes can occur at the same time. Most processes of erosion involve a fluid picking up and transporting material.

- Both air and water move particles by the process of **saltation**, a mechanism of sediment transport in which particles move forward in a series of short hops along arc-shaped paths. Only the smaller sand-sized particles move along the bottom of a stream by saltation, whereas the larger particles move by rolling or sliding. These larger particles are known as the **bed load**; lighter particles carried in suspension form the **suspended load** (see figure), and dissolved ions released by chemical weathering form the dissolved load. The dissolved load may also contain organic matter.

- Air and water carry suspended particles most effectively through turbulent flow. Turbulent flow is dynamic, nonlinear, and generally more effective at picking up particles off the ground. **Glaciers**, however, transport particles by laminar flow, in which the particles move in parallel layers. A glacier is a permanent body of ice consisting largely of recrystallized snow. A glacier plays a significant role in erosion and transport by acting in three ways: as a plow, as a file, and as a sled.

- **Mass wasting** is the en masse downslope movement of rock or regolith under the pull of gravity. In contrast to other types of erosion, in mass wasting the materials moved do not need to be transported by a fluid. **Slope failures** involve downslope movement of relatively coherent masses of rock or regolith. Slope failures can be one of three types: a **fall**, or a sudden vertical drop of rock fragment or debris; a **slide**, which involves rapid displacement of a mass of rock or sediment in a straight line down a steep slope; or a **slump**, a slower type of slope failure that involves rotational movement of rock.

- **Flows** are mixtures of regolith and water or air. Flowing regolith can be wet or dry. Wet flows include slurry flows that can occur rapidly or slowly; a slow slurry flow is also known as solifluction. Granular flows can be rapid, as in debris avalanches, or slow. The most common type of granular flow, which is also the least noticeable because of its slow motion, is called **creep**.

- Both weathering and erosion are controlled by climate and topography. Mass-wasting processes—especially landslides—tend to be particularly frequent along plate boundaries, where earthquakes commonly act as triggering mechanisms.

Bed load and suspended load • Figure 7.14

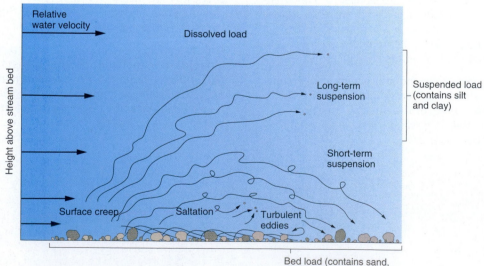

Relative water velocity

Dissolved load

Height above stream bed

Long-term suspension

Suspended load (contains silt and clay)

Short-term suspension

Surface creep

Saltation

Turbulent eddies

Bed load (contains sand, pebbles, boulders)

Key Terms

- bed load   206
- chemical weathering   192
- clay   201
- creep   211
- dissolution   196
- erosion   206
- factor of safety (FS)   211
- flow   208

- glacier   208
- humus   202
- joint   192
- mass wasting   208
- mechanical weathering   192
- regolith   191
- saltation   206
- sand   202

- slope failure   208
- soil   191
- soil horizon   203
- soil profile   203
- suspended load   206
- weathering   190

Critical and Creative Thinking Questions

1. Look around for evidence of mechanical and chemical weathering. How might you determine their relative importance in your area?

2. Many features have recently been discovered on Mars that are suggestive of erosion by water. What is the evidence that Mars once had a hydrosphere? How long ago did it have a hydrosphere? Where did the water go?

3. The Moon lacks an atmosphere, a hydrosphere, and a biosphere, but when the first astronauts landed, they discovered that the Moon has a deep regolith. How might the regolith have been formed, and how does it differ from Earth's regolith?

4. What kinds of mass-wasting processes occur where you live? Can you identify any evidence that suggests how rapidly or how slowly mass wasting is moving regolith downslope? Look especially for signs of creep, which occurs almost everywhere. Some clues are bent tree trunks, curved fences, lobes of soil on grassy slopes, and tilted gravestones.

5. Keep an eye out for the structures in your town used to stabilize slopes or protect property from mass wasting. Are the slopes in your area engineered, or have they been left more or less in their natural state? Where you find retaining walls, do they appear to have stabilized the slope as intended?

What is happening in this picture?

Mudflows are a particularly rapid and dangerous form of mass wasting. In January 2005, 400,000 tons of mud cascaded down on the California town of La Conchita, killing 10 people.

Think Critically

From a geologic point of view, a town should never have been built in this location. Why? What do you think might have caused the mudslide?

Self-Test

(Check your answers in Appendix D.)

1. On this diagram locate and label the following processes in the rock cycle:

   weathering   burial and cementation

   melting   erosion and deposition

   tectonic uplift   intrusion and volcanism

   metamorphism

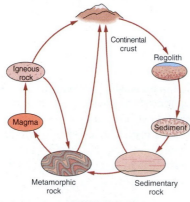

ROCK CYCLE

2. In _____, rock breaks down into solid fragments by physical processes that do not change the rock's chemical composition.

   a. chemical weathering

   b. mechanical weathering

   c. mass wasting

   d. erosion

3. This diagram shows the chemical weathering of a common feldspar mineral. The process shown depicts _____.

   a. a strong chemical reaction

   b. a moderate chemical reaction

   c. dissolution

   d. ion exchange

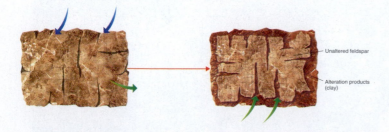

4. Death Valley in California is one of the hottest and driest spots in North America. Summer air temperatures commonly reach 50°C and rainfall averages less than 5 centimeters per year. In this desert environment, what type of weathering would you expect to find?

   a. strong chemical weathering

   b. strong mechanical weathering

   c. moderate chemical weathering

   d. moderate mechanical weathering

   e. very slight weathering, mostly by wind

5. Which of the following is true of sediments found on Earth's surface?

   a. They are one product of the mechanical weathering process.

   b. They are one product of the chemical weathering process.

   c. They are the result of a combination of mechanical and chemical weathering processes.

   d. None of the above statements is correct.

6. A dark-colored layer of mixed mineral and organic matter defines the _____ soil horizon.

   a. O soil horizon.

   b. A

   c. E

   d. B

   e. C

7. This diagram shows three climatic zones and accompanying soil profiles. Draw a line on the illustration, linking each soil profile to the correct climatic zone.

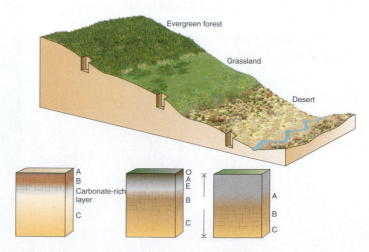

8. _____ is one method modern farmers can use to fight soil erosion.
   a. Uphill plowing
   b. Plowing along natural drainage systems
   c. Contour plowing
   d. Downhill plowing

9. _____ involves the removal and transport of regolith through the combined actions of ice, water, wind, and gravity. This is different from _____, which happens only in place, though both processes can occur at the same time. The process of _____ further alters Earth's surface by the downslope displacement of regolith due to the pull of gravity.
   a. Erosion; mass wasting; weathering
   b. Weathering; erosion; mass wasting
   c. Mass wasting; weathering; erosion
   d. Erosion; weathering; mass wasting
   e. Mass wasting; erosion; weathering

10. _____ is dynamic, nonlinear, and generally more effective at picking up particles off the ground than is _____, in which particles travel in parallel layers.
   a. Viscous flow; laminar flow
   b. Viscous flow; turbulent flow
   c. Laminar flow; turbulent flow
   d. Turbulent flow; laminar flow

11. Air and water tend to carry suspended particles most effectively through _____.
   a. a combination of viscous and laminar flow
   b. turbulent flow
   c. laminar flow
   d. a combination of turbulent and laminar flow

12. _____ transport regolith through laminar flow.
   a. Streams   c. Debris flows
   b. Winds   d. Glaciers

13. _____ are a form of slope failure involving rapid displacement of a mass of rock or sediment in a straight path down a steep or slippery slope.
   a. Rock falls   c. Slides
   b. Slumps   d. Slurries

14. _____ involve rotational movement of rock or regolith.
   a. Rock falls   c. Slides
   b. Slumps   d. Slurries

15. This illustration shows block diagrams of mass wasting of hill slopes through sediment flow. Identify each block diagram based on processes related to rate and degree of wetness from the following list:

| wet, slow | dry, slow |
| wet, fast | dry, fast |

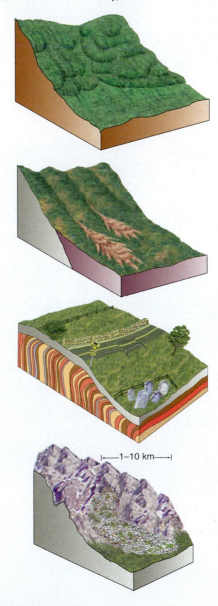

1–10 km

THE PLANNER ✓

Review your Chapter Planner on the chapter opener and check off your completed work.

8 From Sediment to Sedimentary Rock

In this photo, two divers off the island of Rat Cay, in The Bahamas, get ready to explore an underwater cave. They are standing on a pillar of limestone that rises 4 kilometers from the bottom of the ocean. Think of the structure as an underwater skyscraper; the top floor is the Bahama Islands and the lower floors are all submerged. Geologists call the top of the pillar a **carbonate platform**. A photo from space (inset) shows two such platforms in light blue—the islands are brown. The platforms are comparable in area to the peninsula of Florida, at left.

Over millions of years, the pillars grew slowly from the accumulation of coral and other marine debris. The debris combined with minerals from the ocean water and became cemented into limestone.

About 200 million years ago, when the Atlantic Ocean began to form, this region was a shallow sea, a perfect habitat for corals. As the ocean deepened, the coral kept growing new layers. Coral grows faster than the ocean floor sinks, so the carbonate platform remained below sea just the way coral likes it. Like many geologic processes, the accumulation of coral debris and its transformation into rock takes place so slowly it is difficult to imagine.

CHAPTER OUTLINE

CHAPTER PLANNER ✓

❏ Study the picture and read the opening story.
❏ Scan the Learning Objectives in each section:
   p. 220 ❏   p. 223 ❏   p. 229 ❏   p. 240 ❏
❏ Read the text and study all visuals. Answer any questions.

Analyze key features

❏ Geology InSight, p. 221 ❏   p. 223 ❏   p. 224 ❏
❏ Amazing Places, p. 231 ❏
❏ Process Diagram, p. 232 ❏
❏ What a Geologist Sees, p. 235 ❏
❏ Case Study, p. 239 ❏
❏ Stop: Answer the Concept Checks before you go on:
   p. 222 ❏   p. 229 ❏   p. 239 ❏   p. 243 ❏

End of chapter

❏ Review the Summary and Key Terms.
❏ Answer the Critical and Creative Thinking Questions.
❏ Answer What is happening in this picture?
❏ Complete the Self-Test and check your answers.

Sediment

LEARNING OBJECTIVES

1. **Distinguish** three major categories of sediment.

2. **Describe** clastic sediment in terms of size, sorting, and roundness.

3. **Explain** where and how chemical and biogenic sediments are formed.

Nearly every geologic process leaves its mark in the sedimentary record. Tectonic forces raise mountain ranges, which contribute source materials for sediment. Climatic processes control the way rock weathers and how sediment is transported. Both sets of processes influence the characteristics of the sites where sediment is **precipitated**, or **deposited**; we will consider these processes and sites later in the chapter. To understand the complicated record these events leave, we first have to understand the different kinds of sediment. Geologists separate sediment into three broad categories: clastic, chemical, and biogenic.

Clastic Sediment

Clastic sediment derives its name from **clasts**, individual grains of mineral or fragments of rock. Clasts range in size from large boulders down to clay particles finer than flour. In fact, the size of the clasts is the primary basis for classifying clastic sediment (**Figure 8.1**). Other characteristics used to describe clastic sediment include the shape, angularity, and size range of grains (**Figure 8.2**).

> **clastic sediment**
> Sediment formed from fragmented rock and mineral debris produced by weathering and erosion.

Volcaniclastic sediment is another kind of clastic sediment, in which all of the clasts are volcanic in origin. As discussed in Chapter 6, explosive volcanic eruptions blast out large quantities of fragments during an eruption. An old saying describes the uniqueness of volcaniclastic sediment: "Igneous on the way up but sedimentary on the way down." Because the fragments are hot when formed, they are also called **pyroclasts** (from the Greek

From clasts to rock • Figure 8.1

Sediment with different-sized clasts produces sedimentary rock with different names

	Gravel	Sand	Silty mud	Clayey mud
SEDIMENT	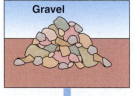	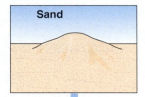	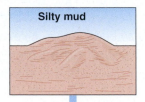	
...WITH COMPRESSION AND TIME, CAN BECOME...	A sediment with pea-sized or larger particles is called *gravel*. When gravel is cemented, the rock so formed is a **conglomerate**.	*Sand* consists of somewhat smaller particles, each about the size of a pinhead. When compacted and cemented, sand becomes **sandstone**.	Sediment with even finer particles, the size of grains of table salt, is called *silt*. The corresponding rock type is **siltstone**.	The finest sedimentary particles, the size of flour or smaller, are called *clay*.* The corresponding rock type is **shale** or **mudstone**.
ROCK				
	This shows a pebble-rich strata of conglomerates interbedded with sandstone; the area photographed is about 1 m wide is exposed in the walls of a canyon eroded by the Tsauchab River in Namibia.	This sandstone from Montana was deposited atop a stratum of pebble-rich conglomerate of Cretaceous age. This sandstone is about 7.5 cm across. The colors of depositional layers vary because of different iron contents.	Siltstone is interlayered with sandstone here, near Adelaide, Australia; the field of view is about 20 cm across.	This shale sample is about 10 cm across. The colors are caused by different contents of organic matter.

*Note that *clay* in this context refers only to particle size, not composition. |

Geology InSight

SORTING

a. In some sediment, all the particles are nearly the same size. Such sediment is said to be well sorted, and it usually has been transported by water or wind. Other sediment, such as sediment transported by ice or by mass wasting, is poorly sorted or even unsorted—a jumble of particles of different sizes.

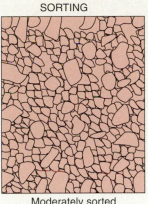
Very poorly sorted

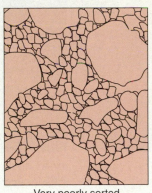
Moderately sorted

Very well sorted

ROUNDNESS

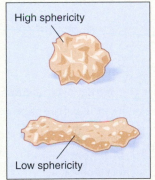

High sphericity

Low sphericity

Angular

High sphericity

Low sphericity

Intermediate

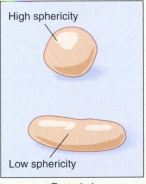

High sphericity

Low sphericity

Rounded

b. Individual particles may take a variety of shapes, from rounded to angular. Note the distinction between roundness and sphericity; even an angular particle can have high sphericity, which simply means that it is not much longer than it is wide.

c. Till, like this deposit from the Exit Glacier in Alaska, is an ice-transported sediment that is usually poorly sorted, of low sphericity, and angular in shape.

d. Quartz sand, such as this (magnified) sample from Wisconsin, tends to be well sorted, with high sphericity and roundness as a result of prolonged weathering and erosion.

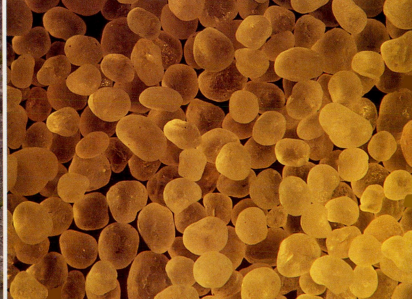

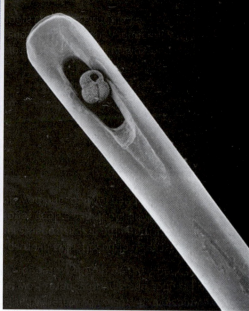

a. Utah's Bonneville Salt Flats is one of the most desolate landscapes on Earth. Chemical sediment containing magnesium and potassium chloride, in addition to ordinary salt, sodium chloride, has been deposited over the past 14,000 years by the evaporation of the prehistoric Lake Bonneville. The Great Salt Lake, a remnant of Lake Bonneville, covers less than one-tenth of the former lake's area.

b. Inside the eye of this needle is the shell of a foraminifer, one-celled plankton that are abundant in the ocean. The shells of foraminifera cover roughly one-third to one-half of the ocean floor. Over time, these biogenic sediment deposits are buried and converted into limestone or chalk.

Chemical and biogenic sediments • Figure 8.3 _____

pyro, meaning "fire"). Pyroclasts are also classified by size as **bombs**, **lapilli**, and **ash**, from largest to smallest.

Chemical and Biogenic Sediments

All surface water and groundwater contains dissolved chemicals; no natural water on or in Earth is completely free from dissolved matter. When dissolved matter precipitates from seawater or lakewater, **chemical sediment** is the result. This precipitation can happen in two ways. First, plants and animals living in the water can alter the chemical balance of the water. For example, increasing the amount of carbon dioxide dissolved in the water will cause calcium carbonate to precipitate. Many limestone deposits are formed in this way.

> **chemical sediment** Sediment formed by the precipitation of minerals dissolved in lakewater, riverwater, or seawater.

Second, if an inland sea is subjected to an increasingly warm and dry climate, or if the inflow of fresh water is restricted for some reason, the rate of evaporation may exceed the rate of input of fresh water into the water body. The sea may then become so shallow and saline that salts that were dissolved in the water will begin to precipitate as solids. Modern examples of this process are found in the Aral Sea (Uzbekistan); Mono Lake, California; and Utah's Great Salt Lake (**Figure 8.3a**).

Biogenic sediment is composed of the remains of plants and animals. This includes the hard parts of animals, such as shells, bones, and teeth, as well as fragments of plant matter, such as wood, roots, and leaves. Two of the most common rock types that come from biogenic sediment are limestone, formed from the calcium carbonate skeletons of marine invertebrates (**Figure 8.3b**), and coal, formed from partially decomposed terrestrial plant material. (Note that limestone can be either biogenic or chemical in origin.)

> **biogenic sediment** Sediment that is primarily composed of plant and animal remains or that precipitates as a result of biologic processes.

CONCEPT CHECK STOP

1. **How** do chemical and biogenic sediments differ from clastic sediment?

2. **What** terms are used to describe the individual particles that make up clastic sediment?

3. **What** are the differences and similarities in the way chemical and biogenic sediments are formed?

Transport and Deposition of Sediment

LEARNING OBJECTIVES

1. **Summarize** the main mechanisms of sediment transport.

2. **Define** deposition and describe the conditions that commonly lead to the deposition of sediment.

3. **Describe** the principal environments where deposition takes place on land.

4. **Describe** the principal environments where deposition takes place in and near the ocean.

As you learned in Chapter 7, particles generated by the weathering of rock can be transported by water, ice, and wind, or they can move downslope under the influence of gravity. The velocity and other characteristics of the transporting medium influence how fast and how far the sediment will be transported. Many of the important mechanisms of sediment transport are summarized in **Figure 8.4**.

Geology InSight Transport of sediment • Figure 8.4 ✓ THE PLANNER

Sediment can be transported by flowing wind, water, and ice. It also can move downslope under the influence of gravity. The major modes of sediment transport are illustrated here.

A dust storm strikes the Darfur region of Sudan, 2007.

Rock falls like this one in the Canadian Rocky Mountains, Alberta, are gravity-driven.

Glacial transport

Stream transport

Mass wasting

Wind transport

Longshore transport

Turbidity currents

Bottom currents

Shelf

Slope

Abyssal plain

Sediment is carried alongshore by waves and currents in Irian Jaya, Indonesia

Ribbons of sediment are transported by the Barnard Glacier, Alaska.

Wherever the carrying capacity of the transporting agent decreases sufficiently for sediment to settle out, **de-position** occurs. Thus, common environments for the deposition of sediment include stream channels and floodplains, lakeshores and lake bottoms, the margins of glaciers, and areas where the wind is intermittently strong, such as beaches and deserts (**Figure 8.5**). Offshore, sediment is deposited where rivers enter the ocean and where oceanic currents move sea-floor sediment.

> **deposition** The laying down of sediment.

Let's look more closely at the principal environments where the deposition of sediment occurs on land, in the ocean, and in coastal zones.

Depositional Environments on Land

There are four principal sites of deposition on land: in stream valleys, lakes, adjacent to glaciers, and wherever winds blow and pick up sediment.

Streams and floodplains Streams are the main transporters of sediment on land. Sediment deposited by streams varies from place to place, depending on the type of stream, the strength of flow, and the sediment load (see Figure 8.5). Generally speaking, stream-carried sediment becomes better sorted and more rounded with distance from the source. Thus, a typical large, smoothly flowing stream may deposit well-sorted coarse and fine particles as it slowly migrates back and forth across its valley. During spring floods, silt and clay are deposited on the floodplain. In contrast, a large mountain stream flowing down a steep valley transports a wide range of particle sizes, including large, angular grains and a high proportion of rock fragments. Such "immature" sediment may contain chemically less-stable minerals, such as amphiboles and feldspars that have not yet had enough time to undergo chemical weathering. When the stream reaches the front of the mountain and is no longer constrained by valley walls, it may spread out into an **alluvial fan**, in which the sediment ranges from coarse, poorly sorted gravel upstream to well-sorted sand downstream.

Lakes Sediment that is deposited in a lake (see Figure 8.5) accumulates on the lakeshore and on the lake floor. The sediment load of a stream entering a lake will be dropped as the stream's velocity and transporting ability suddenly decrease, forming a **delta**. Inclined, generally well-sorted layers on the front of a delta pass downward and outward into thinner, finer-grained, evenly laminated layers on the lake floor. In arid regions, seasonal lakes, or **playas**, are common: They fill up with water for part of the year and then leave deposits of evaporated salts when they dry out.

> **delta** A sedimentary deposit, commonly triangle-shaped, that forms where a stream enters a standing body of water.

Lake deposits may appear similar in many ways to marine sediment. Among the distinguishing characteristics are the generally smaller extent of lake deposits (lakes are typically smaller than seas) and the presence of freshwater fossils instead of marine fossils. As you might expect, lake deposits are less common in the geologic record than marine sediment.

Glaciers Sediment transported by a glacier is either deposited along the glacier's base or released at the margin of the glacier as melting occurs. The sediment may then be subjected to further reworking by running water. Debris that has been deposited directly from ice commonly forms a random mixture of particles that range in size from clay to boulders and consist of all the types of rock over which the ice has passed. This type of sediment is called **till**. Glacial deposits are discussed in greater detail in Chapter 13.

Wind Processes related to the wind are referred to as **eolian** (pronounced "ee-OHL-ee-un"), after Aeolus, the Greek god of wind. **Eolian sediment** tends to be finer than that moved by other erosional agents. Grains of sand are easily transported in places where strong winds are blowing and vegetation is too sparse to stabilize the land surface—for example, seacoasts and deserts. In such places, windblown sand may pile up to form dunes composed of well-sorted grains, with bedding inclined in

> **eolian sediment** Sediment that is carried and deposited by wind.

Geology InSight

Sediment of all sizes has been deposited in this moraine of the Ghiacciaio Dei Forni in the Italian Alps.

A river on Russia's Kamchatka Peninsula contributes sediment that forms a classic fan-shaped delta where the river enters the ocean.

Various depositional environments, both terrestrial and marine, are illustrated here. Each environment has its own unique collection of sediment.

Glacier

Glacial environment

Alluvial fan

Braided meltwater streams

Lake environment

Stream environment

Estuarine environment

Shallow marine environment

Continental shelf

Meandering streams

Playa lake

Sand dunes

Estuary

Delta

Beach

Spit

Bar

Submarine valley

Continental slope

Submarine canyon

Eolian environment

Shore environment

Deep marine environment

Dry lake bed, or playa, in Nevada. Playas may intermittently be covered with water, but normally they have the salt-encrusted appearance seen here.

Sand dunes, like these in the Namib Desert, tend to form in dry environments dominated by aeolian (wind) transport of sediment.

Soil from the last ice age • Figure 8.6

The Loess Hills in western Iowa stand out prominently against the flat Missouri River plain in the distance. The most recent ice age produced copious amounts of glacial sediment, which was transported by wind all over midwestern North America. Here, the loess piled up in large dunes. Though the Loess Hills are too steep for farming, soils developed on deposits like these are highly productive for agriculture, and underlie many of the world's breadbasket regions.

the downwind direction (i.e., the direction toward which the air is flowing). Using these characteristics, geologists can easily identify ancient dune sand in the rock record.

Powdery dust that has been picked up and moved by the wind may travel great distances and be deposited thousands of kilometers away. For example, oceanographers have discovered windblown sediment from the Sahara Desert of North Africa on the other side of the Atlantic Ocean. One important type of windblown sediment is **loess** (a German word meaning "loose" and pronounced "luhss"). Consisting predominantly of yellow-brown silt, loess is windblown dust transported from desert surfaces, glacial sediment, and glacial stream deposits at times of ice-sheet retreat (**Figure 8.6**).

Depositional Environments in and near the Ocean

Sediment can be deposited anywhere in and around the edges of the ocean. In the headings below we discuss the most important sites.

Deltas and estuaries Delta deposits form where streams flow into lakes or seas. Large marine deltas are complex deposits consisting of coarse stream-channel sediment, fine sediment deposited between channels, and still finer sediment deposited on the seafloor.

Where the load of sediment carried by a stream to the sea is smaller, much of it may be trapped in an **estuary** (**Figure 8.7a**; see also Figure 8.5a and b). In estuaries, coarse sediment tends to settle close to land, whereas fine sediment is carried farther seaward. Tiny individual particles of clay carried in suspension settle very slowly to the seafloor. As a transitional environment between land and sea, estuarine sediment

> **estuary** A semi-enclosed body of coastal water, in which fresh water mixes with seawater.

often contains a large amount of organic matter, including fossils of organisms from both land and sea.

Beaches Quartz, the most durable of the common minerals in continental rock, is a typical component of beach sand. However, not all ocean beaches are sandy. Any beach consists of the coarsest rock particles contributed by the erosion of adjacent sea cliffs, together with materials carried to it by rivers or by currents moving along the shore. Beach sediment tends to be better sorted than stream sediment of comparable coarseness. Particles of beach sediment, dragged back and forth by the surf and turned over and over, become rounded by abrasion. Though beach sands are often buff colored, they don't have to be; they may also be white, black, or even green, reflecting the presence of different minerals (**Figure 8.7b**).

a. Where stream meets sea
Where the Tijuana River (on the border between the United States and Mexico) flows into the Pacific Ocean, it deposits sediment—as well as sewage and trash from both sides of the border—into an estuary. Here, seagulls congregate on sandbars formed by sediment.

b. Beach
All beaches contain material that has been transported from somewhere else. The unusual color of Hawaii's Green Sand Beach comes from a high concentration of the mineral olivine in its sand (magnified in the inset). The beach is surrounded on three sides by the eroded walls of a cinder cone called Puu Mahana (on the left in this photo). The basalt in these walls contains olivine, which washes down onto the beach. The glassy-looking fragments of olivine are not very resistant to weathering, so we would only expect to find a green sand beach very close to its volcanic source.

c. Carbonate shelf
Carbonate sediment accumulates in the warm, shallow offshore waters of these islands in The Bahamas. The sediment shows up as white sandbars and consists of fine skeletal debris of tiny sea creatures.

d. Deep-sea turbidites
These deep-sea turbidite beds have been tilted, uplifted, and exposed in a wave-eroded bench along the coast of the Olympic Peninsula in Washington.

Shelves Most of the world's sedimentary rocks originate as strata on continental shelves. As fresh water flows out to sea through an estuary or a river mouth, it continues seaward across the continental shelf. It will deposit most of its sand-sized sediment, whether river derived or formed by erosion of the shore, within about 5 km of the land. However, some sand can be found 100 km or more offshore. Otherwise, continental shelf sediment tends to consist of silty or sandy mud that contains marine fossils.

On the continental shelf of eastern North America, up to a 14-km thickness of fine sediment has accumulated over the past 150 million years. Shelves, in effect, catch weathered continental crust in such a way that it is continually recycled by the processes of plate tectonics and the rock cycle. Because of the abundance of marine life in shallow shelf waters, continental shelf sediment generally contains a high percentage of organic matter.

Carbonate platforms and reefs Where the climate and surface temperature are warm enough to nurture abundant carbonate-secreting organisms, biogenic sediment composed of calcium carbonate may accumulate on continental shelves or broad, flat carbonate platforms that rise from the seafloor (**Figure 8.7c**). A **reef** is a wave-resistant structure built from the skeletons of marine invertebrates. Reefs are generally restricted to warm, sunlit waters of normal marine salinity.

Marine evaporite basins In coastal areas with a sufficiently warm and dry climate, ocean water may evaporate fast enough to leave behind the salts that were dissolved in it, as a marine **evaporite** deposit. These can be distinguished from lake-derived evaporites because they have a different mineral composition. Marine evaporite deposits are surprisingly common (reflecting the many times that shallow seas have flooded large areas of continents in the past) and underlie as much as 30% of the land area of North America.

> **evaporite** A rock formed by the evaporation of lakewater or seawater, followed by lithification of the resulting salt deposit.

Turbidites Thick sediment deposits are found at the foot of the continental slope, at depths as great as 5 km beneath the surface of the ocean. The origin of these deposits was difficult to explain until marine geologists discovered that **turbidity currents**—essentially, underwater landslides that originate on the continental shelf—deposit them. These currents of sediment, sometimes started by the shaking from an earthquake, rush swiftly down the continental slope at velocities of up to 90 km/h. When a turbidity current reaches the ocean floor, it slows down and deposits a graded layer of sediment called a **turbidite**. (To view a turbidity current in a laboratory experiment, see *Where Geologists Click*.) At any site on the continental rise, a major turbidity current is a rare event that happens perhaps once every few thousand years. Nevertheless, over millions of years, turbidites can slowly accumulate and form thick deposits consisting of many layers (**Figure 8.7d**).

> **turbidity current** A turbulent, gravity-driven flow consisting of a mixture of sediment and water, which conveys sediment from the continental shelf to the deep sea.

Where Geologists CLICK
The Geological Society of London

www.geolsoc.org.uk/gsl/education/resources/rockcycle/pid/3660

The Geological Society Web site is a fine source of information of interest to geologists and students of geology. An example of what you will find on the site is the article and video called "How do turbidity currents work?" To view turbidity currents happening in the ocean deeps would be far too dangerous, so geologists wishing to view the process would go to this site to see a turbidity current under laboratory conditions.

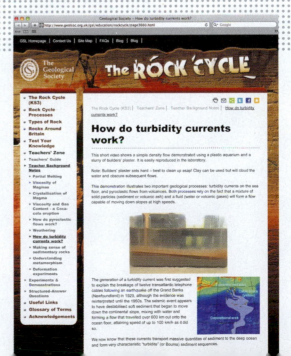

Seafloor As mentioned earlier, a large part of the deep seafloor is covered with biogenic sediment. These deposits come in two principal varieties. **Calcareous ooze**, made of calcium carbonate, comes from the remains of tiny sea creatures. While they are alive, these creatures float, but after they die, their shells drop to the bottom of the ocean. Calcareous ooze forms at low to middle latitudes, where the water is warm. Other parts of the deep ocean floor are mantled with **siliceous ooze** from silica-secreting organisms. (Siliceous ooze is chemically similar to quartz but differs in terms of mineral structure.) This material is most common in the equatorial Pacific and Indian oceans and in a belt encircling the Antarctic continent. These are areas where the biologic productivity of surface waters is high, due partly to the upwelling of deep ocean water that is rich in nutrients.

CONCEPT CHECK STOP

1. **How** do particles transported and deposited by a glacier differ from those transported by wind and water?

2. **Where** in a stream bed would you expect to find each of the following types of sediment: gravel, sand, and mud?

3. **What** kinds of sedimentary deposits would you expect to find at the foot of a mountain range? In a desert?

4. **What** kinds of sedimentary deposits would you expect to find on a beach? On the seafloor?

Sedimentary Rock

LEARNING OBJECTIVES

1. **Relate** the appearance of sedimentary rock to its mode of formation.

2. **Summarize** three processes that lead to the lithification of sediment.

3. **Identify** the most common clastic, chemical, and biogenic sedimentary rock types.

4. **Explain** how features such as ripple marks, cracks, and fossils can tell geologists about the environment in which a rock originated.

5. **Describe** the concept of sedimentary facies.

L ithification, the group of processes by which sediment is transformed into rock, is one of the steps in the rock cycle. Sediment originates from the erosion of preexisting rocks that have been broken down through weathering (Chapter 7). When sediment is transported, deposited, and eventually buried, it lithifies to become sedimentary rock. In this section, we discuss how lithification occurs and how the appearance of the new rock depends on both the materials and the process by which it is created.

> **lithification** The group of processes by which loose sediment is transformed into sedimentary rock.

Rock Beds

When you look at an outcrop of sedimentary rock, such as the one shown in **Figure 8.8**, one of the first things you notice is the **bedding**. As discussed in Chapter 3, the banded appearance comes from the fact that sedimentary

> **bedding** The layered arrangement of strata in a body of sediment or sedimentary rock.

Layers of rock: Bedding • Figure 8.8

The Bungle Bungle Range in northwestern Australia derives its unique coloration from layers of sandstone that have different permeabilities. Algae grow in the more permeable strata, tinting the rock black.

particles are laid down in distinct strata. Over time, the mineral composition of the sediment in a particular location

may change, or the sediment may be transported or deposited in different ways. This causes subsequent strata to look different. The boundary between adjacent strata is called a **bedding surface**. The presence of bedding and bedding surfaces tells geologists that the rock was once sediment.

The appearance of a bed can tell a geologist a great deal about how the sediment was deposited. For exam-

ple, in a **graded bed**, the coarse clasts are concentrated at the bottom, grading up to the finest clasts at the top (**Figure 8.9**). Graded beds often form where a stream or river enters a lake or an ocean.

Turbulent flow in streams, wind, or ocean waves produces a type of bedding called **cross bedding**. The thick strata of sandstone contain many thin beds that are inclined with respect to the stratum in which they occur, as shown in *Amazing Places*. The direction in which the cross bedding is inclined can tell geologists the direction in which the water or air currents were moving when they deposited the sediment.

Clues to a bed's origin • Figure 8.9

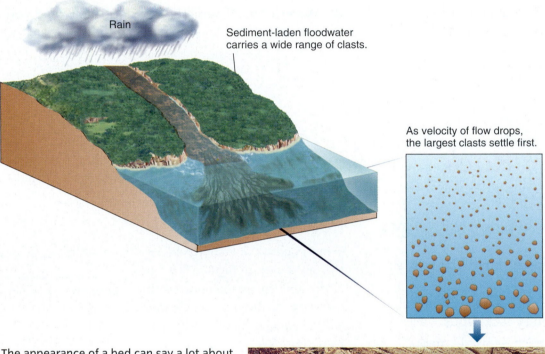

The appearance of a bed can say a lot about its origin. The center stratum in this rock from California is a graded bed, which is produced very quickly—by a single flood. Floods or rainfall cause a swollen river to flow faster and pick up a mixture of fine and coarse sediment. When this rapid, sediment-laden flow reaches a lake or the ocean, its velocity drops, and the larger clasts settle to the bottom first. After the sediment hardens into rock, it forms a graded bed with the largest clasts on the bottom, like the one shown here.

AMAZING PLACES

The Navajo Sandstone

Global Locator

Zion National Park

NATIONAL GEOGRAPHIC

Eolian transport

Sand sea

Stream

Transport

\\\ Appalachians   ⧄ Grenville Mountains

Many of the rock formations in Utah's Zion National Park resemble sand dunes—because that is what they originally were. The imposing cliffs and hills were all once part of a vast sea of sand, larger and thicker than the Sahara Desert is today. The dunes lithified into sandstone and subsequently eroded into the formations you see today. The sedimentary formation they belong to, called the Navajo Sandstone, extends over several states in the American Southwest and attains a thickness of 700 meters in Zion National Park.

But if all of this rock was once sand, where did all the sand come from? The dune patterns indicate that the prevailing winds came from the north, but no mountain ranges of suitable size or age can be found there. The mountain range we call the Rockies did not yet exist in the Lower Jurassic Period, 190 million years ago, when the Navajo sand was being deposited.

However, the Appalachian Mountains and remnants of the more ancient Grenville Mountains did exist then. Lifted up by a plate tectonic collision that began 500 million years ago, they were once as tall as the Himalaya are today and extended all the way into what is now Texas. By the Lower Jurassic, the Appalachians had eroded considerably—and geologists now think that it was their sediment that formed the Navajo Sandstone. Rivers flowing westward carried the erosional sediment toward what is today the center of the continent (inset). As the climate became arid, winds transported the now-dried river sands southward, where they became one of the largest sand seas that has ever existed on Earth.

PROCESS DIAGRAM

Lithification • Figure 8.10

THE PLANNER

① COMPACTION

The weight of accumulating sediment forces the grains together, thereby reducing the *pore space* and forcing water out of the sediment.

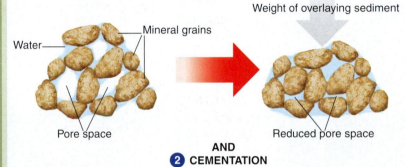

Weight of overlaying sediment

Water — Mineral grains

Pore space

Reduced pore space

AND
② CEMENTATION

Pore water expelled from deeply buried sediments migrates upward toward Earth's surface.

As the water rises and cools, ions dissolved in the water precipitate, forming minerals that cement the grains together.

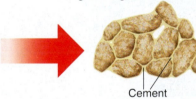
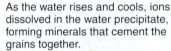

Cement

AND/OR
③ RECRYSTALLIZATION

Pressure causes less stable minerals to rearrange crystals into more stable forms. Aragonite is present in the skeletal structures of living corals and other marine invertebrates.

Over time, aragonite recrystallizes and becomes calcite, which has a different crystalline structure.

Shells made of aragonite

Aragonite in the shells has been transformed into calcite.

Think Critically

Consider the chemical composition of Earth's crust. What would you predict is the most common mineral cement in clastic sedimentary rock?

Lithification

In order for newly deposited, loose sediment to be lithified and turned into rock, the individual particles must somehow be bound together into a cohesive unit. After a layer of sediment is buried, either by the accumulation of more sediment or by tectonic processes, it is placed under higher pressure, leading to **compaction**.

compaction Reduction of pore space in a sediment as a result of the weight of overlying sediment.

Compaction is generally the first step in lithification, which can proceed in several ways. One common process that often happens during lithification is **cementation**, in which the grains of sediment are joined together. Cementation can happen in various ways, such as the evaporation of groundwater under desert conditions. As the water evaporates, chemicals such as silica, calcium carbonate, and iron hydroxide precipitate and cement the grains of sediment together. Another lithification process is **recrystallization**, which is especially common in limestone formed by coral reefs. Calcium carbonate crystals in the fossilized reef change from their original form (aragonite) to the more stable calcite. In the process of recrystallizing, crystals that were separate can grow together.

cementation The process in which substances dissolved in pore water precipitate out and form a matrix in which grains of sediment are joined together.

recrystallization The formation of new crystalline mineral grains from old ones.

Pressure caused by sediment accumulation or tectonic forces initiates the lithification process, as you can see in **Figure 8.10**. The pressures involved in lithification are high by everyday human standards, but they are low pressures by geologic standards, and they are very low in comparison with the pressures that induce metamorphism (Chapter 9). All of the low-temperature, low-pressure changes that happen to sediment after deposition are collectively called **diagenesis**. They include lithification as well as processes involving chemical reactions or microbial activity that are not part of lithification.

Types of Sedimentary Rock

If a particular sedimentary rock is made up of particles derived from the weathering and erosion of igneous rock, it may contain many of the same minerals as the source rock. How, then, can we tell that it is sedimentary rather than igneous? In addition to such clues as bedding, the texture of the rock provides evidence (**Figure 8.11**).

Clastic sedimentary rock When clastic sediment lithifies, it produces a rock whose properties reflect the type of sediment it came from. The four basic classes are conglomerate, sandstone, mudstone, and shale. These are the rock equivalents of gravel, sand, silt, and clay-rich sediment (see Figure 8.1).

To be classified as **conglomerate**, a sedimentary rock must have clasts larger than 2 millimeters. Typically, the large clasts are surrounded by much finer-grained material, called the **matrix**. If the clasts are angular, rather than rounded, the rock is

> **conglomerate**
> Clastic sedimentary rock with large fragments in a finer-grained matrix.

Thinking critically about rock identification: Sedimentary or igneous? • Figure 8.11

Two of these rocks are sedimentary, and one is igneous. How can you tell which is which?

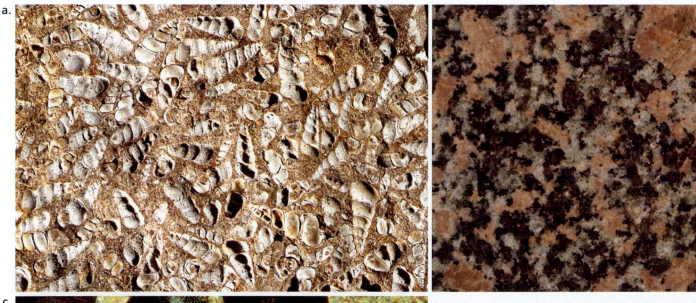

a.

b.

c.

Answers

a. The fossils in this specimen are a give-away. No organism can survive the high temperatures at which igneous rock are formed, so the presence of ancient shells means the rock (limestone) is sedimentary.

b. Looking at this sample, we see interlocking, irregularly shaped crystals without any cement. This rock is granite, an igneous rock.

c. In the magnified view of this sample, we see mineral grains rounded by abrasion. The grains do not abut one another; they have gaps between them that are filled by cement. This rock is sedimentary. (It is sandstone, held together by calcite cement.)

called **breccia**. The presence of angular clasts means that the sediment has been transported only a short distance and was not subjected to a long abrasion process.

sandstone
Medium-grained clastic sedimentary rock in which the clasts are typically, but not necessarily, dominated by quartz grains.

The medium-sized grains in **sandstone** range from 0.05 to 2 mm in size. They are usually dominated by quartz because quartz is a tough mineral that resists weathering. If the sediment has not been transported very far from its source, it may still contain a lot of feldspar or rock fragments. Geologists would call such a sandstone "immature."

Mudstone, often called **mudrock**, consists primarily of silt- and clay-sized particles; these include tiny pieces of rock and mineral grains, as well as clay minerals. Mudrock, as a group, includes the rock types **siltstone**, **mudstone**, and **claystone**—all very fine-grained sedimentary rock types that are distinguished from one another on the basis of their relative proportions of silt-sized to clay-sized particles. Rock that is classified as mudrock breaks into blocky fragments. In contrast, **shale**, a subvariety of mudrock, is **fissile**, which means that it splits into sheet-like fragments.

mudstone A group of very fine-grained, non-fissile sedimentary rock types with differing proportions of silt- and clay-sized particles.

shale Very fine-grained fissile or laminated sedimentary rock, consisting primarily of silt- or clay-sized particles; a fissile mudstone.

Chemical sedimentary rock
Chemical sedimentary rock results from the lithification of chemical sediment. As we have seen, such sediment is formed by the chemical precipitation of minerals from water. Most chemical sedimentary rock types contain only one important mineral, such as calcite, dolomite, gypsum, or halite. These monomineralic compositions, together with their modes of precipitation, provide the basis for classification of these rock types.

Chemical sedimentary rock often forms as **evaporite** deposits. Calcite, gypsum, and halite typically form from the evaporation of seawater, while the evaporation of lakewater may yield more exotic minerals (e.g., sodium carbonate, borax). Many evaporite minerals are mined because they have industrial uses; for example, gypsum is used to make plasterboard. In addition, most of the salt we eat comes from evaporite deposits.

An unusual but economically important kind of chemical sedimentary rock is a **banded iron formation**. Such rock is the source of most of the iron mined today. Not only are banded iron formations valuable for their ore, they also tell the story of a critical period in Earth's history. Almost all banded iron formations are about the same age—1.8 to 2.5 billion years old. This strongly suggests that unique conditions existed on Earth during that period (see *What a Geologist Sees*).

banded iron formation A type of chemical sedimentary rock rich in iron minerals and silica.

Biogenic sedimentary rock
The most abundant biogenic sedimentary rock is **limestone**. It is formed from lithified shells and other skeletal material from marine organisms. Some of these organisms build their shells or skeletons from calcite, but most construct them from the mineral aragonite, which, like calcite, is composed of calcium carbonate. During diagenesis, the aragonite is transformed into the more stable mineral calcite, the main ingredient of limestone.

limestone A sedimentary rock that consists primarily of the mineral calcite.

Calcite is sometimes replaced by the mineral dolomite (a carbonate mineral that contains both magnesium and calcium); the resulting rock is called **dolostone**. Another kind of biogenic rock, which consists of extremely tiny particles of quartz, is called **chert**. The quartz in chert does not come from sand but from the shells of microscopic sea animals.

An important class of biogenic sediment consists of the accumulated remains of terrestrial plants. Over time, and with pressure, this mass gradually becomes **peat**. Eventually, given enough time and pressure, peat may lithify and become **coal**. The process of lithification (in this case, called **coalification**) involves further compaction, release of water, and slow chemical changes that weld the plant fragments together, thereby making the coal relatively lower in water and richer in carbon than the original peat.

peat A biogenic sediment formed from the accumulation and compaction of plant remains.

coal A combustible rock formed from the lithification of plant-rich sediment.

A Change in the Atmosphere

The geologist studying this unusual rock outcrop realizes from the brown color that it is an iron-rich rock. Closer examination (inset) shows that it is composed of thin iron-rich layers alternating with white silica-rich layers. The rock type is called **banded iron formation**. There are no obvious clastic grains present, suggesting that the rock must be the remains of a chemical precipitate. Our geologist hypothesizes that this 2.5-billion-year-old stratum in the Hamersley Range in Australia was formed when iron that was dissolved in seawater precipitated as chemical sediment. Today, seawater contains only slight traces of iron because oxygen in the atmosphere reacts with it to form insoluble iron compounds. If the ocean was once rich in dissolved iron, there must have been very little oxygen in the atmosphere at that time.

The geologist asks the obvious question: How did Earth make the transformation from an oxygen-poor atmosphere 2.5 billion years ago to the oxygen-rich atmosphere of today? In some 2.5-billion-year-old rocks, there are microscopic fossils of cyanobacteria. These bacteria are thought to be the first organisms on Earth to extract energy from sunlight by photosynthesis, a chemical process that releases oxygen. Scientists hypothesize that algae might have oxygenated Earth's atmosphere very rapidly—and possibly more than once, as they proliferated and then poisoned themselves by producing too much oxygen. Each time the oxygen concentration changed, an iron mineral would precipitate out of the seawater, creating a new band of iron-rich sediment. Eventually, about 2.0 to 1.8 billion years ago, the oxygen level of the atmosphere reached a point where the ocean could no longer retain much iron, and banded iron formations could no longer form.

Black layers are rich in reduced iron (Fe^{2+}).

Red layers are rich in oxidized iron (Fe^{3+}).

White layers are rich in silica.

Close-up of layers in a banded iron formation

Think Critically

The photo of the banded iron formation reveals both iron-rich and silica-rich layers. If oxygen from cyanobacteria caused iron to precipitate, what is your hypothesis as to why silica precipitated?

Interpreting Environmental Clues

Just as history books record the changing patterns of civilization, layers of sediment, like pages in a book, record how environmental conditions have changed throughout Earth's history. Geologists are able to "read" this story by interpreting the evidence in the sedimentary rock that forms from the sediment.

We have already seen that the size, shape, and arrangement of particles in sediment (the **texture** of the sediment) provide evidence about the transport of the sediment and the geologic environment in which sediment is deposited and accumulates. These and other clues enable us to demonstrate the existence of past oceans, coasts, lakes, streams, deserts, glaciers, and wetlands.

Ancient and modern features compared • Figure 8.12

a. Ripples are forming in shallow water near the shore of Ocracoke Island, North Carolina.

b. Almost identical ripple marks are exposed on a bedding surface of sandstone at Artist's Point, Colorado National Park, Colorado.

c. Mud cracks are forming on this modern river bed as the river dries up.

d. Similarly shaped mud cracks are preserved on the surface of shale exposed at Ausable Chasm, New York. We can infer that this rock formation was deposited in an intermittently wet environment, such as a seasonal lake bed or a tidal flat. Note that in both **b** and **d**, the present-day environment where the specimens were collected is different from the environment in which the original sediments were deposited.

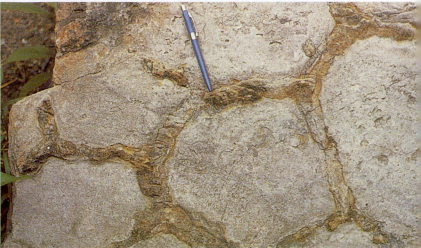

Footprints in the sand • Figure 8.13

a. Seagull footprints are imprinted in sandbars along the Alsek River in Glacier Bay National Park, Alaska.

b. Fossilized footprints are preserved in sandstone near Mount Etjo, Namibia. The animal that left these prints may have been hunting for prey stranded by the falling tide, just like the present-day seagulls in Alaska.

As you will learn in this section, ancient sedimentary deposits bear a striking resemblance to sedimentary environments on Earth today. This is an excellent example of the principle of uniformitarianism, discussed in Chapter 1. Patterns formed by currents of water or air moving across sediment can be preserved and later exposed on bedding surfaces. For example, bodies of sand that are being moved by wind, streams, or coastal waves are often rippled; these wavy structures may be preserved in sandstone as ripple marks (**Figure 8.12a** and **b**). Similarly, mud cracks (**Figure 8.12c** and **d**), fossil tracks (**Figure 8.13**), and even raindrop impacts can be recorded on bedding surfaces, attesting to moist surface conditions at the time they were formed.

Fossils also provide significant clues about former environments. Some animals and plants inhabit warm, moist climates, whereas others can live only in cold, dry climates. Using the climatic ranges of modern plants and animals as guides, we can infer the general character of the climate in which similar ancestral forms lived (**Figure 8.14**). Even microscopic fossils are important; for example, the shells of foraminifera can tell us about former

Plants and climate • Figure 8.14

This fossilized seed fern, from the genus Pecopteris, dates from the Carboniferous Period. Fossils of tropical, moisture-loving plants, such as horsetails, seed ferns, and tree ferns, can be found in Carboniferous-aged sedimentary rock from all over the world, even in areas that now have cooler or drier climates. The particular fern species in this photograph became extinct at the beginning of the Permian Period (about 299 million years ago), so the fossil can be used to establish the age of the sediment in which it was deposited.

temperatures and salinity conditions in the oceans. Fossils are also the basis for determining the relative ages of strata. As you learned in Chapter 3, fossils have played an essential role in efforts to reconstruct the past 635 million years of Earth's history.

The color of fresh, unweathered sedimentary rock can provide clues to the environmental conditions in which the rock originated. The color of a particular rock is determined by the colors of the minerals, rock fragments, and organic matter of which it is composed. Iron sulfides and organic detritus buried with sediment are responsible for most of the dark colors in sedimentary rock. The presence of these materials implies that the sediment was deposited in an oxygen-poor (reducing) environment. Reddish and brownish colors result mainly from the presence of iron oxides, occurring either as coatings on mineral grains or as very fine particles. These minerals point to oxygen-rich (oxidizing) conditions in the environment.

Sedimentary Facies

If you examine a vertical sequence of exposed sedimentary rock, you may notice differences as you move upward from one bed to the next. You may note changes in sedimentary rock type, color, fossil content, and thickness. The differences indicate that the environmental conditions in that location changed over time. If you trace a single bed laterally for a few kilometers, you may also notice changes which indicate that conditions differed from one place to another at any given time during deposition of the sediment.

Changes in the character of sediment from one environment to another are referred to as changes of **sedimentary facies** (pronounced "fay-sheez"). One facies may be distinguished from another by differences in grain size, grain shape, stratification, color, chemical composition, depositional structure, or fossils. Adjacent facies can merge into each other either gradually or abruptly (**Figure 8.15**). For example, coarse gravel and sand on a beach may gradually pass into finer sand, silt, and clay on the floor of the sea or a lake. Coarse, boulder-like glacial sediment, on the other hand, may end abruptly at the margin of a glacier.

By studying the relationships among different sedimentary facies and using these characteristics to identify original depositional settings, we can reconstruct a picture of the environmental conditions that prevailed in a region during past geologic times (see *Case Study*).

Sedimentary facies • Figure 8.15

Each depositional environment leaves its own kind of sedimentary record, which may change over time. This section shows a variety of depositional environments in which distinctive facies are deposited. On the land surface, they lie side by side. In a vertical section, they lie one above another.

Notice the seaward dip of the boundaries between adjacent facies. This indicates that, over time, the boundaries have migrated in the landward direction, due to a rise in sea level.

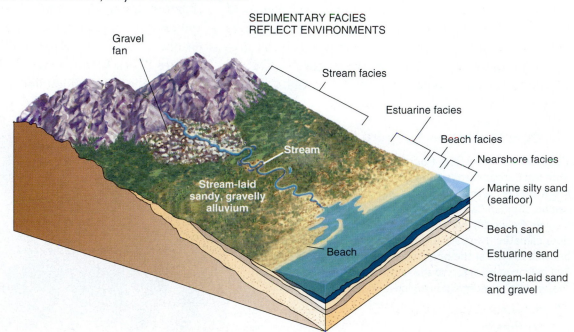

SEDIMENTARY FACIES REFLECT ENVIRONMENTS

Gravel fan

Stream facies

Estuarine facies

Beach facies

Nearshore facies

Stream

Stream-laid sandy, gravelly alluvium

Marine silty sand (seafloor)

Beach sand

Beach

Estuarine sand

Stream-laid sand and gravel

CASE STUDY

Thinking Critically About Sedimentation and Human History

About 30 years ago, in Ethiopia, geologists were looking for fossils in the rocky hills, south of the Awash River near a place called Dikika. Today the region is dry, hot, and seemingly barren (**Figure a**). But the tiny set of bones that emerged from these rocks is the oldest and most complete fossil of a human-like (**hominid**) child ever found (**Figure b**). The paleontologists who found the girl's remains named her Selam—"peace" in the Amharic language. The strata of the Hadar Formation that held her bones reveal much about how Selam lived—and died.

The sedimentary facies revealed that Selam and others of her species, *Australopithecus afarensis* (including the famous fossil "Lucy," found not far away in 1974), lived in an environment of rapidly fluctuating, ephemeral lakes and fast-moving streams, in a deltaic depositional system. Hypothesizing about what might have happened, the paleontologists reviewed what they knew: About 3.3 million years ago, a little girl wandered near a stream in Africa. Sedimentary rock and other fossils have shown that around her were open grasslands and shaded woodlands, home to elephants, hippos, rhinoceroses, and antelopes. Study of the skeleton indicated that although the tiny three-year-old walked upright on two feet, she had strong muscles and gorilla-like shoulders and could climb a nearby tree to escape predators, take shelter, and forage for fruit.

Flooding must have been a common occurrence in this constantly shifting terrain. Perhaps the little girl died by falling into a swiftly flowing stream—her bones were found encased in river channel deposits of gravel and sand, with clear evidence of rapid sedimentation. Her small body would have been buried quickly by sediment-laden water, protected from predators and the elements for millennia.

a. Members of the scientific team sift loose sediment, looking for small fossils. Sedimentary strata of the Ethiopian badlands are in the background.

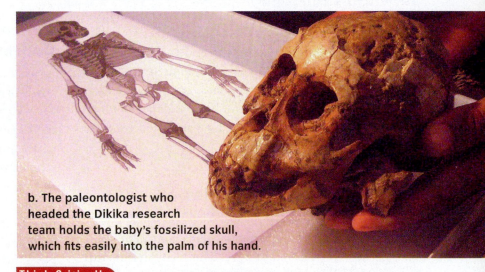

b. The paleontologist who headed the Dikika research team holds the baby's fossilized skull, which fits easily into the palm of his hand.

 Think Critically

The Dikika girl's remains were fossilized because she was quickly covered by sediment, which became sedimentary rock. What geologic processes had to take place after that, in order for her remains to eventually be exposed at the surface?

CONCEPT CHECK 🛑 STOP

1. **What** can cross bedding and graded bedding tell us about the conditions under which a particular sedimentary rock was formed?

2. **How** does sediment become sedimentary rock?

3. **Which** kind of sediment becomes shale when it lithifies?

4. **What** are some clues about past environments that can be preserved in sedimentary rock?

5. **How** do geologists use the concept of sedimentary facies to learn about past environments?

How Plate Tectonics Affects Sedimentation

LEARNING OBJECTIVES

1. **Identify** plate tectonic environments that are favorable for the accumulation of sediment.

2. **Explain** how divergent plate margins influence sedimentation.

3. **Describe** what happens to sediment during continental collisions.

4. **Describe** what happens in subduction zones at convergent plate margins.

Clastic sediment is far more abundant than chemical sediment or biogenic sediment. The locations where clastic sediment is commonly formed and deposited are largely controlled by plate tectonics. Clastic sediment originates from the rock and mineral debris produced by weathering and erosion of continental masses. The sediment accumulates in low-lying areas—troughs, trenches, and basins of various types. **Figure 8.16** summarizes the plate tectonic settings where clastic sediment is most likely to accumulate. Chemical and biogenic sediment are not as strongly influenced by plate tectonics, although the locations of ocean basins where many of these deposits occur, as well as their size and depth, are controlled by plate tectonics.

Divergent Plate Boundaries— Rift Valleys

Low-lying **rift valleys** are formed when a continent splits apart because of tensional forces. The East African Rift

> **rift valley** A linear, fault-bounded valley along a divergent plate boundary or spreading center.

is an example of a rift that is still active in pulling apart a continent; the Red Sea is a young rift where a new ocean is already forming; and the Atlantic Ocean is a mature (i.e., well-developed) rift. A rift valley may eventually become a continental margin bordering an ocean, or the tensional forces may stop at some point, and the valley would then become a failed rift, such as the Newark Basin of New Jersey, or the similarly aged Hartford Basin of Connecticut, down which the Connecticut River flows for much of its way to the sea.

Both the East African Rift and the ancient Newark Basin hold deep wedges of immature clastic sediment deposited by streams. In the Red Sea, marine sediment now covers the clastic sediment that was deposited before the Red Sea was wide enough for the sea to enter and before the evaporite deposits that formed while the seawater was still shallow. On the Atlantic Ocean Margin of North America, sediment that has eroded from the adjacent continent has accumulated over many millions of years to a thickness of over 14 km (see **Figure 8.16a**). Most of the strata deposited on the shelf of a passive continental margin are composed of mature, shallow-water marine sediment. The continental shelf slowly subsides as accumulation takes place, and the pile of sediment grows thicker and thicker. Lithification occurs as sediment is buried progressively deeper in the pile.

Convergent Plate Boundaries— Collisional Type

A variety of low-lying areas are found within and along the edges of high mountain ranges. Many of these are structural basins or troughs caused by faulting and folding of rock associated with the process of mountain building. Coarse, immature (and thus angular and poorly sorted) stream sediment eroded from a rising mountain range accumulates in these low-lying areas as vast thicknesses of conglomerate, gravel, and sand (see **Figure 8.16b**). In the Himalayas, for example, thick sequences of conglomerate and coarse sandstone flank the southern edge of the range, while the finest-grained sediment has been transported all the way to the sea by the many streams that flow to the Indian Ocean.

Convergent Plate Boundaries— Subduction Type

The lowest-lying points on Earth's surface are the long, deep trenches that form on the ocean floor in subduction zones (see **Figure 8.16c**). Subduction zones are often found near continental margins, such as the western margin of South America and the eastern margins of Asia. The elevation plunges from the top of the volcanoes on the edge of the continent (a common feature in subduction zones)

Sites of sedimentation • Figure 8.16

Clastic sediment accumulates in low-lying areas, which occur in specific locations that are strongly controlled by plate tectonics.

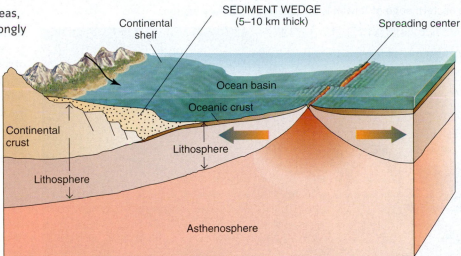

a. Thick sedimentary wedges accumulate in rift valleys and along passive continental margins that are formed when continental crust rifts and a new ocean basin opens.

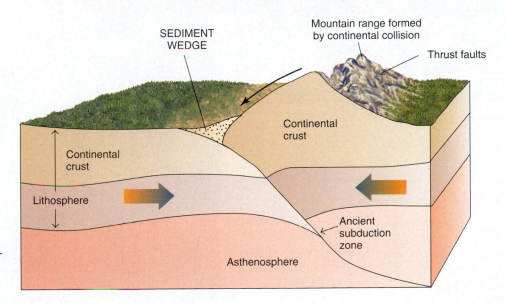

b. Sediment accumulates in structural basins along the edges of mountain ranges thrust up by continental collisions.

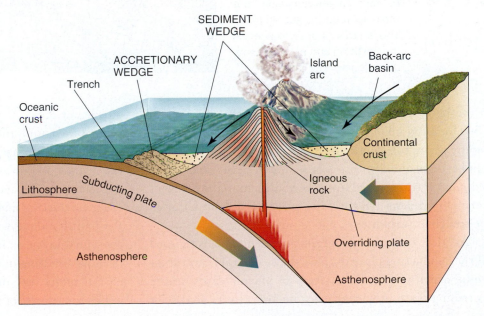

c. Sediment is shed from continents into deep-sea trenches above subduction zones. The sediment forms a wedge that is compressed and crushed as the oceanic plate subducts under the continental plate. The **back-arc basin** that forms on the overriding plate, behind the subduction zone, is another location where sediment is likely to accumulate.

Accretionary wedge • Figure 8.17

The eastern edge of Taiwan is an active accretionary wedge.

a. This chaotic terrain of rock broken during accretion is near Taitung, at the southern end of the wedge. In oceanic subduction zones, the volcanic arc forms a string of islands, typically with a **back-arc basin** between the islands and the mainland. Though not as dramatic topographically as the deep-ocean trenches, these basins are also places where ocean-floor sediment accumulates.

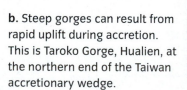

b. Steep gorges can result from rapid uplift during accretion. This is Taroko Gorge, Hualien, at the northern end of the Taiwan accretionary wedge.

to the bottom of the trench, within a lateral distance of a few hundred kilometers. Because erosion proceeds most rapidly where the slopes are steep, we would expect to find rapid erosion and deposition of sediment here—and we do. Sediment is transported from the continent to the adjacent trench by streams and turbidity currents, accumulating in the trench as turbidites.

At some convergent margins, the sediment is scraped off the subducting plate in thick slabs, separated by thrust faults, which pile up like a stack of playing cards on the overriding plate. Adjacent volcanic arcs ensure that a lot of volcanic debris is present in the sediment. The wedge-shaped accumulations of volcaniclastic sediment are called **accretionary wedges** (**Figure 8.17**). For example, the islands of Taiwan and Barbados largely consist of accretionary wedge sediment thrust up above sea level. At other subducting margins, the sediment travels down into the mantle, along with the subducting plate. The water contained in the sediment is released there, and it facilitates wet melting at depths of 100 km and greater. These partial melts fuel the explosive volcanic eruptions that are characteristic of subduction zones.

This chaotic mix of volcaniclastic sediment, fragments of oceanic lithosphere, and rock metamorphosed under the low-temperature, high-pressure conditions of the subduction zone is called a **mélange**, from the French word for "mixture." Mélanges give geologists a close look at processes that occur along subducting margins—geologic environments that are otherwise impossible to access. The fragments of oceanic lithosphere that are sometimes incorporated into the mélange are particularly significant. These fragments, called **ophiolites**, are of immense geologic interest because they expose oceanic crust and underlying mantle at the surface. A complete ophiolite suite typically consists of the mantle rock peridotite at the base, overlain by sheeted complexes of gabbro dikes (remnants of the feeder system for the basaltic magma that formed the oceanic crust), with pillowed oceanic basalt and seafloor sediment at the top. Pillows are balloon-like volcanic structures that form when lavas are extruded underwater. These unusual rock suites provide a rare opportunity for geologists to see oceanic crust from top to bottom, to sample the mantle, and—perhaps most importantly—to study the Moho (crust–mantle boundary) itself.

CONCEPT CHECK

1. **Why** is clastic sediment deposition more strongly influenced by plate tectonics than chemical and biogenic sediment deposition?

2. **What** kind of depositional environment will result if a rift valley continues to widen?

3. **What** happens at the bottom of the pile of sediment that forms on a passive continental margin?

4. **How** does subducted sediment play a role in the igneous rock cycle?

 THE PLANNER

Summary

1 Sediment 220

- Sediment and sedimentary rock provide a record of how climates and environments have changed throughout geologic history. Geologists use three broad categories to distinguish sediment types: clastic, chemical, and biogenic.

- **Clastic sediment** consists of fragmented rock and mineral debris produced by weathering, together with broken remains of organisms. The fragments, called clasts, are classified on the basis of size. Clasts may become rounded and sorted during transport by water and wind, but they typically remain unsorted during transport by glaciers or as a result of mass wasting (see figure). Volcaniclastic sediment is a type of clastic sediment in which the fragments, also called pyroclasts, are volcanic in origin.

- **Chemical sediment** is formed when substances carried in solution in lakewater or seawater are precipitated, generally as a result of evaporation or other processes that concentrate the dissolved substances.

- **Biogenic sediment** is composed of the accumulated remains of organisms. Plants, animals, and microscopic lifeforms may all contribute skeletal and/or organic material to sediment.

From clasts to rock • Figure 8.1

2 Transport and Deposition of Sediment 223

- Sediment can be transported by flowing wind, water, or ice, and it can move downslope under the influence of gravity.

- Stream, lake, glacial, and eolian sediments are common on land. **Eolian sediment**, which is transported by wind, tends to be finer than sediment moved by other erosional agents. We can interpret environmental clues, such as ripple marks, mud cracks, and fossil tracks, to deduce past surface conditions.

- Streams and lakes are common transporters of sediment on land and can form several types of depositional environments, including alluvial fans and **deltas**, triangle-shaped deposits that form where streams enter a standing body of water.

- Other depositional environments are near the ocean (see figure). Some sediment may be trapped in **estuaries**, semi-enclosed bodies of water along the coast, where fresh water and ocean water mix. Beaches tend to comprise finer sediment, as the particles have been repeatedly abraded by surf.

- Sediment is also deposited in the ocean. Reefs are deposits made of the skeletons of marine invertebrates. **Turbidity currents** are underwater landslides that transport sediment from the edge of the continental shelf to the deeper water at the foot of the continental slope. Another type of depositional environment in the ocean is the seafloor, which commonly hosts two types of deposits: calcareous ooze and siliceous ooze.

Depositional environments • Figure 8.5

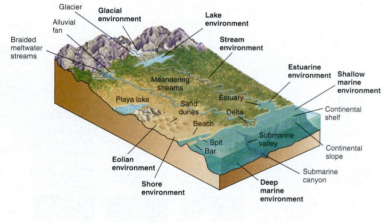

- When sediment is turned into sedimentary rock, the **bedding**, or layered arrangement of the strata, is generally preserved. The presence of bedding and **bedding surfaces**, the boundaries between adjacent strata, indicates that the rock was once sediment. In a graded bed, the coarsest clasts are at the bottom, and the finest are at the top. Cross bedding is produced by turbulent stream or wind flow or by ocean waves.

- **Lithification** is the group of processes that transforms loose sediment into sedimentary rock (see figure). Lithification takes place through a variety of changes, including **compaction**, the reduction of pore space as a result of

increased pressure; **recrystallization**, the formation of new minerals from old ones; and **cementation**, the process by which substances dissolved in pore water precipitate out and cement grains of sediment together. The low-temperature, low-pressure changes that happen to sediment after **deposition** are called diagenesis.

- Clastic sedimentary rock, like sediment, is classified mainly on the basis of clast size. **Conglomerate**, **sandstone**, and **mudstone** are the common rock equivalents of gravel, sand, and mud (silt- and clay-rich sediment), respectively. Conglomerate has large clasts in a fine-grained matrix, sandstone contains medium-sized grains, and mudstone consists of very small particles. **Shale** is mudstone that readily splits into thin layers.

- **Evaporite** is a chemical sedimentary rock formed by evaporation of lake or ocean water. **Banded iron formations**, though uncommon, are significant both economically and for understanding the history of our planet. **Limestone**, **peat**, and **coal** are important kinds of biogenic sedimentary rock. The lithification of peat into coal is known as coalification.

- Sedimentary rocks contain clues that help geologists understand the environments in which they formed. We can interpret clues such as ripple marks, mud cracks, and fossil tracks to deduce past environmental conditions.

- A sedimentary facies is a single geologic unit consisting of sedimentary rock strata deposited at more or less the same time. Geologists learn about the history of a region by studying the way that different sedimentary facies adjoin and grade into each other in space and in time.

Lithification • Figure 8.10

Shells made of aragonite → Aragonite in the shells has been transformed into calcite.

3 Sedimentary Rock   229

How Plate Tectonics Affects Sedimentation   240

- Clastic sediment, which originates from the rock and mineral debris produced by weathering and erosion, is the most abundant type of sediment. Clastic sediment tends to collect in low-lying areas whose location is strongly affected by plate tectonics (see figure). Chemical and biogenic sediment deposition is less strongly influenced by plate tectonics.

- Divergent margins create **rift valleys** and passive continental margins, both of which are common environments in which sediment accumulates.

- Colliding tectonic plates create structural basins and valleys associated with mountain ranges, as a result of the folding and faulting of rock. Low-lying basin topography combined with high mountains that provide a source of sediment typically leads to active sedimentation.

- Subducting plates form deep oceanic trenches along with volcanic arcs that may have back-arc basins. Sediment can be quickly buried in this environment, allowing it to be recycled into the mantle.

Sites of sedimentation • Figure 8.16

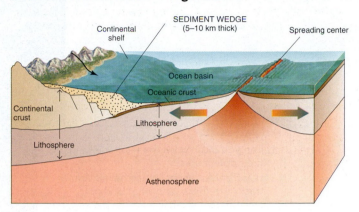

Key Terms

- banded iron formation   234
- bedding   229
- bedding surface   230
- biogenic sediment   222
- cementation   232
- chemical sediment   222
- clastic sediment   220
- coal   234
- compaction   232
- conglomerate   233
- delta   225
- deposition   224
- eolian sediment   225
- estuary   226
- evaporite   228
- limestone   234
- lithification   229
- mudstone   234
- peat   234
- recrystallization   232
- rift valley   240
- sandstone   234
- shale   234
- turbidity current   228

Critical and Creative Thinking Questions

1. Estuaries are generally shallow, yet there are thick accumulations of estuarine sediment in the geologic record. What hypothesis can you suggest to explain this?

2. Do any sedimentary rocks outcrop in the area where you live? If so, see if you can recognize the kinds of rock present and identify the environment in which the sediment was deposited.

3. It is estimated that as much as 25 billion tons of soil are lost through erosion as a result of farming every year. What happens to that soil?

4. Exploration for oil has led to the discovery of up to 14 km of sedimentary rock on the continental shelf of North America.

If the oldest of the strata were deposited 150 million years ago, during the Jurassic Period, and the youngest are still being deposited today, what is the average rate of deposition?

5. Investigate the formation of graded bedding by filling a large beaker with a half-and-half mixture of water and sediment. The sediment should have a variety of grain sizes, including fine clays, sand, and gravel. Shake up the mixture and let it settle quietly until the water is completely clear. Which grains settle first, and which settle last? What does the final sediment deposit look like?

What is happening in this picture?

This slab of rock from the Cambrian Period was collected in Quebec, Canada.

Think Critically

1. Can you distinguish the fossil tracks from the ripple marks?
2. What kind of animal could have made tracks like these, and what kind of environment did it live in? (By the way, no one knows exactly what the animal, called *Climactichnites*, looked like.)

Self-Test

(Check your Answers in Appendix D.)

1. _____ sediment forms from loose rock and mineral debris produced by weathering and erosion.

   a. Clastic   b. Biogenic   c. Chemical

2. These photographs are close-up views of two sediment samples labeled A and B. Which sediment shows a greater degree of rounding?

   a. Sample A

   b. Sample B

   c. Neither sample is well rounded.

   d. Both samples are well rounded.

   a.   **b.**

3. Which of the two sediments in the photographs shows a greater degree of sorting?

   a. Sample A

   b. Sample B

   c. Neither sample is well sorted.

   d. Both samples are well sorted.

4. Examine the middle sedimentary bed depicted in the photograph. This bed is a best described as _____.

   a. cross bedded   b. glacial

   c. graded   d. eolian

5. Three separate processes can lead to the lithification of sediment. During _____, the weight of accumulating sediment reduces pores space and forces out water from sediment. _____ occurs when ions dissolved in solution precipitate out, forming minerals that hold the grains together. Pressure can lead to _____, which causes less-stable minerals to rearrange crystals into more-stable forms.

   a. compaction, Cementation, recrystallization

   b. compaction, Recrystallization, cementation

   c. dehydration, Recrystallization, cementation

   d. dehydration, Cementation, recrystallization

6. Which of the three rock samples in these photographs is *not* a sedimentary rock?

   a. Sample A

   b. Sample B

   c. Sample C

   d. All of the rock samples displayed are sedimentary rock.

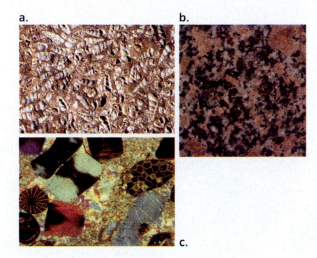

7. _____ is a sedimentary rock that is typically composed entirely of calcite.

   a. Sandstone

   b. Limestone

   c. Shale

   d. Conglomerate

   e. Evaporite

8. _____ is a medium-grained sedimentary rock of clastic origin.

   a. Sandstone

   b. Limestone

   c. Shale

   d. Conglomerate

   e. Evaporite

9. Which one of the following rocks is most likely to have formed in a desert playa?

   a. sandstone

   b. limestone

   c. shale

   d. conglomerate

   e. evaporite

10. If mud cracks are found on a bedding surface of sedimentary rock, what can we deduce about the environmental conditions during deposition of the original sediment?

    a. Deposition occurred during a strong windstorm.

    b. The environment was subject to strong currents.

    c. The environment was moist and then dried out.

    d. Deposition occurred as a result of glaciation.

11. In which of the following sedimentary environments would you expect to find siliceous and calcareous oozes?

    a. shallow marine environment

    b. stream environment

    c. deep marine environment

    d. estuarine environment

12. Given the vertical association of depositional facies presented in this block diagram, what can we deduce about sea level over time?

    a. Sea level was rising during deposition.

    b. Sea level was falling during deposition.

    c. Sea level was stable during deposition.

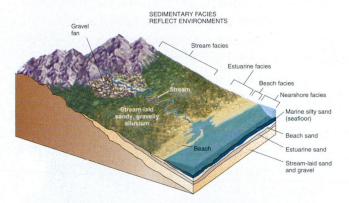

13. In what type of tectonic setting would you expect to find voluminous amounts of arc-derived volcaniclastic sediment?

    a. divergent margin–rift valley

    b. convergent margin–subduction zone

    c. convergent margin–continental collision

    d. passive continental margin

14. Accretionary wedges composed of sediment scraped off the oceanic crust are typical of what type of tectonic setting?

    a. divergent margin–rift valley

    b. convergent margin–subduction zone

    c. convergent margin–continental collision

    d. passive continental margin

15. In what type of tectonic setting would you expect to find sediment in structural basins adjacent to mountain ranges?

    a. divergent margin–rift valley

    b. convergent margin–subduction zone

    c. convergent margin–continental collision

    d. passive continental margin

THE PLANNER

Review your Chapter Planner on the chapter opener and check off your completed work.

Folds, Faults, and Geologic Maps

Moraine Lake is in Banff National Park, located in the beautiful Canadian Rocky Mountains. This view of the lake and the range behind it is looking in a northwesterly direction. It is apparent from the layering that the range is composed of sedimentary strata. Ground studies have shown that strata in the Rockies range in age from early Cambrian to mid-Jurassic. They thin towards the east, so geologists deduce that the source of sediment was in the west. Not apparent in the photo is the fact that the Canadian Rockies consist of a series of parallel ranges that trend north-south. Ground studies have determined that the same strata are present in each of the ranges. The ranges were created by a process that geologists called thin-skinned tectonics, in which compressional forces break and move great slabs of the uppermost crust on thrust faults that dip to the west at shallow angles. The faulting occurred near the end of the Cretaceous, and the sense of movement is from west towards the east—that is towards the viewer's right.

CHAPTER OUTLINE

CHAPTER PLANNER ✔

☐ Study the picture and read the opening story.

☐ Scan the Learning Objectives in each section.
  p. 251 ☐   p. 258 ☐   p. 268 ☐

☐ Read the text and study all visuals. Answer any questions.

Analyze key features

☐ Geology InSight
  p. 252 ☐   p. 260 ☐   p. 268 ☐

☐ Process Diagram, p. 257 ☐

☐ Amazing Places, p. 262 ☐

☐ What a Geologist Sees, p. 264 ☐

☐ Case Study, p. 269 ☐

☐ Stop: Answer the Concept Checks before you go on:
  p. 257 ☐   p. 267 ☐   p. 273 ☐

End of chapter

☐ Review the Summary and Key Terms.

☐ Answer the Critical and Creative Thinking Questions.

☐ Answer What is happening in this picture?

☐ Complete the Self-Test and check your answers.

Rock Deformation

LEARNING OBJECTIVES

1. **Define** three types of stress.
2. **Describe** the differences between elastic, brittle, and ductile deformation.
3. **Explain** how factors such as confining pressure, temperature, strain rate, and rock composition affect deformation.
4. **Define** craton and orogen.
5. **Explain** why orogens are typical settings for highly deformed rock.

A s you learned in Chapters 4 and 5, lithospheric plates are constantly moving around, colliding with one another, and interacting along their margins. Crustal rock is recycled into the mantle at subduction zones, and new crust is created at divergent margins. But what happens at convergent margins in collision zones? Here, enormous forces deform huge masses of continental crust and uplift them into great mountain ranges. We can't actually see the rock being twisted and bent by tectonic forces because this usually happens very slowly and deep underground. However, the aftermath is easily visible in locations where the deformed rock has been uplifted by tectonic forces and exposed by erosion. In **Figure 9.1**, for instance, the originally flat and horizontal beds of sedimentary rock have been folded and tilted into a spectacular zigzag. To understand how folding like this occurs, we rely on both laboratory measurements and field studies of deformed rock.

Stress and Strain

In discussing rock deformation, we use the word **stress** rather than **pressure**. These two words are related in meaning but different in connotation. Both are defined as the force acting on a surface per unit area. The term *pressure*, as used in geology, implies that the forces on a body of rock are essentially uniform in all directions. Sometimes this is also called **uniform stress** or **confining stress**. These are appropriate terms to describe, for instance, the stress on a small body immersed in a liquid such as water or magma.

> **stress** The force acting on a surface, per unit area, which may be greater in certain directions than in others.
>
> **pressure** A particular kind of stress in which the forces acting on a body are the same in all directions.

The core of a former mountain range • Figure 9.1

These tightly folded rock strata in the Himalayan Mountains of southwestern Tibet bear witness to the intense geologic stresses that bent them into a zigzag shape. Such intense deformation is typical for rock that once formed the deep-seated core of a great mountain chain.

Three types of stress • Figure 9.2

The shape of a cube of rock changes, depending on the type of stress applied to it. The arrows indicate tensional, compressional, and shear stress. Rock that is subjected to differential stress—stress that is stronger in one direction than another—typically responds by changing shape, as shown by these blocks.

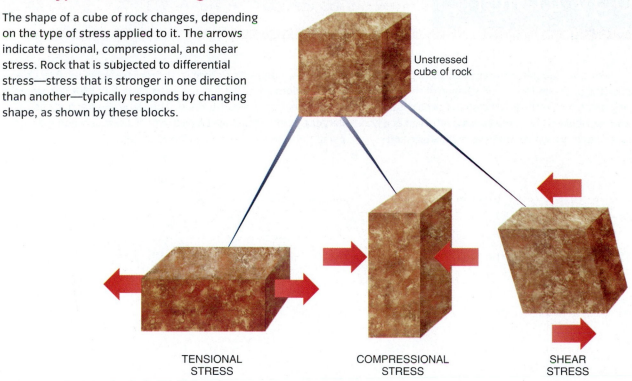

Unstressed cube of rock

TENSIONAL STRESS

COMPRESSIONAL STRESS

SHEAR STRESS

Rock, however, is solid; unlike liquid or gas, solids can resist different pressures in different directions at the same time. For this reason, *stress* is a more versatile term for discussing rock deformation because it does not imply that the forces are necessarily the same in all directions. To be even more precise, we sometimes use the term **differential stress** when the force is greater from one direction than from another. The stresses that cause rock to change shape are differential. They can be classified into three different kinds, as illustrated in **Figure 9.2**: **tension**, **compression**, and **shear**.

In response to stress, rock will experience **strain**. Uniform stress causes a change in volume only, while differential stress may cause a change in shape. For example, if a body of rock is subjected to uniform stress by being buried deep in Earth, its volume will de-

tension A stress that acts in a direction perpendicular to and away from a surface.

compression A stress that acts in a direction perpendicular to and toward a surface.

shear A stress that acts in a direction parallel to a surface.

strain A change in shape or volume of rock in response to stress.

crease; that is, the rock will be compressed. If the spaces (or **pores**) between the grains become smaller as water is expelled from them, or if the minerals in the rock are transformed into more compact structures, the volume change may be quite large.

Brittle and Ductile Deformation

The way rock responds to differential stress depends not only on the amount and kind of stress but also on the nature of the rock itself. For example, rock may stretch like a metal spring and then return to its original shape when the stress is removed. Such a nonpermanent change is called **elastic deformation**. For most solids, including rock, there is a degree of stress—called the **elastic limit**—beyond which the material is permanently deformed. If the rock is subjected to more stress than this, it will not return to its original size and shape when the stress is removed (see **Figure 9.3** on the next page).

elastic deformation A temporary change in shape or volume from which a material rebounds after the deforming stress is removed.

a. A sample of rock tested in a laboratory will exhibit a straight-line relationship between the stress and the strain (change of shape or volume) when being deformed elastically. This straight-line relationship (X to Y) is known as Hooke's Law. If the stress is removed at any point between X and Y, the rock will return to its original size and shape. Beyond the elastic limit (Y) the rock undergoes irreversible ductile deformation. If the stress is removed at point A, for example, the rock returns to an unstressed state along line AA' and has undergone permanent strain equal to XA'. If stress is continued past point A, the rock will eventually break, at point Z.

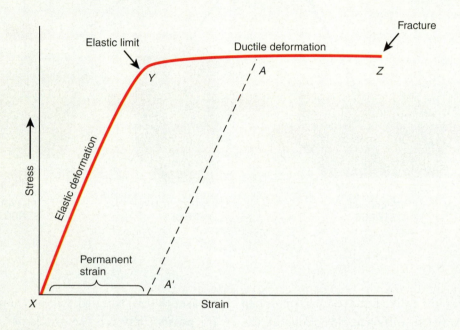

b. A paper clip that is holding only 3 sheets of paper is deformed elastically and returns to its original shape after the papers are removed. This corresponds to a position anywhere along the line XY in **a.**

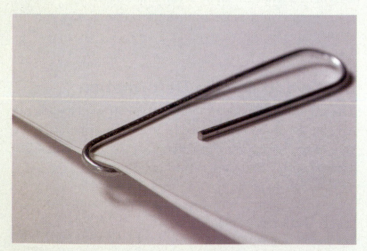

c. The same paper clip becomes permanently bent when used to hold together a 50-page document. Its elastic limit is somewhere between 3 and 50 pages. Beyond the elastic limit, the clip has suffered permanent strain. This corresponds to point A' in **a.**

d. At room temperature, glass deforms in a brittle manner, as shown by this lightbulb shattering when it hits the floor. Very little ductile deformation occurred, which means the elastic limit and the point of fracture are very close together. This corresponds to point Z in **a**.

e. If glass is heated slowly over a flame, it can bend and flow in a ductile manner, as in this glassmaker's studio. This means that at high temperature, glass has an extensive range of ductile deformation. This corresponds to any point along the line YZ in **a**.

f. This rock responded to stress by folding and flowing, a ductile deformation, so the deformation occurred between points Y and Z. At room temperature, the deformation is permanent, and this corresponds to point A' in **a**.

g. This rock exhibits both brittle fracture and ductile deformation. The central yellow-colored rock has been deformed by brittle fracture, corresponding to point Z in **a**. The white rock enclosing the yellow fragments and the blue rock below have not fractured but have flowed and been deformed by ductile deformation, corresponding to a point somewhere along the line YZ. Each of the three rock types—yellow, white, and blue—were subjected to the same stress regime, so composition must play a role in how rock deforms under a given stress regime.

When rock is stretched past its elastic limit, it can deform in two different ways. **Ductile deformation**, also called **plastic deformation**, is one type of permanent deformation in rock (or other solid) that has been stressed beyond its elastic limit. Alternatively, the rock may undergo **brittle deformation**.

A brittle material deforms by fracturing, whereas a ductile material deforms by changing its shape. Drop a piece of chalk on the floor, and it will break; drop a piece of modeling clay and it will bend or flatten instead of break. Under the conditions of room temperature and atmospheric pressure, chalk is brittle, and modeling clay is ductile. Similarly, some rock behaves in a brittle manner and some in a ductile manner (see Figure 9.3). However, rock that is brittle in one set of conditions may be ductile in different conditions.

The main factors that affect how rock deforms are temperature, confining pressure, rate of deformation, and composition of the rock. Let's look briefly at each of these.

ductile deformation A permanent but gradual change in shape or volume of a material, caused by flowing or bending.

brittle deformation A permanent change in shape or volume, in which a material breaks or cracks.

Temperature As shown in Figures 9.3d and e, glass is brittle at room temperature but becomes ductile at a higher temperature. Rock is like glass in this respect; it is brittle at the surface but becomes ductile deep inside the planet, where temperatures are higher.

Confining pressure The effect of (uniform) confining pressure on deformation is not familiar from our everyday experience. High confining pressure reduces the brittleness of rock because it hinders the formation of fractures. **Figure 9.4** shows the results of a series of experiments that demonstrate this effect. Near Earth's surface, where confining pressure is low, rock exhibits brittle behavior and develops many fractures. At great depth, however, where confining pressure is high, rock tends to be ductile and deforms by flowing or bending.

Rate of deformation The rate at which stress is applied to a solid is another important factor in determining how a material will deform. If you take a hammer and suddenly whack a piece of ice, it will fracture. But if you apply stress to the ice little by little over a long period, it will sag, bend, and behave in a ductile manner. The same is true of

Rock deformation under pressure • Figure 9.4

These cylinders show the results of a series of experiments on the effects of confining pressure on rock.

a. This is an undeformed cylinder of rock.

b. This cylinder was subjected to high confining pressure (uniform in all directions) and, at the same time, compression from above. It deformed in a ductile manner, becoming shorter and fatter.

c. An identical cylinder was subjected to the same amount of compression from above, but this time with a lower confining pressure. It deformed in a brittle manner, with many large fractures.

Silly Putty® • Figure 9.5

The effects of different strain rates can be observed with Silly Putty.

a. At very slow strain rates, Silly Putty flows like a liquid. Starting from an apparently solid ball shape, it changes to the shape of a liquid drop over a few hours.

b. At medium strain rates, such as when stretched by human hands, Silly Putty deforms in a ductile manner.

c. At high strain rates, Silly Putty deforms elastically. (Bouncing is an elastic effect, caused by the material's tendency to return to its original shape.)

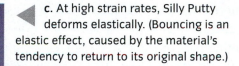

d. At extremely high strain rates, such as when shot by a bullet, Silly Putty shatters like glass.

rock. If stress is applied quickly, the rock may behave in a brittle manner, but if small stresses are applied over a very long period, the same rock may behave in a ductile manner. The term **strain rate** refers to the rate at which rock is forced to change its shape or volume. The lower the strain rate, the greater the tendency for ductile deformation to occur (**Figure 9.5**).

To summarize, low temperature, low confining pressure, and high strain rates tend to enhance the brittle behavior of rock. Low-temperature and low-pressure conditions are characteristic of Earth's crust, especially the upper crust. As a result, fracturing is common in upper-crustal rock. High temperature, high confining pressure, and low strain rates, which are characteristic of the deeper crust and the mantle, reduce the brittle properties of rock and enhance the ductile properties. The depth below which ductile properties predominate is referred to as the **brittle–ductile transition**. Fractures are uncommon deep in the crust and in the mantle because rock at great depths (below about 10–15 kilometers) tends to behave in a ductile manner.

Composition The composition of a material determines the exact point at which its brittle–ductile transition will occur. For example, both chalk and modeling clay are brittle at −50° Celsius. When warmed to room temperature, the modeling clay behaves in a ductile manner, but the chalk is still brittle. This is because they have different chemical compositions and, therefore, different properties. The same is true of different rocks and their mineral constituents. Some minerals—notably quartz, garnet, and olivine—are strong and brittle, as are the rocks that contain these minerals (e.g., sandstone and granite). For the most part, quartz-bearing rocks control the depth of the brittle–ductile transition. Other minerals—notably mica, calcite, and gypsum—are more often ductile under natural conditions. Thus, the rocks that contain them (e.g., limestone, marble, shale, slate) also tend to deform in a ductile manner.

Water is another component that enhances the ductile properties of rock. It reduces the friction between mineral grains and dissolves material at points of high stress, which permits the material to move to places where the stress is lower. Trace amounts of water can also enter strong minerals such as quartz and olivine and significantly weaken them, through a process called **hydrolytic weakening**.

Where Rock Deformation Occurs

In which tectonic environments can the different types of stress—tension, compression, and shear stress—be expected to occur? Tensional stress involves a pulling-apart motion; it is characteristic of environments in which divergent motion is occurring, such as a continental rift or mid-ocean ridge. Shear stress occurs where lithospheric plates are sliding past one another, such as along the boundary between the Pacific Ocean plate and the North American plate. Compression is characteristic of environments in which convergent motion is occurring, such as subduction zones and continent–continent collisions.

The deformation of rock in continent–continent collisions provides some of the best evidence that Earth has been tectonically active for billions of years. As discussed in Chapter 4, the conclusive evidence for plate tectonics—seafloor spreading—occurs in oceanic crust. However, all of today's oceanic crust is relatively young—less than 200 million years old. Therefore, evidence for earlier plate tectonics comes from continents and how they are put together (**Figure 9.6**), including the placement of **cratons** and **orogens**. Some orogens, including the Alps and the Himalayas, are still active today. Others, such as the Appalachians, are inactive and now highly eroded. Through radiometric dating, some orogens have been found to be as old as 4 billion years.

> **craton** A region of continental crust that has remained tectonically stable for a very long time.
>
> **orogen** An elongated region of crust that has been deformed and metamorphosed through a continental collision.

Cratons and orogens • Figure 9.6

Even though all of North America lies on one lithospheric plate today, a closer look at the rocks shows that it has been assembled like a jigsaw puzzle from older parts. The core of a continent is formed of cratons, which represent the very oldest tectonic units. Between the cratons lie elongated regions called orogens. These contain highly deformed rock strata and lots of metamorphic rock (see Chapter 10)—both signs that these were once sites of intense mountain-building activity.

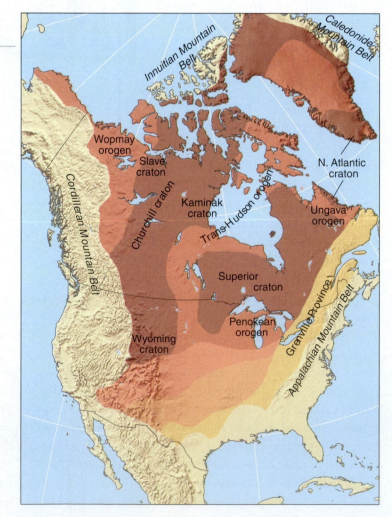

Billions of years old

> 2.5	1.8–1.7	1.2–1.0
1.9–1.8	1.7–1.6	< 1.0

Adjusting the height of continents • Figure 9.7

The rigid lithosphere "floats" on the ductile asthenosphere.

a. In a collision of two cratons, the lithosphere thickens—both upward and downward—and the lithosphere–asthenosphere boundary is pushed down, creating a mountain belt (an orogen) with a deep root.

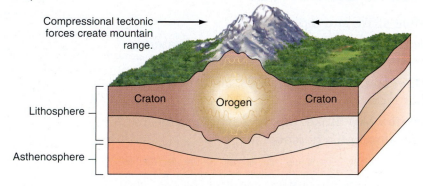

Compressional tectonic forces create mountain range.

Lithosphere
Craton   Orogen   Craton

Asthenosphere

Think Critically

Which orogens in North America would you expect to have the deepest roots, and how could you test your hypothesis?

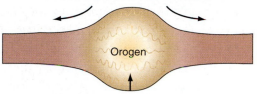

Erosion wears down the mountains.

Orogen

b. As erosion slowly reduces the height of the mountains in the orogen, the root slowly rises to compensate for the removal of mass. This continuing adjustment to maintain buoyancy, like a cork finding its level in water, is called isostasy.

Orogen bobs upward, maintaining isostasy and preserving mountains even after the tectonic forces that created them are gone.

Orogens such as the Appalachians still display some topographic relief even after hundreds of millions of years of erosion because of a phenomenon called **isostasy** (**Figure 9.7**). Isostasy is the process in which a floating object automatically adjusts to a position of equilibrium with the medium in which it is floating. From everyday experience, you know that a block of wood floating in water will sink if a weight is put on top of it, and it will bob up if the weight is removed. The same thing happens with icebergs floating in the ocean. Icebergs are partially visible above the ocean surface, but every iceberg has a deep root that keeps it floating in balance. As the exposed top of an iceberg melts, the entire block of ice will bob upward to maintain this flotational balance.

isostasy The flotational balance of the lithosphere on the asthenosphere.

Isostasy plays an important role in shaping Earth's topography, particularly in orogens. Continental collisions cause the entire continental crust to thicken; as a result, beneath every great mountain range is a root of thickened continental crust that dips down into the asthenosphere, similar to the root of an iceberg. This root keeps the entire orogen in flotational balance with respect to the underlying hot, weak asthenosphere. As erosion weathers away the top of the mountain range, the root pushes upward, like the root of a melting iceberg. But unlike an iceberg, which floats in seawater, an orogen is solid rock that floats in the hot, ductile asthenosphere, which makes the process extremely slow. The Appalachian Mountains are still adjusting isostatically after more than 300 million years of weathering.

CONCEPT CHECK   STOP

1. **How** does shear stress differ from compressional stress?

2. **Why** is rock buried deep underground more likely to undergo ductile deformation than brittle deformation?

3. **How** would slate and granite respond differently to stress?

4. **Why** do orogens preserve a record of ancient deformations long after you might expect them to have eroded away?

5. **When** observing a very large area of highly deformed rock, why might a geologist draw the conclusion that it is part of an orogen?

Structural Geology

LEARNING OBJECTIVES

1. **Define** strike and dip.

2. **Identify** which kinds of stress are associated with various types of faults.

3. **Define** and describe synclines, anticlines, and other types of folds.

4. **Describe** how strike, dip, and simple folds are represented on a geologic map.

he concepts of stress and strain play a central role in **structural geology**. Structural geologists attempt to decipher the geologic history of a region by identifying and mapping deformational features in rock. These features are also important from a practical perspective. For example, faults control the locations of certain types of ore deposits. Other rock structures can affect slope stability, influence the flow of groundwater, or trap oil and natural gas deep underground (**Figure 9.8**).

> **structural geology** The study of stress and strain, the processes that cause them, and the deformation and rock structures that result from them.

Many practical applications, such as oil prospecting and groundwater flow mapping, require geologists to make inferences about structures that lie underground, hidden from view. Most of their information comes from observations made at the surface, with additional information from drilling and other methods of subsurface study. A systematic program of measurements and mapping can give structural geologists a very good idea of what might lie beneath the surface.

Strike and Dip

The **principle of original horizontality** (see Chapter 3) tells us that sedimentary strata are horizontal when they are first deposited. Where such rocks are tilted, we generally can assume that deformation has occurred. To describe this deformation, a geologist starts by measuring the orientation of the tilted rock layer. This information is given by the **strike**. It is a basic fact of geometry that a sloping plane—in this case, the tilted

> **strike** The compass orientation of the line of intersection between a horizontal plane and a planar feature, such as a rock layer or fault.

Structural geology and oil • Figure 9.8

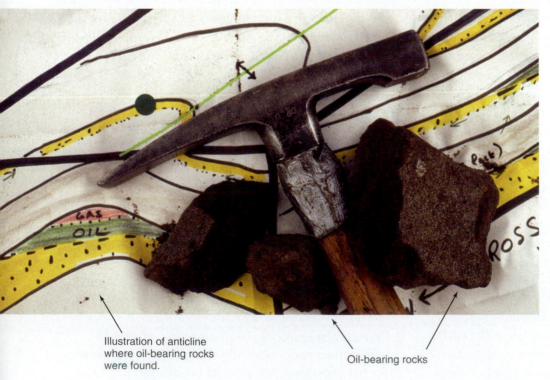

Illustration of anticline where oil-bearing rocks were found.

Oil-bearing rocks

The rock samples in this photograph, which contain oil, are from an exploratory oil well site on the North Slope of Alaska. The diagram underneath the rock pick shows a typical rock formation that harbors oil. A layer of porous rock (shown in yellow) forms an arch-shaped fold with a layer of impervious rock such as shale overlying it. Oil and natural gas, because they are less dense than the water in the pore spaces of the rock, gradually rise through the porous rock until they are trapped at the top of the arch.

Strike and dip • Figure 9.9

Geologists use the terms strike and dip to describe the orientation of a tilted layer of rock.

a. The strike is formed by the intersection of a rock layer and a (sometimes imaginary) level plane. The dip is the angle of tilt of the rock stratum, measured from the horizontal plane.

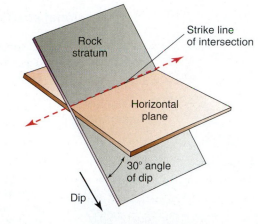

b. In this drawing, the top of the water provides a suitable horizontal plane. The shoreline (the intersection of the water with the rock surface) indicates the strike. The dip is the angle at which the rock layers are tilted. In this drawing, they are tilted 30° from the horizontal. The patterns shown in the diagram are commonly used to distinguish different rock types: The brick-like pattern represents limestone, the tiny dots represent sandstone, and the dashes represent shale.

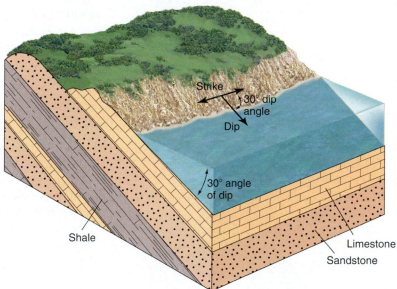

rock layer—intersects a horizontal plane to form a line (**Figure 9.9a**). The strike of the rock layer is the compass direction of that line (measured in degrees north, south, east, or west).

We need another measurement to fully describe the orientation of a tilted rock layer. **Dip** is measured as an

> **dip** The angle between a tilted surface and a horizontal plane.

angle downward from the horizontal plane in degrees, using an instrument similar to a protractor. Dip is always measured in a direction perpendicular to strike. In **Figure 9.9**, the rock strata are tilted at 30°.

If they were dipping more shallowly, the angle would be smaller. If they were dipping more steeply, the angle would be greater. The maximum possible dip is 90°, which represents a layer that has been tipped so that it is completely vertical. When rock layers are tilted even further than the vertical orientation, we say they are **overturned**.

It is important to remember that strike and dip describe the orientation and slope of the rock layer, specifically its **bedding surface**, not the ground itself. Strike and dip are also used to describe the orientation of other types of planar structural features, including faults.

Tensional stress

a. When the crust is stretched by tension, normal faults occur. The block on the overhanging part of the fault (called the **hanging-wall block**) moves down relative to the block underneath the fault (called the **footwall block**). The motion may expose a cliff-like landform at the surface called a **scarp**.

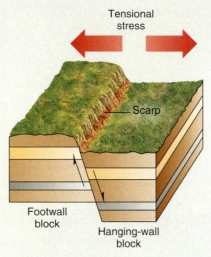

Tensional stress

Scarp

Footwall block

Hanging-wall block

b. Normal faults often occur in pairs. In a **graben**, the two faults dip toward each other, and the block between them drops down. In a **horst**, the faults dip away from each other, and the block between them rises. Although the relative motion of the blocks is nearly vertical, keep in mind that it is the horizontal stretching of the crust, due to tensional stress, that causes it.

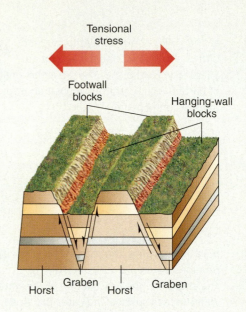

Tensional stress

Footwall blocks

Hanging-wall blocks

Horst   Graben   Horst   Graben

Compressional stress

c. In a reverse fault, compressional stress pushes the hanging-wall block up and over the footwall block. The direction of movement along the fault is opposite that on a normal fault.

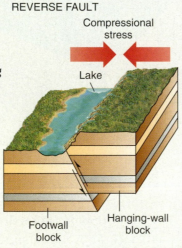

REVERSE FAULT

Compressional stress

Lake

Footwall block

Hanging-wall block

d. A reverse fault with a very shallow dip is called a thrust fault. Geologists indicate this with a row of triangles pointing toward the hanging-wall block.

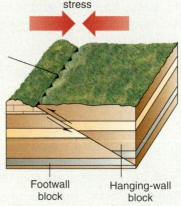

THRUST FAULT

Compressional stress

Symbols used on a map to indicate a thrust fault

Footwall block

Hanging-wall block

Shear stress

e. In a strike-slip fault, the movement is mostly horizontal and parallel to the strike of the fault. Such faults are created by shear stresses.

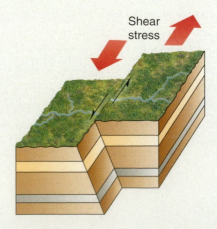

Shear stress

Faults and Fractures

Fractures, or cracks, are characteristic of brittle rock deformation. Fractures occur in all sizes. Some are so tiny that you would need a microscope to see where an individual mineral grain has cracked. Cracks this small are sometimes called **microfractures**. A **fault**, as defined in Chapter 4, is a fracture in rock along which movement has occurred along the fracture surface. Some faults are small, only meters long, but others are very large.

There are many types of faults, caused by different kinds of stress. Tensional (or extensional) stress, stress that stretches or pulls apart the crust, causes **normal faults** (**Figure 9.10a**). Normal faults sometimes occur in pairs, as shown in **Figure 9.10b**, creating a distinctive pairing of uplifted and down-dropped blocks called horsts and grabens. For example, the East African Rift Valley is a huge system of roughly parallel faults that extends for more than 6000 km through the countries of East Africa (**Figure 9.11**). In North America, the region lying between the Sierra Nevada and the Rocky Mountains, known as the Basin and Range Province, is also made up of alternating horsts and grabens.

Compressional stress is responsible for **reverse faults** and **thrust faults** (**Figure 9.10c** and **d**). In reverse faults, the hanging-wall block is pushed over the footwall block, shortening and thickening the crust. The reverse fault in Figure 9.10c dips steeply. When a reverse fault dips more shallowly, less than 45°, it is called a thrust fault (see Figure 9.10d). Thrust faults are common in mountain chains along convergent plate boundaries. The ranges in the Canadian Rockies, described in *Amazing Places* (on the next page), were formed by thrust faults.

In large thrust faults, the hanging-wall block may move thousands of meters, coming to rest on top of much younger rock in the footwall block. This seems to contradict the principle of superposition. However, that principle says that in any undisturbed sequence of strata, the younger strata are deposited on top. A thrust fault represents a disturbance of the original stratigraphic sequence.

Shear stress typically creates **strike-slip faults**, along which two adjacent blocks are displaced horizontally relative to one another. The name comes from the fact that the blocks slip in a direction parallel to the strike of the fault (not the strike of the rock layers), as shown in **Figure 9.10e**.

> **normal fault** A fault in which the block of rock above the fault surface moves downward relative to the block below.

> **reverse fault** A fault in which the block on top of the fault surface moves up and over the block on the bottom.

> **thrust fault** A reverse fault with a shallow angle of dip.

> **strike-slip fault** A fault in which the direction of the movement is mostly horizontal and parallel to the strike of the fault.

The Great Rift Valley • Figure 9.11

The Great Rift Valley in Kenya is under tensional stress because it is a divergent plate margin that is stretching the crust beneath the African and Somali plates. A series of horsts and grabens can be seen in this photograph.

Horst   Graben   Horst   Graben

AMAZING PLACES

The Canadian Rockies

Just to the north of Glacier National Park, the Canadian Rockies in western Alberta and eastern British Columbia offer many beautiful views of folding and faulting.

a. This dramatic fold at Mt. Kidd marks the end of the Lewis Thrust fault into folded rock.

b. Visible from the Kananaskis Highway west of Calgary, a peak in the Opal Range, known locally as Opal Mountain, is a combination of folded and faulted strata.

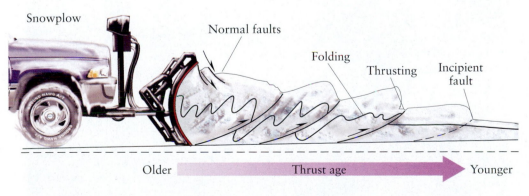

Snowplow

Normal faults

Folding

Thrusting

Incipient fault

Older ← Thrust age → Younger

c. The Canadian Rockies comprise what geologists call a fold-thrust belt, resulting from thin-skinned tectonics (see the chapter-opening photo). A good way to visualize the relationship between the folds and the faults in thin-skinned tectonics is to imagine a snowplow pushing a layer of snow. The snow compresses horizontally and thickens vertically. In the process, a series of thrust faults form in front of the plow, and the snow is highly folded between the faults.

The San Andreas fault system • Figure 9.12

The San Andreas fault is part of a complicated system of faults along which the Pacific plate is moving in a northwesterly direction relative to the North American plate. This map shows the major faults in the vicinity of San Francisco along which motion has occurred in the past 10,000 years.

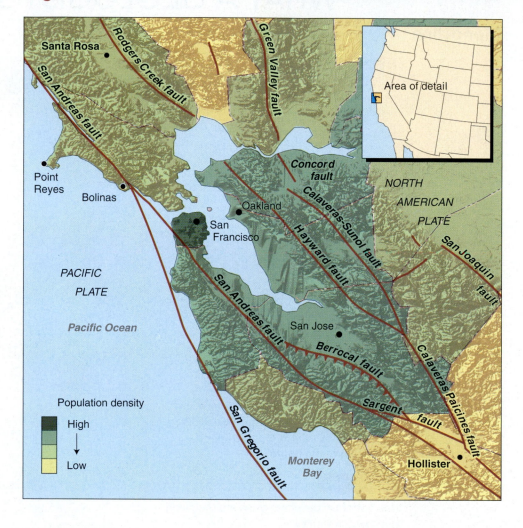

One strike-slip fault is so famous that almost everyone has heard of it: the San Andreas fault in California. Along this fault, the Pacific plate is moving toward the northwest relative to the North American plate (**Figure 9.12**). The word *relative* is very important here. In fact, both the Pacific plate and the North American plate are moving in a roughly northwesterly direction, but the Pacific plate is moving more quickly, about 10 centimeters/year, like a fast runner overtaking a slower one. Over the past 15 million years, the Pacific plate may have moved more than 600 km northwest relative to the North American plate, which is moving at about 5 cm/yr.

Strike-slip faults can be described according to the direction of relative horizontal motion as follows: To an observer standing on either block, the movement of the other block is **left lateral** if it has moved to the left, and

right lateral if it has moved to the right. The San Andreas fault is right lateral because to an observer standing on the Pacific plate, the North American plate appears to be moving to the right (i.e., toward the southeast). Note that it does not matter which plate you stand on; if you stand on the North American plate, the Pacific plate will still appear to be moving toward your right (i.e., toward the northwest), so the San Andreas is still a right-lateral fault.

Folds

When rock deforms in a ductile manner, it bends and flows, creating **folds**. A fold may be a broad, gentle warping over many hundreds of kilometers, a tight flexing of microscopic size, or anything in between.

> **fold** A bend or warp in layered rock.

Monocline

a. A geologist viewing this scene in southern Utah would see that the reddish-colored strata are bent, because the flat-lying strata at the top of the photo dip downward, toward the right-hand side of the photo.

Think Critically

Which strata in the photograph are the oldest, and how can you tell?

b. The geologist prepares a schematic drawing of the area in the photo. You can see that the strata become nearly vertical on the right side of the monocline. The land surface cuts across the rock strata on the right, but it lies parallel to the rock strata on the left.

Approximate area of photo

The simplest type of fold is a **monocline**, a local steepening in otherwise uniformly dipping strata (see *What a Geologist Sees*). An easy way to visualize a monocline is to lay a book on a table and drape a handkerchief over one side of the book. So draped, the handkerchief forms a monocline. However, most folds are more complex than monoclines. In fact, a

> **anticline** A fold in the form of an arch, with the rock strata convex upward and the older rock in the core.
>
> **syncline** A fold in the form of a trough, with the rock strata concave upward and the younger rock in the core.

monocline that is visible at the surface may be just one part of a larger and more complicated fold that is partially hidden underground. Folds are often combinations or variations of two basic types: **anticlines** and **synclines**. If you push the edge of a rug with your foot, it will form a series of anticlinal and synclinal folds.

Several measurements are needed to describe the geometry and orientation of a fold. Imagine a plane dividing the fold in half, as symmetrically as possible (**Figure 9.13**). This is the **axial plane** of the fold. The two halves of the fold, on either side of the axial plane, are the **flanks**, or **limbs**. Axial planes can be vertical, as in Figure 9.13b , or they can be tilted (e.g., the axial planes of the folds in Figure 9.1 are tilted). The axial planes of simple folds are planar, but in a complicated fold, the axial plane can be a curved surface.

Notice that the axial planes in Figure 9.13b connect the most strongly curved parts of each rock layer—in other words, the axial plane passes through the crests of the rock layers in the anticline and the troughs of the rock layers in the syncline. The line where the axial plane intersects the fold is called the **axis**, or **hinge**. The axis of a simple fold may be horizontal (as in the anticline on the left in Figure 9.13b), but some fold axes are tilted. A fold in which the axis is not horizontal is said to be *plunging* (like the anticline on the right in Figure 9.13b).

Maps are two-dimensional representations of three-dimensional geologic structures, so we must use lines to represent axial planes. The projection of a fold axis or an axial plane onto a horizontal plane is called the **axial trace**. To fully represent the geometry of a fold, the map needs to show the following elements: (1) a line representing the axis and the axial plane (i.e., the axial trace); (2) a symbol to indicate the type of fold (anticlines are illustrated using two arrows pointing away from the axial trace; synclines are illustrated using two arrows pointing toward the axial trace); (3) an arrow to show the direction in which the

Simple folds • Figure 9.13

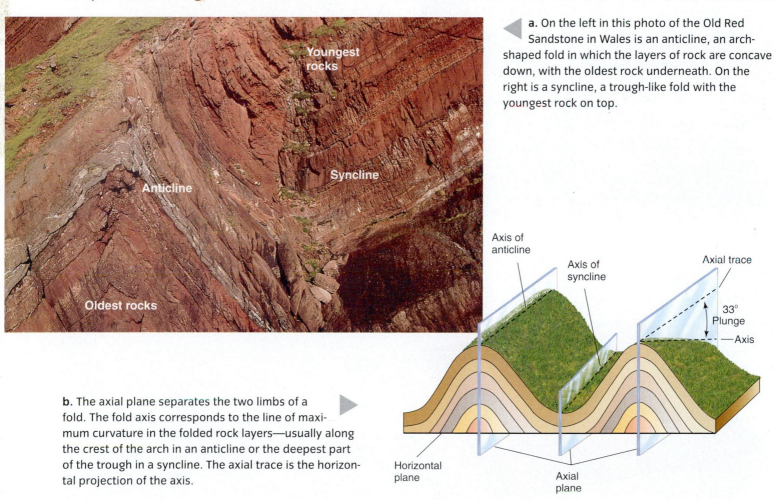

a. On the left in this photo of the Old Red Sandstone in Wales is an anticline, an arch-shaped fold in which the layers of rock are concave down, with the oldest rock underneath. On the right is a syncline, a trough-like fold with the youngest rock on top.

b. The axial plane separates the two limbs of a fold. The fold axis corresponds to the line of maximum curvature in the folded rock layers—usually along the crest of the arch in an anticline or the deepest part of the trough in a syncline. The axial trace is the horizontal projection of the axis.

axis is plunging; and (4) a number (measured in degrees from horizontal) to show the plunge angle. **Figure 9.14** gives an example of how these symbols are used on a map.

Note that synclines do not always form valleys, and anticlines do not always form ridges. Sometimes an anticline may expose a rock stratum that is more susceptible to erosion than the surrounding layers; in that case, the anticline will erode away quickly and form a valley. In general, the contours of the land do not always follow the bedding surfaces. The shapes of the bedding surfaces matter most for geology because they record the long-term deformational history of the rock.

Showing folds on maps • Figure 9.14

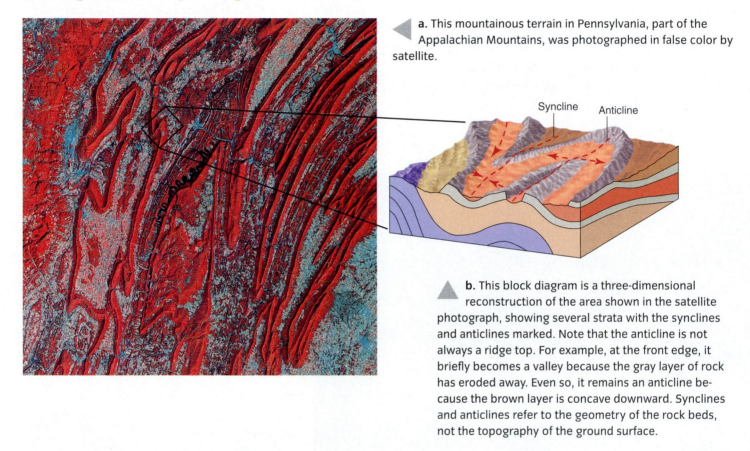

a. This mountainous terrain in Pennsylvania, part of the Appalachian Mountains, was photographed in false color by satellite.

b. This block diagram is a three-dimensional reconstruction of the area shown in the satellite photograph, showing several strata with the synclines and anticlines marked. Note that the anticline is not always a ridge top. For example, at the front edge, it briefly becomes a valley because the gray layer of rock has eroded away. Even so, it remains an anticline because the brown layer is concave downward. Synclines and anticlines refer to the geometry of the rock beds, not the topography of the ground surface.

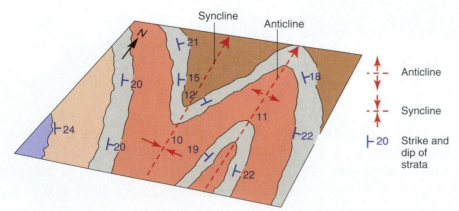

c. This is a map of the area in the block diagram, using geologic symbols for the anticlines, synclines, strike, and dip.

Domes and basins • Figure 9.15

Upwarping of the crust causes domes to form. When domes erode, they expose older rock at their centers.

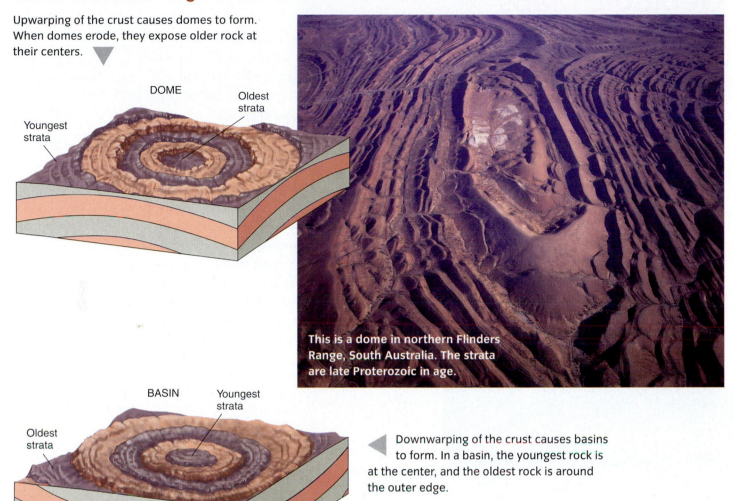

DOME

Youngest strata

Oldest strata

This is a dome in northern Flinders Range, South Australia. The strata are late Proterozoic in age.

BASIN

Oldest strata

Youngest strata

Downwarping of the crust causes basins to form. In a basin, the youngest rock is at the center, and the oldest rock is around the outer edge.

Folds can also take more complicated forms than the ones we have discussed. Sometimes a large area of crust undergoes upwarping or downwarping, which forms broad, gentle folds. Upwarping of strata forms **domes**, while downwarping forms large, bowl-like **basins** (**Figure 9.15**). Folds can also be **asymmetrical**, with one limb dipping more steeply than the other, or even **overturned**, with one limb tilted so far over that it is upside down. Folds that are so strongly overturned that they are almost lying flat are called **recumbent**. This is another way in which rock deformation can change the normal sequence of strata.

CONCEPT CHECK   STOP

1. **What** information do the strike and dip of a rock layer provide?

2. **What** is the difference between a normal fault and a reverse, or thrust, fault? Which one changes the usual sequence of rock strata?

3. **What** process could create a pair of scarps with a valley between them? What is such a formation called?

4. **Why** is the shape of the ground not necessarily a reliable guide to the shape of the underlying strata?

Geologic Maps

LEARNING OBJECTIVES

1. **Distinguish** between a topographic map and a geologic map.
2. **Describe** how a geologist makes a geologic map.
3. **Explain** what kind of information can be shown on a geologic cross section.

G eologists cannot see all the structural details of deformed rock strata in a given area; soil, water, vegetation, and buildings cover much of the evidence. They must gather geologic information from **outcrops**—places where bedrock is exposed at the surface (**Figure 9.16**). At outcrops, geologists note the type of rock present, the orientation of the layers, and the presence of structural features such as fractures or folds. They extrapolate from these data—perhaps using additional sources of information, such as drilling or seismic studies—to determine what lies beneath the soil, vegetation, water, and buildings in the areas between the outcrops. The final result is a **geologic map**.

The oldest-known geologic map—possibly the first one ever made—was drawn on a papyrus scroll in ancient Egypt. Evidently, it was used to show the locations of certain rock types for quarrying purposes. Probably the most important geologic map ever produced was made by an English surveyor, William Smith (see *Case Study*). Today, geologic maps still help geologists interpret the geologic history of an area. Geologists from mining companies, oil companies, engineering firms, environmental agencies, consulting firms, and government agencies refer to such maps regularly.

> **geologic map**
> A map that shows the locations, kinds, and orientations of rock units, as well as structural features such as faults and folds.

Geology InSight   Making a geologic map • Figure 9.16

✓ THE PLANNER

a. This block diagram shows a landscape with tilted rock strata. Much of the rock is covered at the surface by grass and soil, but there are some outcrops where bedrock is exposed. ▼

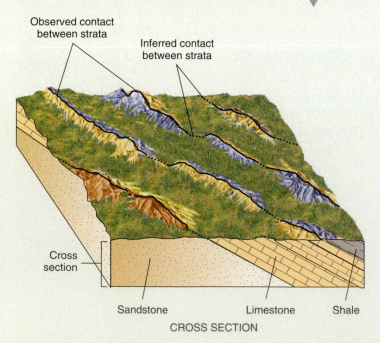

Observed contact between strata

Inferred contact between strata

Cross section

Sandstone   Limestone   Shale

CROSS SECTION

b. The geologist has transferred the information about rock types and boundaries (**geologic contacts**) between rocks onto a map of the area. In areas where the bedrock is covered, the geologist makes an educated guess about the types of rock and the locations of boundaries. This educated guess can be used to construct a vertical profile of the strata, shown as the front panel in the block diagram; it is called a **geologic cross section**. ▼

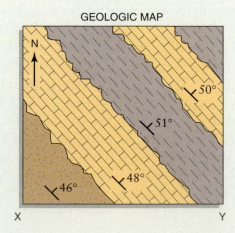

GEOLOGIC MAP

N

50°

51°

48°

46°

X   Y

The Map That Changed the World

In 1815, the English geologist and canal surveyor William "Strata" Smith published this immense geologic map of England, which has been described in a recent history as "the map that changed the world." It was a given this impressive name because it was a major step in the emergence of the modern science of geology. Smith used 23 different colors to depict where each stratum (from "London Clay" to "Granit, Syenit and Gneifs") was exposed at the surface.

The map is literally a work of art. It was printed in 15 sections, measures 2 m by 3 m, and includes more than 1,000 place names—many of which had never been shown on a map before. Every copy of the map was hand-painted with watercolors, and only 400 copies were ever printed. (Only 43 are known to still exist.)

But even more important than the map's size and detail was its organization. The map's legend (inset) shows that Smith understood that all 23 strata lie in a definite vertical sequence, and the sequence is the same anywhere in England. In other words, he had discovered the key to understanding all geologic processes: the geologic column. His map also enabled geologists to predict where certain economically important deposits, such as coal, could be found.

Modern-day geologic maps still show strata, although they now use standardized symbols instead of watercolors. They also depict three-dimensional structural information, such as the locations and slopes of faults and folds, which the geologists of Smith's day did not yet understand.

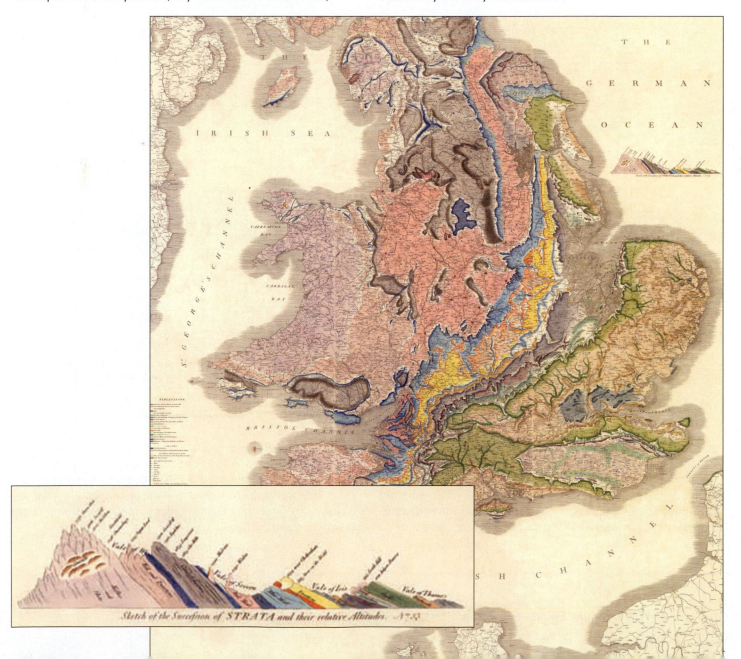

Creating a topographic profile • Figure 9.17

Topographic maps contain information about the ups and downs of the land. Contour lines indicate points of equal height; for example, every point on the line labeled "1100" is 1100 feet (335 m) above sea level.

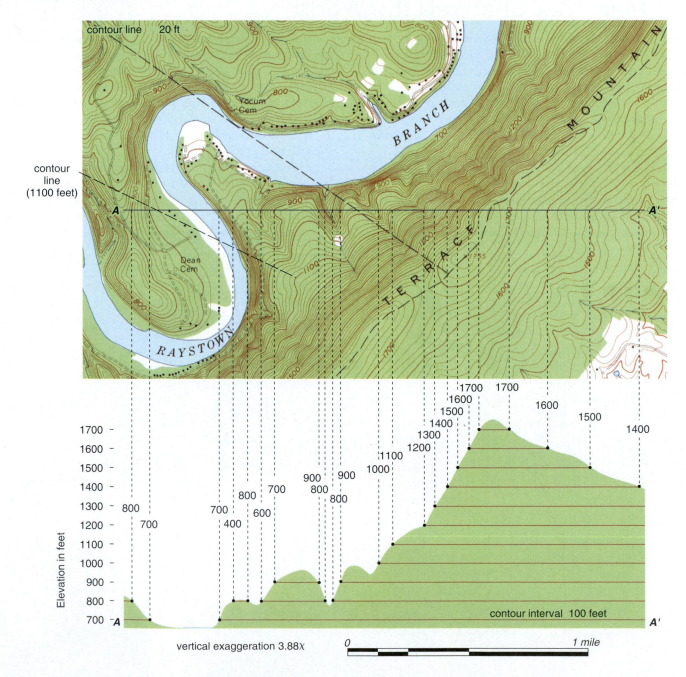

You can use the height information from a topographic map to plot a **topographic profile** of the landscape. First, pick a line along which you would like to construct the topographic profile (e.g., the line labeled A–A' on the map). Next, set up your topographic profile: Use the line A–A' as the horizontal axis and draw and label a vertical axis to represent height or elevation. Directly below each point where A–A' crosses a contour line on the map, place a dot at the appropriate height on the profile. Finally, connect the dots with a smooth curve. This curve represents the topography of the mountain. (Here we have exaggerated the height in order to make the differences in elevation stand out.)

Making and Interpreting Geologic Maps

Commonly, a **topographic map** is used as the base for a geologic map (**Figure 9.17**). You may have used a topographic map if you have ever gone backpacking or hiking. An important aspect of topography is **relief**, the difference between the lowest and highest elevations in the area. A mountainous region has high relief, whereas a flat plain has low relief. Topographic maps use **contour lines**, lines of equal elevation, to portray topography.

topographic map
A map that shows the shape of the ground surface, as well as the location and elevation of surface features, usually by means of contour lines.

However, topography is not geology. A geologist also needs to know about the rocks that form the topography, as well as the rocks beneath the surface. A **formation** is a unit of rock that can be distinguished from units above and below it on the basis of rock type and recognizable boundaries, or geologic contacts, with other rock units.

The colors of the rock units do not reveal whether the underlying rock strata are vertical or horizontal or tilted in one direction or another. This information can be indicated using symbols that show the locations and orientations of folds, faults, and other geologic features. The finished product looks like the map in **Figure 9.18**.

WILEY PLUS | NATIONAL GEOGRAPHIC | Video

Interpreting geologic maps • Figure 9.18

This is part of a geologic map that shows the Canmore Quadrangle in Alberta, Canada. Note the colors, which indicate the age and type of rock, and the contour lines, which show the elevation. In the inset, a magnified part of the map, you can see the symbols for thrust faults, strike and dip, synclines, and anticlines.

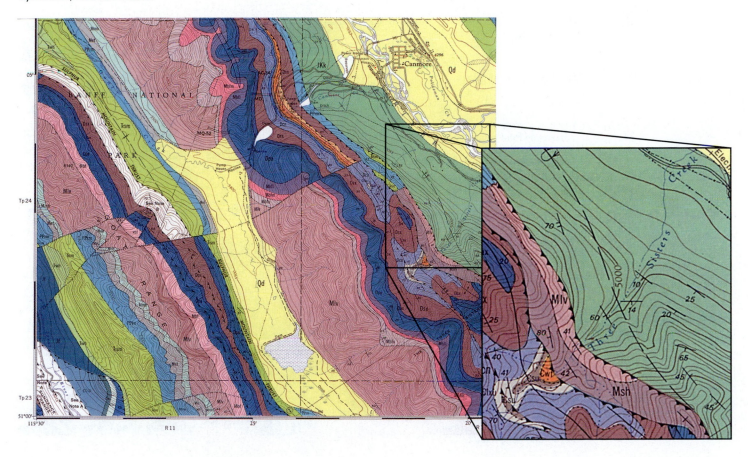

The colors and symbols most commonly used to portray rock formations on maps are shown in **Figure 9.19**.

Geologic Cross Sections

A geologic map shows the locations of all rock outcrops on the ground surface, as well as the orientations of layering, structures, and other geologic features. It also shows the geologist's educated guess as to what lies under the soil, vegetation, and buildings between the outcrops. A geologic map can be used to make inferences about what happens to the rock layers just under the ground. Do they bend completely around and come back to the surface? Do they level out and become flat? Do they grade into a different type of rock? To answer these questions, we must try to visualize the area in three dimensions, even though we may

Symbols commonly used on geologic maps • Figure 9.19

Representative patterns are commonly, but not universally, used to show various kinds of rock in geologic maps and cross sections.

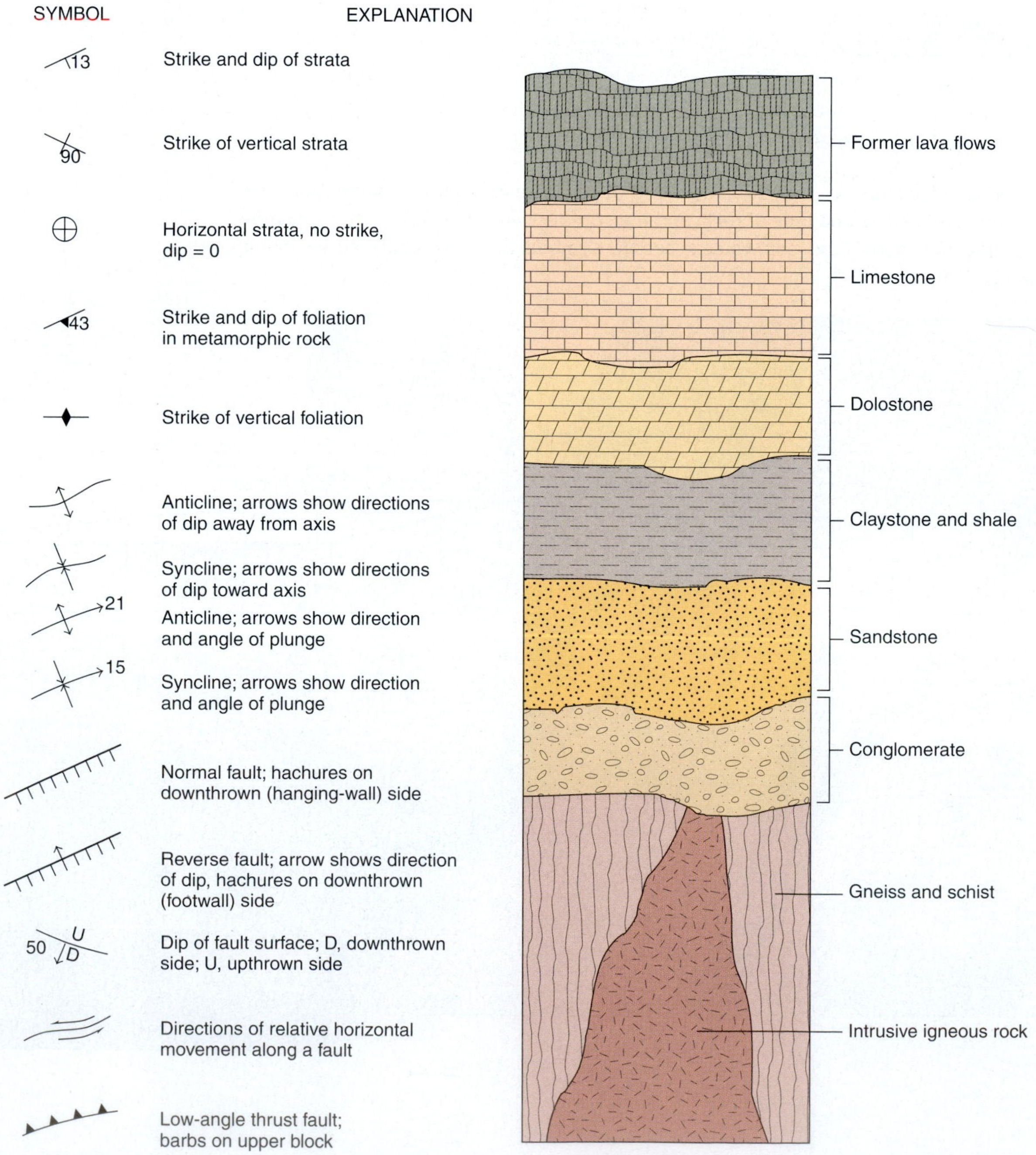

SYMBOL   EXPLANATION

13 — Strike and dip of strata

90 — Strike of vertical strata

⊕ — Horizontal strata, no strike, dip = 0

43 — Strike and dip of foliation in metamorphic rock

— Strike of vertical foliation

— Anticline; arrows show directions of dip away from axis

— Syncline; arrows show directions of dip toward axis

21 — Anticline; arrows show direction and angle of plunge

15 — Syncline; arrows show direction and angle of plunge

— Normal fault; hachures on downthrown (hanging-wall) side

— Reverse fault; arrow shows direction of dip, hachures on downthrown (footwall) side

50 U/D — Dip of fault surface; D, downthrown side; U, upthrown side

— Directions of relative horizontal movement along a fault

— Low-angle thrust fault; barbs on upper block

— Former lava flows

— Limestone

— Dolostone

— Claystone and shale

— Sandstone

— Conglomerate

— Gneiss and schist

— Intrusive igneous rock

Underneath the Alps • Figure 9.20

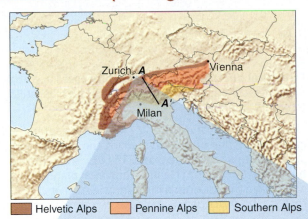

Helvetic Alps Pennine Alps Southern Alps

a. The map shows part of the Alps, a great mountain chain that borders Italy, France, and Switzerland, which was formed by crustal deformation and compressive forces. The points labeled A and A' are the beginning and end of the cross section shown here.

b. The cross section A–A' reveals the intense folding and faulting that have resulted in the mountainous landscape of the Alps. Features you should be able to spot in the cross section include several faults, an enormous anticline-like structure (called a **nappe**), and the remnants of a subduction zone.

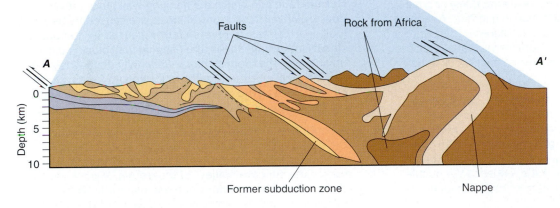

Faults

Rock from Africa

Former subduction zone

Nappe

Where Geologists CLICK

National Geologic Maps Database

http://ngmdb.usgs.gov/MapProgress/MIP_home.html

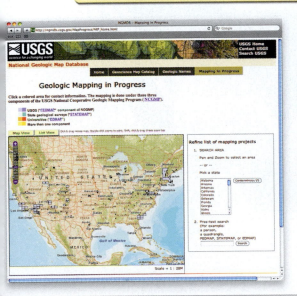

For a more extensive discussion of geologic maps, go to the National Geologic Maps database, maintained by the USGS, where you can view geologic maps in progress, and see areas where geologic mapping is still being done.

only have information concerning the rocks at the surface (see *Where Geologists Click*).

Geologists do this three-dimensional visualization by constructing a **geologic cross section** like the one shown in **Figure 9.20**. This is a bit like a topographic map to which a geologist adds his or her best guess as to how the strata fold and fault underneath the ground. As you can see, the layering can become very complicated.

> **geologic cross section** A diagram that shows geologic features that occur underground.

CONCEPT CHECK   STOP

1. **What** types of information are included on geologic maps, that are not included on topographic maps?

2. **What** are the basic data that a geologist records on a geologic map?

3. **How** can a geologist illustrate the three-dimensional distribution of rock strata under the surface?

Summary

1 Rock Deformation 251

- In response to **stress**, a rock may undergo **strain**; that is, it may change its shape, volume, or both. A nonpermanent change is **elastic deformation**. A permanent change that involves folding or flowing is **ductile (or plastic) deformation**. A permanent change that involves fracturing is **brittle deformation**.

- Rock deformation results from **pressure** or stress placed on rock by the movements and interactions of lithospheric plates.

- Stress can be uniform (the same in all directions) or differential (stronger in one direction than in another). As shown in the figure, **compression** results from forces that squeeze a body of rock; **tension** results from forces that stretch the rock, or pull it apart; and **shear** stress causes rock to be twisted and to change shape.

- Low temperature, low confining pressure, and high rate of strain enhance the brittle properties of rock. Failure by fracture is common in upper-crustal rock, where temperature and pressure are low. High temperature, high confining pressure, and low rate of strain, which are characteristic of the deeper crust and the mantle, enhance the ductile behavior of rock. The composition of rock also influences its deformational properties.

- Deformed rock is commonly found in **orogens**, where former tectonic plates collided. Today's continents are complicated assemblages of orogens and **cratons**—regions of ancient crust eroded to near sea level. Orogens are collision zones between cratons, and, because of **isostasy**, they remain subject to vertical movement long after the collisions that produced them have ended.

Three types of stress • Figure 9.2

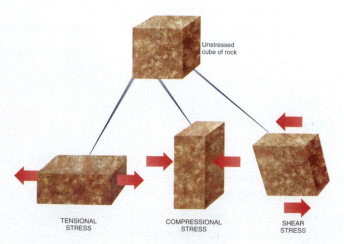

TENSIONAL STRESS   COMPRESSIONAL STRESS   SHEAR STRESS

2 Structural Geology 258

- **Structural geology** deals with the processes and structures associated with stress and strain, such as faults, fractures, and **folds**.

- The **strike** of a stratum, a fault, or another geologic surface is the orientation of the line marking the intersection of the surface with a horizontal plane. The **dip** is the angle between the tilted surface and a horizontal plane, measured down from the horizontal in degrees. Any planar features in rock can be described using these two measurements.

- **Normal faults**, in which the hanging-wall block moves down relative to the footwall block, are caused by tensional (pull-apart) stress. In **reverse faults**, caused by compressional (squeezing) stress, the hanging-wall block moves up and over the footwall block. Shallowly dipping reverse faults are called **thrust faults**. In **strike-slip faults**, caused by shear stress, the movement is mainly horizontal and parallel to the strike of the fault (as shown in the figure).

- An **anticline** is an upward fold in the form of an arch, with the oldest rock in the core of the arch. A **syncline** is a downward fold in the form of a trough, with the youngest rock in the concave "valley" of the trough. Many folds are combinations or variations of these two basic types. To fully portray the geometry of a fold, we must describe the fold axis and (if necessary) the direction in which the axis is plunging, as well as the dip of the plunging axis.

Stresses and the faults they cause • Figure 9.10

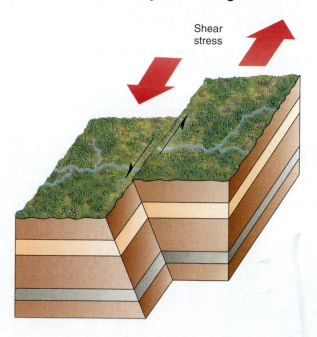

Shear stress

3 Geologic Maps 268

- Geologists use **geologic maps** to portray rock units and structures found at Earth's surface. It is common to use a **topographic map** (see figure at right) as the base for a geologic map. Topographic maps give information about the relief, or "ups and downs," of the land surface, using contour lines to connect points of equal elevation.

- Geologic maps provide information about the distribution, locations, and orientations of rock formations and structures at the surface, using standardized symbols, colors, and fill textures to portray various geologic features.

- Sometimes it is important to be able to visualize an area in three dimensions in order to understand what happens to the geologic structures in the subsurface and to interpret the geologic history of the area. This is done by constructing a **geologic cross section**. Geologic cross sections are based partly on the surface geology, as revealed by geologic mapping, and partly on other sources of information about underground geology, such as drilling or seismic studies.

Creating a topographic profile • Figure 9.17

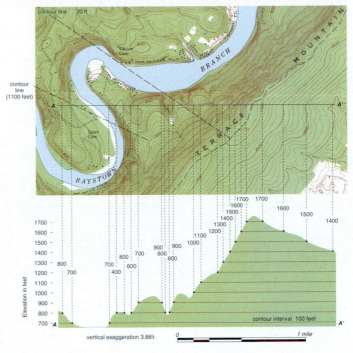

Key Terms

- anticline 264
- brittle deformation 254
- compression 251
- craton 256
- dip 259
- ductile deformation 254
- elastic deformation 251
- fold 263
- geologic cross section 273

- geologic map 268
- isostasy 257
- normal fault 261
- orogen 256
- pressure 250
- reverse fault 261
- shear 251
- strain 251
- strike 258

- strike-slip fault 261
- stress 250
- structural geology 258
- syncline 264
- tension 251
- thrust fault 261
- topographic map 271

Critical and Creative Thinking Questions

1. Find real examples of plate boundaries along which each of the following types of stress predominates: (a) compression, (b) tension, and (c) shearing. Try to find examples different from those used in the text.

2. What might happen to a rubber ball or a banana or Silly Putty when they are frozen in liquid nitrogen? Can you find other common materials or objects that make a transition from ductile to brittle (or vice versa) at different temperatures, pressures, or strain rates?

3. To construct geologic cross sections, geologists use the information on topographic maps, often supplemented with information about the subsurface provided by remote study techniques. What are some of the techniques geologists use to study rock structures hidden beneath Earth's surface? (Hint: Look back at Chapter 5.)

4. In this chapter we mentioned the role of structural geology in determining the locations of mineral ores, oil, and natural gas. Try to find out about some other practical uses and applications of structural geology, geologic maps, and cross sections.

5. Imagine that you are taking a field course in the Alps. From geologic cross sections, you expect to find some very large folds (see *What is happening in this picture?*) where the rock layers have been bent over backward. What kinds of evidence could you look for that would help you determine whether a particular sequence of rock is right-side up or upside down? In other words, how would you determine whether a trough-shaped fold was a right-side up syncline or an upside down (i.e., overturned) anticline?

What is happening in this picture?

Chief Mountain, in Glacier National Park in Montana, is a huge chunk of Precambrian rock that lies on top of a layer of much younger Cretaceous shale and sandstone (underneath the arrow).

Think Critically

1. How is such a formation possible?
2. The arrow points to a fault. Can you guess what kind of fault it is?

Self-Test

(Check your answers in the Appendix D.)

1. _____ is a type of stress that acts in a direction perpendicular to and away from a fault surface.

   a. Shear stress

   b. Compression

   c. Tension

2. _____ is a type of stress that acts in a direction parallel to and along a fault surface.

   a. Shear stress

   b. Compression

   c. Tension

3. In _____ deformation, rock will bend as long as stress is applied to the crust but will resume its original shape if the stress is released.

   a. elastic

   b. brittle

   c. ductile

4. What type of crustal deformation is depicted in this photograph?

   a. elastic

   b. brittle

   c. ductile

5. The structure depicted in the photograph in question 4 could have formed as a result of stress, under conditions that have any combination of _____.

   a. high confining pressure, low temperature, and high strain rate

   b. low confining pressure, low temperature, and high strain rate

   c. high confining pressure, low temperature, and high strain rate

   d. high confining pressure, high temperature, and low strain rate

6. A(n) _____ is an elongate region of crust that has been deformed and metamorphosed by a continental collision. By contrast, a(n) _____ is a region of continental crust that has remained undeformed for a very long time.

   a. syncline; anticline

   b. anticline; syncline

   c. orogen; craton

   d. craton; orogen

7. On this block diagram of tilted strata, draw symbols to show the direction of strike and dip and indicate the 30° angle of dip.

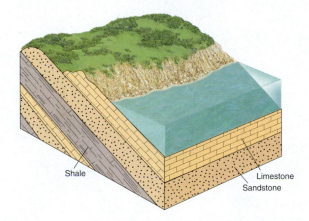

Shale   Limestone   Sandstone

8. This diagram shows a faulted block of Earth's crust. What type of fault is depicted in the diagram?

   a. normal fault

   b. thrust, or reverse, fault

   c. strike-slip fault

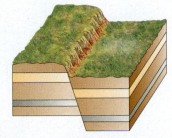

9. The fault depicted in the diagram in question 8 must have formed in response to what kind of stress?

   a. tension

   b. compression

   c. shear

10. A _____ fault is a product of compression of Earth's crust, whereas a _____ fault is a product of tension. _____ faults result from shear stress.

   a. strike-slip; normal; Thrust (reverse)

   b. normal; thrust (reverse); Strike-slip

   c. thrust (reverse); strike-slip; Normal

   d. thrust (reverse); normal; Strike-slip

   e. strike-slip; thrust (reverse); Normal

11. Label this diagram with the following terms:

   anticline   axial plane

   syncline   plunge

   axial trace

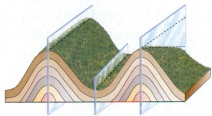

12. A(n) _____ is a local steepening in otherwise uniformly dipping strata.

   a. anticline

   b. syncline

   c. monocline

   d. dome

   e. basin

13. A _____ would allow a geologist to locate the steepest cliff face in a region.

   a. geologic map

   b. topographic map

   c. geologic cross section

14. A _____ can be used to show the subsurface structure of an orogen.

   a. geologic map

   b. topographic map

   c. geologic cross section

15. The areal distribution of rocks and structures in an area is best depicted on a _____.

   a. geologic map

   b. topographic map

   c. geologic cross section

THE PLANNER

Review your Chapter Planner on the chapter opener and check off your completed work.

Metamorphism: New Rock from Old

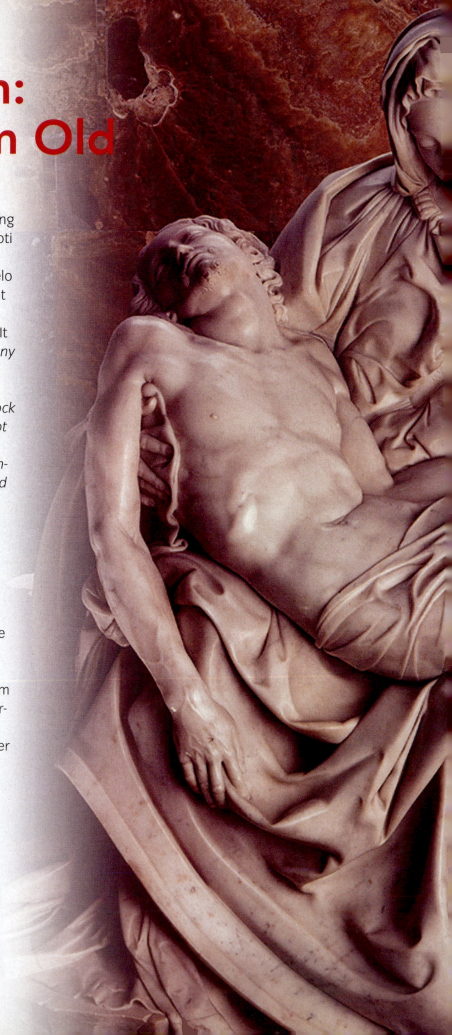

To carve the *Pietà*, a statue portraying Mary holding the lifeless body of Jesus, Michelangelo Buonarroti needed a block of perfect marble 2 meters wide. The year was 1498; the place was Rome. After Michelangelo had resigned himself to a visit to the quarries in distant Carrara, word reached him that a block had been cut on order and shipped as far as Rome but not paid for. It would go no further. As Irving Stone writes in *The Agony and the Ecstasy*:

> He watched the rays of the rising sun strike the block and make it transparent as pink alabaster, with not a hole or hollow or crack or knot in all its massive white weight. It tested out perfect against the hammer, against water, its crystals soft and compacted with fine graining. His *Pietà* had come home.

White marble starts out as marine sediment composed of fragments of shells, most of them made of calcite. The sedimentary rock that forms from this sediment as a result of diagenesis is limestone. When subjected to further heat and pressure, the calcite recrystallizes, all traces of the shells disappear, and the limestone becomes a crystalline white marble. This transformation is called metamorphism.

Which is more remarkable: the metamorphism from calcareous sediment to pristine marble, or the transformation from a raw quarried marble block to a finished sculpture? We are amazed by them both. In this chapter you will learn about the processes of change during metamorphism and the many kinds of metamorphic rock and where they are found.

CHAPTER OUTLINE

CHAPTER PLANNER ✓

❏ Study the picture and read the opening story.

❏ Scan the Learning Objectives in each section:
   p. 280 ❏   p. 289 ❏   p. 293 ❏   p. 298 ❏

❏ Read the text and study all visuals.
   Answer any questions.

Analyze key features

❏ Geology InSight, p. 282 ❏

❏ What a Geologist Sees, p. 285 ❏

❏ Process Diagram, p. 286 ❏

❏ Case Study, p. 287 ❏

❏ Amazing Places, p. 291 ❏

❏ Stop: Answer the Concept Checks before you go on:
   p. 288 ❏   p. 292 ❏   p. 297 ❏   p. 299 ❏

End of chapter

❏ Review the Summary and Key Terms.

❏ Answer the Critical and Creative Thinking Questions.

❏ Answer What is happening in this picture?

❏ Complete the Self-Test and check your answers.

What Is Metamorphism?

LEARNING OBJECTIVES

1. **Explain** how metamorphism differs from diagenesis and lithification.
2. **Describe** the temperature and pressure conditions required for metamorphism.
3. **Define** the grade of metamorphism.
4. **Describe** how pore fluid, stress, and time affect the results of metamorphism.
5. **Define** foliation.

Perhaps the best analogy for **metamorphism** is simple, everyday cooking—a process that takes, for instance, flour, salt, sugar, yeast, and water and transforms them into a completely new substance called bread, or turns raw meat into a medium-rare steak (**Figure 10.1a**). Like cooking and like diagenesis (see Chapter 8), metamorphism involves heat and pressure, but the temperatures and pressures involved are considerably higher. Some of the same physical processes that occur in diagenesis, such as compaction and recrystallization, continue during metamorphism—but in a more intense

> **metamorphism**
> The mineralogical, textural, chemical, and structural changes that occur in rock as a result of exposure to elevated temperatures and/or pressures.

way. In metamorphism, these processes are accompanied by chemical reactions that change the rock's mineral assemblage and texture and sometimes its chemical composition as well. Metamorphism also takes place at greater depths than diagenesis—typically between about 5 and 40 kilometers, which is most of the extent of the thickness of the crust.

There is one important kind of change that does not take place during metamorphism: The rock does not melt and that is what makes metamorphic rock so interesting. The rock may have been squeezed, stretched, heated, and altered in complex ways (like the gneiss in **Figure 10.1b**), but it has remained solid. Solids, unlike liquids and gases, preserve clues about the events that changed them. For example, if you throw a stone into a pond, the splash and resulting ripple soon disappear. Throw a stone at a window, however, and the result is permanently cracked glass. Metamorphic rocks preserve a record of all the heatings, stretchings, and grindings that have happened to them throughout geologic history. When we encounter an area with a lot of metamorphic rock, we can be certain that it has been geologically busy. Decoding the record of that activity is challenging, but the rewards are great because the decoding can tell us about long-ago collisions of continents, the break-up of ancient supercontinents, and whether rich mineral deposits might be present as a result of such activity.

Metamorphism and cooking • Figure 10.1

a. In the thermal metamorphism of a roast, the outer layers of meat have been subjected to higher temperatures, and therefore a higher degree of metamorphism, than the inner portion, including changes in texture and color. The outermost layer is well-done; moving in, the lighter-colored meat is medium-rare; and in the center, the reddest meat is in the rare zone.

b. High temperatures and pressures deep in the crust converted the ingredients in this mix of minerals into a new metamorphic rock called gneiss. The extreme folding indicates that the rock underwent ductile deformation during metamorphism. This photograph, shows an outcrop in the Sand River, South Africa.

a. This diagram shows schematically where in the crust metamorphism and melting occur.

b. Four colored bands illustrate the temperature and pressure conditions for diagenesis, low-grade and high-grade metamorphism, and melting. Note that the vertical scale is given in units of both pressure and depth beneath the surface. The cross-hatched diagonal band indicates the pressures and temperatures most commonly found in continental crust.

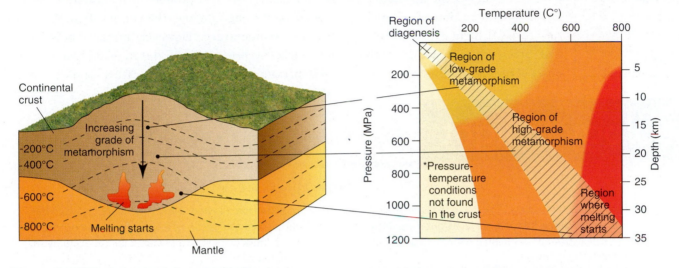

The Limits of Metamorphism

The two most important factors in metamorphism are heat and pressure. The heat that causes metamorphic reactions is Earth's internal heat. We know from drilling deep gas and oil wells, and from deep gold mines, that temperature in the continental crust increases with depth at a rate of about 30°C/km. At a depth of about 5 km, the temperature is about 150°C. This temperature represents the dividing line between the processes of diagenesis (Chapter 8), which change sediment into sedimentary rock (below 150°C), and the processes of metamorphism (above 150°C). Geologists have settled on 150°C as the "official" boundary between diagenesis and metamorphism, but the actual transition from diagenesis into metamorphism is more gradational; some rocks begin to show distinct signs of metamorphic changes at 150°C, but others do not.

The upper temperature limit of metamorphism is about 800°C. The onset of melting above this temperature marks the transition from metamorphic processes to magmatic processes. As discussed in Chapter 6, the beginning of melting can occur at a variety of temperatures, depending on the pressure, the composition of the rock, and the amount of fluid present, particularly water and carbon dioxide. Thus, 800°C is only a generalization of the temperature at which fractional melting typically begins in the crust.

Just as pressure influences the temperature at which melting begins, it also influences the kind of metamorphic changes that occur at temperatures below the onset of melting. The air pressure at sea level on Earth's surface is defined to be 1 atmosphere (atm)—about half the pressure of the air inside an automobile tire. However, rock is heavier than air. In the crust, pressure increases with depth at a rate of about 300 atm per kilometer, or about 30 megapascals per kilometer. Therefore, the pressure 5 km below the surface is 1500 times greater than the surface atmospheric pressure—that is, 1500 atm, or 150 MPa. This is the depth at which both temperature and pressure are high enough for recrystallization and growth of new minerals to start.

Figure 10.2 shows the range of temperatures and pressures over which metamorphism occurs. In **Figure 10.2b**, pressure is represented on the vertical axis, increasing with depth. Temperature is represented on the horizontal axis, increasing from left to right. The pressure and temperature conditions under which sediment is formed and changed into sedimentary rock during diagenesis occupy the upper-left corner of the diagram. Below this, and down to

low-grade Rock metamorphosed under temperature and pressure conditions up to 400°C and 400 MPa.

high-grade Rock metamorphosed under temperature and pressure conditions higher than about 400°C and 400 MPa.

a depth of about 15 km, where the pressure reaches 400 MPa (4000 atm) and the temperature is about 400°C, metamorphism occurs but is not very intense. In **low-grade** metamorphic rock, the minerals have changed, but the appearance is still similar to that of sedimentary rock. At higher temperatures and pressures, extending to the onset of melting, **high-grade** meta-

morphic rock is formed. In the lower-right corner of the diagram, representing the hottest temperatures and highest pressures, rock may begin to melt and would no longer be considered metamorphic.

Factors Influencing Metamorphism

As in cooking, the end product of metamorphism is controlled by the ingredients (the composition of the rock) and the temperature. However, metamorphism is also influenced by pressure, the duration and rate at which high pressures and temperatures are applied, and the

Geology InSight   From shale to schist • **Figure 10.3**

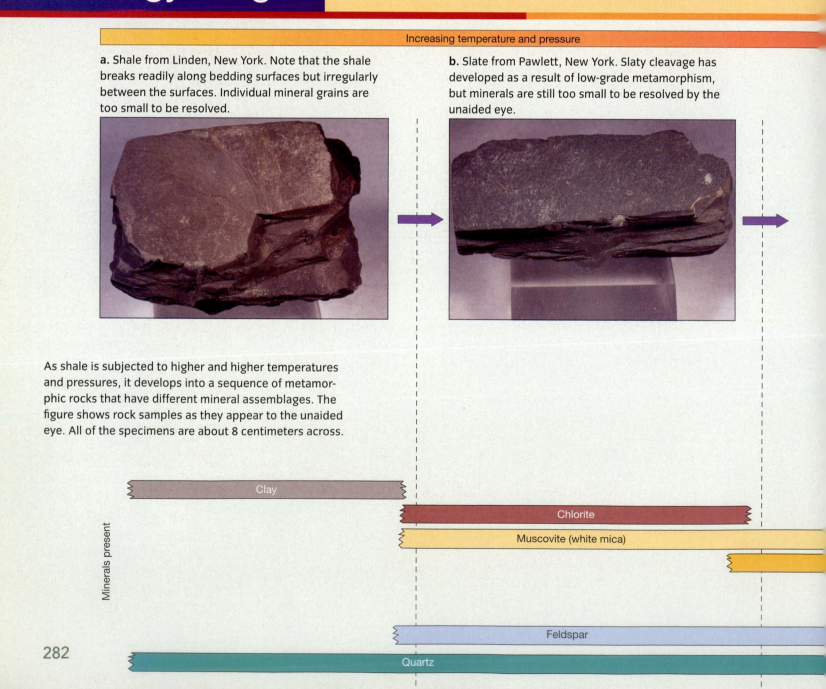

Increasing temperature and pressure

a. Shale from Linden, New York. Note that the shale breaks readily along bedding surfaces but irregularly between the surfaces. Individual mineral grains are too small to be resolved.

b. Slate from Pawlett, New York. Slaty cleavage has developed as a result of low-grade metamorphism, but minerals are still too small to be resolved by the unaided eye.

As shale is subjected to higher and higher temperatures and pressures, it develops into a sequence of metamorphic rocks that have different mineral assemblages. The figure shows rock samples as they appear to the unaided eye. All of the specimens are about 8 centimeters across.

Minerals present

Clay

Chlorite

Muscovite (white mica)

Feldspar

Quartz

presence or absence of fluid. The directionality of pressure also plays an important role in metamorphism. As explained in Chapter 9, geologists use the term *stress* to refer to pressures that are greater in one direction than in others. Many metamorphic rocks clearly show the effects of directional stress. Let's look at all these factors.

Temperature and pressure When rocks are heated, some of the original minerals recrystallize but don't change composition; others are involved in chemical reactions that form new minerals. **Figure 10.3** shows the effect of increasing temperature and pressure on shale, the most common variety of sedimentary rock. Notice that as the grade of metamorphism increases, completely new minerals appear—and, in some cases, disappear again at higher grades. Thus, identifying the component minerals is a good way to tell these metamorphic rocks apart. Notice also that as the temperature and pressure of metamorphism increase, the minerals grains become progressively larger. Rock subjected to low-grade metamorphism tends to be fine-grained, while high-grade metamorphism produces coarse-grained rock.

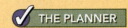
THE PLANNER

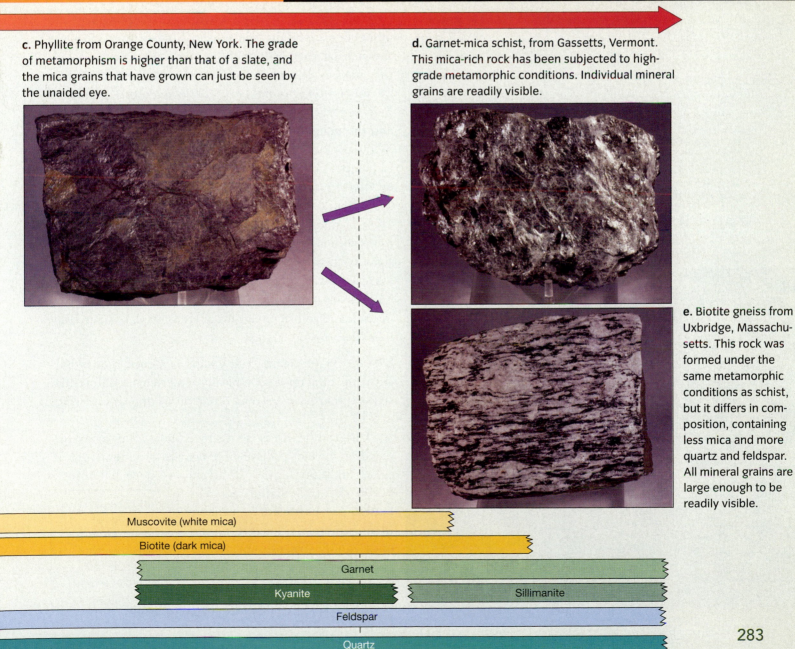

c. Phyllite from Orange County, New York. The grade of metamorphism is higher than that of a slate, and the mica grains that have grown can just be seen by the unaided eye.

d. Garnet-mica schist, from Gassetts, Vermont. This mica-rich rock has been subjected to high-grade metamorphic conditions. Individual mineral grains are readily visible.

e. Biotite gneiss from Uxbridge, Massachusetts. This rock was formed under the same metamorphic conditions as schist, but it differs in composition, containing less mica and more quartz and feldspar. All mineral grains are large enough to be readily visible.

Muscovite (white mica)

Biotite (dark mica)

Garnet

Kyanite   Sillimanite

Feldspar

Quartz

283

Pore fluid Another factor that affects metamorphism is the presence of open spaces, called **pores**. The same term is used for any small, intergranular spaces in rock or sediment, such as those between the grains in a sedimentary rock, as well as small fractures in igneous and metamorphic rock.

Most pores are filled with either a watery fluid or a gaseous fluid, such as carbon dioxide. The watery, or aqueous, fluid is never just pure water; it has small amounts of gases and salts dissolved in it, along with traces of all the mineral constituents present in the enclosing rock. At high temperatures, the pore fluid is likely to be entirely gaseous. Regardless of its specific characteristics, pore fluid plays a vital role in metamorphism.

Pore fluid enhances metamorphism in two ways. The presence of pore fluid permits material to dissolve from one place, move quickly via the fluid, and be precipitated in another place. In this way, pore fluid speeds up recrystallization and generally leads to the formation of rock with relatively large mineral grains. Pore fluid also enhances metamorphism by acting as a reservoir during the growth of new minerals. When the temperature and pressure of a rock undergoing metamorphism change, so does the composition of the pore fluid. Some of the dissolved constituents move from the fluid to the new minerals growing in the metamorphic rock. Other constituents move in the other direction, from the minerals to the fluid. Pore fluid speeds up chemical reactions in the same way that water in a stew pot speeds up the cooking of a tough piece of meat. Because of this, metamorphism proceeds rapidly when pore fluid is present, but when pore fluid is absent or present in tiny amounts, metamorphic reactions occur very slowly.

As pressure increases and metamorphism proceeds, the amount of pore space decreases, and the pore fluid is slowly driven out of the rock. The escaping pore fluid carries with it small amounts of dissolved minerals. As the fluid flows through fractures in the rock, some of the dissolved minerals may precipitate, creating veins (**Figure 10.4**). Veins are quite common in metamorphic rock.

The presence of pore fluid also greatly influences the onset of rock melting. The effect of fluid on rock is to lower the melting temperature of the rock. The upper temperature limit of metamorphism therefore depends in part on the amount of pore fluid present. When a tiny amount of fluid is present, only a small amount of melting occurs, and the magma remains trapped in small pockets in the metamorphic rock. When the rock cools, so do the pockets of magma. The result is a composite rock, part igneous and part metamorphic, called **migmatite**.

When abundant pore fluid is present and large volumes of magma develop, the magma rises and intrudes into overlying metamorphic rock. In such cases, we observe batholiths of granitic rock (see Chapter 6) closely associated with large volumes of metamorphic rock. This igneous–metamorphic rock association occurs along subduction and collision margins of tectonic plates (see *What a Geologist Sees*).

Stress As you learned in Chapter 9, rock that is buried deep underground typically experiences differential stresses that have a strong effect on its appearance (**Figure 10.5**).

Differential stress causes rock undergoing metamorphism to develop a distinctive layered or planar texture called **foliation**. **Figure 10.5a** shows a nonfoliated rock; **10.5b** shows a foliated rock. Foliation is particularly evident when minerals belonging to the mica family are present. Micas are minerals in which the silicate anions link together (polymerize) in flat sheets. (See Chapter 2

> **foliation** A planar arrangement of textural features in metamorphic rock that give the rock a layered or finely banded appearance.

Quartz vein • Figure 10.4

This sample of gneiss contains many quartz veins (white) that crisscross the rock. The quartz precipitated out of pore fluid that was expelled from the rock during metamorphism.

WHAT A GEOLOGIST SEES

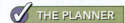

At the Roadside

A geologist who looks at this road cut, **(Figure a)**, in central Connecticut sees dark-colored rock with a nearly vertical direction of breakage. The geologist therefore suspects metamorphic rocks that have suffered ductile deformation, with the principal direction of compressive stress perpendicular to the plane of breakage. Looking more closely **(Figure b)**, the geologist sees rock in which all the mineral grains are large enough to be readily visible, so he concludes that he is looking at high-grade metamorphic rock. Examination of the outcrop reveals the pre-sence of quartz, feldspar, amphibole, and biotite, so he conclu-des that the rock is a gneiss.

Cutting through the gneiss are two white dikes. On close examination **(Figure c)**, it is apparent that the dikes consist main-ly of quartz and feldspar, with minor amounts of biotite, and that the texture is that of an igneous rock—there is no platy fabric or plane of breakage as there is in the gneiss. The geologist conclu-des that the two dikes of granite were intruded after the main phase of metamorphic deformation.

Effects of uniform and differential stress • Figure 10.5

The two rock samples in these photos have similar mineral assemblages but look very different because of their different stress histories.

a. This granite consists of quartz (glassy), feldspar (white), and biotite (dark), which crystallized from magma (a liquid) under conditions of uniform stress. Note that the biotite grains are randomly oriented.

b. This gneiss, a high-grade metamorphic rock, contains the same minerals as the granite, but they developed entirely in the solid state and under differential stress. The biotite grains are aligned, giving the rock a pronounced layered texture.

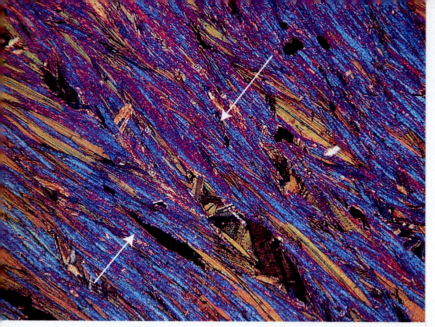

Foliation in schist seen under a microscope • Figure 10.6

This image was created by gluing a rock chip to a glass microscope slide and grinding the chip down until it was thin enough to let light pass through, then viewing it in polarized light. This sample of foliated metamorphic rock (schist) consists mostly of muscovite (a mica; colored grains in the photo) and quartz (the black grains). The direction of maximum stress is indicated by the arrows.

to review the structure of mica and other minerals mentioned here.) Under differential stress, mica grows so that the sheets are perpendicular to the direction of maximum stress (**Figure 10.6**).

Low-grade metamorphic rock tends to be so fine grained that the new mineral grains can be seen only under a microscope (see the slate specimen in Figure 10.3). Foliation in these rocks produces a distinctive style of fracture, called **slaty cleavage**. The orientation of the cleavage planes says a great deal about the conditions under which metamorphism occurred, as explained in **Figure 10.7**.

> **slaty cleavage**
> Foliation in low-grade metamorphic rock, which causes the rock to break into flat, plate-like fragments.

PROCESS DIAGRAM

Development of slaty cleavage • Figure 10.7

☑ THE PLANNER

a. In low-grade metamorphism, the maximum stress typically is from the weight of overlying rock and, therefore, is perpendicular to the bedding. This creates foliated rock with cleavage planes parallel to the bedding.

b. During a collision of tectonic plates, strata are squeezed and folded, causing a ripple-like appearance. The maximum stress is from the compression of the colliding plates; it is horizontal, as shown by the arrows. This causes the rock to develop cleavage in the vertical direction, at an angle to the bedding.

c. In this photograph from the Martinsburg Formation in Pennsylvania, the slaty cleavage has developed at an angle to the bedding, indicating that this region was under compression when the rock was metamorphosed.

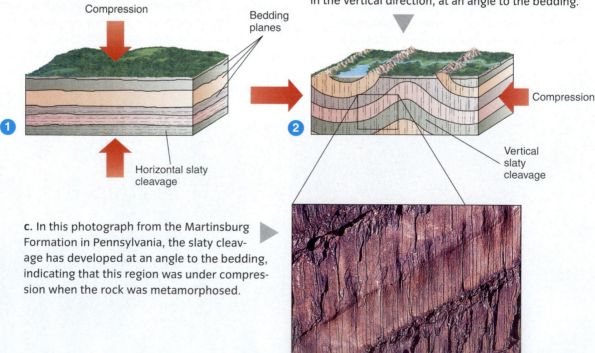

CASE STUDY

Metamorphism and Billiards

The modern game of billiards is played on tables made from meta-morphic rock. Early billiard tables resembled modern ones in that they were rectangular, surrounded by cushions, and fitted with six pockets. However, they were made of wood, which had a tendency to vibrate and thus interfere with the accuracy of the players. In addition, the wood warped after only a few years. Around 1825, at the height of the Industrial Revolution, an Englishman named John Thurston invented the slate bed. Manufacturers of billiard tables discovered that if slate was used for a table (with the slaty cleavage surface forming the table top), they could ensure a surface that was very smooth and free of vibrations. The slate table was also much sturdier and less prone to wear and tear than a wooden table.

a.

Metamorphism converts shale to slate.

Slate comes from the low-grade metamorphism of shale (**Figure a**). In North America, a belt of slate runs along the eastern seaboard from Georgia to Maine. Thus, the collision of continents that created the supercontinent of Pangaea 400 million years ago also created the rock (**Figure b**) that enable these shepherds in western China (**Figure c**) to work on their skills at the billiard table while their flocks graze in the background.

b.

c.

Slaty cleavage is similar to cleavage in a mineral (see Chapter 2) but with a very important difference. A mineral grain is a single crystal, and cleavage occurs because the bonds between atoms in the crystal are weaker in some directions than in others. Cleavage in a rock, in contrast, involves a great many crystal lattices, and it happens because the crystals themselves are roughly parallel. The flat cleavage planes that result make slate an excellent material for roofing and paving tiles, black-boards, billiard tables, and other uses that require a durable natural material with a flat surface (see the *Case Study*).

Schistosity versus slaty cleavage • Figure 10.8

a. Slate

a. Slaty cleavage in a red slate from Granville, New York. This view shows a cleaved surface.

b.

b. Specimen a viewed perpendicular to the cleavage direction. Note the numerous small parallel steps and ridges; they are parallel to the direction of slaty cleavage.

c. Mica schist

c. Schistosity in mica schist from Manhattan, New York. The view is looking at the surface of schistosity.

d.

d. Specimen c viewed perpendicular to the direction of schistosity.

Another kind of foliation, **schistosity**, forms under conditions of high-grade metamorphism, which causes mineral grains to grow large enough to be seen with the naked eye. This kind of foliation differs from slaty cleavage primarily in the size of the mineral grains. Another difference is that schistose rock generally breaks along wavy or distorted surfaces, while slaty cleavage is strictly planar (**Figure 10.8**).

> **schistosity** Foliation in coarse-grained metamorphic rock.

Duration and rate of metamorphism Chemical reactions can occur rapidly or slowly. Some reactions, such as the burning of natural gas to produce carbon dioxide and water, happen so fast that they can cause an explosion if not handled carefully. At the other end of the scale are reactions that take thousands or even millions of years to complete. Most of the reactions that happen during metamorphism are the latter kind.

Despite the slowness of most metamorphic reactions, scientists have been able to use laboratory experimentation to demonstrate that high temperatures, high pressures or stresses, abundant pore fluid, and long reaction times produce large mineral grains. Coarse-grained rock—with mineral grains the size of a thumbnail or larger—is formed under long-sustained metamorphic conditions (possibly over millions of years) with high temperature, high stress, and abundant pore fluid. On the other hand, fine-grained rock—with mineral grains the size of a pinhead or smaller—is either produced under conditions involving lower temperature and lower stress or under conditions where pore fluid was scarce or reaction times were short.

CONCEPT CHECK   STOP

1. **What** is the approximate depth where diagenesis ceases and metamorphism starts?

2. **What** three variables, besides rock composition, are most important in causing metamorphism?

3. **How** do temperature and pressure affect the grade of metamorphism?

4. **What** does the presence of a vein in a rock reveal about the way the rock formed?

5. **Why** are some metamorphic rocks foliated, while others are not?

Metamorphic Rock

LEARNING OBJECTIVES

1. **Identify** the common metamorphic products of shale.
2. **Identify** the common metamorphic products of basalt.
3. **Explain** how basalt can be metamorphosed in different ways.
4. **Identify** several common types of foliated and nonfoliated metamorphic rock.

T he names of metamorphic rocks are mostly based on their textures and mineral assemblages (see *Where Geologists Click*). The most widely used names apply to metamorphic rock derived from the sedimentary rock shale, sandstone, and limestone, and from the igneous rock type basalt. This is because shale, sandstone, and limestone are the most abundant sedimentary rock, and basalt is the most abundant igneous rock. We will first describe the metamorphic rock types that have some degree of foliation; then we will move on to describe metamorphic rocks that tend to occur without foliation.

Rock with Foliation

As mentioned earlier in this chapter, the metamorphic products of shale form a sequence of rock types, beginning with slate, a low-grade product. As the metamorphism proceeds to higher grades, the rock looks less and less like its parent rock, shale, and develops larger and larger mineral grains.

There are some subtleties of terminology in the sequence of rock derived from shale. The names **slate** and **phyllite** describe textures. They are usually used without adding mineral names as adjectives because their mineral assemblages are not easy to see. On the other hand, the mineral grains in **schist** and **gneiss** are large enough to be identified, so geologists usually add the mineral assemblage to the name, as in "garnet-mica schist."

Metamorphic rock doesn't always develop from sedimentary rock; igneous rock also can undergo metamorphism. The names given to metamorphosed igneous rocks differ from those used for the metamorphic rocks that develop from shale and other sedimentary rocks. Basalt—the most abundant

slate A very fine-grained, low-grade metamorphic rock with slaty cleavage; the metamorphic product of shale.

phyllite A fine-grained metamorphic rock with pronounced foliation, produced by further metamorphism of slate.

schist A high-grade metamorphic rock with pronounced schistosity, in which individual mineral grains are visible.

gneiss (pronounced "nice") A coarse-grained, high-grade, strongly foliated metamorphic rock with separation of dark and light minerals into bands.

Where Geologists CLICK

Smithsonian Institution's National Museum of Natural History

http://collections.mnh.si.edu/search/ms/

At this site you can search an extensive database of more than 350,000 metamorphic—and other—rocks and minerals, including meteorites, from the collections of the Smithsonian Institution's National Museum of Natural History. Use the site's "Quick Browse" to look for images of Gems and Minerals, Meteorites, or Rocks and Ores.

Metamorphism of basalt • Figure 10.9

When basalt is subjected to metamorphism under conditions in which pore water can enter the rock, distinctive mineral assemblages develop.

a. Low-grade metamorphism produces a rock called green-schist, so named because of the presence of a green, mica-like mineral called chlorite. Like phyllite, greenschists are fine grained, with pronounced foliation. This specimen is from Minnesota.

b. Under high-grade metamorphism, chlorite is replaced by amphibole, and the resulting high-grade rock is an amphib-olite. Foliation is present in amphibolites but is not pronounced because micas are usually absent. Notice that the elongate crystals of hornblende in this specimen from North Carolina are all parallel.

c. Under conditions of very high stress, as in subduction zones, the low-grade metamorphic product of basalt is blue-schist, which owes its color to a mineral called glaucophane. This specimen is from Greece.

d. At even higher pressures and moderate temperatures, blueschist is replaced by the high-grade metamorphic rock eclogite, which contains such minerals as jadeite and garnet. In this specimen from Greece, the red garnets are obvious; the dark green mineral surrounding the garnets is a pyroxene.

e. Jadeite is a pyroxene that is formed by metamorphic transformation of sodium-rich plagioclase in subduction zones. The photo from an outcrop near specimen **d** contains large green grains of jadeite. The color and toughness of jadeite make it a desirable material for jade carvings. Jadeite is a semi-precious stone; the value of jadeite lies mainly in the exquisite workmanship of jade carvings.

Metamorphic Processes

LEARNING OBJECTIVES

1. **Identify** two types of physical changes that occur in rock during metamorphism.

2. **Describe** contact, burial, and regional metamorphism.

3. **Identify** the tectonic settings where the different types of metamorphism are likely to occur.

4. **Define** metasomatism and explain how it differs from metamorphism.

The processes that cause changes in texture and mineral assemblages in metamorphic rock are **mechanical deformation** and **chemical recrystallization**. Mechanical deformation includes grinding, crushing, bending, and fracturing. **Figure 10.11a** shows a conglomerate in which the pebbles have not been mechanically deformed. By contrast, **Figure 10.11b**, which shows flattened pebbles in a conglomerate, is an example of mechanical deformation. Chemical recrystallization includes changes in mineral composition, growth of new minerals, recrystallization of old minerals, and changes in the amount of pore fluid due to chemical reactions that occur when rock is heated and squeezed. Metamorphism typically involves both mechanical deformation and chemical recrystallization processes, but their relative importance varies greatly.

Types of Metamorphism

We can distinguish among several kinds of metamorphism on the basis of the dominant processes (mainly the relative importance of mechanical deformation and chemical recrystallization) and the tectonic environments in which they occur.

Mechanical deformation • Figure 10.11

a. This undeformed conglomerate contains pebbles rounded from stream transport.

b. In the deformed conglomerate, differential pressure has squeezed the once-rounded pebbles to such an extent that they are now flat.

Contact metamorphism Where hot magma intrudes into cooler rock, high temperatures cause chemical recrystallization in surrounding rock. The magma itself may also release pore fluid—chiefly heated water—as it cools. The fluid, in turn, transports materials and accelerates the growth of new minerals. **Contact metamorphism** is primarily temperature driven; mechanical deformation—and therefore foliation—are minor.

> **contact metamorphism**
> Metamorphism that occurs when rock is heated and chemically changed adjacent to an intruded body of hot magma.

An igneous intrusion is commonly surrounded by a zone or an aureole of contact metamorphism (**Figure 10.12**). If the rock is relatively impervious, such as shale, or if the magma contains little water, then the aureole may extend only a few centimeters. However, a large intrusion contains more heat energy than a small one and may release a large volume of pore fluid as it cools. When intrusions are very large (the size of stocks and batholiths), and when they intrude into highly reactive and relatively pervious rock, such as limestone, the effects of contact metamorphism may extend for hundreds of meters. Unlike other kinds of metamorphism, contact metamorphism is not associated with any particular tectonic setting; it may occur wherever magma intrusions occur.

Burial metamorphism The first stage of metamorphism to occur in sedimentary rock after diagenesis is **burial metamorphism**. The metamorphic processes caused by burial begin at about 150°C, when a sedimentary rock has been buried at a depth of about 5 km. The maximum stress exerted during burial metamorphism tends to be vertical; foliation, if present, is thus parallel to bedding. Burial metamorphism requires a relatively thick layer of overlying rock or sediment, so it is usually observed in deep sedimentary basins such as those found along passive margins or fore-arc basins (see Chapter 8). Because the temperature of the rock in these basins seldom exceeds 300°C, burial metamorphism tends to be low grade. As temperature and pressure increase beyond this point, or as sediment on a continental margin is deformed during a tectonic collision, burial metamorphism grades into regional metamorphism.

> **burial metamorphism**
> Metamorphism that occurs after diagenesis, as a result of the burial of sediment in deep sedimentary basins.

Contact metamorphism • Figure 10.12

A layer of limestone undergoes contact metamorphism when granite magma intrudes into it. The intrusion is surrounded by an aureole of altered rock, with several distinct zones. In the outermost zone, farthest from the heat of the granite intrusion, the limestone has been metamorphosed to marble. Inside the marble is a zone where chemicals dissolved in the pore fluid have reacted with the limestone to form chlorite and serpentine. Closest to the magma is a zone of high-grade garnet and pyroxene.

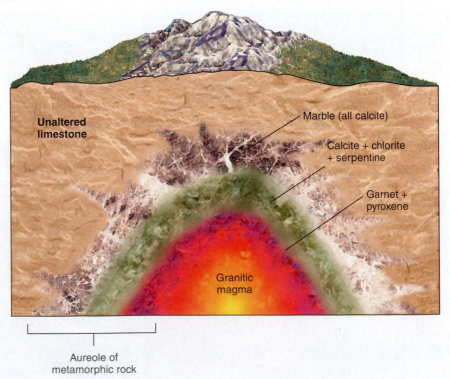

Unaltered limestone

Marble (all calcite)

Calcite + chlorite + serpentine

Garnet + pyroxene

Granitic magma

Aureole of metamorphic rock

Tectonics and metamorphism • Figure 10.13

The theory of plate tectonics provides a unified view of burial metamorphism (1), regional metamorphism in a subduction zone (2), and regional metamorphism in a collision zone (3). Contact metamorphism (4) can occur adjacent to an igneous intrusion in any tectonic setting. The dashed lines are **isotherms** (lines denoting equal temperature).

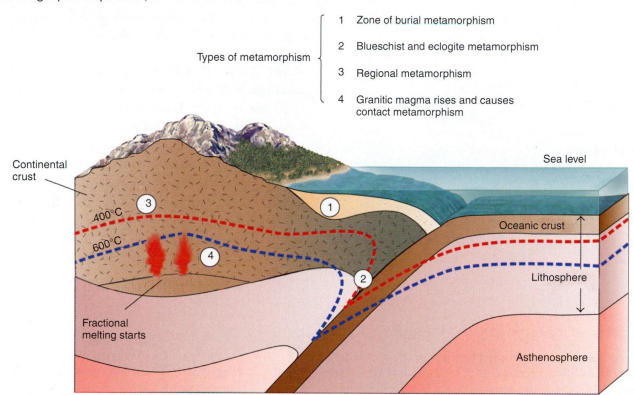

Types of metamorphism

1 Zone of burial metamorphism

2 Blueschist and eclogite metamorphism

3 Regional metamorphism

4 Granitic magma rises and causes contact metamorphism

Continental crust

Sea level

400°C

600°C

Oceanic crust

Lithosphere

Fractional melting starts

Asthenosphere

Regional metamorphism The most common kinds of metamorphic rock in the continental crust—slate, phyllite, schist, and gneiss—are found in areas extending over tens of thousands of square kilometers. They are formed by a process called **regional metamorphism**, which occurs only at convergent plate margins. Regionally metamorphosed rock is found in mountain ranges and in the eroded remnants of former mountain ranges (**Figure 10.13**). It is formed as a result of subduction or through collisions between masses of continental crust. Rock at convergent margins is subjected to intense differential stresses; the foliation that is characteristic of regionally metamorphosed rock is a consequence of such stresses, although, as discussed

regional metamorphism
Metamorphism of an extensive area of the crust, associated with plate convergence, collision, and subduction.

above, the full development of foliation depends on the composition of the rock.

Regionally metamorphosed rock can also be found in places where no plate collisions are occurring today. We described these regions (called orogens) in Chapter 9. These are regions where plates have collided in the past, and their fragments have been assembled into the lithospheric plates we see today. Orogens provide some of our best evidence that plate tectonics and regional metamorphism have been active on Earth for billions of years.

A special kind of regional metamorphism occurs at subduction margins. In region 2 of Figure 10.13, you can see the isotherms bend down steeply, indicating that these rocks are much cooler than the surrounding crust. This phenomenon occurs because the solid, cold oceanic crust dives rapidly (by geologic standards) into the hot, weak asthenosphere. The pressure on the subducting rock rises

Quick pressure, slow heat • Figure 10.14

When a block of rock is squeezed, the entire block "feels" the stress immediately. When the rock is heated on one side, however, it takes a long time for the other side to get hot. A thermometer on the side away from the heat source will still register a low temperature after a fairly extended period of heating.

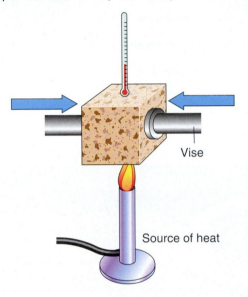

Vise

Source of heat

quickly, but the temperature of the rock cannot rise at the same rate. This conclusion follows from a basic property that can be observed in a laboratory: Rock transmits pressure immediately but conducts heat slowly (**Figure 10.14**). Therefore, in a subduction zone, metamorphism occurs at very high pressures but low to moderate temperatures. This produces a distinctive series of rocks, including blueschist and (at higher grades) eclogite.

Other types of metamorphism

Occasionally metamorphism occurs in other geologic settings where high temperature and/or elevated pressure or stress exist, even if only briefly or sporadically. These include fault zones, meteorite impact craters, and even the sites of large fires. As geologists have become more aware of the role of meteorite impacts in our planet's geologic history, the unique kinds of metamorphism produced by these rare events have become important diagnostic tools for detecting highly eroded ancient impact sites (**Figure 10.15**).

Instantaneous metamorphism • Figure 10.15

a. Nördlingen Cathedral in Germany sits in the center of an ancient impact crater. The cathedral was constructed of local rock that was metamorphosed by the heat and high pressure associated with the meteorite impact.

b. This is a microscopic view of shocked quartz of the type found in Nördlingen Cathedral building stone. The lines in the quartz record a kind of very high-pressure deformation that is found only at impact craters and atomic bomb blast sites. The stones of the cathedral also contain tiny diamonds, which ordinarily form deep underground under high pressure but were formed here instantaneously by the blast wave from the exploding meteorite.

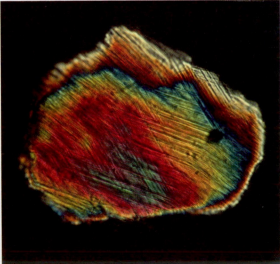

Metasomatism

Metamorphic processes may dramatically change the appearance of a rock and may cause its chemical constituents to move and crystallize into new mineral assemblages. However, they have little effect on the overall chemical composition of the rock. This is one reason geologists can tell the identity and composition of metamorphic parent rock.

There is an important exception to this general rule. Pore fluids, chiefly water and carbon dioxide, can be squeezed into or out of a body of rock during metamorphism. Most metamorphic environments have a small fluid-to-rock ratio, which means the amount of fluid present compared to the amount of rock is less than about 1:10. That is enough fluid to facilitate metamorphism but not enough to dissolve much of the rock and thus change its composition noticeably.

In a few circumstances, though, large fluid-to-rock ratios of 10:1 or even 100:1 can occur. One example of such a circumstance is a large, open rock fracture through which a lot of fluid flows. The rock adjoining the fracture can be drastically altered by the addition of new material, the removal of material into the fluid solution, or both. The term **metasomatism**, which comes from the Greek words for "change" and "body," is applied in such cases. Note that *metamorphism* refers to a change of form (*morph* is the Greek word for "shape," or "form"), often accompanied by a redistribution of chemical constituents. *Metasomatism*, on the other hand, refers to a change in the chemical composition of the rock as a whole.

> **metasomatism** The process whereby the chemical composition of a rock is altered by the addition or removal of material by solution in fluids.

Figure 10.16 is a photograph of a contact metamorphic rock that was originally a limestone. Without the addition of new material, the limestone would have become a marble, dominated by calcite. Through metasomatism, however, it gained the constituents needed to crystallize garnet and a pyroxene called diopside, in addition to calcite. Metasomatic fluids may also carry valuable minerals in solution, which can lead to the formation of ore deposits, a topic discussed in Chapter 16.

Metasomatism • Figure 10.16

Metasomatism of limestone produced this colorful rock from the Gaspé Peninsula in Quebec. The white is calcite, the red is garnet, and the dark brown is pyroxene. The sample is 2.5 by 1.5 centimeters.

CONCEPT CHECK

1. **What** physical and chemical changes happen in rock undergoing metamorphism?

2. **What** distinguishes burial metamorphism from regional metamorphism?

3. **How** does regional metamorphism in a subduction zone differ from regional metamorphism in a collision zone?

4. **What** process changes the chemical composition of a rock, rather than just its texture or mineral assemblage?

Metamorphic Facies

LEARNING OBJECTIVES

1. **Define** metamorphic zones and metamorphic facies.

2. **Explain** how metamorphic facies correlate with the temperature and pressure the rock is exposed to.

Much of the early research on metamorphism and metamorphic rock was done in the Scottish Highlands in the late 1800s. Geologists discovered that mineral assemblages varied a great deal from place to place but that the overall chemical composition was essentially the same as that of shale. The rocks thus differed in terms of the grade of metamorphism they had been subjected to and the mineral assemblages they had developed, but not in the composition of their source rock. The geologists identified characteristic **index minerals** that marked the appearance of each new mineral assemblage in a progression from low-grade to higher-grade metamorphic rock. They drew lines, called **isograds** (lines of equal grade), connecting the map locations where a given index mineral first appeared (**Figure 10.17**). Rock lying between one isograd and the next would thus have similar mineral assemblages, indicating that they formed under similar metamorphic conditions; they were defined as belonging to the same **metamorphic zone**. This method of mapping, first used in Scotland over a century ago, has been successfully tested on different types of rock around the world.

Regional metamorphism in Scotland and Michigan • Figure 10.17

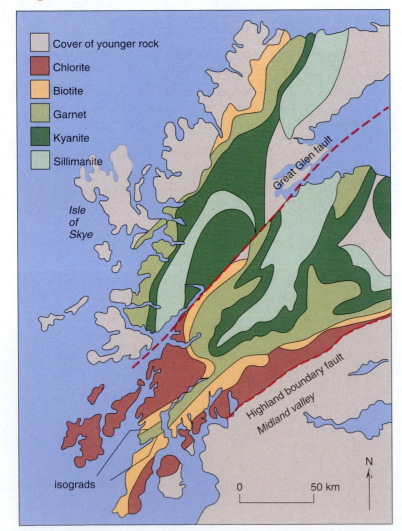

isograds

a. This map shows isograds in a body of regionally metamorphosed rock in Scotland. The isograds mark the first appearances of various index minerals. Rock lying between two isograds is said to be in a particular metamorphic zone. For example, rock between the biotite and garnet isograds would contain the index mineral biotite but would not yet have reached the metamorphic conditions in which garnet would appear; it would be said to be in the "biotite zone."

b. Isograds in a regionally metamorphosed region of the Upper Peninsula of Michigan.

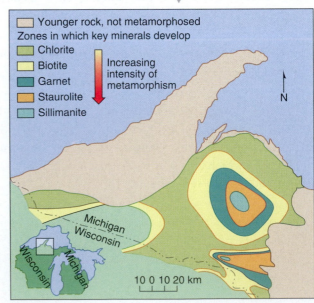

Metamorphic facies • Figure 10.18

This graph shows the regions of pressure and temperature that characterize the different metamorphic facies. The zeolite facies is characteristic of burial metamorphism, and the hornfels facies is characteristic of contact metamorphism. Blueschist and eclogite facies are typical for subduction zones. Greenschist, amphibolite, and granulite facies occur in the regional metamorphism of continental crust thickened by plate collisions. The lines indicate how pressure and temperature change under conditions of (a) contact metamorphism, (b) regional metamorphism in continental plate collisions, and (c) regional metamorphism in subduction zones.

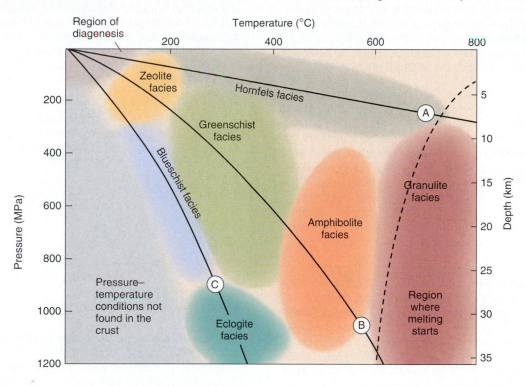

The metamorphic rock in Scotland came from source rock of similar composition (shale) that was exposed to a range of metamorphic conditions. It is also possible to group together metamorphic rock of differing compositions that have all been exposed to the same temperature and stress conditions; such a grouping is called a **metamorphic facies**. Each of the metamorphic facies has been named, as you can see in **Figure 10.18**. The names typically reflect some obvious feature, such as the color of a common rock (e.g., blueschist, greenschist); the presence of a distinctive mineral (e.g., zeolite, amphibole); an unusual type of rock (e.g., eclogite); or a distinctive texture (e.g., granulite). Each facies is representative of a specific set of metamorphic conditions. For example, the blueschist facies represents relatively high-pressure, low-temperature

metamorphic facies The set of metamorphic mineral assemblages that form in rock of different compositions under similar temperature and stress conditions.

metamorphism, while the hornfels facies represents relatively high-temperature, low-pressure metamorphism (i.e., contact metamorphism).

The pressure–temperature diagram in Figure 10.18 shows the characteristic pressure and temperature ranges for each metamorphic facies and how they relate to specific tectonic settings. For any given composition of source rock, it is possible to prepare a pressure–temperature grid like the one in Figure 10.18 that shows the specific mineral assemblages characteristic of each metamorphic facies.

CONCEPT CHECK   STOP

1. **What** is a metamorphic facies?

2. **What** temperature and pressure conditions characterize each of the following facies: hornfels, blueschist, amphibolite, granulite?

Summary

1 What Is Metamorphism? 280

- New rock textures and new mineral assemblages develop when rock is subjected to elevated temperatures and stresses. **Metamorphism** is a term that describes all such processes that occur at higher temperatures and pressures than diagenesis (Chapter 8) but without melting the rock.

- The transition from diagenesis to metamorphism is gradual, but the dividing line between the two is customarily taken to be a temperature of around 150°C, which typically occurs at a depth of about 5 km. The temperature and pressure are both high enough at that depth to initiate the formation of new minerals. Between 5 and 15 km beneath the surface, rock is subjected to **low-grade** metamorphism. The region of **high-grade** metamorphism lies from 15 km to the depth at which melting occurs, often between 30 and 40 km. The kind of metamorphism that occurs depends on temperature and pressure as well as composition.

- Pore fluid enhances metamorphism by permitting material to dissolve, move around, and be precipitated somewhere else in the rock, as well as speeding up some chemical reactions. Pore fluid is driven out of rock as metamorphism progresses. Veins in metamorphic rock mark the passageways through which pore fluid once flowed. At the high-temperature limit of metamorphism, pore fluid can reduce the melting point of rock to the point where small pockets of magma form. The resulting rock, with small pockets of igneous rock surrounded by metamorphic rock, is called migmatite.

- Differential stress during metamorphism produces a distinctive texture known as **foliation**, marked by parallel cleavage planes and plates formed by crystals that are all aligned in the same direction. It is particularly noticeable in rock containing minerals of the mica family. Foliation developed in low-grade metamorphic rock is termed **slaty cleavage**; in coarser-grained, higher-grade metamorphic rock, it is called **schistosity** as seen in the photo. The orientation of

the foliation is perpendicular to the direction of maximum stress. Under conditions of low-grade metamorphism, the maximum stress is usually downward, and the slaty cleavage is horizontal and parallel to bedding. Under conditions of compression, produced by plate tectonics, the maximum stress is horizontal, and the slaty cleavage is vertical.

- High temperature, high pressure, abundant pore fluid, and long reaction times tend to produce metamorphic rock with large mineral grains. Metamorphic rock containing small mineral grains usually forms at lower temperatures and pressures but can also form if pore fluid is scarce or reaction time is short.

Foliation in schist seen under a microscope
• Figure 10.6

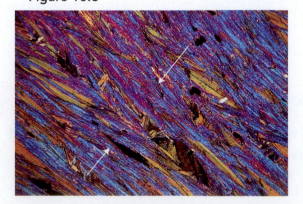

2 Metamorphic Rock 289

- The names of metamorphic rocks are based partly on texture and partly on composition. Foliated metamorphic rocks derived from shales are **slate**, **phyllite**, **schist**, and **gneiss**, in order of increasing metamorphic grade. The first two terms refer to textures of fine-grained rock. In schist and gneiss, where the mineral grains are large enough to see, it is customary to identify rock further by its mineral composition, as in "biotite schist" or "quartz-biotite-amphibole gneiss." Gneiss is characterized by a coarsely foliated texture, with alternating bands of mica-rich and mica-poor material, giving the rock a banded appearance.

- Foliated rock can derive from igneous rock, the most common of which is basalt. Rock derived from basalts include greenschists (low-grade metamorphic rock) and amphibolites (high-grade metamorphic rock). Under conditions of high pressure and moderate temperature, which occur in subduction zones, a different series of metamorphic rock forms from basalts. These are blueschists (low grade) and eclogites (high grade).

- Two common kinds of metamorphic rock have no foliation. These are **marble** (see photo), a metamorphic product of limestone, and **quartzite**, a metamorphic product of sandstone. These tend to be monomineralic, or nearly so; the main mineral present in marble is calcite, and the main mineral present in quartzite is quartz.

Non foliated metamorphic rock • Figure 10.10a

- Sedimentary rock that is subjected to sufficiently high temperatures and pressures due solely to the weight of overlying strata undergoes **burial metamorphism**. Burial metamorphism is a low-grade metamorphic process. If the resulting rock is foliated, the foliation is usually parallel to the bedding planes.

- **Regional metamorphism** occurs under the conditions of differential stress and elevated temperature that result from collisions between tectonic plates. A special kind of regional metamorphism occurs at subduction zones, where pressures are high but temperatures are moderate, due to the very rapid burial of the rock.

- Very high pressures and temperatures caused by meteorite impacts produce a unique kind of high-pressure metamorphism, similar to that caused by underground atomic bomb explosions.

- **Metasomatism** occurs when a large volume of fluid flows into or out of a rock. It is a process that changes the overall chemical composition of the rock by moving chemical components into or out of the rock rather than just altering the texture or mineral assemblage.

3 Metamorphic Processes   293

- The processes that are primarily responsible for changes in texture and mineral assemblages in metamorphic rock are mechanical deformation and chemical recrystallization. Typically both sets of processes are involved in metamorphism, but their relative importance depends on the specific pressure and temperature conditions of metamorphism.

- **Contact metamorphism**, shown in the diagram, is caused by the intrusion of hot magma or fluid (primarily water) into cooler rock. Because contact metamorphism is caused by high temperatures, the main process involved is chemical recrystallization, and there is little mechanical deformation of the rock.

Contact metamorphism • Figure 10.12

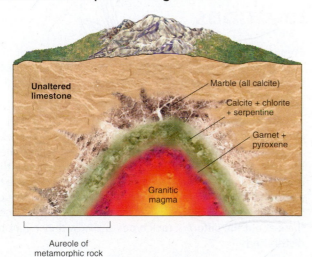

Unaltered limestone

Marble (all calcite)

Calcite + chlorite + serpentine

Garnet + pyroxene

Granitic magma

Aureole of metamorphic rock

- For a given rock composition, the assemblages of minerals that are formed under a given set of temperature and stress conditions as shown in the diagram, are the same, regardless of where in the world the metamorphism happens.

- Rock of different chemical compositions that are metamorphosed under the same temperature and stress conditions are said to belong to the same **metamorphic facies**, even though they will develop different mineral assemblages.

- Each metamorphic facies is associated with a particular tectonic setting. Blueschist facies and eclogite facies conditions occur in the high-pressure environment characteristic of subduction zones. Greenschist, amphibolite, and granulite facies conditions occur along convergent margins where continental masses collide. The zeolite facies is characteristic of burial metamorphism, and the hornfels facies is characteristic of contact metamorphism.

Metamorphic facies • Figure 10.18

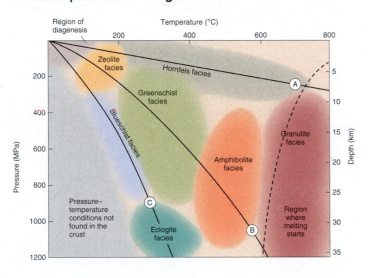

Key Terms

- burial metamorphism 294
- contact metamorphism 294
- foliation 284
- gneiss 289
- high-grade 282
- low-grade 282

- marble 292
- metamorphic facies 299
- metamorphism 280
- metasomatism 297
- phyllite 289
- quartzite 292

- regional metamorphism 295
- schist 289
- schistosity 288
- slate 289
- slaty cleavage 286

Critical and Creative Thinking Questions

1. Briefly describe how pressure and temperature might change over time in rock being subjected to contact metamorphism, burial metamorphism, and subduction-related regional metamorphism.

2. Suppose a large meteorite struck a pile of clastic sedimentary rock. How would pressure and temperature change in the rock? Would the effects be the same for rock on the Moon as for rock in a sedimentary basin on Earth?

3. Examining the texture of a rock is an important process for geologists because the texture can reveal whether the rock has been metamorphosed and under what conditions. Explain why. Would this process work for both foliated and nonfoliated rock? Why (or why not)?

4. Compare the concept of metamorphic facies to that of sedimentary facies (see Chapter 8). In what ways are they similar? In what ways are they different?

5. Let's say that you have bought a property in a hilly area, and you are going to build a new house on the property. You like the idea of building on a hill slope with a view, but you are concerned about safety issues. One of the hills on the property is composed of gneiss, another is slate, and a third is a marble in which there are caves. Which one of these hill slopes do you think would provide the most stable location for your home, and why? Try sketching the hill slope to help think about the different rock types and how they might be distributed.

What is happening in this picture?

This roof in France was made from a commonly occurring planar metamorphic rock.

Self-Test

(Check your answers in Appendix D.)

1. How does metamorphism differ from diagenesis and lithification?

   a. Diagenesis and lithification occur at higher temperatures and pressures.

   b. Diagenesis and lithification occur at lower temperatures and pressures.

   c. Metamorphism requires melting of preexisting rock.

   d. Diagenesis and lithification do not cause any significant changes in the rock or sediment.

2. Under which of the following conditions does metamorphism occur?

   a. high temperature and high pressure

   b. high temperature and low pressure

   c. low temperature and low pressure

   d. All of the above answers are correct.

3. For this diagram locate and label the areas on the temperature–pressure diagram that correspond to regions of Earth's crust listed below.

region of diagenesis region of high-grade metamorphism

region where melting starts region of low-grade metamorphism

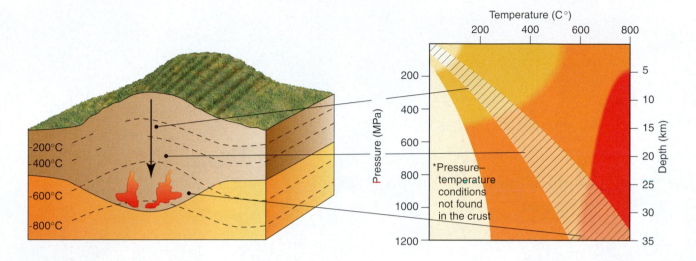

4. Pore fluids enhance metamorphism by _____.

a. permitting material to dissolve

b. permitting chemical components to move around and be precipitated somewhere else

c. speeding up some chemical reactions

d. All of the above answers are correct.

5. _____ is the planar alignment of minerals due to differential stress during metamorphism.

a. Ionization

b. Metasomatism

c. Foliation

d. Diagenesis

6. Foliated metamorphic rock derived from shales, in order of increasing metamorphic grade, are _____.

a. phyllite, slate, and schist or gneiss

b. phyllite, gneiss, and schist or slate

c. slate, phyllite, and schist or gneiss

d. slate, gneiss, and phyllite or schist

7. Basalt can be metamorphosed to form a series of foliated rocks ranging from greenschists (low grade) to amphibolites (high grade). It is also possible for a different series of metamorphic rocks to form, ranging from blueschists (low grade) to eclogites (high grade). Under what conditions does this latter series form?

a. low pressure and moderate temperature

b. low pressure but high temperature

c. high pressure but moderate temperature

d. high pressure and high temperature

8. The two rock samples shown in these photographs are nonfoliated metamorphic rock. Sample A is composed of calcite, and Sample B was formed through the metamorphism of sandstone. Which of the following correctly identifies the two samples?

a. Sample A is quartzite, and Sample B is schist.

b. Sample A is quartzite, and Sample B is marble.

c. Sample A is marble, and Sample B is quartzite.

d. Sample A is marble, and Sample B is schist.

9. Metamorphism involves two main groups of processes: _____.

   a. fractional melting and chemical recrystallization

   b. fractional melting and fractional crystallization

   c. fractional melting and mechanical deformation

   d. mechanical deformation and chemical recrystallization

10. _____ occurs over an extensive area of the crust and is associated with plate convergence, collision, and subduction.

   a. Regional metamorphism

   b. Burial metamorphism

   c. Contact metamorphism

   d. Diagenesis

11. _____ occurs when rocks are heated and chemically changed adjacent to an intruded body of hot magma.

   a. Regional metamorphism

   b. Burial metamorphism

   c. Contact metamorphism

   d. Diagenesis

12. On this diagram, label the areas designated 1 through 4 from the following choices:

   regional metamorphism

   blueschist- and eclogite-facies metamorphism

   burial metamorphism

   zone where fractional melting starts, granite magma rises, and contact metamorphism occurs

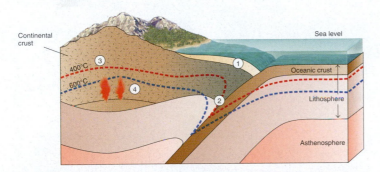

13. Metasomatism is a process by which the chemical composition of rock is altered. What is the mechanism of this chemical alteration?

   a. addition or removal of material by solution in fluids

   b. loss of chemical material during fractional melting

   c. loss of chemical material during fractional crystallization

   d. None of the above answers are correct.

14. Label this diagram with the proper metamorphic facies.

   zeolite facies   granulite facies

   blueschist facies   greenschist facies

   amphibolite facies   hornfels facies

   eclogite facies

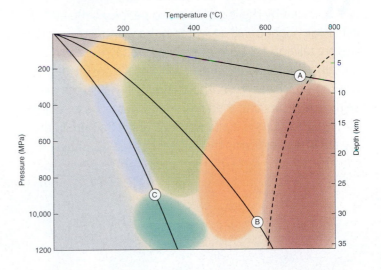

15. In the diagram in question 14, the three lines labeled A, B, and C represent temperature and pressure pathways for different types of metamorphism. Which of these three pathways describes the suite of metamorphic facies that you would expect to see in an area of regional metamorphism due to continental collision?

   a. A

   b. B

   c. C

THE PLANNER ✓

Review your Chapter Planner on the chapter opener and check off your completed work.

Water On and Under the Ground

Mosi-oa-Tunya, or "the smoke that thunders," as it is called by the local Makololo people, is the largest single sheet of falling water in the world—over 100 meters tall and 1.5 kilometers wide. Today, we know this curtain of water as Victoria Falls. The falls are formed as the Zambezi River, in southern Africa, flows across basaltic lava plains and then plummets into a chasm about 120 m wide, carved by its waters along a fracture in the basalt. The river forms the border between Zambia on the north (to the right in the photograph) and Zimbabwe to the south. The waterfall converts the calm river into a ferocious torrent. The roar of the water can be heard from 40 km away. Enough water passes over these falls every hour to supply a city of 2 million people for a year.

Humans have learned to harness the power of water for electricity, as well as for drinking and other uses. The first power station was set up at Victoria Falls in 1938. In North America, the Niagara Power Project exploits the 52-m drop over Niagara Falls, plus a further 55-m drop down the Niagara Gorge, by withdrawing water

Author Skinner (dark helmet) views the Falls from above.

upstream of the waterfalls and carrying it through tunnels to generating plants located downstream. It's easy to imagine how the immense force of these waterfalls can mold the land below. In this chapter, we will explore the many ways in which water shapes our land and in which human activities affect our water.

Global Locator

Africa
Victoria Falls

NATIONAL GEOGRAPHIC

CHAPTER PLANNER ✔

- ❑ Study the picture and read the opening story.
- ❑ Scan the Learning Objectives in each section.
  p. 308 ❑   p. 312 ❑   p. 320 ❑   p. 329 ❑
- ❑ Read the text and study all visuals. Answer any questions.

Analyze key features

- ❑ Process Diagram, p. 308 ❑
- ❑ Geology InSight, p. 312 ❑, p. 326 ❑
- ❑ What a Geologist Sees, p. 318 ❑
- ❑ Case Study, p. 328 ❑
- ❑ Amazing Places, p. 338 ❑
- ❑ Stop: Answer the Concept Checks before you go on:
  p. 311 ❑   p. 320 ❑   p. 325 ❑   p. 337 ❑

End of chapter

- ❑ Review the Summary and Key Terms.
- ❑ Answer the Critical and Creative Thinking Questions.
- ❑ Answer What is happening in this picture?
- ❑ Complete the Self-Test and check your answers.

The Hydrologic Cycle

LEARNING OBJECTIVES

1. **Define** and describe the hydrologic cycle.
2. **Identify** the main pathways in the hydrologic cycle.
3. **Identify** the main reservoirs in the hydrologic cycle.

Four great reservoirs make up the Earth system (see Chapter 1): the geosphere, hydrosphere, atmosphere, and biosphere. Water moves—and helps move materials—among all four spheres. Water vapor is an important part of the atmosphere. Water is a constituent of many common minerals (e.g., micas, clays) in the geosphere, where it is tightly bonded in their crystal structures. And, of course, water is a fundamental component of living things in the biosphere.

Water in the Earth System

The **hydrologic cycle**, also called the **water cycle**, describes how water moves among these four reservoirs (**Figure 11.1**), and the scientific study of water is called **hydrology**.

Water moves through the hydrologic cycle along numerous pathways and processes. These include **evaporation** and **transpiration**, both of which are powered by energy from the Sun. Depending on local conditions of temperature, pressure, and humidity, some of the water vapor in the atmosphere will undergo **condensation**,

> **evaporation** The process by which water changes from a liquid into a vapor.
>
> **transpiration** The process by which water taken up by plants passes directly into the atmosphere.
>
> **condensation** The process by which water changes from a vapor into a liquid or a solid.

PROCESS DIAGRAM

☑ THE PLANNER

The hydrologic cycle • Figure 11.1

a. The hydrologic cycle as part of the Earth system.

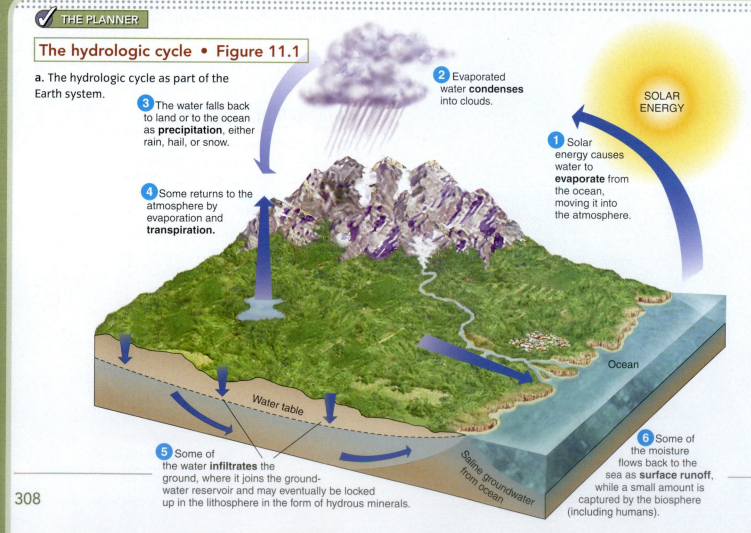

3 The water falls back to land or to the ocean as **precipitation**, either rain, hail, or snow.

2 Evaporated water **condenses** into clouds.

SOLAR ENERGY

1 Solar energy causes water to **evaporate** from the ocean, moving it into the atmosphere.

4 Some returns to the atmosphere by evaporation and **transpiration.**

Water table

Ocean

5 Some of the water **infiltrates** the ground, where it joins the groundwater reservoir and may eventually be locked up in the lithosphere in the form of hydrous minerals.

Saline groundwater from ocean

6 Some of the moisture flows back to the sea as **surface runoff**, while a small amount is captured by the biosphere (including humans).

precipitation The process by which water that has condensed in the atmosphere falls back to the surface as rain, snow, or hail.

surface runoff Precipitation that drains over the land or in stream channels.

infiltration The process by which water works its way into the ground through small openings in the soil.

changing to a liquid or a solid. It will then fall back to the land or ocean as rain, snow, or hail via the process of **precipitation**. Some precipitation becomes **surface runoff**, and some trickles directly into the ground via **infiltration**.

The schematic representation of the hydrologic cycle (**Figure 11.1b**) will look familiar to you if you have read Chapters 1, 4, and 7, which introduced the tectonic cycle and the rock cycle. The water cycle is linked to both of these other cycles. Like them, it is a closed cycle of open systems.

Because it is a closed cycle, the total amount of water is fixed. However, all the local reservoirs within the cycle, such as rivers and trees, are free to gain or lose water. They sometimes do so quite dramatically, as during a flood or drought.

Pathways and Reservoirs

Unlike many aspects of the tectonic cycle and the rock cycle, most of the water cycle is easily observable to us. We can measure the amount of global precipitation; using satellite monitoring, we can even measure the amount of evaporation. With these measurements, along with the overall mass balance of the water cycle, geologists can roughly deduce how much water is exchanged along each of the pathways shown in Figure 11.1b.

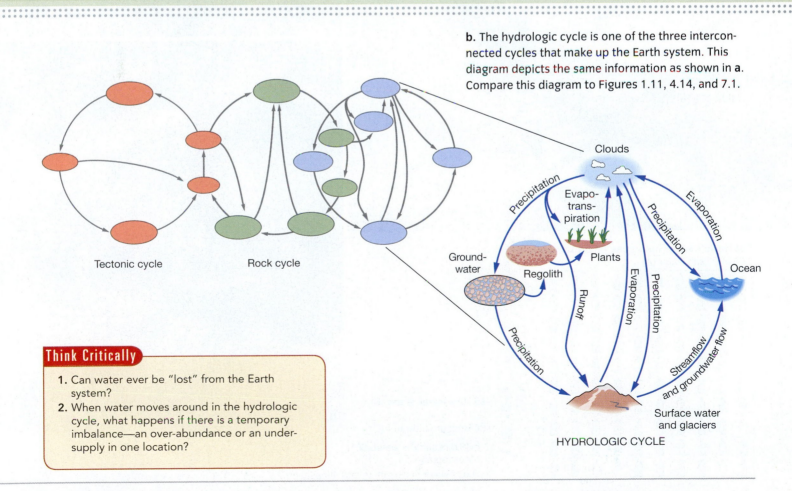

b. The hydrologic cycle is one of the three interconnected cycles that make up the Earth system. This diagram depicts the same information as shown in **a**. Compare this diagram to Figures 1.11, 4.14, and 7.1.

Tectonic cycle

Rock cycle

Clouds

Precipitation

Evapo-transpiration

Precipitation

Evaporation

Ground-water

Regolith

Plants

Runoff

Evaporation

Precipitation

Ocean

Precipitation

Streamflow and groundwater flow

Surface water and glaciers

HYDROLOGIC CYCLE

Think Critically

1. Can water ever be "lost" from the Earth system?
2. When water moves around in the hydrologic cycle, what happens if there is a temporary imbalance—an over-abundance or an under-supply in one location?

Reservoirs in the hydrologic cycle • Figure 11.2

The vast majority of Earth's water is either salty (**a**), frozen (**b**), or underground (**c**). The most visible everyday sources of fresh water, such as rivers (**d**), lakes, and the atmosphere, together comprise less than one-hundredth of a percent of Earth's water budget.

a.

b.

c.

d.

The world's water resources (in proportion):
97.5 liters salt water (A)

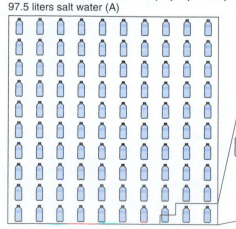

 1.85 liters frozen water (B)

0.64 liters groundwater (C)

10 milliliters surface water (D)*
(= 2 teaspoons)

*includes water in biosphere and atmosphere

It is also instructive to compare the sizes of each of the reservoirs in the water cycle (**Figure 11.2**). The largest reservoir in the hydrosphere—by far—is the world ocean, which holds 97.5% of Earth's water. Thus, the vast majority of Earth's water is saline (salty), not fresh. This has important consequences for humans because we depend on fresh water as a resource for drinking, agriculture, and industrial use.

The other numbers in Figure 11.2 may surprise you. Most of Earth's fresh water (almost 74%) is locked up in polar ice sheets, where it is almost inaccessible to humans. The vast majority of unfrozen fresh water (98.5%) lies underground. Historically, human settlement has concentrated along lakes or rivers, where the other 1.5% of the world's fresh water—namely, surface water—is readily available (**Figure 11.3**). But in most rural areas and quite a few urban areas as well, underground water sources are much more plentiful.

There is a correlation between the size of a reservoir and the **residence time**, the average length of time that a water molecule spends in that reservoir. Residence times in the large-volume reservoirs, such as the ocean and ice sheets, are several thousands of years. The time spent by water in the groundwater system may amount to hundreds of years or more. In small-volume reservoirs, the residence time of water is much shorter—weeks in streams and rivers, days in the atmosphere, and hours in some living organisms.

Although water is continuously cycling from one reservoir to another, the total volume of water in each reservoir is approximately constant over short time intervals.

Living on the water's edge • Figure 11.3

Many of the world's great cities are built on riverbanks or coastlines. St. Louis has long benefited from its proximity to the Mississippi River. In the foreground, a towboat pushes barges upstream; in the background, the *Mississippi Queen* riverboat brings tourists to the Gateway Arch.

Where Geologists CLICK

Exploring the Water Cycle

http://ga.water.usgs.gov/edu/watercycle.html

The U.S. Geological Survey website provides an interactive summary of the water cycle, with extensive information about groundwater, surface water, and all of the various transfer processes in the hydrologic cycle. You can also find specific data about groundwater, surface water, and water use in the area where you live, as well as a water map of the world.

However, the volume of water in each reservoir can change dramatically over longer intervals. During glacial ages, for example, vast quantities of water evaporate from the ocean and are precipitated on land as snow. The snow slowly accumulates to build ice sheets that are thousands of meters thick and cover vast areas. At such times, the amount of water removed from the ocean is so large that the global sea level can fall by many meters, and the expanded glaciers increase the ice-covered area of Earth. *Where Geologists Click* provides some additional resources for exploring the water cycle with the U.S. Geological Survey.

CONCEPT CHECK

1. **What** are the major reservoirs in the hydrologic cycle?

2. **How** does water move among these reservoirs?

3. **How** does the residence time of water correlate to the size of a reservoir?

How Water Affects Land

LEARNING OBJECTIVES

1. **Identify** the basic characteristics of streams.
2. **Explain** how straight, braided, and meandering channels form.
3. **Identify** three common land formations made by stream deposits.
4. **Define** drainage basin.
5. **Describe** how lakes form and disappear.

I f you stand outside during a heavy rain, you can see that, initially, water tends to move downhill in a process called **overland flow** (or **sheet flow** because the flowing water often takes the form of a thin, broad sheet). After traveling a short distance, overland flow begins to be concentrated into well-defined passageways, thereby becoming **streamflow**. Overland flow and streamflow together constitute surface **runoff**, one of the main pathways in the water cycle. Let's look more closely at streams and streamflow, the passageways of streams, and the interactions of streams with the land.

Streams and Channels

Every **stream** or river has a **channel** and several factors affect the shape of the channel and the landforms it creates. Streams create landforms through two processes: erosion (see Chapter 7) and deposition (see Chapter 8). Both processes go on throughout a stream's existence and along its length, but one or the other may predominate at a particular location or during a particular time, depending on a variety of factors.

The three most important factors controlling a stream channel are the **gradient**, **discharge**, and **load**. The three factors are interrelated. For example, if the gradient of a stream becomes steeper along a particular stretch of channel, the velocity of flow is likely to increase as well. If the velocity is high, a greater load can be carried.

> **stream** A body of water that flows downslope along a clearly defined natural passageway.
>
> **channel** The clearly defined natural passageway through which a stream flows.
>
> **gradient** The steepness of a stream channel.
>
> **discharge** The amount of water passing by a point on a channel's bank during a unit of time.
>
> **load** The suspended and dissolved sediment carried by a stream.

If the discharge increases, the channel must handle more water in a given period; as a result, both the velocity of flow and the depth of the water in the stream will increase. (Note that this is the first of two different meanings of *discharge* used in this chapter.) In some cases, the channel

Geology InSight

Channel patterns and features associated with them. Pools are places along a channel where the water is deepest. Arrows indicate the direction of streamflow and trace the path of the deepest water.

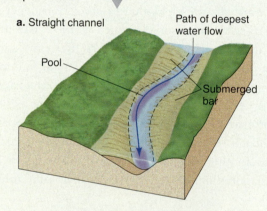

a. Straight channel

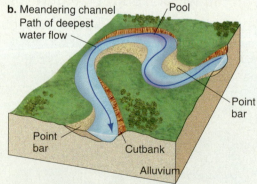

b. Meandering channel

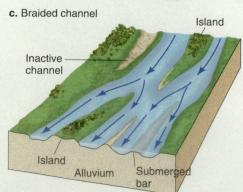

c. Braided channel

itself will increase in width and depth as the water scours the banks and the bottom. This scouring, in turn, adds new sediment to the load. When the velocity of the water eventually decreases, the sediment settles out and allows the channel to return to its original size.

While gradient, discharge, and load are the most important determinants of a channel's size and shape, other factors are also important. Topography, for example, determines the gradient, and climate determines the amount of precipitation, and hence the discharge. The geologic characteristics of the underlying rock are also important. For example, a stream may suddenly bend, or its gradient may increase when it passes from erosion-resistant rock into rock that is easily eroded. Because these factors interact in different ways, no two stream channels are exactly alike. Nevertheless, we can classify them into three broad categories, as shown in **Figure 11.4**.

Three kinds of streams • Figure 11.4

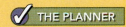 THE PLANNER

a. Straight channels, such as this stream that drains a glacier in Alaska, usually occur only in relatively short stretches. They are often found where streams have a high gradient (near the stream's headwaters), and they generally have a classic V-shaped valley shaped by mass wasting and overland flow.

b. A meandering stream near Phnom Penh, Cambodia, shows several features typical of a meandering river: a low gradient, light-colored sandy bars on the inside edge of the bends, and nearby oxbow lakes marking previous watercourses that have been cut off.

c. Note the intricate braided pattern of the Tasman River, below Tasman Glacier in New Zealand's Southern Alps. It has the classic profile of a braided stream, with a low gradient and a large and variable load of sediment—produced in this case by the action of glaciers.

A river changes shape • Figure 11.5

In the background of this photograph, taken several months after the Mt. St. Helens eruption, you can see the braided form of the Toutle River. In the foreground, engineers have built an artificial levee to straighten the river's course and control erosion.

Braided flow created by high bed load upstream

Muddy water due to high rate of erosion upstream

Straight channel flow created by artificial levees

Unlike engineered aqueducts and canals, natural streams are never really straight from start to finish. **Straight channels** may occur over short distances, particularly in **upstream** areas (that is, near the **headwaters,** or **source**, of the stream), where the gradient is high and the channel deeply incised. Even in a "straight" channel, close examination will show that the deepest part of the channel oscillates from side to side.

Meandering channels tend to develop where the stream gradient is low, typically in the lower, or **downstream,** parts of a stream system (close to the **mouth**, where the stream empties into another surface water body). The erosion in a meandering stream concentrates along the sides of the channel rather than the bottom. As water sweeps around a bend, it flows more rapidly along the outer bank, undercutting and steepening it to form a **cut bank**. Meanwhile, along the inner side of each meander, where the water is shallow and velocity is low, sediment accumulates to form a **point bar**, as shown in **Figure 11.4c**. Thus, meanders slowly change shape and shift their position along a valley as the stream erodes material from one bank and deposits sediment on the other. Sometimes the water finds a shorter route downstream, bypassing a meander by cutting across the narrow part of the loop. As sediment is deposited along the banks of the new channel route, the former meander is cut off and converted into a curved **oxbow lake**.

Braided channels arise when a stream's ability to move its sediment load varies over time. At times of high flow, a stream can carry more sediment. If the discharge decreases but the load does not, the stream deposits the excess sediment in its own channel as bars or islands. These variations in flow over time cause the channel to repeatedly divide and reunite, as shown in **Figure 11.4d**. Braided patterns tend to form in streams with highly variable discharge and large loads of coarse sediment. Braided patterns can form in streams with discharge that varies seasonally (e.g., following rapid snowmelt) and in streams with easily eroded banks.

As an extreme example of dramatic changes in a river channel, Washington's Toutle River changed almost overnight from a meandering stream to a braided stream in 1980, when Mt. St. Helens erupted (**Figure 11.5**). The debris avalanche that resulted from the eruption deposited huge amounts of loose and easily eroded pyroclasts and volcanic ash into the Toutle Valley, dramatically increasing the river's load. In addition, there were no longer any living trees to hold the old riverbanks in place. Changes in an ancient river's flow patterns (as revealed in sedimentary strata) can tell geologists about the geologic history of a region.

Stream valley deposits • Figure 11.6

a. Some of the landforms created by sediment in stream valleys are floodplains, natural levees, and alluvial fans.

Oxbow lake After the cutoff, silt and sand are deposited across the ends of the abandoned channel, producing an oxbow lake. The oxbows fill with fine sediment and organic matter produced by aquatic plants and eventually turn into swamps.

Floodplain The meandering river channel dominates the floodplain, along with stretches of abandoned channels. These were abandoned after cutoff events, when the river cuts across a meander loop to create a shorter, more direct path.

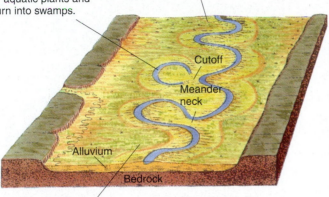

Natural levees are created during overbank flooding, when sand and silt are deposited next to the channel creating belts of higher land on either side of the channel. Deposition is heavier closest to the channel, so the levee surface slopes away from the channel.

b. Where this stream emerges from the mountains into Death Valley, California, it abruptly slows and deposits its sediment load. This has created a symmetrical alluvial fan, covered by a braided system of channels. The stream was dry at the time the photograph was taken.

c. Where the Nile River empties into the Mediterranean Sea, the sediment it deposits has formed a fan-shaped delta that supports the green vegetation seen in this satellite image. Its triangular shape, similar to the Greek letter delta (Δ), gave rise to the term *delta* that we use today.

Stream Deposits

Stream deposits form along channel margins, valley floors, mountain fronts, and at the stream's mouth, where it opens into the ocean or a lake. These are all places where the stream loses energy and, therefore, its ability to carry a load. A point bar is one example of a stream deposit.

When a stream rises during a flood, water overflows the banks of the stream's channel and inundates the **floodplain** (**Figure 11.6a**). As sediment-laden water flows out of the channel, its depth, velocity, and turbulence

> **floodplain** The relatively flat valley floor adjacent to a stream channel, which is inundated when the stream overflows its banks.

decrease abruptly at the margins of the channel. This results in sudden, rapid deposition of the coarser part of the load along the margins, which builds up a broad, low ridge of **alluvium** atop each bank, called a **natural levee**. Farther away, the finer particles settle out in the quiet water covering the valley. This creates the broad, flat, fertile land that is typical of floodplains.

> **alluvium** Unconsolidated sediment deposited in a recent geologic time by a stream.

Another kind of alluvial structure develops where a stream draining a steep upland region suddenly emerges onto the floor of a much broader lowland valley. The stream will slow down and lose some of its ability to carry sediment. It will deposit the coarser part of its load (the part that can no longer be transported) in an **alluvial fan** (**Figure 11.6b**). Such fans are typical of semi-arid conditions where vegetation is sparse and infrequent rainfall creates streams that are heavily laden with sediment.

> **delta** A sedimentary deposit, commonly triangular, that forms where a stream enters a standing body of water.

A similar situation occurs when a stream flows into a standing body of water, such as the ocean or a lake. It quickly loses velocity and fans out, dropping its sediment load (the heaviest particles first and the finer particles farther seaward). Over time, the sediment builds up a deposit called a **delta** (so named because of its shape, which is commonly triangular, resembling the Greek letter delta, Δ). Most of the world's great rivers, including the Nile, Ganges-Brahmaputra, Huang He, Amazon, and Mississippi, have built massive deltas (**Figure 11.6c**).

Disappearing coastline • Figure 11.7

The taming of the Mississippi River with levees and artificially modified channels has slowed the deposition of new sediment in its delta downstream, where the river flows into the Gulf of Mexico. As a result, the entire Mississippi Delta has been shrinking. Marshes give way to open water, ponds turn into lakes, and barrier islands shrink. The top image shows the extent of the delta in 1839, the middle image shows its extent in 1993, and the bottom image shows a projection of its extent in 2090. Land areas are shown in brown colors, marshes and wetlands are in green, and open water is in blue tones. The degradation of the delta is at least partly responsible for the increased vulnerability of New Orleans to hurricane damage.

DECONSTRUCTING A COAST

1839

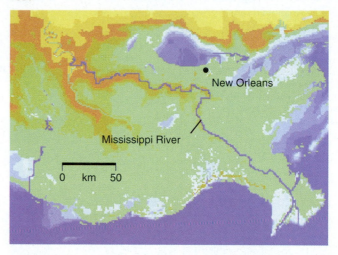

1993

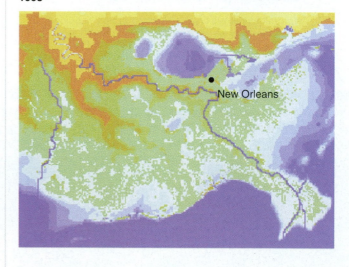

2090 *projection*

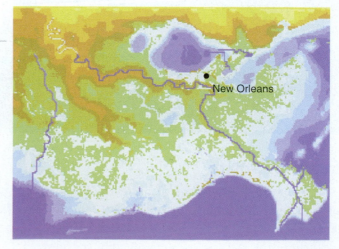

Prominent deltas do not form in places where strong wave, current, or tidal action redistributes sediment as quickly as it reaches the coast. However, if the rate of sediment supply exceeds the rate of coastal erosion, a delta will form. The converse holds, too: If the rate of deposition slows down, the delta will disappear (**Figure 11.7**).

Large-Scale Topography of Stream Systems

Streams are governed by a simple principle: Water flows downhill. Rainwater that falls on the land surface will move from higher to lower elevations under the influence of gravity. A stream's headwater region is the area of relatively higher elevation from which streams have their source. Small, high-gradient **tributary** streams carry water downslope from the headwater region, combining their flow to form a larger stream. The gradient gradually decreases toward the low-lying region of the stream's mouth.

Every stream is surrounded by a **drainage basin** (sometimes called a **catchment** or **watershed**). Drainage basins range in size from less than a square kilometer to areas the size of subcontinents. In general, the greater a stream's annual discharge, the larger its drainage basin. The vast drainage basin of the Mississippi River encompasses more than 40% of the total area of the contiguous United States (**Figure 11.8**). From an environmental perspective,

> **drainage basin**
> The total area from which water flows into a stream.

The Mississippi River drainage basin • Figure 11.8 _____

The drainage basin of the Mississippi River encompasses most of the midwestern United States and extends into southern Canada. In this diagram, the widths of the rivers are exaggerated to represent the discharge in cubic meters per second.

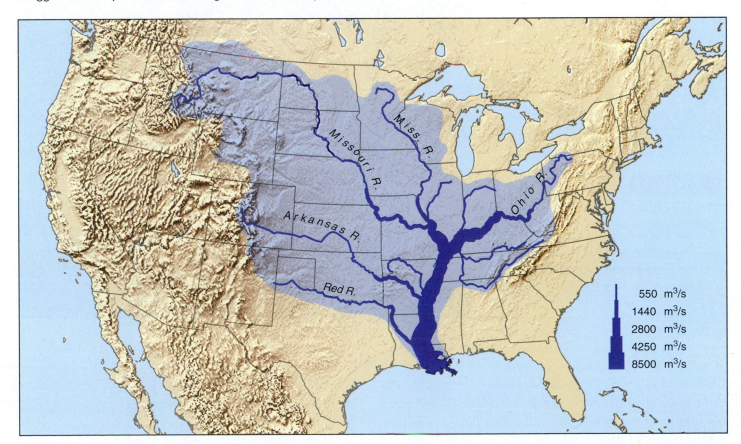

550	m³/s
1440	m³/s
2800	m³/s
4250	m³/s
8500	m³/s

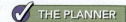
Drainage Basins

The drainage basins of many rivers are hard to see from a satellite because they are covered by vegetation. However, a geologist looking at this space shot of Wadi Al Masilah in South Yemen, adjacent to the Rubh-al-Khali, a desert, would not have any trouble drawing in the drainage basin of one tributary of the river. The geologist has used a dashed line to outline the boundary—that is, the divide—that surrounds and defines the drainage basin of the tributary. He identified the divide by looking to see where streams flowed in opposite directions. Note that this basin would itself be a part of the drainage basin of the larger river at the top of the photograph.

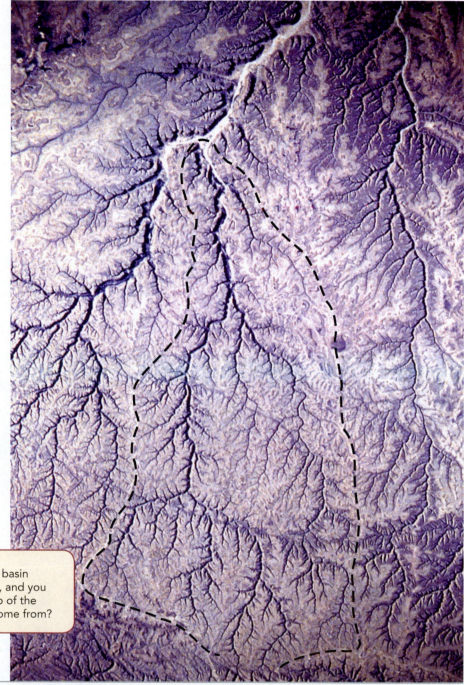

Think Critically

The stream channels in the outlined drainage basin appear to be completely dry. But look closely, and you can see that the main river channel, at the top of the photo, has water in it. Where did the water come from?

a drainage basin is a more natural geographic entity than a country or a state because issues of water supply, pollution, and wildlife management that affect one part of a watershed are likely to affect it all. (See *What a Geologist Sees.*)

If you have driven across North America, you may have seen highway signs marking the **continental divide.** This major **divide** separates streams that drain toward one side of the continent from streams that drain toward the other side. The continental divide of western North America lies along the length of the Rocky Mountains

divide A topographic high that separates adjacent drainage basins.

(**Figure 11.9**). Any two adjacent drainage basins are separated by a divide that directs runoff in one direction or another, even if streams flowing from the basins ultimately flow into the same ocean.

Lakes

Lakes are standing bodies of water that have open surfaces, in direct contact with the atmosphere. Water enters lakes from streams, overland flow, and groundwater, and it exits either by evaporation or by flowing through an outlet. All freshwater lakes have outlets; however, some saline lakes lack outlets and therefore lose water only through evaporation, which inevitably leads to a buildup of salt. The Great Salt Lake in Utah is an important example of an inland saline lake.

Lakes are important to us as sources of fresh water and food, as well as for transportation and recreation. The majority of lakes in the United States have been altered, and in some cases, even created, by humans. Sometimes the changes are deliberate, as when a **reservoir** lake is created by a dam or a wetland is drained to make it accessible for human development. In other cases, the human effects are inadvertent. For example, runoff of sewage or fertilizer

Continental drainage in the Americas • Figure 11.9

This map of North and South America shows the location of continental drainage basins and divides.

A dying river • Figure 11.10

This river in Florida has turned green and mucky because of the growth of algae stimulated by excessive nutrients (possibly from sewage or fertilizer). Eventually, the algae will use up all the oxygen in the water, making it impossible for other life to survive in the river. This is called eutrophication.

gravel deposits. Kettle lakes, a common landform in New England, form when these pits fill with water.

An important characteristic of lakes is that they are short-lived features on the geologic time scale. They disappear through one of two processes or a combination of both. First, lakes that have stream outlets will be gradually drained as the outlets are eroded to lower levels. Second, lakes accumulate inorganic sediment carried by streams entering the lake and organic matter produced by plants within the lake. Eventually, they fill up, forming boggy **wetland** with little or no free water surface.

In arid climates, many lake beds are either dry or only intermittently filled with shallow water. Streams bring dissolved salts to these ephemeral, or short-lived, lakes. Because evaporation removes only pure water, the salts remain behind, and salinity levels increase. Eventually the salts may be precipitated as solid evaporites, some of which have economic value.

into a lake can cause **eutrophication**, which can kill most of the life in the lake (**Figure 11.10**). Eutrophication also occurs as a natural part of the process of swamp formation.

Lakes can form as a result of several different geologic processes. Crustal faulting creates many large, deep lakes. A lava flow may form a dam in a river valley, causing water to back up as a lake. Landslides suddenly create lakes by blocking valleys. Throughout formerly glaciated regions of North America and Europe, plains of glacial sand and gravel contain natural pits and hollows left by the melting of stagnant ice masses that were buried in the sand and

CONCEPT CHECK STOP

1. **What** do geologists mean by the term "stream"?

2. **How** do the three main types of stream channels develop?

3. **What** are the major types of stream deposits?

4. **What** separates a drainage basin from adjacent drainage basins?

5. **Why** are lakes geologically short-lived?

Water as a Hazard and a Resource

LEARNING OBJECTIVES

1. **Describe** how floods occur and what factors may make them worse.

2. **Define** recurrence interval and show how this interval is used to predict floods.

3. **Explain** why flood prevention efforts sometimes actually make flooding worse.

4. **Explain** why interbasin transfer of water can be a flawed solution to water scarcity.

T hough water is a vital resource, it can also be a dangerous force. Uneven distribution of rainfall through the year causes some water bodies to dry up, but others rise and overflow their banks, creating hazardous circumstances for people who live in the area.

flood An event in which a water body overflows its banks.

Floods

Under natural circumstances, all water bodies undergo changes in the volume of water they hold or transport. From time to time, when the discharge or water level becomes too much to handle, the water body will **flood** (**Figure 11.11**).

Mississippi River flood • Figure 11.11

a. A pair of satellite images shows the region where the Missouri River joins the Mississippi River at St. Louis, Missouri. The photo on the left shows a dry summer with low flows (July 1988). The photo on the right shows the same region in July 1993. Weeks of rain hundreds of kilometers away caused the rivers to overflow their levees. Numerous towns, along with 44,000 km² of farmland in nine states, were flooded by an estimated 3 km³ of floodwater.

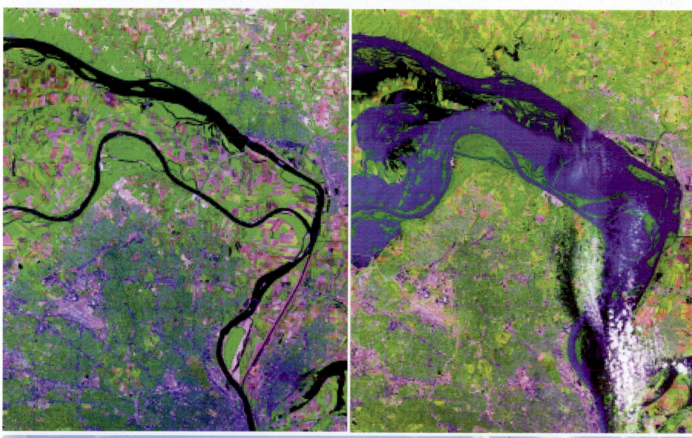

b. Flooding of the Mississippi River occurred again in 2011. Here, on May 3, 2011, flood waters have swamped the region just north of New Madrid, Missouri.

During a stream flood, the "extra" water flowing in the channel, contributed by excess precipitation, is called **storm runoff.** The peak discharge of a flood usually comes well after the rains that produced it. Geologists record the development of a flood with a **hydrograph,** a graph that shows the stream's discharge as a function of time. **Figure 11.12** shows an example of a hydrograph in which a passing storm generated a brief interval of intense rainfall. As the runoff moved into the stream channel, the discharge quickly rose. The **crest** of the resulting flood—when the peak flow passed the hydrologic station where the measurements were made—occurred about two hours after the storm. It took another eight hours for the flood runoff to pass through the channel and for the discharge to return to its normal level.

A hydrograph of stream discharge • Figure 11.12

This diagram illustrates the hydrograph of a stream after a brief, intense storm.

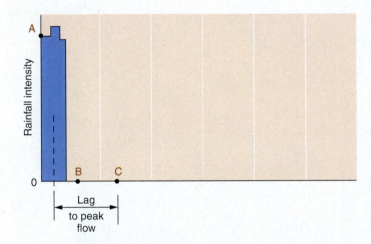

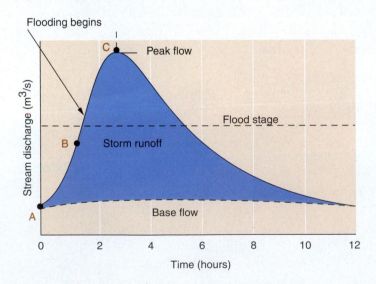

Base flow is the "normal" flow of water in a stream, contributed by groundwater.

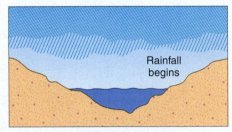

A. Onset of storm (0 hours):
The peak discharge is delayed as the runoff collects and runs down the stream channel.

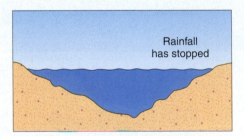

B. 1 hour:
One hour after the cloudburst, the stream can still contain the increased volume.

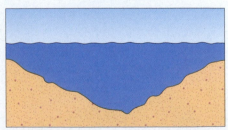

C. 2 hours:
After two hours, the stream reaches its peak flow and cannot be contained by its banks any more.

Coastal flood • Figure 11.13

As Hurricane Katrina moved toward New Orleans from the Gulf of Mexico in August 2005, high waters breached the city's protective levees. The subdivision shown here in St. Bernard Parish was constructed on sand dredged from the Mississippi River and was protected by a levee. But the levee was not sufficient to withstand the storm surge. Ten days after the storm, when this photo was taken, the subdivision was still under almost 2 m of water.

Lakes also flood, as do oceanic coastal zones. In the case of coastal flooding, it may be the inflow of water from the ocean, rather than the runoff of water from the land, that does most of the damage. The **storm surge** associated with Hurricane Katrina in 2005, in which water from the Gulf of Mexico was blown onshore by high winds of the hurricane, temporarily raised the water level near New Orleans by 6 m or more. The storm surge breached the levees even before the main force of the storm winds and rain hit the city (**Figure 11.13**).

Though both coastal flooding and stream flooding often take people by surprise, geologists view them as normal and inevitable events. The geologic record shows that floods have been occurring throughout Earth's history, for as long as there has been a hydrosphere. Even though flooding is a natural geologic process, it can quickly become a human catastrophe when it affects population centers. The Huang He in China, sometimes called the Yellow River because of its heavy load of yellowish-brown silt, has a long history of catastrophic floods. In 1887, the river inundated 130,000 km² and swept away many villages in the heavily populated floodplain. In 1931, another Huang He flood killed a staggering 3.7 million people. Yet these same floods help replenish the soil in the floodplain, which explains why people keep moving back to the area.

Human activity sometimes increases the chance that flooding will occur (instead of decreasing it). Urban development can exacerbate the problem of flooding in a variety of ways. Urban construction on compressible sediment, often accompanied by withdrawal of groundwater, can lead to **subsidence** (a drop in the surface of the land), which increases the danger of flooding (as it did in New Orleans). The impermeable ground cover associated with urbanization can add substantially to surface runoff in urban areas. Storm sewers can contribute to flooding because they allow the runoff from paved areas to reach the river channel more quickly. Floods in urbanized basins often have higher peak discharges and reach their peaks more quickly than do floods in undeveloped basins. This quicker and higher crest means that people living in a flood-prone area must be able to move very quickly or be able to predict when a flood might strike.

Because floods can be so damaging, predicting and preparing for them is essential (**Figure 11.14**). To do this, scientists plot the frequency of past floods of different sizes on a graph to produce a **flood–frequency curve**. The average time interval between two floods of the same magnitude is called the **recurrence interval**. For example, a "10-year flood" has a recurrence interval of 10 years, which means that there is a 1-in-10 (or 10%) chance that such a flood will occur in any given year. A flood with an even greater discharge having a recurrence interval of 50 years would be termed a "50-year flood" for this particular stream, and there would be a 1-in-50 (2%) chance of such a flood occurring in any given year. Regional planners should (and do) keep these intervals in mind when planning development on or near a flood plain.

Predicting floods • Figure 11.14

a. The graph below shows the frequency of floods of different sizes on the Skykomish River at Gold Bar, Washington. A flood with a discharge of 1750 m³ per second has a recurrence interval of 10 years, and hence a 1-in-10 chance of occurring in any given year.

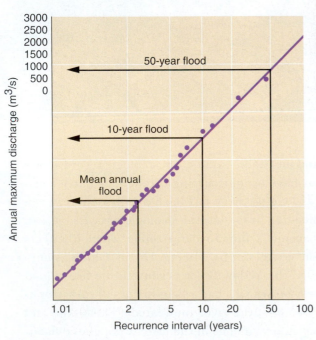

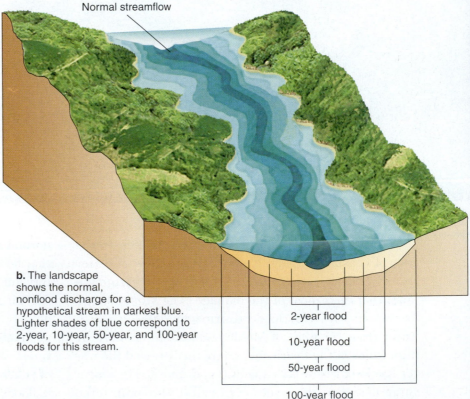

Normal streamflow

b. The landscape shows the normal, nonflood discharge for a hypothetical stream in darkest blue. Lighter shades of blue correspond to 2-year, 10-year, 50-year, and 100-year floods for this stream.

2-year flood

10-year flood

50-year flood

100-year flood

Another aspect of flood prediction is real-time monitoring of storms and water levels. Hydrologists can combine information about the weather with their knowledge of a river basin's geology and topography to forecast the peak height of a flood and the time when the crest will pass a particular location. Such forecasts, which are often made with the aid of computer models and **geographic information systems (GIS)**, can be very useful for planning evacuation or defensive measures.

Understandably, many people throughout history have been unsatisfied with simply predicting floods and have attempted to prevent them. River channels are often modified, or "engineered," for the purposes of flood control and protection as well as to increase access to floodplain lands, facilitate transport, enhance drainage, and control erosion. The modifications usually consist of some combination of widening, deepening, straightening, clearing, or lining of the natural channel. All of these approaches are collectively called **channelization**.

Like dams, channelization projects can contribute to the economic well-being of a community—but at a price. Channel modifications interfere with natural habitats and ecosystems. Such modifications can degrade the aesthetic value of a river and aggravate water pollution. Projects sometimes control flooding in the immediate area but contribute to more intense flooding downstream. Perhaps most importantly, any modification of a channel's course or cross section renders invalid the hydrologic data collected there in the past. During the Mississippi River floods of 1973 and 1993, experts could not account for water levels that were higher than predicted by the historical data; the likely cause was extensive upstream modifications of the river channel by humans.

Surface Water Resources

A reliable water supply is critical—not only for human survival and health but also for the role it plays in industry, agriculture, and other economic activities (**Figure 11.15** on pages 326–327). According to the United Nations, 43 countries worldwide, with a total population of almost 700 million people, are today designated as **water-scarce**; by 2025 this could increase to 1.8 billion people. The lack of water in these countries places serious constraints on agricultural production, economic development, health, and environmental protection.

Globally, crop irrigation accounts for about 73% of the demand for water, industry for about 21%, and domestic use for the remaining 6%, though the proportions vary from one region to another. Demand in each of these sectors has more than quadrupled since 1950. Population growth is partly responsible for the increasing demand, but improvements in standards of living around the world have also contributed to the large increase in water use per capita over the past few decades. The total amount of water being withdrawn (i.e., diverted from rivers, lakes, and groundwater) for worldwide human use is about eight times the annual streamflow of the Mississippi River.

Sometimes, because of population growth and development, regions with the greatest demand for water do not have abundant and readily available supplies of surface water. For this reason, surface water is often transferred from one drainage basin to another, sometimes over long distances. Besides raising political issues related to water rights, such **interbasin transfer** can have negative environmental impacts (see *Case Study* on page 328).

CONCEPT CHECK

1. **What** factors contribute to flooding of a particular stream?

2. **How** is the recurrence interval of a flood calculated?

3. **What** is channelization, and how does it influence flooding?

4. **What** are some consequences of transferring water from one drainage basin to another?

The large map shows the world's main watersheds in terms of the amount of renewable water available per person, per year. The bottom map and graphics show access to improved drinking water, and the main purposes of water withdrawals in different world regions.

Primary watersheds
Annual renewable water
(cubic meters per person, 2000)

- More than 100,000
- 10,000 to 100,000
- 4,001 to 10,000
- 1,701 to 4,000
- 1,001 to 1,700
- Less than 1,000
- No data
- ▲ Water related conflict in the last 100 years
- – Large dam - volume (in thousands) greater than 38,000 cu m (50,000 cu yds)

Percent of total population using improved drinking water sources, 2000
- More than 90%
- 76% - 90%
- 51% - 75%
- 25% - 50%
- Less than 25%
- No data available

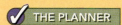

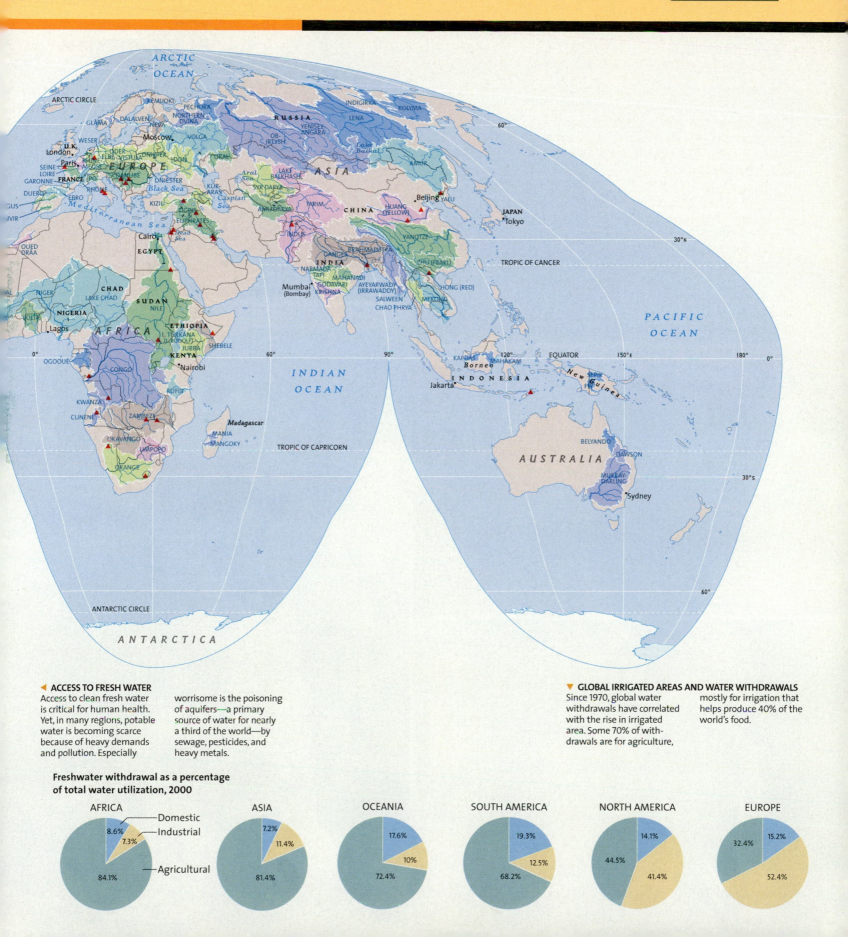

◀ **ACCESS TO FRESH WATER**
Access to clean fresh water is critical for human health. Yet, in many regions, potable water is becoming scarce because of heavy demands and pollution. Especially worrisome is the poisoning of aquifers—a primary source of water for nearly a third of the world—by sewage, pesticides, and heavy metals.

▼ **GLOBAL IRRIGATED AREAS AND WATER WITHDRAWALS**
Since 1970, global water withdrawals have correlated with the rise in irrigated area. Some 70% of withdrawals are for agriculture, mostly for irrigation that helps produce 40% of the world's food.

Freshwater withdrawal as a percentage of total water utilization, 2000

AFRICA
- Domestic: 8.6%
- Industrial: 7.3%
- Agricultural: 84.1%

ASIA
- 7.2%
- 11.4%
- 81.4%

OCEANIA
- 17.6%
- 10%
- 72.4%

SOUTH AMERICA
- 19.3%
- 12.5%
- 68.2%

NORTH AMERICA
- 14.1%
- 41.4%
- 44.5%

EUROPE
- 15.2%
- 52.4%
- 32.4%

CASE **STUDY**

Mono Lake and the Los Angeles Water Supply

California's Mono Lake, in the Sierra Nevada Mountains, has been the site of a collision between human needs for reliable supplies of fresh water and the needs of a unique habitat and its wildlife.

Seventy years ago, the water level of the lake was 1956 m above sea level, and these calcium carbonate spires (**Figure a**) were underwater. But in 1941, the Los Angeles Aqueduct began diverting water from four of the six streams that empty into Mono Lake. With insufficient input to make up for its evaporation losses, the water level dropped, the lake shrank to half its original volume, and the salinity doubled. Migratory birds, such as the nation's second-largest colony of California gulls, were placed in jeopardy. Their food supply (brine shrimp) was dying because of the high salinity of the water, and the islands on which the birds nested were in danger of being connected to the mainland because of the retreating water, making the birds vulnerable to predators, mainly coyotes.

By the late 1980s, the lake and its ecosystem were nearing collapse. In 1994, after a court battle between the city of Los Angeles and environmental groups, both sides agreed to a plan to raise the water level by 5 m and compensate Los Angeles for the loss of part of its water supply.

By May 2011, water levels in the lake had risen to 1945 m, about halfway to the target of 1948 m set by the agreement. The lake is expected to reach its target level in 15 to 20 years. Dams that once diverted streams into the aqueduct (**Figure b**) have been reengineered to do the opposite: They maintain a steady flow into Mono Lake and divert water into the aqueduct only if there is an overflow. For the California gulls (**Figure c**), the change may have come just in time.

a.

b.

c.

Global Locator

Mono Lake

NATIONAL GEOGRAPHIC

Fresh Water Underground

LEARNING OBJECTIVES

1. **Define** water table.

2. **Explain** how porosity and permeability of rock affect the motion of groundwater.

3. **Identify** two types of aquifers.

4. **Explain** why some wells require pumping, while others flow unaided.

5. **Describe** how subsidence relates to groundwater.

6. **Describe** how a cave forms.

 ess than 1% of all water in the hydrosphere cycle is **groundwater**. Although this sounds small, the volume of groundwater is 40 times greater than that of all the water in freshwater lakes and streams. Water can be found everywhere beneath the land surface, even beneath parched deserts. About half of it is near the surface, no more than 750 m below ground. At greater

groundwater
Subsurface water contained in pore spaces in regolith and bedrock.

depths, the pressure exerted by overlying rock reduces the pore space, making it difficult for water to flow freely. In addition, water that occurs at great depths tends to be briny, that is, rich in dissolved mineral salts, and not well suited for human use. Therefore, from a practical perspective, we can think of groundwater as the water found between the land surface and a depth of about 750 m, even though an equally large amount of water is present at greater depths.

The Water Table

Much of what we know about groundwater has been learned from the accumulated experience of generations of people who have dug or drilled millions of wells. This experience tells us that a hole penetrating the ground ordinarily first encounters a zone in which the spaces between the grains in regolith or bedrock are filled mainly with air, although the material may be moist to the touch. This is the **zone of aeration**, also known as the **vadose zone** (**Figure 11.16**).

After passing through the zone of aeration, the hole reaches the **water table** and enters the

water table The top surface of the saturated zone.

Water under the ground • Figure 11.16

A well first passes through the zone of aeration, where pores in the soil are filled with both air and water. Eventually, it reaches the water table, where the pore spaces are completely filled with water. Underground water exists everywhere, and water flows out wherever the ground surface intersects the water table.

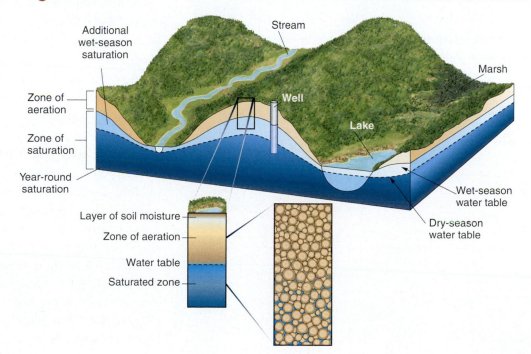

Porosity in sediment and rock • Figure 11.17

In these examples, all of the pore spaces are filled with water, as they would be in the saturated zone.

a. The porosity is about 30% in this sediment with particles of uniform size.

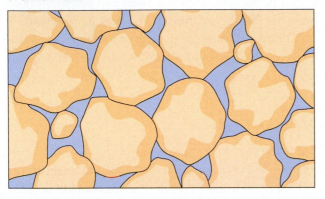

b. This sediment, in which fine grains fill the space between larger grains, has a lower porosity, around 15%.

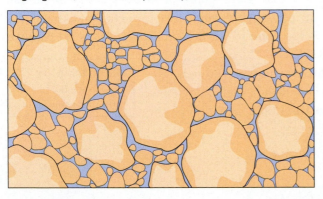

c. In sedimentary rock, the porosity may be reduced by cement that binds the grains together and fills the pores.

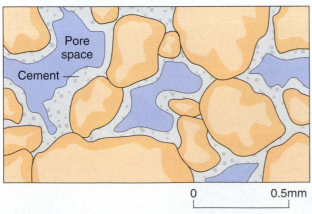

Pore space

Cement

0   0.5mm

saturated zone, also known as the **phreatic zone**, in which all openings are filled with water. The water table is high beneath hills and low beneath valleys. This may seem surprising, because the surface of a glass of water or a lake is always level. But water underground flows very slowly and is strongly influenced by surface topography. If all rainfall were to cease, the water table would slowly flatten. Seepage of water into the ground would diminish and then stop entirely, and streams would dry up as the water table fell. During droughts, the depression of the water table is evident from the drying up of springs, stream beds, and wells. Repeated rainfall, which soaks the ground with fresh supplies of water, maintains the water table at a normal level and keeps surface water bodies replenished.

Whether it is deep or shallow, the water table marks the upper limit of readily usable groundwater. For this reason, a major aim of groundwater specialists and well drillers is to determine the depth and shape of the water table. To do this, they must first understand how groundwater moves and what forces control its distribution underground.

How Groundwater Moves

Most groundwater is in motion. Unlike the swift flow of rivers, however, which is measured in kilometers per hour, the movement of groundwater is so slow that it is measured in centimeters per day or meters per year. The reason is simple: Whereas the water of a stream flows through an open channel, groundwater must move through small, constricted passages. Therefore, the rate of groundwater flow depends on the nature of the rock or sediment through which the water moves, and especially on the porosity and permeability.

Porosity and permeability

Porosity determines the amount of fluid a sediment or rock can contain. The porosity of sediment is affected by the size and shape of the particles and the compactness of their arrangement (**Figure 11.17a, b**). The porosity of a sedimentary rock is also affected by the extent to which the pores have been filled with cement (**Figure 11.17c**). Plutonic igneous rock and metamorphic rock, which consist of many closely interlocked crys-

> **porosity** The percentage of the total volume of a body of rock or regolith that consists of open spaces (pores).

tals, generally have lower porosities than do sediment and sedimentary rock. However, joints and fractures may increase their porosity.

A rock with low porosity is likely also to have low **permeability**. However, high porosity does not necessarily mean high permeability because both the sizes and continuity of the pores (i.e., the extent to which the pores are interconnected) influence the ability of fluids to flow through the material.

> **permeability** A measure of how easily a solid allows fluids to pass through it.

Percolation After water from a rain shower soaks into the ground, or infiltrates, some of it evaporates, while some is taken up by plants. The remaining water continues to **percolate** under the influence of gravity until it reaches the water table. The "perc test" that must be carried out when a new septic system is being installed is a measure of percolation. The movement of groundwater in the saturated zone is similar to the flow of water that occurs when you gently squeeze a

> **percolation** The process by which groundwater seeps downward and flows under the influence of gravity.

water-soaked sponge. Water moves slowly through very small pores along threadlike paths. The water flows from areas where the water table is high toward areas where it is lower. In other words, it generally flows toward surface streams or lakes (**Figure 11.18**). Some of the flow paths turn upward and enter the stream or lake from beneath, seemingly defying gravity. This upward flow occurs because groundwater is under greater pressure beneath a hill than beneath a stream or lake. Because water tends to flow toward points where pressure is low, it flows toward bodies of water at the surface.

Recharge and discharge Recharge of groundwater occurs when rainfall and snowmelt infiltrate the surface and percolate downward to the saturated zone (see Figure 11.18). The water then moves slowly along its flowpath toward zones where **discharge** occurs. (Note that earlier in the chapter, we used the term *discharge* for a

> **recharge** Replenishment of groundwater.
>
> **discharge** The process by which subsurface water leaves the saturated zone and becomes surface water.

How groundwater flows • Figure 11.18

This diagram illustrates the flow paths of water as it seeps in through the ground surface, flows downward to the zone of saturation, and then percolates along curved pathways under the influence of gravity, eventually emerging at the surface once again.

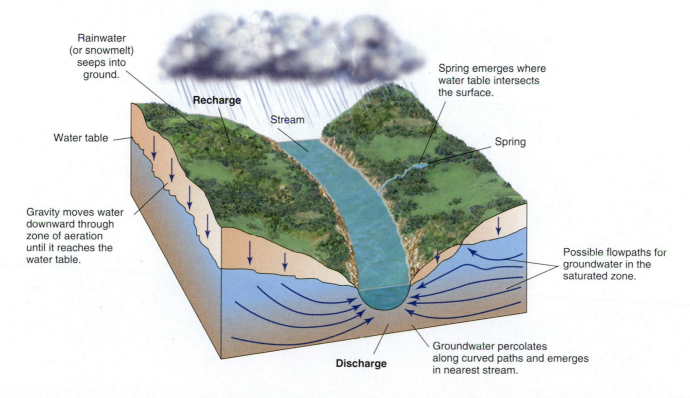

Rainwater (or snowmelt) seeps into ground.

Spring emerges where water table intersects the surface.

Recharge

Stream

Spring

Water table

Gravity moves water downward through zone of aeration until it reaches the water table.

Possible flowpaths for groundwater in the saturated zone.

Discharge

Groundwater percolates along curved paths and emerges in nearest stream.

somewhat different concept—the flow of water along a stream channel.) In discharge zones, subsurface water either flows out onto the ground surface as a **spring** or joins bodies of water such as streams, lakes, ponds, swamps, or the ocean. Groundwater discharge maintains the base flow of a stream. Pumping groundwater from a well also creates a point of discharge. The amount of time water takes to move through the ground to a discharge area depends on distance and rate of flow. It may take as little as a few days or as long as thousands of years.

spring A natural outlet for groundwater that occurs where the water table intersects the land surface.

aquifer A body of rock or regolith that is water saturated, porous, and permeable.

Where Groundwater Is Stored

When we wish to find a reliable supply of groundwater, we search for an **aquifer** (Latin for "water carrier"). An aquifer is not a body of water—it is a body of rock or regolith that is water saturated, as well as porous and permeable. Gravel and sand generally make good aquifers; many sandstones are also good aquifers, as are fractured or cavernous limestone and granite. An aquifer in which the water is free to rise to its natural level is called an **unconfined aquifer** (**Figure 11.19**). In a well drilled into an unconfined aquifer, the water will rise to the level of the surrounding water table. Bringing it to the surface requires a pump or a bucket.

A **confined aquifer** is overlain by impermeable rock units, called *confining layers*, or **aquicludes** (see Figure 11.19 and **Figure 11.20**). Common aquicludes are shale and clay layers, as well as unfractured, nonporous, impermeable bedrock. The water in a confined aquifer is held in place by the overlying impermeable unit, and its recharge zone may be many kilometers away at a higher elevation. If a well is drilled into the aquifer, the high water

aquiclude A layer of impermeable rock.

Aquifers, confined and unconfined • Figure 11.19

An unconfined aquifer is open to the atmosphere through pores in the rock and soil above the aquifer. In contrast, the water in a confined aquifer is trapped between impermeable rock layers.

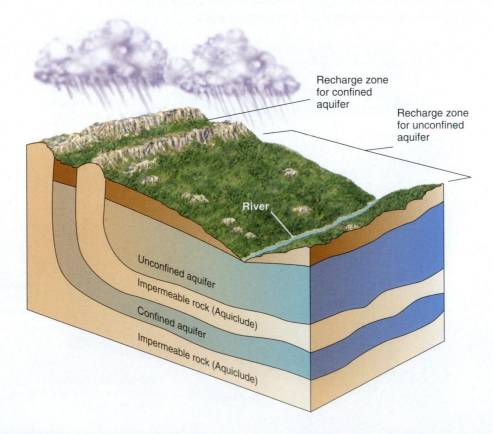

Artesian water • Figure 11.20

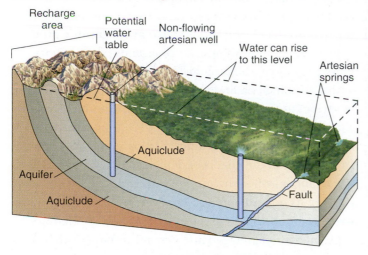

a. This diagram illustrates the geologic conditions that produce artesian wells or springs. Note that the water enters the aquifer at a higher elevation than the well. As a result, the water flows or gushes out of the well with positive water pressure.

b. This flowing artesian well in Letchworth State Park, New York, freezes during the winter to form a core of ice that sometimes grows as high as 20 m.

pressure due to the elevation of the recharge zone will cause the water to rise or even flow out of the well without having to be pumped. This is called an **artesian well**. A fault can also serve as a natural conduit for artesian water.

A change in permeability of rock at ground level may give rise to a spring, which is a natural analogue of an artesian well. Such a change may be due to the presence of an aquiclude (**Figure 11.21**), or it may happen along the trace of a fault (**Figure 11.20a**).

What causes springs? • Figure 11.21

a. This spring in the Grand Canyon is fed by water from the porous Redwall and Muav Limestones. These cavernous limestones are the water source for many springs. The impermeable shale unit beneath them, the aquiclude, is the Bright Angel Shale.

b. Water flows from a spring in a limestone aquifer underlain by an impermeable shale aquiclude.

Groundwater Depletion and Contamination

A well will supply water if it is deep enough to intersect the water table. As shown in **Figure 11.22**, a shallow well may become dry during periods when the water table is low, whereas a deeper well may yield water throughout the year. When water is pumped from a well, a **cone of depression** (a cone-shaped dip in the water table) will form around the well. In most small domestic wells, the cone of depression is hardly discernible. Wells pumped for irrigation and industrial uses, however, sometimes withdraw so much water that the cone may become very wide and steep and can lower the water levels in surrounding wells. When large cones of depression from pumped wells overlap, the result is regional depression of the water table.

If the rate of withdrawal of groundwater regularly exceeds the rate of natural recharge, the volume of stored water steadily decreases; this is called **groundwater mining**. It may take hundreds or even thousands of years for a depleted aquifer to be replenished. The results of excessive withdrawal include lowering of the water table, drying up of springs and streams, compaction of the aquifer, and subsidence, a decline in land surface elevation. Sometimes it is possible to recharge an aquifer by pumping water into it. In other cases, the effects of depletion may be permanent. For example, when an aquifer suffers **compaction**—that is, when its mineral grains collapse on one another because the pore water that held them apart has been removed—it is permanently damaged and may never be able to hold as much water as it originally held.

Year-round and seasonal wells • Figure 11.22

Seasonal changes affect the height of the water table. A well will produce year-round only if it extends into the year-round zone of saturation.

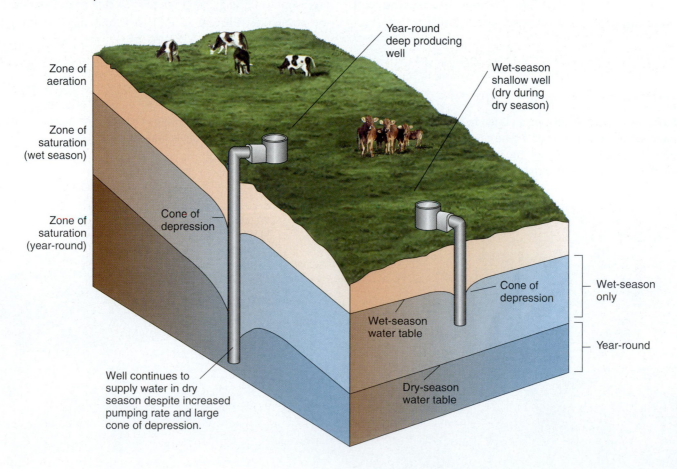

In Galveston, Texas, these homes had to be abandoned because pumping of groundwater lowered the level of the land, allowing seawater to flow in from the Gulf of Mexico. The snapshot in the foreground shows the same neighborhood, high and dry, years earlier.

Urban development can increase the rate of depletion of groundwater, not only by increasing the demand for water but also by increasing the amount of impermeable ground cover in the area.

When roads, parking lots, buildings, and sidewalks cover a recharge area, the rate of infiltration and groundwater recharge can be substantially reduced. Subsidence caused by withdrawal of groundwater (**Figure 11.23**) is a problem in many urban areas, such as Mexico City, Bangkok, and Venice. The weight of buildings also contributes to the compaction of compressible sediment.

Laws and policies relating to water rights are very complicated, and the application to groundwater is even more complicated than for surface water. Because groundwater is hidden from view, it is difficult to monitor its flow and regulate its use. If you drill a well into an aquifer underlying your property, are you entitled to withdraw as much water as you need from that well? Should you withdraw water only for your own purposes, or should you be permitted to withdraw the water and sell it elsewhere? What happens if withdraw-

ing the groundwater depletes the aquifer, and your neighbor's well runs dry? Similar problems arise when a landowner's actions cause an aquifer to become contaminated. In some jurisdictions, it is technically legal to contaminate groundwater; it becomes illegal only when the contaminated groundwater flows into a neighboring property. To make matters worse, because of the slow movement of groundwater, those responsible for the problem may be long gone before their actions begin to affect their neighbors.

Many of the types and sources of contaminants that affect surface water also cause groundwater contamination. Because of its hidden nature, however, groundwater contamination can be much more difficult to detect, control, and clean up than surface water contamination. The most common source of water pollution in wells and springs is untreated sewage. Agricultural pesticides and fertilizers, significant sources of surface water pollution, are also common contaminants of groundwater. Harmful chemicals leaking from waste disposal facilities can also infiltrate groundwater reservoirs and contaminate them. Probably the most serious groundwater contamination problem in

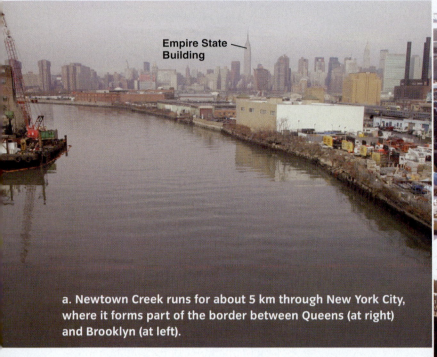

Empire State Building

a. Newtown Creek runs for about 5 km through New York City, where it forms part of the border between Queens (at right) and Brooklyn (at left).

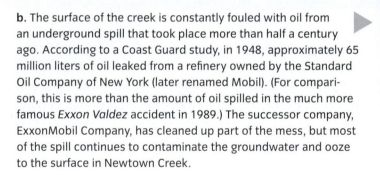

b. The surface of the creek is constantly fouled with oil from an underground spill that took place more than half a century ago. According to a Coast Guard study, in 1948, approximately 65 million liters of oil leaked from a refinery owned by the Standard Oil Company of New York (later renamed Mobil). (For comparison, this is more than the amount of oil spilled in the much more famous *Exxon Valdez* accident in 1989.) The successor company, ExxonMobil Company, has cleaned up part of the mess, but most of the spill continues to contaminate the groundwater and ooze to the surface in Newtown Creek.

North America is caused by leaking underground storage tanks at gas stations, refineries, and other industrial settings (**Figure 11.24**).

When Groundwater Dissolves Rock

In regions underlain by rock types that are highly susceptible to chemical weathering (see Chapter 7), groundwater creates extensive systems of underground caverns. In such areas, a distinctive landscape forms on the surface. **Karst topography**, named for the Karst region of the former Yugoslavia (now Slovenia), is characterized by many **sinkholes** (small, closed basins) and disrupted drainage patterns (**Figure 11.25**). Streams disappear into the ground and join the groundwater. Large springs form downstream, where the water table intersects the land surface, or along favorable routes of escape, such as faults. Karst is most typical of regions underlain by soluble carbonate rock (limestone or dolostone), although it can also occur in regions with extensive evaporite (salt) deposits. In carbonate terrains with karst topography, the rate of dissolution is faster than the average rate of erosion of surface materials by streams and mass wasting.

a. These sinkholes in Florida are part of a karst terrain. Most of Missouri and large parts of Kentucky, Tennessee, Texas, and New Mexico also have carbonate karst landscapes.

b. This is a karst terrain near Guilin, China. White limestone is visible in the pillars that remain after a cavern collapse created the valley in which the village is located. The pillars are riddled with caves and passageways.

Caves and Sinkholes

Caves and **caverns** are formed when circulating groundwater at or below the water table dissolves carbonate rock.

> **cave** An underground open space; a cavern is a system of connected caves.

The process begins with dissolution along interconnected fractures and bedding planes. A cave passage then develops along the most favorable flow route.

The development of a continuous passage by slowly moving groundwater may take up to 10,000 years, and the further enlargement of the passage by more rapidly flowing groundwater needed to create a fully developed cave system may take an additional 10,000 to 1 million years. Finally, the cave may become accessible to humans after the water table drops below the floor of at least some of the chambers. The deepest cave yet discovered in the United States is Lechuguilla Cave, in Carlsbad Caverns National Park in New Mexico, where there are more than 100 known caves (see *Amazing Places* on page 338).

In the parts of the cave that lie above the water table (in the vadose zone), groundwater continues to percolate downward, dripping from the ceiling to the floor. Calcium carbonate dissolved in the water precipitates out of solution and builds up beautiful icicle-like decorations on the cave walls, ceilings, and floor. These include **stalactites** (hanging from the ceiling) and **stalagmites** (projecting upward from the floor), as well as columns, draperies, and flowstones. In the saturated zone, below the water table, water movement is largely horizontal, creating tubular passages.

Caves are dissolution cavities that are closed to the surface or have only a small opening. In contrast, a sinkhole is a dissolution cavity that is open to the sky. Some sinkholes are formed very abruptly, when the roofs of caves collapse. Most sinkholes, though, develop much more slowly and less catastrophically, simply growing wider over time as the carbonate bedrock slowly dissolves.

CONCEPT CHECK   STOP

1. **What** lies above and below the water table?
2. **How** does permeability differ from porosity?
3. **What** is the geologic difference between an unconfined and a confined aquifer?
4. **What** is an artesian well, and how is it related to a spring?
5. **How** and **why** can ground subsidence be related to groundwater withdrawal?
6. **How** and **why** do caves and sinkholes develop in certain regions?

AMAZING PLACES

Lechuguilla Cave

In 1986, cavers (also known as spelunkers) discovered the deepest-known cave in the United States, called Lechuguilla Cave, in Carlsbad Caverns National Park. Its entrance had been known for decades, but it had been considered a dead end until cavers dug through the floor to a huge network of passages on the other side.

Lechuguilla Cave has now been explored to a depth of 475 m, and it has almost 160 km of mapped passages. It is as spectacular as it is deep, but it is closed to the public to preserve its unusual formations.

a. A "bush" made of fragile aragonite pokes out of a stalagmite made of calcite. Aragonite and calcite are polymorphs of calcium carbonate (see Chapter 2).

b. "Soda straws" reach down from the ceiling in this small chamber. Water flows down through the center of each straw. If water starts flowing down the outside, it will build a stalactite.

c. The origin of this rare formation, called "pool fingers," is a mystery, perhaps related to bacterial activity. The "fingers" crystallized in a pool of water and were left behind when the water retreated.

d. Gypsum crystals provide a clue to this cave's unusual history. Unlike most other limestone caves, which are formed by carbonic acid in rainwater, Lechuguilla formed from the bottom up. Hydrogen sulfide from lower-lying oil deposits percolated up into the groundwater and formed sulfuric acid, which dissolved away the rock. As the water table dropped, gypsum (calcium sulfate) deposits precipitated out of the acidic water.

Summary

1 The Hydrologic Cycle 308

- The water cycle, or hydrologic cycle, describes the movement of water from one reservoir to another in the hydrosphere as shown in the figure. The ocean is the largest reservoir, followed by the polar ice sheets. The largest reservoir of unfrozen fresh water is groundwater. Surface water bodies, the atmosphere, and the biosphere are much smaller water reservoirs.

The hydrologic cycle • Figure 11.1

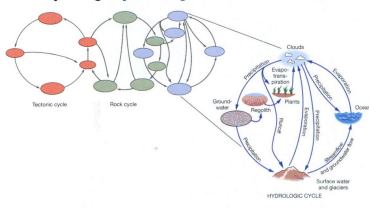

- The pathways or processes by which water moves from one reservoir to another include **evaporation**, **condensation**, **precipitation**, **transpiration**, **infiltration**, and **surface runoff**.

- Like the tectonic and rock cycles, the hydrologic cycle is a closed cycle of open systems. Because it is closed, the global hydrologic cycle maintains a mass balance.

2 How Water Affects Land 312

- **Streams** and rivers flow downslope along a clearly defined natural passageway, the **channel**. Interrelated factors that influence the behavior of a stream include the **gradient**, **discharge**, **load**, and velocity of the water.

- Streams create landforms through erosion and deposition. Straight, braided, and meandering channels, as shown in the diagram, and oxbow lakes are erosional landforms. Point bars, alluvial fans, **deltas**, and **floodplains** are depositional landforms made of recently deposited sediment, or **alluvium**.

Three kinds of streams • Figure 11.4

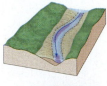

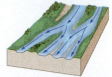

- Every stream is surrounded by its **drainage basin**, the total area from which water flows into the stream. The topographic "high" that separates adjacent drainage basins is a **divide**. Streams originate in a headwater region, where they collect water from a large number of smaller tributary streams. Water exits a stream system from the stream's mouth, which may empty either into a larger river, a lake, or the ocean.

- Lakes form in topographic basins created by faults, glacial debris, landslides, and lava flows, as well as in areas of poor drainage. Lakes are usually short-lived geologic features because they tend to disappear by erosion of the outlet or by silt deposition.

3 Water as a Hazard and a Resource 320

- A **flood** occurs when a stream's discharge becomes so great that it exceeds the capacity of the channel, causing the stream to overflow its banks. The risk of floods is increased by subsidence, and flooding can be exacerbated by human activity.

- Prediction of flooding is based on analysis of the frequency of occurrence of past events as shown in the diagram, and on real-time monitoring of storms using a hydrograph. River channel modifications made for purposes of flood control and protection, as well as for navigation and other purposes, are collectively known as channelization.

Predicting floods • Figure 11.14

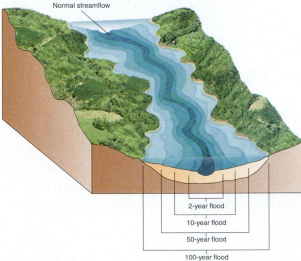

- Agriculture is by far the largest consumer of fresh water. Surface water can be transported from one drainage basin to another, but only at the risk of long-term environmental changes in the affected basins.

4 Fresh Water Underground 329

- **Groundwater** is subsurface water contained in spaces within bedrock and regolith. In the zone of aeration (unsaturated zone or vadose zone), water is present but does not completely saturate the ground. In the phreatic or saturated zone, all openings are filled with water. The top of the saturated zone is the **water table**.

- The rate of groundwater flow is dependent on the characteristics of the rock or sediment through which the water must move. **Porosity** is the percentage of the total volume of a body of rock or regolith that consists of open spaces. **Permeability** is a measure of how easily a solid allows fluids to pass through it. Groundwater in the saturated zone moves slowly by **percolation** through very small pores from areas where the water table is high to where it is lower.

- Groundwater **recharge** occurs when rainfall and snowmelt infiltrate and percolate downward to the saturated zone. **Discharge** occurs where subsurface water leaves the saturated zone and becomes surface water in a stream, lake, or **spring**.

- An **aquifer** is a body of permeable rock or regolith in the zone of saturation. An unconfined aquifer is in contact with the atmosphere through the pore spaces of overlying rock or regolith, while a confined aquifer lies between layers of impermeable rock, called **aquicludes** as shown in the diagram. The water pressure in a confined aquifer may be high enough to force the water partway or all the way to the surface when a well is drilled into it. Such a well is called artesian. Excessive withdrawal from an aquifer can cause the water table to drop, springs and streams to dry up, and regolith to compact and subside.

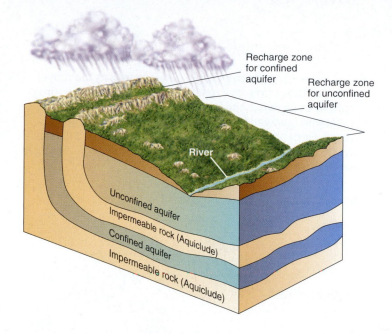

- Many of the types and sources of contaminants that affect surface water also cause groundwater contamination. These include untreated sewage, agricultural pesticides and fertilizers, and leaks or spills of chemicals from commercial and industrial sites.

- **Caves**, caverns, sinkholes, and karst topography are formed when rock—most commonly carbonate rock—is dissolved by circulating groundwater, creating underground cavities. Stalactites and stalagmites are built up from calcium carbonate precipitated from percolating groundwater.

Key Terms

- alluvium   316
- aquiclude   332
- aquifer   332
- cave   337
- channel   312
- condensation   308
- delta   316
- discharge (1)   312
- discharge (2)   331
- divide   318

- drainage basin   317
- evaporation   308
- flood   320
- floodplain   315
- gradient   312
- groundwater   329
- infiltration   309
- load   312
- percolation   331
- permeability   331

- porosity   330
- precipitation   309
- recharge   331
- spring   332
- stream   312
- surface runoff   309
- transpiration   308
- water table   329

Critical and Creative Thinking Questions

1. List as many ways as you can in which we depend on the availability of fresh water in our daily lives.

2. Investigate the ways in which water shortages, floods, or other processes associated with water (e.g., erosion and deposition of sediment) have affected human history. How did societies respond, and what effects did those responses have?

3. The concept of residence time is very important in geology. It applies not only to water but also to any substance that moves from one reservoir to another in the Earth system. What other substances, besides water, move around in the Earth system? Why would we want to monitor these substances and keep track of their residence times?

4. What evidence has been gathered for water-shaped landscapes on Mars or elsewhere in the solar system? How are these landforms similar to or different from their analogues on Earth?

5. Where does your community obtain its water supply? Is it a groundwater or a surface freshwater source? Is either the quantity or the quality of the water threatened?

6. Visit a stream before and after an intense rainfall. Observe the gradient, discharge, load, and velocity of the streamflow. What changes do you notice?

What is happening in this picture?

This 100-m-wide sinkhole opened up one day in Winter Park, Florida, and grew to the point where it eventually swallowed up part of a house, six commercial buildings, and the municipal swimming pool.

Think Critically

1. What could have caused this to happen?
2. What kind of rock do you think might underlie a location that is prone to this sort of collapse?

Self-Test

(Check your answers in Appendix D.)

1. On this diagram, label each stage of the hydrologic cycle (1 through 6) using the following terms:

   precipitation   cloud formation through condensation

   surface runoff   surface evaporation and transpiration

   infiltration   evaporation from the ocean

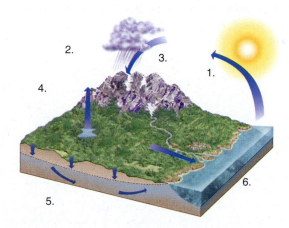

2. The diagram with question 1 depicts Earth's hydrologic cycle. Which reservoir in the hydrologic cycle holds *most* of the freshwater resources of the planet?

   a. the ocean

   b. ice sheets

   c. lakes

   d. groundwater

3. The _____ of a stream will have a large influence over channel development and the evolution of associated land-forms.

   a. discharge

   b. gradient

   c. sediment load

   d. All of the above answers are correct.

4. _____ channels will form in streams when there is a low slope gradient and large and variable sediment load.

   a. Straight

   b. Meandering

   c. Braided

5. Lakes are ephemeral features that disappear by _____.

   a. the accumulation of inorganic sediment carried in by streams

   b. the accumulation of organic matter produced by plants within the lake

   c. gradually being drained by stream outlets eroding to lower levels

   d. All of the above answers are correct.

6. On this diagram, label each of the landforms associated with a meandering stream system:

   alluvium   cutoff

   oxbow lake   meander neck

   floodplain   natural levees

7. A(n) _____ is the total area from which water flows into a stream.

   a. alluvial fan

   b. braided channel

   c. water cycle

   d. drainage basin

8. Urban development can lead to increased risk of flooding by _____.

   a. compressing underlying sediment, causing subsidence

   b. increasing surface runoff

   c. channeling runoff more quickly to rivers through storm drains

   d. All of the above answers are correct.

9. There is a _____ chance of a "50-year flood" occurring in any given year.

   a. 1%

   b. 2%

   c. 5%

   d. 10%

   e. 50%

10. Though the proportions vary from one region to another, what accounts for the greatest demand for water globally?

    a. industry

    b. domestic use

    c. crop irrigation

11. The _____ controls the porosity of an aquifer.

    a. size of the particles in the material

    b. shapes and uniformity of the material

    c. amount of cementing agent in the pore space of the material

    d. All of the above factors control the porosity of an aquifer.

12. A(n) _____ is an underground reservoir of water that is overlain by impermeable rock units.

    a. aquiclude

    b. aquifer

    c. confined aquifer

    d. unconfined aquifer

13. In a well drilled into a(n) _____, the water will rise to the level of the surrounding water table.

    a. aquiclude

    b. aquifer

    c. confined aquifer

    d. unconfined aquifer

14. Placing a well in a position where Earth's surface intersects the water table will result in a well that does not require pumping to produce water.

    a. True

    b. False

15. On this diagram, label each groundwater feature using the following terms:

dry-season water table	additional wet-season saturation
wet-season water table	year-round saturation
zone of saturation	

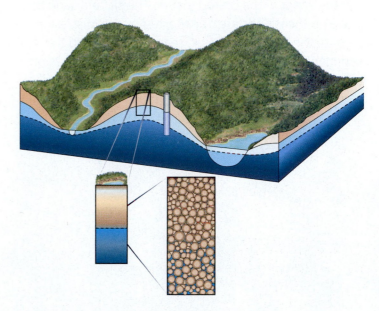

THE PLANNER ✓

Review the Chapter Planner on the chapter opener and check off your completed work.

The Ocean and the Atmosphere

Sculpted by winds and waves, Louisiana's Chandeleur Islands are some of the many barrier islands that line the coast of the United States. The silt and sand that make up the islands are transported by the Mississippi River to the Gulf of Mexico. Ocean currents build the sediment into sandbars and spits. Hurricanes occasionally carve channels through the islands or wash sand over the top, creating dunes on the seaward side (left in this photo) and flat, marshy ground on the landward side (right). Like other barrier islands, the Chandeleurs and their wetlands protect the shore against the battering of storm waves, filter and store groundwater, and provide habitat for countless species.

Barrier islands are constantly reshaped, with the erosive effects of weather and waves balanced by deposition of new sediment. Human intervention can alter this balance. The Chandeleurs have been eroding for half a century, as levees altered the flow of the Mississippi River and deprived the islands of sediment.

In 2005, Hurricane Katrina struck a devastating blow, submerging the entire north end of the island in the foreground of this pre-2004 photo. The islands lost 50% of their overall area (inset). In 2010, the islands were struck again when the BP Deepwater Horizon exploded, spilling 4.9 million barrels of oil. Some of this oil washed up onto the beaches of the Chandeleur Islands. We will explore the fate of the oil, and its impacts on the Gulf and the Chandeleur Islands, later in this chapter.

CHAPTER OUTLINE

CHAPTER PLANNER ✔

- ❑ Study the picture and read the opening story.
- ❑ Scan the Learning Objectives in each section.
  p. 346 ❑   p. 352 ❑   p. 366 ❑   p. 374 ❑
- ❑ Read the text and study all visuals. Answer any questions.

Analyze key features

- ❑ Geology InSight, p. 346 ❑   p. 366 ❑
- ❑ What a Geologist Sees, p. 354 ❑
- ❑ Case Study, p. 363 ❑
- ❑ Process Diagram, p. 369 ❑
- ❑ Amazing Places, p. 364 ❑
- ❑ Stop: Answer the Concept Checks before you go on:
  p. 351 ❑   p. 365 ❑   p. 374 ❑   p. 379 ❑

End of chapter

- ❑ Review the Summary and Key Terms.
- ❑ Answer the Critical and Creative Thinking Questions.
- ❑ Answer What is happening in this picture?
- ❑ Complete the Self-Test and check your answers.

The Ocean

LEARNING OBJECTIVES

1. **Explain** where Earth's ocean water came from and what controls the shape and distribution of ocean basins.

2. **Summarize** the processes that add and remove salts from seawater.

3. **Identify** the ocean's three layers and describe the differences between them.

4. **Describe** the overall pattern of surface and deep-water currents.

I n Chapter 11 we introduced the hydrologic cycle and described the events in this cycle that take place on land. If we stopped there, we would be left with a rather distorted view of the hydrologic cycle as a whole. The **world ocean** is by far Earth's largest reservoir of water. The atmosphere, a much smaller reservoir than either the ocean or fresh water on land, also plays a critical role in redistributing Earth's water. In fact, it is best to think of the ocean and the atmosphere as a linked system. Atmospheric circulation drives ocean waves and currents; in turn, the ocean regulates the temperature and humidity of the lower atmosphere. The atmosphere is also the conveyor that transfers water vapor from the ocean to the land.

We will discuss the atmosphere and ocean individually, and we will then consider how they are linked as one great system.

Ocean Basins

Water has been present on Earth's surface since its earliest days. In fact, we do not have any geological record of a time before water. The oldest rock so far discovered on Earth (about 4.0 billion years old) is gneiss that was once sedimentary strata. These strata were deposited in water and are similar to strata being deposited today. We can be reasonably certain, as a result, that the ocean formed sometime between 4.56 billion years ago (when Earth was formed) and 4.0 billion years ago.

It is not as certain where the water in the ocean came from (**Figure 12.1**). One hypothesis is that the water originated in Earth's interior. A more recent hypothesis suggests that the water arrived from outer space via cometary impacts. Possibly both processes occurred at the same time.

Geology InSight

Where did the water come from?

Scientists do not fully agree on where Earth got its water. Some water was present from the beginning, trapped inside the planet along with other volatile elements among the materials from which Earth was formed. In the first few hundred million years after the planet formed, water was released in the form of steam from volcanoes. Early in Earth history, though, the surface of the planet was too hot for liquid water to persist.

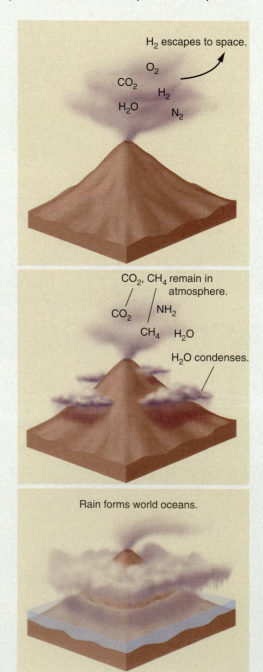

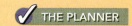

Possible external sources for Earth's water

Additional water may have been added to Earth's chemical mix by comets coming in from the outer regions of the solar system after the planet had already formed. As the surface slowly cooled, liquid water began to be retained in low-lying basins.

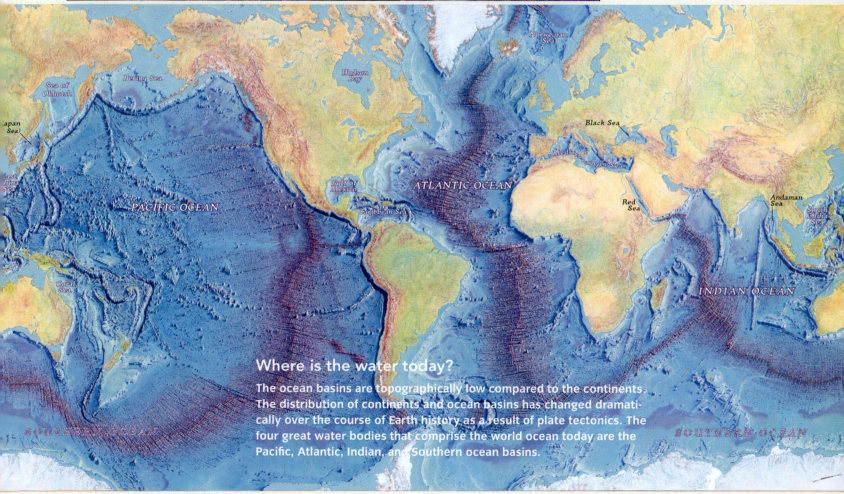

Where is the water today?

The ocean basins are topographically low compared to the continents. The distribution of continents and ocean basins has changed dramatically over the course of Earth history as a result of plate tectonics. The four great water bodies that comprise the world ocean today are the Pacific, Atlantic, Indian, and Southern ocean basins.

Before we consider the characteristics of the water in the ocean, let's have a look at the "containers" that hold this water. Most of the water on the surface of our planet is contained in four huge interconnected ocean basins (**Figure 12.1c**). The Pacific, Atlantic, and Indian oceans are connected with the Southern Ocean, the body of water that encircles Antarctica. (The Arctic Ocean is considered a northern extension of the Atlantic Ocean.) Collectively, these four bodies of water, together with a number of smaller ones, such as the Mediterranean Sea, Hudson Bay, and the Persian Gulf, cover 71% of Earth's surface.

The ocean basins are topographically low, separated by the continents, which are topographically high. Recall from Chapter 4 that the ocean basins are underlain by the relatively dense rock, basalt, whereas the continents consist of relatively less-dense rock that is approximately granitic in composition. The shape of the ocean basins is controlled by plate tectonics. An ocean begins as a rift, widening into a full oceanic basin as plates diverge along midocean ridges. On the other hand, ocean basins can close and disappear altogether when two continents converge and collide.

From subduction zone trenches—the deepest locations on our planet's surface—to the chains of volcanic mountains that snake the lengths of entire ocean basins, to the flat expanses of the abyssal plains, the ocean floor is characterized by a wide variety of dramatic topographic features, as shown in Figure 12.1c. The volcanic island of Mauna Kea extends 10,203 meters from its base on the ocean floor to its summit, making it the tallest mountain on Earth (considerably taller than Mount Everest, which at 8,840 m above sea level is the highest location on Earth).

The Composition of Seawater

If you have ever swum in the ocean, you know that seawater contains a lot of salt, which makes it not only unpalatable but also dangerous for human consumption. (If you tried living on salt water, you would become dehydrated, as your body would try to rid itself of the excess of ions such as sodium and potassium.)

The **salinity** of seawater ranges between 3.3 and 3.7%. This may seem quite small, but it is

> **salinity** A measure of the salt content of a solution.

about 70 times the salinity of tap water. Not surprisingly, when seawater is evaporated, more than three-quarters of the dissolved matter that remains behind is sodium chloride—that is, table salt. Seawater contains most of the other natural elements as well, but it contains many of them in such low concentrations that only extremely sensitive analytical instruments can detect them.

The elements dissolved in seawater come from several sources. Chemical weathering of rock releases soluble materials such as salts of sodium, potassium, and sulfur. The soluble compounds are leached out of the weathered rock, and become part of the dissolved load in river water flowing to the sea. Volcanic eruptions, both on land and beneath the sea, also contribute soluble compounds via volcanic gases and hot springs. The "black smoker" vents found at spreading centers (see Chapter 4) also add dissolved minerals to the water. Two other processes, evaporation of surface water and freezing of seawater, tend to make seawater saltier because they remove fresh water while leaving the salts behind.

Why, then, doesn't the sea continue to become saltier and saltier? The answer is that it is an open system and constantly receives fresh water from precipitation and river flow (**Figure 12.2**). Also, aquatic plants and animals withdraw some elements, such as silicon, calcium, and phosphorus, to build their shells or skeletons. Other elements precipitate out in mineral form and settle to the sea bottom. All of these processes balance each other, keeping the composition of seawater essentially unchanged. This is the problem that John Joly ran into when he tried to use the salinity of seawater to estimate the age of Earth (Chapter 3); he assumed, incorrectly, that the salinity of seawater had been changing more or less continuously since the planet was formed.

Layers in the Ocean

The salinity and temperature of seawater together control its density: Cold, salty water is dense and will sink, whereas warm, less salty water is less dense and will rise. Figure 12.2 shows this very clearly; the warm and less salty water from the Amazon River water stays on top of the cold, salty Atlantic Ocean water for hundreds of kilometers.

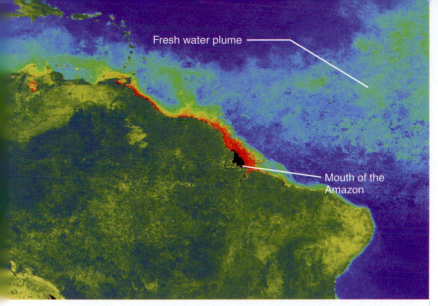

Fresh water plume

Mouth of the Amazon

A river in the ocean • Figure 12.2

These satellite photos use false colors to represent the amount of living plant material (chlorophyll) at the sea surface, where fresh water flows from the Amazon River into the Atlantic Ocean. The freshwater "plume" (shown here in blue-green) extends hundreds of kilometers off the coast of Brazil. (Source of photo: SeaWiFS Project and GeoEye, Scientific Visualization Studio, Goddard Space Flight Center, NASA.)

Ocean scientists have discovered that there are three major layers, or zones, in the ocean, in which the density of the water differs. The differences are caused by changes in both temperature and salinity, with temperature being the major factor (**Figure 12.3**).

Water in the **surface layer** is the warmest because it is directly exposed to sunlight. The depth to which the surface layer extends varies geographically; in tropical regions, where the sun shines directly on the ocean surface, the surface layer is thicker than in polar regions, where the sun's energy is less intense and the air temperature is very cold. Below the surface layer is a transitional zone called the **thermocline**, in which the temperature drops rapidly to just above the freezing point of water. In the **deep layer**, extending to the ocean floor, the water temperature is cold and nearly constant. The layers ordinarily do not mix, but in certain places—called **upwellings** and **downwellings**—they do. Those locations are very important for driving the global circulation of seawater, as you will learn.

Structure of the ocean • Figure 12.3

a. The three major density zones in the ocean are the surface layer, the transitional zone or thermocline, and the deep zone.

b. Change of water temperature with depth in the ocean

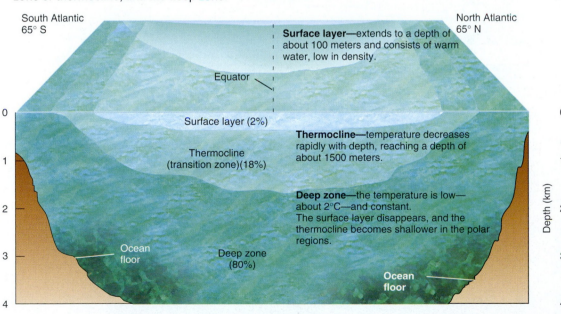

South Atlantic 65° S

North Atlantic 65° N

Surface layer—extends to a depth of about 100 meters and consists of warm water, low in density.

Equator

Surface layer (2%)

Thermocline—temperature decreases rapidly with depth, reaching a depth of about 1500 meters.

Thermocline (transition zone)(18%)

Deep zone—the temperature is low— about 2°C—and constant. The surface layer disappears, and the thermocline becomes shallower in the polar regions.

Ocean floor

Deep zone (80%)

Ocean floor

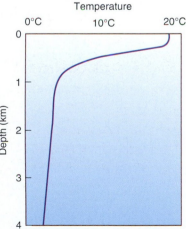

Ocean Currents

In 1492, when Christopher Columbus set sail across the Atlantic Ocean in search of China, he took an indirect route. On the outward voyage, he sailed southwest toward the Canary Islands and then west on a course that carried him to the Caribbean Islands, where he first sighted land. In choosing this course, he was following the path of the prevailing winds and surface ocean currents instead of fighting the **westerly** winds (i.e., flowing from the west) and currents at 40°N latitude. On the return trip, he sailed a more northerly route in order to take advantage of the westerly winds and currents at that latitude. His route is shown by the dotted black arrows in the top portion of **Figure 12.4**.

Surface ocean currents, like those Columbus followed, are set in motion by the prevailing winds. The surface of the ocean creates friction when the wind passes over it, with the result that the wind literally "drags" the surface water along. For this reason, these currents only extend to about 50 to 100 m deep and are confined to the surface zone.

The pattern of water flow in the deep ocean is quite different from that of surface currents and needs to be pictured in three dimensions (**Figure 12.5**). Differences in temperature and salinity—which change the water's density—drive this circulation pattern, and the current directions are controlled by the Coriolis force (discussed later in the chapter). Water off the coast of Greenland is very cold; it is also salty because it has had fresh water removed from it to form sea ice. This water is therefore dense, and it sinks to form a large mass of water called the **North Atlantic Deep Water**.

Surface ocean currents • Figure 12.4

Sailors have long been aware of the prevailing currents in the ocean. Christopher Columbus used the North Atlantic current to good advantage in 1492. Surface ocean currents in the northern hemisphere tend to curve to the right (clockwise), whereas in the southern hemisphere, they curve to the left (counterclockwise).

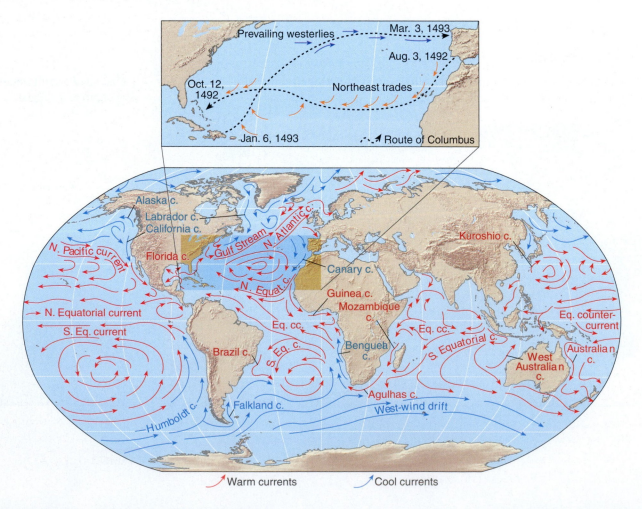

The ocean conveyor belt • Figure 12.5

a. As in a conveyor belt, the deep ocean currents of the thermohaline circulation flow in the opposite direction from the warmer surface currents. Vertical motions and differences in temperature, salinity, and density drive the conveyor. Cold, salty water sinks in a few locations near Antarctica and Greenland. These cold, dense water masses combine and flow into the Pacific, where they eventually well up, becoming warmer and fresher. It takes about 1000 years for water to complete one circuit of this global flow pattern.

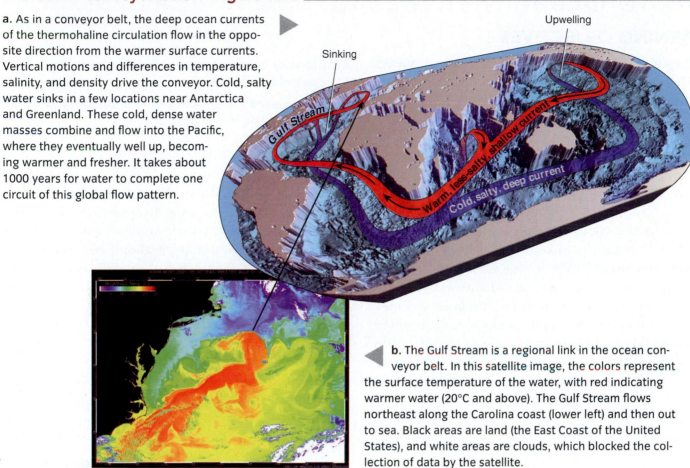

b. The Gulf Stream is a regional link in the ocean conveyor belt. In this satellite image, the colors represent the surface temperature of the water, with red indicating warmer water (20°C and above). The Gulf Stream flows northeast along the Carolina coast (lower left) and then out to sea. Black areas are land (the East Coast of the United States), and white areas are clouds, which blocked the collection of data by the satellite.

The mass of cold, salty, dense water from the North Atlantic moves slowly along the bottom and mixes in the South Atlantic with another mass of cold, salty, dense water, known as the **Antarctic Bottom Water**. The current of combined water masses flows across the seafloor and northward into the Pacific. Eventually, in the northern Pacific, the cold water wells up, becoming shallower and warmer.

The shallower, warmer currents eventually flow back into the Atlantic and toward the north, where they are largely responsible for maintaining a mild climate in Europe. The warm **Gulf Stream** current is part of this northward flow. By the time the water reaches the North Atlantic, it begins to cool and sink, creating a downwelling. Because of its continuous, repetitive nature, this circuit has been called "the great ocean conveyor belt." Its scientific name is the **thermohaline circulation** because it is driven by variations in ocean temperature (*thermo*), and salinity (*haline*).

thermohaline circulation The deep-ocean global "conveyor belt" circulation, driven by differences in water temperature, salinity, and density.

CONCEPT CHECK STOP

1. **What** are two possible origins of Earth's water?

2. **What** are two processes that add salts to and two processes that remove salts from seawater?

3. **What** is the name of the zone in the ocean that is transitional between the surface and deep zones?

4. **How** do surface currents differ from deep-ocean currents?

Where Ocean Meets Land

LEARNING OBJECTIVES

1. **Explain** why worldwide sea levels do not always stay the same.

2. **Explain** what causes tides.

3. **Describe** how waves and longshore currents transport sediment.

4. **Identify** common types of shorelines and coastal landforms.

5. **Summarize** the process of reef building and the status of the world's coral reefs.

The majority of the world population lives within 100 kilometers of a coastline. This reflects our dependence on the ocean, especially the rich resources in coastal zones. However, the concentration of large numbers of people in coastal areas means that the coastal environment must absorb the impacts of a wide range of human activities. It also means that human vulnerability to hazards can be particularly high in coastal zones; for example, infrequent events such as large storms can cause major loss of life and damage to property.

If you visit almost any coastline on two occasions one year apart, you will see changes. Sometimes the changes are small, but often they are substantial. Large sand dunes may have shifted. Sand may have built up behind barriers or may have been eroded away. Steep sections of coastline may have collapsed. Channels may have broken through from the sea to lagoons on the landward side, where there were no channels before. The energy driving these continual changes comes from storms and the waves that accompany them.

Changes in Sea Level

Water plays a powerful role in shaping coastlines. Before describing the landforms that result from the water's action, we will describe the processes themselves, which operate on very different time scales. On a very long time

The Bering land bridge • Figure 12.6

a. About 20,000 years ago, during the most recent ice age, sea levels were about 120 m lower than they are today. This had important geological ramifications, including the emergence of a "land bridge" between Asia and North America. Because there is no definitive evidence of humans in America before that time, it has been suggested that humans migrated across the "land bridge" into the western hemisphere. The green regions are above sea level.

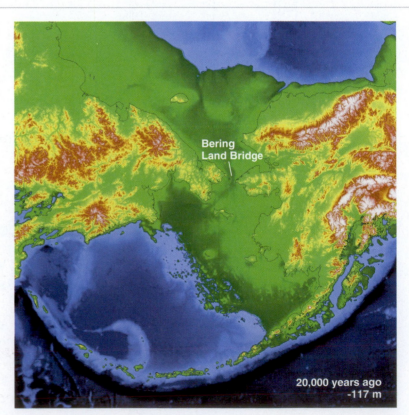

Bering Land Bridge

20,000 years ago
-117 m

scale, over many thousands of years, sea level can rise or fall by hundreds of meters. On a shorter time scale—twice a day—tides produce changes in water level that can amount to several meters in certain places. On the shortest time scale, ocean waves constantly stir up the sediment along the coast, and, especially during storm season, waves batter the coastline and can produce dramatic effects.

Global volume changes

One of the most important factors in determining the long-term position of the shoreline is the volume of the ocean. This in turn is strongly tied to global climatic change. When the climate warms, water stored on continents in glaciers and ice caps melts and returns to the sea. This causes a worldwide rise in sea level. Conversely, during cold climatic periods, glaciers and ice caps expand, and water is withdrawn from the ocean and stored on land (in the form of ice); this results in a drop in sea level. Currently, global warming is suspected to be responsible for a worldwide rise in sea level of approximately 2.4 millimeter/year. Such a trend is nearly imperceptible on the scale of a human lifetime, but over geological time it may account for great change in the position of a shoreline (**Figure 12.6**).

Tides

From a human perspective, a very obvious change in sea level occurs due to the effect of the **tides**. The gravitational attraction of the Moon makes ocean water bulge toward the Moon on the side of Earth nearest the Moon. There is also a bulge away from the Moon on the opposite side, whose origin is the inertial force (the same force that pulls the string of a yo-yo tight if you swing it around your finger). The Sun also affects the tides, but because it is so much farther away than the Moon, its tide-producing force is only about half as strong.

To visualize how tides work, consider the tidal bulges oriented with their maximum height lying along a line running through the center of Earth and the center of the Moon. Whereas Earth rotates around its axis, the tidal bulges remain stationary, opposite the Moon. Thus, any given coastline will move eastward through both tidal bulges each day. Every time a landmass encounters a tidal bulge, the water level along the coast rises. As Earth rotates, the coast passes through the highest point of the tidal bulge (**high tide**) and the water level begins to fall until it passes through the lowest point (**low tide**). Along most coastlines, two high tides and two low tides are observed each day.

> **tides** A regular, daily cycle of rising and falling sea level that results from the gravitational action of the Moon, the Sun, and Earth.

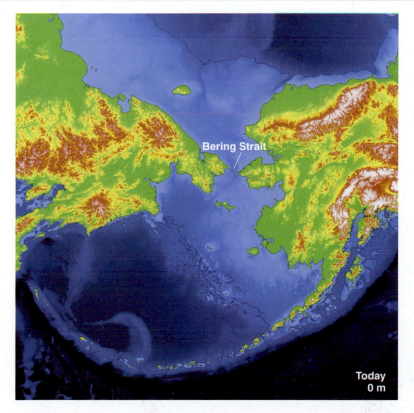

b. Today Asia and America are separated by an 85-km-wide stretch of sea called the Bering Strait.

Bering Strait

Today
0 m

High-tide water level

What Causes Tides?

The tidal range in the Bay of Fundy in eastern Canada is one of the largest in the world (Figure a). These fishing boats are grounded at low tide. At high tide, the water will rise up to the level where the posts in the photograph change color.

A geologist visiting the Bay of Fundy would realize that the Moon's gravitational attraction and the inertia of the rotating Earth–Moon system combine to produce this tidal variation. As shown in this greatly exaggerated sketch (Figure b), the two forces stretch Earth into an oblong shape, with bulges directed toward and away from the Moon. The bulges remain essentially stationary while Earth rotates through them, creating two high tides and two low tides per day.

The geologist would also know that local topography strongly affects both the timing and size of the tides in any particular place. The tides in the Bay of Fundy are so high because the frequency of the tides closely matches the natural period of oscillation of water in the long, thin bay. It is like sloshing the water in a bathtub: If you move back and forth at just the right frequency, the water will spill over the sides. Other places in the world have almost no tides, or even once-a-day tides instead of twice-a-day tides, depending on the local conditions.

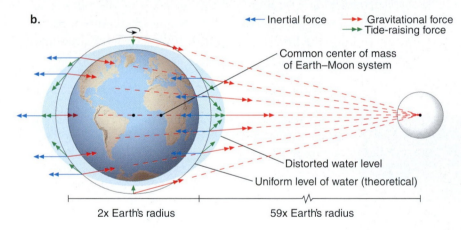

b.

◄◄— Inertial force   ►►— Gravitational force
►►— Tide-raising force

Common center of mass of Earth–Moon system

Distorted water level
Uniform level of water (theoretical)

2x Earth's radius   59x Earth's radius

Think Critically

Does Earth cause tides on the Moon?

In the open sea, the effect of tides is small. However, the shape of a coastline can greatly influence the tidal **runup height**, the highest elevation reached by the incoming water. Narrow openings into bays, rivers, estuaries, and straits can amplify normal tidal fluctuations. At the Bay of Fundy in Nova Scotia, a **tidal range** (the difference between high and low tide) of up to 16 m has been reported (see *What a Geologist Sees*). The bay is very long and narrow, which causes the incoming tide to rush in, forming a steep-fronted wall of water called a

tidal bore that can move faster than a person in average physical condition can run.

In most places, though, the tides act too slowly to have much geological impact. Perhaps their most important effect is global in scale. Tides gradually slow down Earth's rotation because the Earth's tidal bulge tends to get ahead of the Moon. This means that the Moon pulls slightly backward on it. The effect is very small, but significant on a geological time scale. In the early Phanerozoic Eon, the day was about 22 hours long, and there were nearly 400 days in a year.

Waves

Like surface ocean currents, ocean waves receive energy from winds. The size of the largest wave depends on how fast, how far, and how long the wind blows across the water surface. A gentle breeze blowing across a bay may ripple the water or form low waves less than 1 m high. In contrast, storm waves whipped up by intense winds over several days may tower over ships unfortunate enough to be caught in them (**Figure 12.7**). The waves generated by a storm can travel great distances. For example, surfers in California keep an eye on the weather report for storms far away in the tropical Pacific because they know that a good storm creates a "swell" that will arrive at their shores several days later.

Wave action along coastlines As a wave approaches the shore, it undergoes a rapid transformation. In deep water, a buoy or a parcel of water makes circular loops

Rogue waves generated by storms • Figure 12.7

Aug 20, 1996
Overhead view of ocean surface (satellite radar)

a. In a storm, the highest waves can wash over the deck of a ship. For centuries, sailors have told tales about rogue waves as high as a 10-story building, which are big enough to capsize a ship. For years scientists did not believe such stories.

b. However, evidence such as the radar image, taken from a satellite in 1996, has removed all doubt. In the center of the white bar, you can see a dark trough that goes down more than 10 m below sea level, next to a bright crest that reaches more than 15 m above sea level. A ship caught in the trough would have seen a wall of water more than 25 m high (inset). Note that the rogue wave is quite localized; all around it are waves that are less than 10 m high from trough to crest.

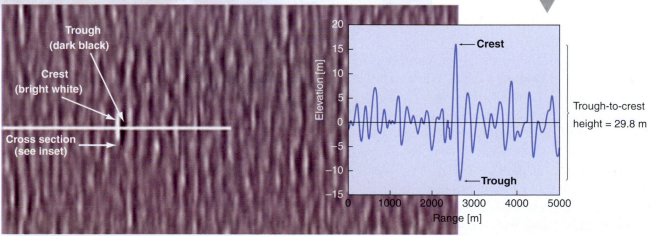

Trough (dark black)

Crest (bright white)

Cross section (see inset)

Trough-to-crest height = 29.8 m

−15 −10 −5 0 5 10 15
Elevation [m]

How waves change near the shore • Figure 12.8

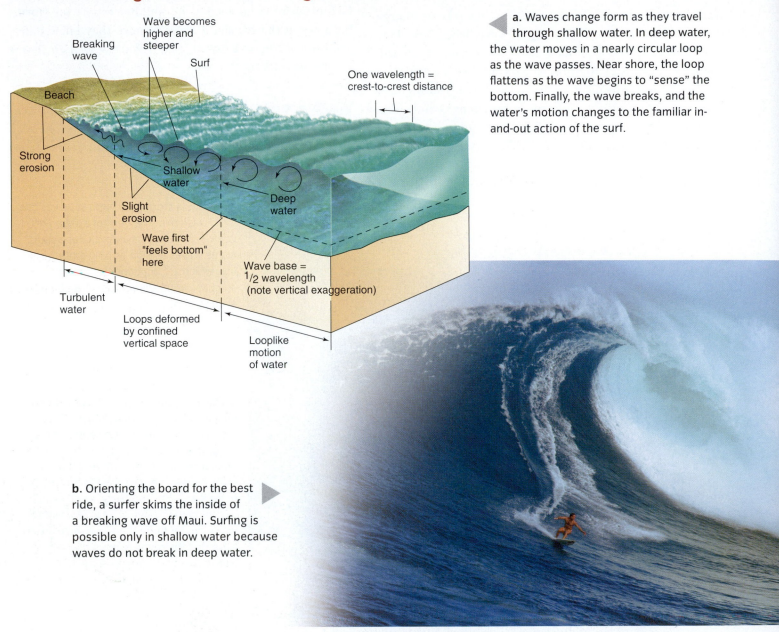

a. Waves change form as they travel through shallow water. In deep water, the water moves in a nearly circular loop as the wave passes. Near shore, the loop flattens as the wave begins to "sense" the bottom. Finally, the wave breaks, and the water's motion changes to the familiar in-and-out action of the surf.

Wave becomes higher and steeper
Breaking wave
Surf
Beach
One wavelength = crest-to-crest distance
Strong erosion
Shallow water
Deep water
Slight erosion
Wave first "feels bottom" here
Wave base = 1/2 wavelength (note vertical exaggeration)
Turbulent water
Loops deformed by confined vertical space
Looplike motion of water

b. Orienting the board for the best ride, a surfer skims the inside of a breaking wave off Maui. Surfing is possible only in shallow water because waves do not break in deep water.

as the waves pass by. But in shallower water, the circular loops become flatter (**Figure 12.8a**). Where water depth becomes less than half a wavelength, the increasingly shallow seafloor interferes with wave motion and distorts the wave's shape. The wave height increases, and the wavelength decreases. Now the front of the wave is in shallower water and is also steeper than the rear. Eventually, the front becomes too steep to support the advancing wave. As the rear part continues to move forward, the wave collapses, or breaks—hence the term **breakers**, referring to a line of breaking waves (**Figure 12.8b**).

When a wave breaks, the motion of the water instantly becomes turbulent, like that of a swift river. **Surf** is found in the **surf zone** between the line of breakers and the shore. Each wave finally dashes against the rocks or rushes up a sloping beach until its energy is expended; then the water flows back toward the open sea. Water that has piled up against the shore returns seaward in an irregular and complex way, partly as a broad sheet along the bottom and partly in localized narrow channels known as **rip currents** that can sweep unwary swimmers out to sea. Experienced swimmers know that a rip current is not wide; by swimming sideways to the current, rather than against it, they can get out of trouble.

surf The "broken," turbulent water found between a line of breakers and the shore.

A wave approaching a coast generally does not encounter the bottom simultaneously all along its length. When a segment of a wave touches the seafloor, that part of the wave slows down. Gradually the trend of a wave becomes realigned to parallel the contours of the seafloor. This process, **refraction**, causes a series of waves approaching a shoreline at an angle to change their direction of movement. (Note that the word *refraction*, in this context, means "bending"—just as it does in the context of seismic-wave refraction or light-wave refraction.)

Erosion and transport of sediment by waves

Surf is a powerful erosive force because it possesses most of the original energy of the waves that created it. Wave erosion takes place not only at sea level but also below sea level and—especially during storms—above sea level. In the surf zone, rock particles are worn down, becoming smoother, rounder, and smaller. At the same time, through continuous rubbing and grinding with these particles, the surf scours and deepens the bottom. Onshore, the surf acts like a saw cutting horizontally into the land. Its energy is eventually consumed in turbulence, in friction at the bottom, and in the movement of sediment thrown up from the bottom. Two processes, one on land and one underwater, transport the sediment: the longshore current and beach drift.

Most waves reach the shore at an oblique angle (**Figure 12.9a**). Part of the force of an incoming wave is oriented perpendicular to the shore; this produces the crashing surf. Another component of the wave motion is oriented parallel to the shore. The parallel component sets up a **longshore current**. While surf erodes sediment at the shore, the longshore current moves the sediment longitudinally in the surf zone.

> **longshore current**
> A current within the surf zone that flows parallel to the coast.

Longshore current and beach drift • Figure 12.9

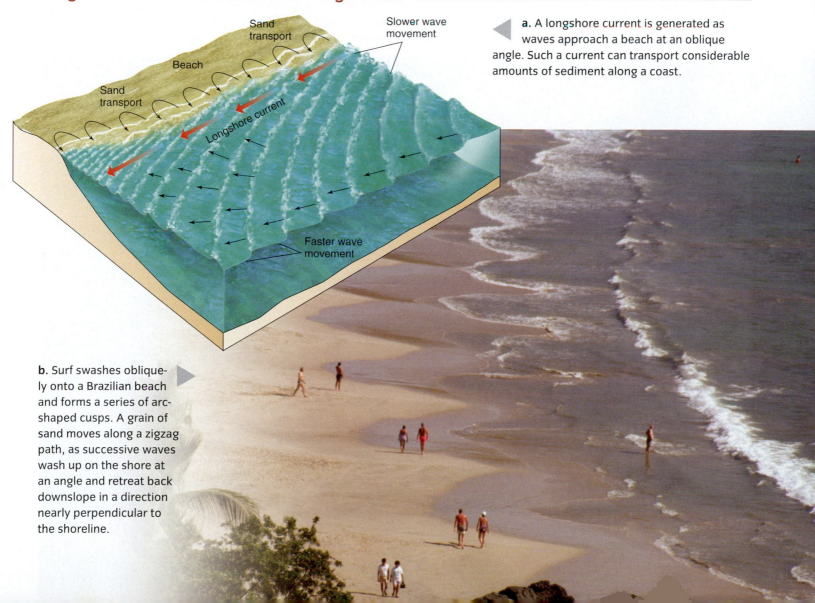

Sand transport

Beach

Sand transport

Longshore current

Slower wave movement

Faster wave movement

a. A longshore current is generated as waves approach a beach at an oblique angle. Such a current can transport considerable amounts of sediment along a coast.

b. Surf swashes obliquely onto a Brazilian beach and forms a series of arc-shaped cusps. A grain of sand moves along a zigzag path, as successive waves wash up on the shore at an angle and retreat back downslope in a direction nearly perpendicular to the shoreline.

Meanwhile, on the exposed beach, incoming waves produce an irregular pattern of water movement along the shore. Because waves generally strike the beach at an angle, the **swash** (uprushing water) of each wave travels obliquely up the beach before gravity pulls the water down (**Figure 12.9b**). This zigzag movement of water carries sand and pebbles first up and then down the beach slope. Successive movements of this type gradually transport sediment along the shore, causing **beach drift**. The greater the angle of waves to the shore, the greater the rate of drift. Marked pebbles have been observed to drift along a beach at a rate of more than 800 m per day.

The processes of erosion and deposition along a coast are anything but steady. A single winter storm can erode cliffs or beaches more

beach drift The movement of particles along a beach as they are driven up and down the beach slope by wave action.

than a full year's worth of ordinary surf. The balance between erosion and deposition can change with the seasons. Along parts of the Pacific coast of North America, winter storm surf tends to carry away fine sediment, and the beach becomes narrow and steep. In calm summer weather, fine sediment drifts in and the beach assumes a gentler profile.

Human actions can also shift the balance between erosion and deposition, as you learned at the beginning of this chapter in the case of the Chandeleur Islands. When a seawall interrupts the flow of sediment along the coastline, the beach builds up on one side of the seawall but erodes away on the other side because the longshore current no longer brings in enough material to replenish the eroded sediment (**Figure 12.10**).

Engineering the shoreline • Figure 12.10

In Ocean City, Maryland (right), the construction of a jetty has caused sand to accumulate and make the beach wider. Just south of Ocean City, Assateague Island in Virginia (left) has been deprived of the sand that the longshore current would have deposited on its shore. Thus the beach on Assateague is retreating inland.

Ocean City

Direction of longshore current

Jetty

Sand accumulates on the up-current side of the jetty

Assateague

Wave-cut benches • Figure 12.11

The coast of New Zealand at Tongue Point has two terrace-like landforms called **benches**. They are both former seafloor and were elevated above sea level by two stages of tectonic uplift.

In this photograph you can see two other common features of rocky coasts: **headlands** that jut out into the sea, and **pocket beaches** that form in the bays between the headlands.

Pocket beach

Upper bench

Headland

Lower bench

Shorelines and Coastal Landforms

The end result of the constant interplay between erosional and depositional forces along coastlines is a wide variety of shorelines and coastal landforms. Their forms depend on the geological processes at work, the susceptibility to erosion of coastal rock, and the length of time these processes have been operating. Changes in sea level can also influence the development of coastal features. Many coastal landforms show clear signs of different sea levels at different times in the past (**Figure 12.11**).

Despite the variability of coasts and shorelines, three basic types are most common: the rocky (cliffed) coast, the lowland beach and barrier island coast, and the coral reef. Each one has a particular set of erosional and depositional landforms.

Rocky coasts The most common type of coast, comprising about 80% of ocean coasts worldwide, is a rocky, or cliffed, coast. When a cliffed coast is seen in profile, the usual elements are a vertical **wave-cut cliff** and a horizontal wave-cut bench at its base, both products of erosion. As the upper part of the cliff is undermined, it collapses, and the resulting debris is redistributed by waves. An undercut cliff that has not yet collapsed may have a well-developed notch at its base. The bench may be covered by sand, or the bedrock may be exposed, especially at low tide.

> **wave-cut cliff** A coastal cliff cut by wave action at the base of a rocky coast.

If the coast has been uplifted by tectonics, a wave-cut bench and its sediment cover can be lifted up out of the water and become a marine terrace. In some locations, you can see two or more terraces ascending out of the ocean like a giant staircase (see Figure 12.11).

Expensive follies • Figure 12.12

Millionaire's folly

Cliff collapses when undercut rock can no longer support it

Undercutting by wave action

Typically this kind of slope failure occurs because waves have undercut the base of the cliff. Unfortunately, this scenario will continue to repeat itself as long as people choose home sites based on the desire for "million-dollar views" rather than on sound geological reasoning.

Area of collapse

Houses dangle on the edge of a cliff in Pacifica, California, after winter storms caused a section of cliff to collapse. Note the rubble on the beach and the scar on the cliff where the debris slide occurred.

The rocky character of cliffed coasts may be misleading to people looking for a dramatic home site. They see a rocky cliff as a sign of permanence and stability, whereas in fact quite the reverse is the case. Shorelines with cliffs are susceptible to frequent landslides and rock falls as erosion eats away at the base of a cliff. Roads, buildings, and other structures built too close to such cliffs can be damaged or destroyed when sliding occurs (**Figure 12.12**).

Beaches and barrier islands

Beaches are a striking feature of many coasts. Most people think of a beach

> **beach** Wave-washed sediment along a coast.

as the sand surface above the water along a shore. Actually, a **beach** also includes sediment in the surf zone, which is underwater and therefore continually in motion. At low tide,

when a large part of a beach is exposed, onshore winds may blow beach sand inland to form belts of coastal dunes. A landform commonly associated with beaches is the **barrier island** (**Figure 12.13**

> **barrier island** A long, narrow, sandy island lying offshore and parallel to a lowland coast.

and chapter-opener). Sand dunes are typically the highest topographical points on a barrier island. Barrier islands are found along most lowland coasts. For example, the Atlantic and Gulf coasts of the United States consist mainly of a series of barrier beaches ranging from 15 to 30 km in length and 1.5 to 5 km in width, located 3 to 30 km offshore. Examples include Coney Island, New York, the long chain of islands known as the Outer Banks of North Carolina, and the Chandeleur Islands in Louisiana.

a. This aerial view of the Outer Banks in North Carolina shows a series of barrier islands. The action of wind and waves constantly pushes the sediment on the island toward the mainland (top left in this photo).

b. Salishan Spit in Oregon, shown here, is attached to the mainland at one end, while the other end terminates at a tidal inlet.

c. Spits and barrier islands often create a sheltered lagoon on the landward side. This lagoon is found on Cayo Costa Island in Florida.

spit

tidal inlet

Barrier islands are topographically low, so they are very susceptible to flooding. During a major storm, surf washes across the low places and erodes them, cutting **tidal inlets** that may remain open permanently. At such times, fine sediment is washed between the barrier island and the mainland. Because of this deposition and erosion of sediment, the length and shape of barrier islands is always changing. Unfortunately, the ever-changing nature of barrier island coasts conflicts with our desire to erect permanent buildings on them. Property owners often protect their properties with artificial seawalls, which, though they may serve as a local fix, only hasten the erosion or deposition processes elsewhere on the coast.

Barrier island beaches typically exhibit depositional landforms such as **spits** (elongated ridges of sand or gravel that project from land into the open water of an embayment along the coast), **tombolos** (spit-like ridges of sand and gravel that join an island to the mainland), and **bay barriers**, which may completely close off the mouth of a small bay. A well-known example of a large, complex spit is Cape Cod, Massachusetts.

The elongated bay lying inshore from a barrier island or other low, enclosing strip of land (e.g., a coral reef) is called a **lagoon** (**Figure 12.13c**). Lagoons are commonly fed by **estuaries**, the wide, fan-shaped mouths of rivers in the tidal zone where fresh water and salt water meet. Lagoons and estuaries are important habitats for a wide variety of plants and animals. They also play an important role in the protection of mainland shorelines because they serve as buffers against storm waves. Human activities can also adversely affect these sensitive environments (see the *Case Study*).

Coral reefs Many of the world's tropical coastlines consist of limestone **reefs** built by vast colonies of organisms, principally corals, which secrete calcium carbonate (the main chemical constituent of limestone) as their skeletal material. Reefs are built up very slowly over thousands of years. Each of the tiny coral animals deposits a protective layer of calcium carbonate; over time the layers build up, forming a complex reef structure. **Fringing reefs** form coastlines that closely border the adjacent land. **Barrier reefs** are separated from the land by a lagoon, as in the case of the Great Barrier Reef off Queensland, Australia. Reefs are highly productive ecosystems that support a diversity of marine life forms (**Figure 12.14**). They also perform an important role in the recycling of nutrients in shallow coastal environments. They provide physical barriers that dissipate the force of waves, protecting the ports, lagoons, and beaches that lie behind them, and are an important aesthetic and economic resource.

> **reef** A hard structure on a shallow ocean floor, usually but not always built by coral.

Coral reef hosts bountiful life • Figure 12.14

Reef-building corals provide complex habitat choices and typically host an abundant diversity of life.

Thinking Critically About the Effects of Oil Spills

The Chandeleur Islands can't win. After decades of sediment-starvation from engineering of the Mississippi River, and massive storm-surge damage by Hurricane Katrina in 2005, the barrier islands were hit again on April 20, 2010, when the BP Deepwater Horizon oil rig exploded in a fiery inferno. The damaged well eventually released almost 5 million barrels (about 780 million liters) of crude oil into the Gulf of Mexico.

It took months to stop the leaking oil, which caused extensive damage to coastal and marine plants, wildlife, habitat, and fishing and tourism. The cleanup effort, deploying more than 28,000 people, involved controlled burn-offs; skimming and suctioning of oil from the surface; chemical dispersants; floating containment booms; and sand barriers along shorelines.

What was the ultimate fate of those 5 million barrels of oil? A team of scientists has been trying to answer this. Some oil washed up on nearby shores, including the Chandeleur Islands (Figure a and Figure b). Some was collected or dispersed by various cleanup methods. A large portion evaporated or sank to the bottom of the Gulf, where its fate remains uncertain. The pie chart (Figure c) shows an early estimate of the fate of the oil. Uncertainties involved in this calculation include the lack of standardized sampling protocols for surface and subsurface oil; variation in droplet sizes of dispersed oil; and lack of understanding of the impacts of longer-term processes, such as biodegradation of the oil.

Crude oil varies widely in composition. Thousands of hydrocarbon compounds, many toxic, make up each batch of crude oil. Spilled oil behaves differently depending on weather, wave and wind action, depth and temperature of the water, nature of the shoreline, and many other factors. Add a few technical complexities like the great depth of the well head in this case, throw in the urgency associated with possible damage to fragile ecosystems, and you get an idea of how extremely challenging large oil spills can be.

a. This is a photo of the Chandeleur Islands, with oil from the BP oil spill of April 2010 visible just offshore.

b. Oil washed up on the beaches of the Chandeleur Islands after the oil spill.

Deepwater Horizon Oil Budget
Based on estimated release of 4.9m barrels of oil

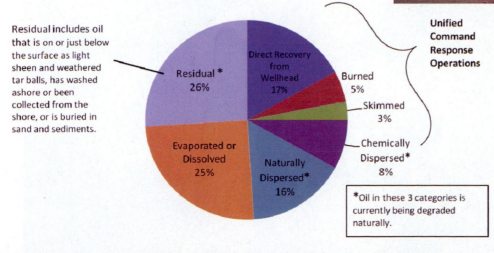

Residual includes oil that is on or just below the surface as light sheen and weathered tar balls, has washed ashore or been collected from the shore, or is buried in sand and sediments.

Residual * 26%

Direct Recovery from Wellhead 17%

Burned 5%

Skimmed 3%

Chemically Dispersed* 8%

Naturally Dispersed* 16%

Evaporated or Dissolved 25%

Unified Command Response Operations

*Oil in these 3 categories is currently being degraded naturally.

c. This pie chart shows the fate of the almost 5 million barrels of spilled oil, estimated as of August 2010 by the NOAA Office of Response and Recovery, Deepwater Horizon/BP Oil Spill.

Corals require shallow, clear water in which the temperature remains above 18°C but does not exceed 30°C. Reefs therefore are formed only at or close to sea level and are characteristic of warm, low latitudes. They also favor places where there is deep-sea upwelling that provides abundant nutrients, normal salinity—not too salty, not too fresh—and an absence of sediment being deposited from large streams.

Because of their very specific requirements, coral reefs are highly susceptible to damage from human activities as well as from natural causes such as tropical storms. Recently, scientists have suspected that the absorption of atmospheric carbon dioxide by the ocean is changing the composition—mainly the acidity—of ocean water. This may be having a negative impact on corals, which are showing signs of damage in many parts of the world ocean. The decline of coral health in the Florida Keys is discussed in *Amazing Places*.

AMAZING PLACES

The Florida Keys Reef

a. The world's third-longest coral barrier reef, and the only living coral reef in the continental United States, runs parallel to the Florida Keys. It stretches for 320 km and in most places is just a short boat ride offshore. Unlike other animals, whose skeletons become stone only after thousands of years of burial, corals have skeletons that are limestone from day one, and they gradually add to the landmass of the continent.

In the Florida Keys Reef, you can find a variety of corals, such as the fire coral and the mustard hill coral in the foreground, and the brain coral in the background. Unfortunately, though, the coral in the Florida Keys has been in declining health for a long time.

CONCEPT CHECK STOP

1. **What** processes can cause sea level to change?

2. **Why** are the heights of the tides not the same in every location?

3. **How** do waves in deep water behave differently from waves in shallow water?

4. **Why** do jetties and seawalls on one part of a beach often have adverse effects on other parts?

5. **What** are some of the beneficial effects of coral reefs?

b. In this series of photos, you can see how part of Carysfort Reef progressed from a healthy state in 1975 to a sick one in 1985, and by 1995 it was almost completely dead. By 2000, Carysfort Reef had lost 90% of its original coral cover. The causes of coral decline are numerous and not completely understood. Coral polyps can be killed by natural causes, such as hurricanes, but the reef usually recovers from them over the long term. Anthropogenic injuries, such as scarring from boats that run aground on the reef, are much harder for coral to recover from.

c. In recent years, a pervasive problem has been **coral bleaching**. When the water gets too warm, coral polyps expel the symbiotic algae that ordinarily give them their bright colors. The bleached coral is not necessarily dead, but without its algae, it will die soon. The white parts of the staghorn coral in this photo are bleached, whereas the brown parts are still healthy (but probably in danger).

The Atmosphere

LEARNING OBJECTIVES

1. **List** the main chemical constituents of Earth's atmosphere.

2. **Identify** the four layers of Earth's atmosphere.

3. **Explain** how the atmosphere protects Earth from radiation and contributes to the warming of Earth's surface.

4. **Describe** how the Sun's heat and Earth's rotation affect the movement of air.

5. **Relate** the movement of air masses to climate zones.

Atmosphere is the generic term for a gaseous layer that surrounds a planet or other celestial body. When Earth first formed some 4.6 billion years ago, it was surrounded by an envelope of gaseous components; additional volatiles were trapped in the interior of the planet. This **primary atmosphere** was lost early in Earth history—stripped away by the intense solar wind associated with the young Sun. Earth evolved a new, **secondary atmosphere** through volcanic outgassing of some of the volatile constituents from its interior, with some additional material from incoming comets (**Figure 12.15**).

The atmosphere and hydrosphere have evolved chemically over Earth history, in close partnership with the biosphere and the geosphere. The early atmosphere had a very different composition and was not "breathable" in the sense that we know it today. The advent of a breathable atmosphere with available oxygen took hundreds of millions of years, and depended on the activity of photosynthetic

Geology InSight

The atmosphere • **Figure 12.15**

Where did the air come from?
Earth's primary atmosphere was lost—blown away by a strong solar wind. Volcanic emissions brought to the surface volatile constituents that had been trapped inside the planet. Eventually the planet cooled sufficiently that it was able to retain this envelope of gases, the secondary atmosphere.

Breathable air
It took billions of years of change before the atmosphere contained enough "free" oxygen to be "breathable" in the sense that we now know it. This change involved both the removal of carbon and the production of oxygen, both of which depended on the evolution of life. Mats of photosynthetic algae, like the one shown here, were key players in the production of a breathable atmosphere.

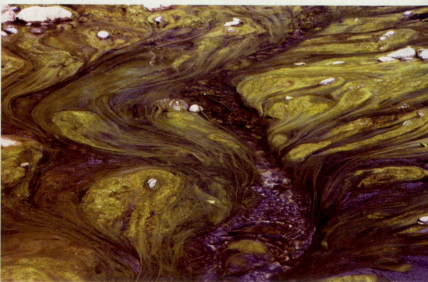

organisms like algae. You will learn more about these chemical changes and about how the atmosphere and hydrosphere were influenced by the origin and evolution of life in Chapter 15.

Composition of Earth's Atmosphere

Many planets and even some moons have atmospheres, but **air** is the gaseous envelope that surrounds one planet in particular: Earth (see **Figure 12.15**).

air A mixture of 78% nitrogen, 21% oxygen, and trace amounts of other gases, found in Earth's atmosphere.

Air is an invisible, normally odorless mixture of gases and suspended particles. Two components of air are highly variable in concentration: aerosols and water vapor. **Aerosols** are liquid droplets or solid particles that are so small that they remain suspended in the air. Water droplets in fogs are liquid aerosols. Some examples of solid aerosols are tiny ice crystals, smoke particles from fires, and sea-salt crystals from ocean spray.

Water is always present in the air, but in varying amounts expressed by the **humidity** of the air. On a hot, humid day in the tropics, as much as 4% of the air by volume may be water vapor. On a crisp, cold day, less than 0.3% may be water vapor. (Note that 100% **relative humidity**, a term that you might hear in a weather report, does not mean the air is 100% water vapor! It simply means that the air contains as much water vapor as it can carry at that particular temperature. The temperature at which the relative humidity is 100% is called the **dew point**.)

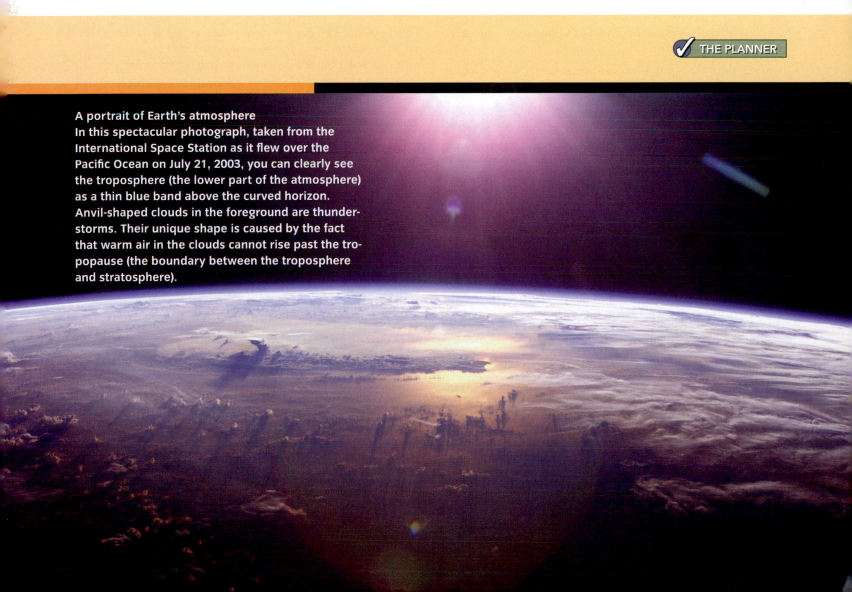

✓ THE PLANNER

A portrait of Earth's atmosphere
In this spectacular photograph, taken from the International Space Station as it flew over the Pacific Ocean on July 21, 2003, you can clearly see the troposphere (the lower part of the atmosphere) as a thin blue band above the curved horizon. Anvil-shaped clouds in the foreground are thunderstorms. Their unique shape is caused by the fact that warm air in the clouds cannot rise past the tropopause (the boundary between the troposphere and stratosphere).

What air is made of • Figure 12.16

Air contains two substances whose concentration varies from place to place and time to time: water vapor and aerosols. The rest of the atmosphere consists primarily of nitrogen and oxygen, with small amounts of other gases.

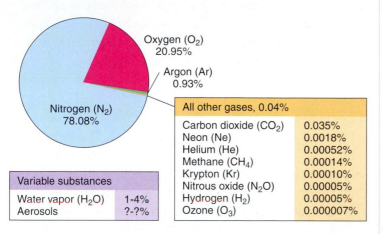

All other gases, 0.04%	
Carbon dioxide (CO_2)	0.035%
Neon (Ne)	0.0018%
Helium (He)	0.00052%
Methane (CH_4)	0.00014%
Krypton (Kr)	0.00010%
Nitrous oxide (N_2O)	0.00005%
Hydrogen (H_2)	0.00005%
Ozone (O_3)	0.000007%

Variable substances	
Water vapor (H_2O)	1-4%
Aerosols	?-?%

Because the water vapor and aerosol contents of air vary so widely, the relative amounts of the remaining gases are reported as if the air were entirely lacking in water vapor and aerosols. When these two components are ignored, the relative proportions of the remaining gases in the air—termed **dry air**—are essentially constant. As shown in **Figure 12.16**, three gases—nitrogen (78%), oxygen (21%), and argon (0.93%)—make up 99.96% of dry air by volume. The remaining gases (carbon dioxide, 0.035%; neon, 0.0018%; and six others) are present in very small quantities. However, some of these minor gases are profoundly important for life on Earth because they absorb certain wavelengths of sunlight. They act both as a warming blanket and as a shield against deadly ultraviolet radiation.

The oxygen and carbon dioxide in the atmosphere are constantly being removed or replenished by plants and animals, by chemical processes such as the weathering of rock and formation of soil. Gases also cycle from the atmosphere to the ocean and back again, passing through the ocean surface. Further recycling of volatile constituents occurs through geological processes such as volcanism and subduction, which link the atmosphere and hydrosphere to the geosphere. If anything happens to affect the rates of any of these processes, the chemical makeup of the atmosphere will change.

Layers in the Atmosphere

The characteristics of the atmosphere are not constant throughout its depth. When sunlight hits the top of the atmosphere, about 30% of it is reflected back into space, but the remaining 70% is absorbed by the atmosphere itself, and by the ocean, land, and biosphere. The absorbed solar energy warms Earth's surface, which in turn warms the bottom portion of the atmosphere. Eventually, all of this energy is re-emitted in the form of long-wavelength infrared radiation, back to outer space. However, as the sunlight passes through the atmosphere, water vapor and some of the minor gases absorb certain wavelengths of solar radiation, which raises the temperature of the air.

Scientists have identified four major atmospheric layers with distinct temperature profiles (**Figure 12.17**). The layers are separated from each other by boundaries called **pauses**.

Four major layers • Figure 12.17

Temperature varies with altitude in the atmosphere. In the lowest level, the troposphere, the temperature drops rapidly with increasing altitude because the troposphere is warmed from below, by Earth's surface. In the next layer, the stratosphere, the reverse is true. Two more reversals occur at higher altitudes. The air in the outermost layer is so tenuous that this zone is defined by rocket scientists (though not by geologists) to be part of "space." NASA awards a space-flight badge to anybody who flies above 80 km.

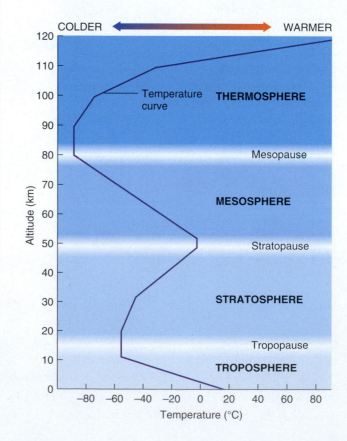

How a greenhouse works • Figure 12.18

The glass in a greenhouse (**Figure a**) works by trapping warm air underneath the glass. Radiatively active, or "greenhouse," gases in the troposphere (**Figure b**) work in a similar way, but instead of physically trapping the outgoing warmed air, the gases absorb infrared radiation, warming the air and the surface.

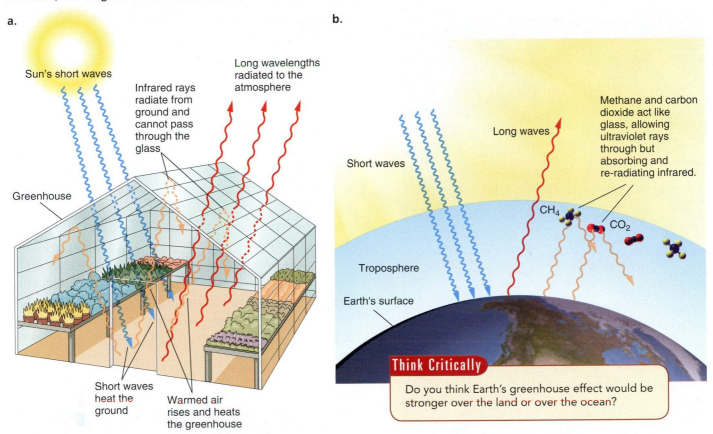

a.

Sun's short waves

Infrared rays radiate from ground and cannot pass through the glass

Long wavelengths radiated to the atmosphere

Greenhouse

Short waves heat the ground

Warmed air rises and heats the greenhouse

b.

Short waves

Long waves

Methane and carbon dioxide act like glass, allowing ultraviolet rays through but absorbing and re-radiating infrared.

CH_4

CO_2

Troposphere

Earth's surface

Think Critically

Do you think Earth's greenhouse effect would be stronger over the land or over the ocean?

The troposphere and greenhouse effect Most humans, with the exception of military pilots and astronauts, never fly outside the troposphere. The **troposphere** contains 80% of the actual mass of the atmosphere, including almost all of the water vapor and clouds. The thin blue layer of atmosphere shown in Figure 12.15c is mainly the troposphere. This is also where almost all weather-related phenomena originate. Although very little mixing occurs between the troposphere and the stratosphere, the troposphere itself is constantly moving and thoroughly mixed by winds.

The troposphere contains most of the heat-absorbing gases (also called **radiatively active gases**, or **greenhouse gases**) that play a role in warming Earth's surface. The most important greenhouse gas is water vapor. Another very

troposphere
The lowest layer of Earth's atmosphere, extending (variably) to about 15 km in altitude.

important greenhouse gas is carbon dioxide, and another is methane, the main constituent of natural gas. Greenhouse gases intercept and absorb some of the outgoing infrared terrestrial radiation and prevent it from escaping, in much the same way that the glass of a greenhouse does (**Figure 12.18**), so this natural atmospheric process is called the **greenhouse effect**. Without the greenhouse effect, the surface of Earth would be a cold and inhospitable place. However, it is a matter of serious concern that the concentration of greenhouse gases in the atmosphere is steadily increasing as a result of human industrial activity. We will explore this phenomenon and its potential impact on the global climate system in greater detail in Chapter 14.

greenhouse effect
The absorption of long-wavelength (infrared) energy by radiatively active gases in the atmosphere, causing heat to be retained near Earth's surface.

The Atmosphere **369**

A shield against radiation • Figure 12.19

Ultraviolet radiation from the Sun can be harmful or lethal; generally speaking, the shorter the wavelength, the more harmful the radiation. Fortunately, the atmosphere protects us from almost all these rays because they are absorbed by three kinds of oxygen—O, O_2, and O_3 (ozone).

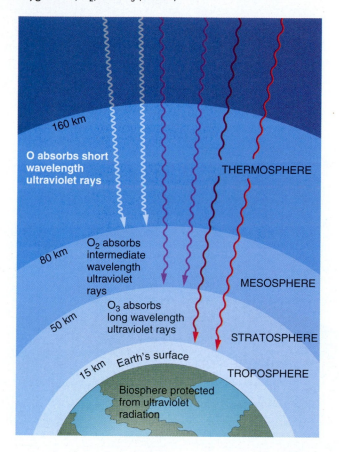

160 km

O absorbs short wavelength ultraviolet rays

THERMOSPHERE

80 km

O_2 absorbs intermediate wavelength ultraviolet rays

MESOSPHERE

50 km

O_3 absorbs long wavelength ultraviolet rays

STRATOSPHERE

15 km   Earth's surface

TROPOSPHERE

Biosphere protected from ultraviolet radiation

The stratosphere and the ozone layer

The **stratosphere** contains 19% of the atmosphere's total mass, which means that the troposphere and the stratosphere combined contain 99% of the atmosphere. Beyond the stratosphere are the **mesosphere** and **thermosphere**, which together comprise only 1% of the mass of the atmosphere. The three outer layers of the atmosphere play very important roles for life on Earth despite their relatively small masses (when compared to the troposphere). Oxygen in various forms absorbs harmful ultraviolet rays coming from the Sun (**Figure 12.19**). The thermosphere absorbs short ultraviolet wavelengths, the mesosphere absorbs intermediate wavelengths, and the stratosphere absorbs long wavelengths. The gas ozone (O_3) in the stratosphere, in tiny but vital amounts, absorbs the most dangerous of the ultraviolet rays.

> **stratosphere**
> The layer of Earth's atmosphere above the troposphere, extending to about 50 km altitude.

In recent years scientists have become very concerned about the destruction of the stratospheric **ozone layer** by chemical pollutants (**Figure 12.20**). Among these pollutants are chlorofluorocarbons (CFCs), which were formerly an ingredient in aerosol sprays but were banned by an international treaty in 1996. Although it is still too early to be certain, there are signs that the destruction of ozone is abating as a result of this action taken by the

> **ozone layer** A zone in the stratosphere where ozone is concentrated.

The ozone "hole" • Figure 12.20

In 1985, scientists discovered that a previously unnoticed gap in the ozone layer was forming over Antarctica during the southern spring. (A smaller hole also formed over the Arctic Ocean.) For many years the hole (seen here in blue) grew larger, and the depletion at its center grew more severe. This image shows the ozone hole in September 2006—the largest ever recorded. (Source: NASA Ozone Hole Watch.)

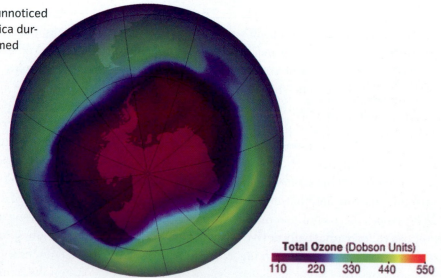

Total Ozone (Dobson Units)

110   220   330   440   550

Where Geologists CLICK

Ozone Hole Watch

http://ozonewatch.gsfc.nasa.gov/

Visit NASA's Ozone Hole Watch Web site to view an extremely interesting series of maps and animations that show the history of the Antarctic ozone hole. Compare the ozone hole in this image from November of 2010 with the ozone hole in Figure 12.20 from September of 2006. Do you notice any differences? Visit this Web site to learn more about the factors that cause the ozone hole to vary from season to season and year to year.

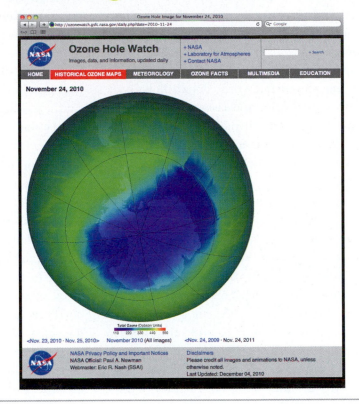

---

global community. Perhaps, in the future, this will be cited as an example of a successful environmental intervention. To see maps and animations showing changes in the ozone hole over time, see *Where Geologists Click*.

Movement in the Atmosphere

Two things energize the atmosphere: the Sun's heat and Earth's rotation. Because Earth is a sphere, the Sun does not warm every place on Earth equally. In places where the Sun is directly overhead, the incoming rays are perpendicular to the surface, and a maximum amount of heat is received per unit area (**Figure 12.21**). Because of the curvature of Earth's surface, at all other locations the surface is at an angle to the incoming rays; therefore, these locations receive less heat per unit of surface area. The atmosphere and ocean redistribute this uneven heat, mainly through winds and ocean currents, which move heat from the equator—where the input of solar heat is greatest—toward the poles, where it is least.

Who gets the most sunlight?
• Figure 12.21

At all times, the point on Earth's surface where the Sun is directly overhead receives the maximum amount of solar energy (about 1366 watts per square meter). At the spring and fall equinoxes (center), the maximum energy falls on the equator. In December (right), because of the 23.5° tilt of Earth's axis of rotation to the plane of its passage around the Sun, solar input is most intense at 23.5°S, the Tropic of Capricorn. In June (left), the incoming solar energy is most intense at 23.5°N, the Tropic of Cancer.

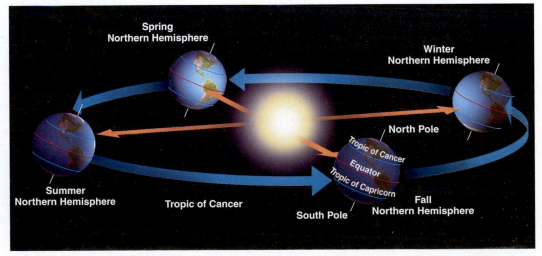

Heating of Earth's surface causes convection currents in the atmosphere. Heated air near the equator expands, becomes lighter (less dense), and rises. Near the top of the troposphere, it spreads outward toward the poles. As the upper air travels northward and southward toward the poles, it gradually cools, becomes heavier, and sinks. Upon reaching the surface, this cool air flows back toward the equator, warms up, and rises again, thereby completing a convective cycle. In reality, it is not quite so simple: The global circulation actually organizes itself into three major sets of convection cells, which interlock like gears (**Figure 12.22**).

When air rises, it leaves a region of low **atmospheric pressure** at the surface. Atmospheric pressure, or **air pressure**, is a measure of how much of the mass of the atmosphere overlies a particular location. A low-pressure area has less atmospheric mass overlying it; a high-pressure area has more atmospheric mass overlying it. In the parts of convection cells where the air is denser and is descending from aloft, a region of high atmospheric pressure is formed at the surface. This difference between high-pressure and low-pressure areas, caused by rising and falling air, adds to the dynamic nature of the lower atmosphere. This is because air always moves away from a zone of high atmospheric pressure and toward a zone of low atmospheric pressure. This **pressure gradient** is thus a force that causes air to move, creating **wind**.

> **wind** Air in motion.

The Coriolis force If Earth did not rotate, convection currents in the atmosphere would simply flow north and south, moving from the equator to the poles in a simple convection cell. But Earth does rotate, and its rotation complicates the convection currents in the atmosphere (as well as in the ocean). The **Coriolis force** causes anything that moves freely with respect to the rotating Earth (including water and air, as well as planes or missiles) to veer off a straight path. It is similar to trying to throw a ball to a friend when you are both on a spinning merry-go-round. In the northern hemisphere, the Coriolis force causes a moving mass to veer toward the right, and in the southern hemisphere toward the left. The effect strongly influences both the global pattern of wind systems (see **Figure 12.22**) and the pattern of ocean currents, including the deep-ocean thermohaline "conveyor belt" circulation.

> **Coriolis force**
> An effect due to Earth's rotation, which causes a freely moving body to veer from a straight path.

Global atmospheric circulation • Figure 12.22

Huge convection cells transfer heat from the equatorial regions, where the input of solar energy is greatest, toward the poles, where the solar input is least. Because Earth is rotating, the flow of air toward the poles and the return flow toward the equator are constantly deflected sideways, creating the circulating volumes of air, or **air masses**, that you see outlined on weather maps. The convection cells are permanent features of Earth's atmosphere and therefore have a great influence on both day-to-day weather and long-term climate.

Think Critically

1. In what other contexts have we discussed the process of convection?
2. What are the similarities and differences between those examples and convection in the atmosphere?

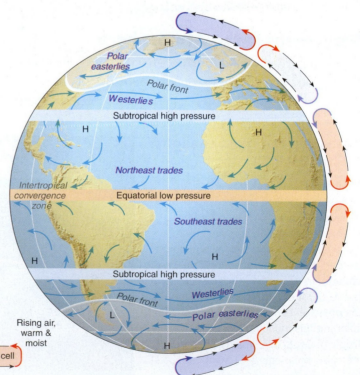

Descending air, cold & dry — Polar cell
Rising air, warm & moist — Ferrel cell
Descending air, cool & dry — Hadley cell
Rising air, warm & moist

Wind systems The Coriolis force breaks up the flow of convective air between the equator and the poles into belts. For example, a large belt, or cell, of circulating air lies between the equator (0°) and about 30° latitude in both the northern and southern hemispheres. Warm air rises at the equator, creating a low-pressure zone called the **intertropical convergence zone** or **ITCZ** (**Figure 12.23a**), which is characterized by cloud cover. The air rises to the top of the troposphere and begins to flow toward the poles, but it veers off course as a result of the Coriolis force. By the time it reaches latitude 30° (N and S), the high-altitude air mass has cooled, and therefore sinks. The descending air flows back across Earth's surface toward the equator. As it flows, the land and sea warm the air so that it eventually becomes warm enough to rise again.

The low-latitude cells (from 0° to 30°N and 0° to 30°S) created by this circulation pattern are called **Hadley cells**. The returning surface winds of Hadley cells are deflected by the Coriolis force so that they are **northeasterlies** in the northern hemisphere (i.e., flowing from the northeast toward the southwest), whereas in the southern hemisphere they are **southeasterlies**. Turning to flow along the equator, these winds are called **trade winds** because their consistent direction and flow carried trade ships across the tropical ocean at a time when winds were the chief source of navigational power (**Figure 12.23b**).

Another belt of convecting air cells, called the **polar cells**, lies over the polar regions. In a polar cell, frigid air flows across the surface, away from the pole and toward the equator, slowly warming as it moves. When the polar air reaches about latitude 60°N or 60°S, it has warmed sufficiently to rise convectively high into the troposphere and flow back toward the pole, where it cools and descends again, thereby completing the convection cell. Because of the Coriolis force, the cold air that flows away from the poles is deflected to the right, giving rise to a wind system called the **polar easterlies**.

Trade winds • Figure 12.23

a. The intertropical convergence zone, where the trade winds of the northern and southern hemispheres merge, shows up very clearly in this satellite photograph. Warm, rising air near the equator causes ocean water to evaporate and form a nearly perpetual band of storm clouds.

b. The continual blowing of the east-to-west (easterly) trade winds has caused this tree, near the southern tip of Hawaii, to lean far to one side.

Between the Hadley cells and the polar cells, a third, but less well-defined, set of convection cells is found in the mid-latitudes, between about latitude 30° to 60°N and 30° to 6°S. In these mid-latitude cells, called the **Ferrel cells**, deflection of air by the Coriolis force creates winds that blow from the west in the northern hemisphere. That is why weather systems in the continental United States and southern Canada generally move in from west to east. These circulation patterns will become clearer if you take some time to study Figure 12.23.

CONCEPT CHECK   STOP

1. **What** are the two components of air that vary significantly in concentration from day to day and place to place?

2. **What** are the two lowest layers of Earth's atmosphere? Describe their main characteristics.

3. **Why** does the ozone hole allow more ultraviolet radiation from the Sun to reach Earth's surface?

4. **Why** does the Sun's heat not warm Earth's surface uniformly?

5. **What** are trade winds, westerlies, and polar easterlies, and what are their physical causes?

Where Ocean Meets Atmosphere

LEARNING OBJECTIVES

1. **Describe** some of the major interactions between the atmosphere and the ocean.

2. **Summarize** how the ocean and the atmosphere regulate climate.

3. **Explain** how hurricanes and other tropical cyclonic storms result from atmosphere–ocean interactions.

4. **Outline** what happens during an El Niño event.

The surface of the ocean is an interface—a boundary between the two great fluid reservoirs of this planet. Through this boundary the atmosphere and ocean are linked, acting in tandem as one great system to control both climate and weather. **Climate** is an average of weather patterns over a long period, generally on a regional or global scale. The processes we think of as **weather**—wind, rain, snow, sunshine, storms, and even floods and droughts—are temporary local variations against the more stable, longer-term background of climate.

We will first consider the roles of the atmosphere and the ocean in regulating climate and driving weather systems. Then we will look at two important weather phenomena that illustrate how this interaction operates.

Ocean–Atmosphere–Climate Interactions

Much happens at the ocean's surface, the dynamic interface between atmosphere and ocean. Of course, the water is constantly in motion, broken by waves and moved by winds and currents. People and animals use this surface for transport and for gathering resources. Materials also move from one reservoir to the other across the interface. For example, sea spray allows salt particles to move from the ocean to the atmosphere, where they are transported by wind and eventually deposited on land. Dust and particulate pollutants similarly are blown by the wind and deposited from the atmosphere onto the surface of the ocean.

Some of the most important material transfers across the ocean–atmosphere interface occur in the vapor phase. For example, water leaves the ocean and moves to the atmosphere by evaporation—the breaking away of water molecules from the ocean's surface. Carbon dioxide (along with other gases) is absorbed by ocean water across this interface, which allows the ocean to act as a sink for excess carbon dioxide from the atmosphere. This is a crucial process in controlling the carbon content of the atmosphere, and therefore plays a crucial role in the greenhouse effect. As mentioned previously, it also has implications for the composition (mainly the acidity) of ocean water.

Heat also is exchanged across the ocean–atmosphere interface. This exchange is extremely important in the regulation of Earth's global climate system since these two fluid bodies are responsible for transporting heat from the equator to the poles. The ocean differs greatly from the land in the amount of heat it can store. When the Sun's rays strike the land, only the top meter or so is warmed. Below the top few meters, rock temperatures are controlled by the geothermal gradient; at any given depth, the temperature is roughly constant year round. In the ocean, however, mixing by currents and waves means the Sun's heat affects much more than just the surface layer.

Water has a higher **heat capacity** (an indicator of a material's ability to store heat) compared to many terrestrial materials, which contributes to its temperature stability. Both the ocean and the atmosphere are also affected by water's **latent heat**—the energy that the ocean absorbs when water evaporates, and releases when water freezes. When water evaporates from the ocean, it carries some of the Sun's heat energy with it into the atmosphere, leaving the ocean water slightly cooler and the lower part of the atmosphere slightly warmer. The reverse happens when water freezes: It releases its latent heat and maintains the temperature of the water around it.

These processes, together, reduce the amount of variation in ocean temperatures (**Figure 12.24**). The temperature and moisture of air masses are strongly influenced by

The global air temperature distribution • Figure 12.24

This map shows the range in average daily air temperature over both land and sea. Notice the much broader range of temperatures over land (in red, orange, and yellow), as compared to the range over water (in blue). There is also a pronounced seasonal change in land temperatures between summer and winter, whereas over the ocean there is little seasonal difference (especially in the tropics).

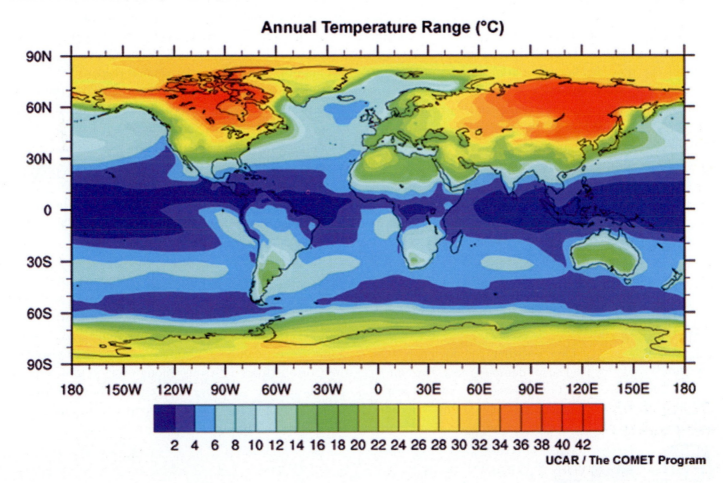

Annual Temperature Range (°C)

2 4 6 8 10 12 14 16 18 20 22 24 26 28 30 32 34 36 38 40 42

UCAR / The COMET Program

the characteristics of the surface over which the air is flowing. The ocean influences the temperature and moisture content of the lower atmosphere, which in turn moderates the climate of coastal areas on land. Along the Pacific coast of Washington and British Columbia, for example, winter air temperatures seldom drop to freezing, whereas inland temperatures plunge to −30°C or lower. The global patterns of air circulation that we discussed previously in this chapter are principally responsible for controlling the variety and distribution of Earth's climatic zones.

Even if you live far from the ocean rather than in a coastal zone, you feel the ocean's influence on a regular basis, courtesy of atmospheric circulation. For example, in eastern North America, the muggy weather and thunderstorms of summertime originate with water that evaporated off the Gulf of Mexico. The monsoon rains in India, nor'easters off the coast of New England, and winter storms in the Pacific Northwest and the Prairies—all of these are weather systems that are transported by the atmosphere, from the ocean to the land.

One of the most important natural influences on climate is the oceanic thermohaline circulation. For example, as mentioned earlier, the Gulf Stream current—the segment of the thermohaline circulation that returns warm water to the North Atlantic—is largely responsible for the moderate climate in coastal regions of eastern North America and western Europe. Because the ocean's conveyor belt circulation redistributes heat from the tropics to the poles, any weakening of this circulation would be expected to have a dramatic effect on the temperature distribution. For example, during the last ice age, the polar ice caps may have cut off the thermohaline circulation in the North Atlantic, which likely increased the length and severity of the glaciation. We will explore this and other factors that play a role in Earth's climate in greater detail in Chapter 14.

Tropical Cyclones

cyclone A wind system that is circulating around a low-pressure center.

One weather phenomenon that clearly demonstrates the interaction between ocean and atmosphere is cyclonic storms. There are many different types of **cyclones**,

distinguished by where they form, how they develop, and their characteristics of temperature, air pressure, wind speed, and geographic extent. The most intense and typically the most damaging are **tropical cyclones**, which are called **cyclones** if they form over the Pacific Ocean, **typhoons** if they form over the western Pacific or Indian Ocean, and **hurricanes** if they form over the Atlantic Ocean.

Cyclones circulate counterclockwise in the northern hemisphere and clockwise in the southern hemisphere, as dictated by the twin forces of air pressure gradient and Coriolis force. Air naturally moves into a low-pressure center; in general, the greater the pressure drop, the more vigorous the wind speeds. Meanwhile, the Coriolis force deflects the pathway of the air to the right in the northern hemisphere, and to the left in the southern hemisphere. The end result is a cyclonic wind system, rotating around the low-pressure center.

One type of cyclone is an **extra-tropical** ("outside of" the tropics) or **mid-latitude cyclone**. These form between 30° and 60° N and S, as a result of the interaction between warm and cold air along the polar fronts. They are large in geographic extent—up to 2000 km across—and tend to last many days. This type of cyclone is responsible for most weather events in the mid-latitude regions of the world. They also can spawn more local, intense weather systems, such as **thunderstorms** and **tornadoes**.

Hurricanes, typhoons, and cyclones are intense tropical cyclonic storms that start as  wind systems gently circulating over warm ocean water. Hurricanes generally start to form in the eastern Atlantic, near Africa (**Figure 12.25**). They require a warm sea-surface temperature (at least 26.5°C) to develop. Drawn upward into the low-pressure center, the warm ocean water evaporates and then condenses. This releases latent heat and energizes the storm, which may then develop into a hurricane. Hurricanes and other cyclonic storms can be generated only in latitudes where the Coriolis force is strong enough for cyclonic circulation to develop—that is, higher than about latitude 5° N or S. The Coriolis force is zero at the equator, which means that hurricanes cannot form there.

On the track of a hurricane • Figure 12.25

Hurricanes draw their energy from warm surface water in the ocean.

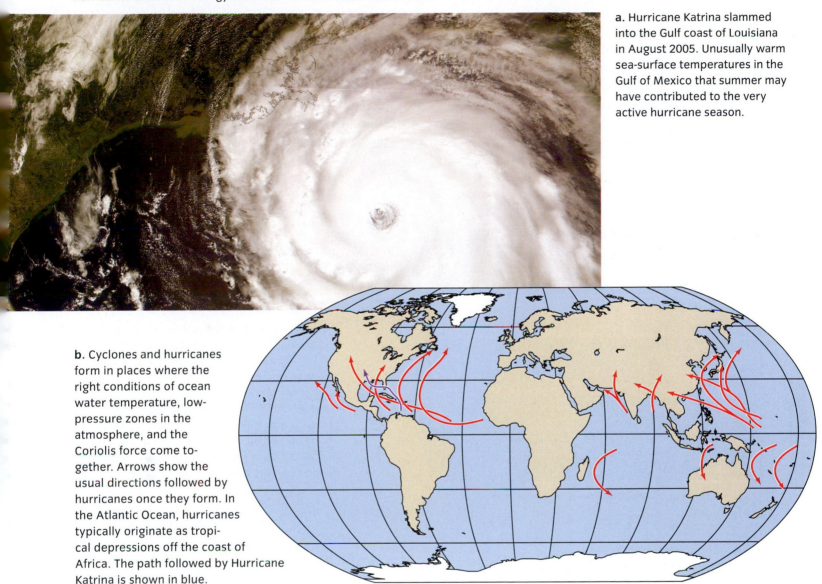

a. Hurricane Katrina slammed into the Gulf coast of Louisiana in August 2005. Unusually warm sea-surface temperatures in the Gulf of Mexico that summer may have contributed to the very active hurricane season.

b. Cyclones and hurricanes form in places where the right conditions of ocean water temperature, low-pressure zones in the atmosphere, and the Coriolis force come together. Arrows show the usual directions followed by hurricanes once they form. In the Atlantic Ocean, hurricanes typically originate as tropical depressions off the coast of Africa. The path followed by Hurricane Katrina is shown in blue.

A hurricane, by definition, has wind speeds greater than 119 km/hour. (At lower wind speeds, it would be called a **tropical storm** or **tropical depression**.) Because a hurricane draws its energy from warm ocean water, its wind speeds will diminish, and the hurricane will dissipate once it moves onshore, away from its energy source. For this reason, most hurricane wind damage occurs near the coast. The winds are usually accompanied by torrential rain and sometimes by **storm surge**, a lo- cal, exceptional flood of ocean water onto coastal areas. The center (or **eye**) of a hurricane is a region of very low air pressure, which causes the surface of the ocean to bulge upward locally. Hurricane-force winds drive the high seas onshore, and extensive flooding can result. Storm surge, combined with failure of the levee system, is what led to the massive devastation in New Orleans and coastal areas like the Chandeleur Islands during Hurricane Katrina.

El Niño and La Niña

Another important weather phenomenon that highlights the complex nature of the interactions between atmosphere and ocean is **El Niño** (**Figure 12.26**).

> **El Niño** A regional weather system that involves unusual warming of equatorial Pacific surface water.

Roughly every four years, a mass of unusually warm water appears off the coast in the Pacific Ocean. Fish populations decline, warm-water organisms foreign to these waters begin to appear, and the trade winds slacken. Peruvians refer to this as El Niño because it commonly appears at Christmas time (*El Niño* refers to "the Christ child"). Coincident with Peruvian El Niño conditions, very heavy rains fall in normally arid parts of Peru and Ecuador; Australia experiences drought conditions; anomalous cyclones appear in Hawaii and French Polynesia; the seasonal rains of northeastern Brazil are disrupted; and the Indian monsoon may fail to appear. During exceptional El Niño years, weather patterns over much of Africa, eastern Asia, and North America are affected.

El Niño has been experienced by generations of Peruvians, but its broader significance as a coupled

La Niña and El Niño • Figure 12.26

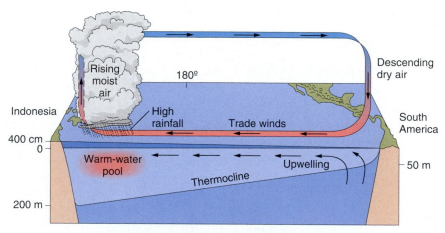

Normal conditions in the tropical Pacific

a. In a "normal" year, persistent trade winds blow westward across the tropical Pacific from the zone of upwelling water off Peru (moving from a high-pressure zone in the eastern Pacific to a low-pressure zone in the western Pacific). The warm water that is pushed westward by the trade winds collects in a large pool of warm water above the thermocline in the western Pacific. The warm water causes the moist maritime air to rise and cool, bringing abundant rainfall to Indonesia. When this "normal" situation is expressed unusually strongly, it is referred to as **La Niña**.

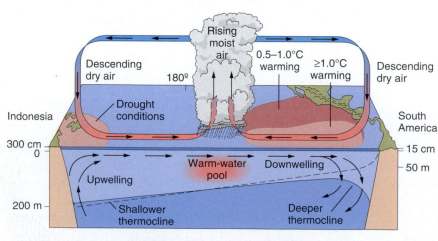

El Niño conditions in the tropical Pacific

b. During an El Niño event, the relative air pressure between the eastern and western Pacific weakens, or even reverses (i.e., the low-pressure zone shifts to the east and the high-pressure zone to the west). As a result, the trade winds weaken. The pool of warm water is not pushed westward but remains in the central Pacific. Rising moist air above the warm-water pool greatly increases rainfall in the mid-Pacific, while descending cool, dry air brings drought conditions to Indonesia and Australia. Surface waters in the eastern Pacific become unusually warm, shutting off the supply of nutrients from deep, cold water. This adversely affects the fishery off Peru. During a strong El Niño year, such as 1997–1998, the weather is disrupted over more than half of the planet.

atmosphere–ocean phenomenon has been recognized only recently. Normally there is a gradient from a high-pressure zone in the eastern tropical Pacific to a low-pressure zone in the western Pacific. This pressure difference causes the trade winds to blow toward the west (thus they are **easterlies**, blowing from the east) because air always moves from an area of high atmospheric pressure toward an area of lower atmospheric pressure. In the western Pacific the warm, moist air rises, causing cloudiness and rain in the area of Indonesia. During these normal (or **La Niña**) years, the easterly trade winds push the warm surface water toward the western Pacific, allowing deep, cold water from the Southern Ocean to well up to the surface along the coast of South America (**Figure 12.26a**).

When the atmospheric pressure difference at the surface is large (high in the east, low in the west), the trade winds blow strongly toward the west; when the pressure difference is small (almost the same from east to west), the trade winds weaken. An El Niño event begins when the air pressure difference becomes small and the trade winds weaken, or even reverse. Without the strong easterly trade winds to push warm surface water away from the equatorial coast of South America, the upwelling of cold, deep water is suppressed (**Figure 12.26b**). Anomalously warm surface water accumulates in the central and eastern

Pacific, near South America. The zone of high rainfall that is normally situated near Indonesia shifts toward the central Pacific, bringing drought conditions to Indonesia.

The effects of major El Niño events are felt over at least half of the planet. Therefore, a major area of current research interest is to identify the factors that trigger El Niño events or at least to detect their onset as early as possible. We now know that La Niña, too, has worldwide impacts on weather systems. Understanding El Niño and La Niña events requires scientists to consider the atmosphere and ocean as one great, interlocked system.

CONCEPT CHECK STOP

1. **What** are three ways that the atmosphere and ocean exchange materials?

2. **How** does the thermohaline circulation affect climate?

3. **Where** do hurricanes and other tropical cyclonic storms get their energy?

4. **Why** does the upwelling of cold, deep water from the Southern Ocean fail to occur along the Pacific coast of South America during an El Niño event?

✓ THE PLANNER

Summary

1 The Ocean 346

- There has been an ocean on this planet for at least 4.0 billion years. The water in the ocean may have condensed from steam produced by primordial volcanic eruptions, or some of the water may have been delivered to the planet's surface via cometary impacts.

- Most of Earth's water is contained mainly in four great, interconnected basins, which form the world ocean. The size, distribution, and topography of these basins are largely controlled by plate tectonic processes.

- Seawater ranges in **salinity** from 3.3 to 3.7% and varies measurably from place to place. Freezing and evaporation make the water saltier, whereas rain, snow, and river flow make it less salty.

- The water in the ocean forms layers based on density, which is controlled by temperature and salinity. The water in the

surface layer, from the surface to a depth of about 100 m, is relatively warm and moves in broad, wind-driven currents. The water in the deep layer moves in great, slow, global currents of the **thermohaline circulation**, driven by differences in temperature and salinity as shown in the diagram.

The ocean conveyor belt • Figure 12.5

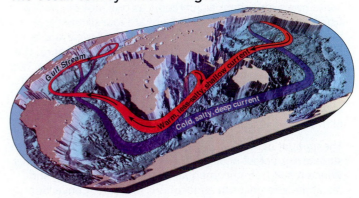

2 Where Ocean Meets Land   352

- The ocean's level varies over a wide range of time scales and for various reasons. Worldwide changes in sea level due to the melting or growth of polar ice sheets take place over thousands of years. Local changes due to the **tides** occur twice a day along most coasts.

- When a wave moves onto the shore, its motion is distorted, as the shallow bottom interferes with the circular motion of water in the wave. It eventually breaks, creating turbulent **surf**. Much of the resulting erosion and transport of sediment is accomplished by **longshore currents** in the surf zone, as shown in the diagram, and by **beach drift** (on land).

Longshore current and beach drift • Figure 12.9

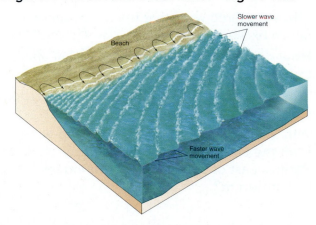

- Shorelines are highly variable, but three basic types are most common: rocky coasts and **wave-cut cliffs**, lowland **beaches** and **barrier islands**, and coral **reefs**. Shorelines and coastal landforms are shaped by a combination of erosive and depositional processes. Storms and artificial structures often accelerate the process of coastal erosion.

3 The Atmosphere   366

- The composition of **air**, excluding water vapor and aerosols, is 99.96% by volume nitrogen, oxygen, and argon.

- Air temperature changes markedly as one moves upward from the surface. As a result, there are four distinct layers in the atmosphere, each with a distinct temperature profile. From the bottom up, they are the **troposphere**, the **stratosphere**, the mesosphere, and the thermosphere. The troposphere contains most of the gases that play a role in Earth's **greenhouse effect**. The **ozone layer**, which protects life on Earth by absorbing harmful incoming ultraviolet radiation, is a concentration of O_3 in the stratosphere.

- The amount of heat energy from the Sun reaching Earth's surface is greatest near the equator and least near the poles. **Winds** result from the unequal distribution of energy, and resulting differences in atmospheric pressure. Convection

currents move heat from the equator toward the poles as shown in the diagram. Earth's rotation produces the **Coriolis force**, which causes winds to veer to the right in the northern hemisphere and to the left in the southern hemisphere.

Global atmospheric circulation • Figure 12.22

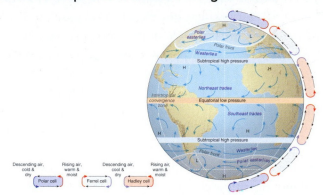

4 Where Ocean Meets Atmosphere   374

- Climate is the average, or "normal," weather for a particular location over a long time.

- The heat-absorbing capacity of the ocean combines with circulation in the atmosphere to control the variety and distribution of Earth's climatic zones.

- **Cyclones** are wind systems that circulate around centers of low atmospheric pressure. Tropical cyclones draw their energy from warm surface water, illustrating the interaction between ocean and atmosphere.

- **El Niño** is an important weather system that involves complex interactions between the ocean and the atmosphere. El Niño events, as shown in the diagram, which can affect the weather over half of the globe, are triggered by variations in atmospheric pressure, which cause weakening of the trade winds and anomalies in equatorial sea surface temperatures.

La Niña and El Niño • Figure 12.26

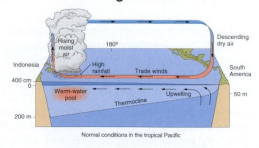

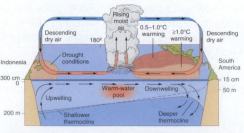

Key Terms

- air 367
- barrier island 360
- beach 360
- beach drift 358
- Coriolis force 372
- cyclone 376
- El Niño 378

- greenhouse effect 369
- longshore current 357
- ozone layer 370
- reef 362
- salinity 348
- stratosphere 370
- surf 356

- thermohaline circulation 351
- tides 353
- troposphere 369
- wave-cut cliff 359
- wind 372

Critical and Creative Thinking Questions

1. Compare Earth's atmosphere with what is known about the atmospheres of other planets. How does its composition differ, and why? Why do some planets have dense atmospheres, and others have nearly none? Does Jupiter's Great Red Spot have any analogues on Earth?

2. During the early Tertiary Period, North and South America were not connected. How might that have changed the global pattern of ocean circulation? How would this in turn have affected global climate? Do some library research and suggest some tests for your hypothesis.

3. Visit a shoreline. If possible, visit the same spot on the shoreline on two or more occasions. What kinds of coastal landforms do you observe? What kinds of changes can you notice from one visit to the next?

4. Rock of Permian and Triassic age in both North America and Western Europe record the existence of desert conditions. What might have caused the deserts to form?

5. Geological evidence indicates that the atmosphere of early Earth was free of oxygen. What constraints might that have had on early life?

What is happening in this picture?

This is a satellite image of tropical cyclone Edzani, in January 2010.

Think Critically

Can you tell whether Edzani was in the northern or southern hemisphere, just by looking at the picture? (Look it up online to see if you guessed the hemisphere correctly.)

Self-Test

(Check your answers in Appendix D.)

1. Which of the following processes increases the salinity of seawater?

   a. freezing

   b. rain, snow, and river runoff

   c. evaporation

   d. Both a and c are correct.

2. Which of the following processes decreases the salinity of seawater?

   a. freezing

   b. rain, snow, and river runoff

   c. evaporation

   d. Both a and c are correct.

3. The water in the _____ extends to a depth of 100 m, is relatively warm, and moves in broad, wind-driven currents.

   a. deep zone

   b. surface layer

   c. thermocline

   d. transition zone

4. Deep-water ocean currents are more influenced by the _____ of the water than by wind directions.

   a. density

   b. temperature

   c. salinity

   d. All of the above answers are correct.

5. In coastal regions, the ocean can act to _____.

   a. increase average temperatures of the land

   b. decrease average temperatures of the land

   c. induce ice formation

   d. moderate temperatures of the land

6. Which of the following statements about global sea level is correct?

   a. Global sea level is constant over both human and geological time.

   b. Global sea level can vary tens to hundreds of meters over geological time.

   c. Large variations in global sea level can be an effect of climate change.

   d. Both b and c are correct.

7. This diagram depicts the forces involved with the generation of tides on Earth. Label the diagram with the following terms:

   distorted water level

   inertial force

   gravitational force

   common center of mass of Earth–Moon system

   uniform level of water (theoretical)

   tide-raising force

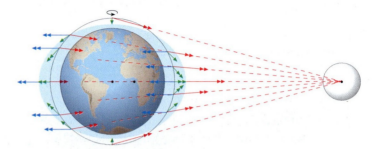

8. Shorelines are highly variable, but three basic types of coastlines are most common. Which of the following is *not* a common type of coastline?

   a. rocky coasts and wave-cut cliffs

   b. trenches

   c. lowland beaches and barrier islands

   d. coral reefs

9. This block diagram shows the beach and shore environment. Place arrows on the diagram, showing the direction of longshore current and sand transport. Be sure to label each of the sets of arrows.

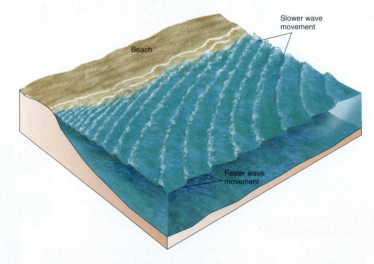

10. _____ is the most abundant chemical constituent of Earth's atmosphere.

   a. Oxygen

   b. Carbon dioxide

   c. Argon

   d. Nitrogen

   e. Water vapor

11. This diagram shows a graph of temperature variations vertically through the atmosphere. Label the diagram with the following terms:

   thermosphere   tropopause

   stratosphere   mesopause

   troposphere   stratopause

   mesosphere

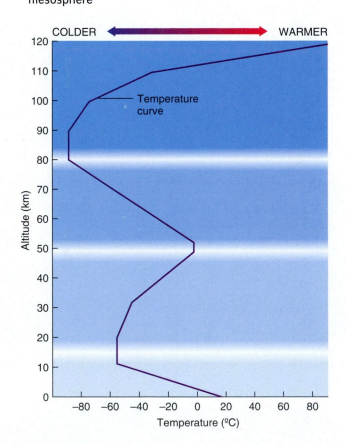

12. The atmosphere contributes to the warming of Earth's surface _____.

   a. by trapping ultraviolet rays in the troposphere

   b. by focusing the Sun's rays through refraction

   c. through evaporation, which causes heating of the surface

   d. by absorbing and reradiating infrared radiation

13. The atmosphere can shield organisms from harmful _____ rays from the Sun, as they are absorbed by O, O_2, and O_3 (ozone).

   a. infrared

   b. ultraviolet

   c. gamma

   d. neutron

14. The movement of air masses is a result of _____.

   a. differential heating of the curved Earth by the Sun

   b. the deflection of moving air masses by Earth's rotation

   c. the seasonal melting of ice in the polar regions

   d. Both a and b are correct.

   e. Both b and c are correct.

15. Subtropical high pressure is created by _____.

   a. descending cool and dry air at the juncture of a Ferrel cell and Hadley cell

   b. rising warm and moist air at the juncture of a polar cell and Ferrel cell

   c. rising warm and moist air at the juncture of a Ferrel cell and Hadley cell

   d. descending cool and dry air at the juncture of a polar cell and Ferrel cell

THE PLANNER ✓

Review your Chapter Planner on the chapter opener and check off your completed work.

Climatic Extremes: Deserts and Glaciers

The Mandara Lakes in Libya are stunning (see inset). The water is as flat as a mirror, reflecting a turquoise sky. The palm trees are dwarfed by the nearby dunes of the Sahara Desert. It is easy to see why early explorers were drawn to the harshness of the desert: Its vastness, its sense of danger, and the life-giving beauty of its oases make human undertakings seem trivial by comparison.

Explorers of another stripe, like this mountaineer exulting at the summit of Aurora Peak in Alaska, are fascinated by the world's icy wildernesses. In this case, the mountaineer is looking at Black Rapids Glacier. Almost every description of the Sahara Desert could also apply to this landscape: abstract, beautiful, immense, dangerous, remote, and vulnerable.

For geologists, deserts and glaciers are linked by more than beauty; they are the results of extreme climates—the extremes of dry and cold, linked together through the processes of our global climate system. The erosional power of wind and ice, which play lesser roles elsewhere, are dominant in these regions. In this chapter, we will look closely at the geological processes that operate in the locations that are characteristic of these climatic extremes.

CHAPTER OUTLINE

CHAPTER PLANNER ✓

- ❏ Study the picture and read the opening story.
- ❏ Scan the Learning Objectives in each section:
  p. 386 ❏   p. 395 ❏   p. 398 ❏
- ❏ Read the text and study all visuals.
  Answer any questions.

Analyze key features

- ❏ Process Diagram, p. 389 ❏   p. 391 ❏   p. 404 ❏
- ❏ Geology InSight, p. 392 ❏   p. 400 ❏   p. 408 ❏
- ❏ Case Study, p. 396 ❏
- ❏ Amazing Places, p. 402 ❏
- ❏ What a Geologist Sees, p. 410 ❏
- ❏ Stop: Answer the Concept Checks before you go on:
  p. 394 ❏   p. 398 ❏   p. 411 ❏

End of chapter

- ❏ Review the Summary and Key Terms.
- ❏ Answer the Critical and Creative Thinking Questions.
- ❏ Answer What is happening in this picture?
- ❏ Complete the Self-Test and check your answers.

Deserts and Drylands

LEARNING OBJECTIVES

1. **Identify** five different geological settings for deserts.

2. **Explain** how the wind shapes desert landforms through abrasion and deflation.

3. **Describe** the structure of a sand dune.

4. **Identify** five types of sand dunes.

onvection (see Chapter 12) creates global belts of rising and falling air masses. This results in three belts of high rainfall and four of low rainfall (**Figure 13.1**). High-rainfall belts are regions of **convergence**, where warm, moist air masses meet and rise. These belts lie along the equator and along the polar fronts, at 50°N and 50°S latitudes, resulting in warm-humid (equatorial) and cold-humid (polar front) climate zones. Low rainfall belts are regions of **divergence**, where cool, dry air masses sink downward. These belts lie in the two polar regions and in the subtropical regions at 30°N and 30°S latitudes creating two dry climate subtropical regions, and two dry, cold polar climates.

Types of Deserts

The word *desert* literally means a deserted region. However, irrigation is changing the meaning by making deserts suitable for agriculture and therefore habitable. **Desert** is now defined in terms of precipitation: Deserts form in **arid** lands, which receive less than 250 millimeters of precipitation per year. Deserts total 25% of the land area outside the polar regions. In addition, there is a smaller percentage of **semi-arid** land in which the annual rainfall ranges between 250 and 500 mm.

> **desert** An arid land that receives less than 250 mm of rainfall or snow equivalent per year and is sparsely vegetated unless it is irrigated.

The world's deserts • Figure 13.1

This map shows the distribution of arid and semi-arid climates and the major deserts associated with them. Many of the world's great deserts are located where belts of dry air descend along the 30°N and 30°S latitudes. Notice also that regions of cold, descending air also surround both of the poles. Despite being covered by ice, polar regions receive little precipitation and are considered to be frozen deserts; indeed, Antarctica is the world's largest desert.

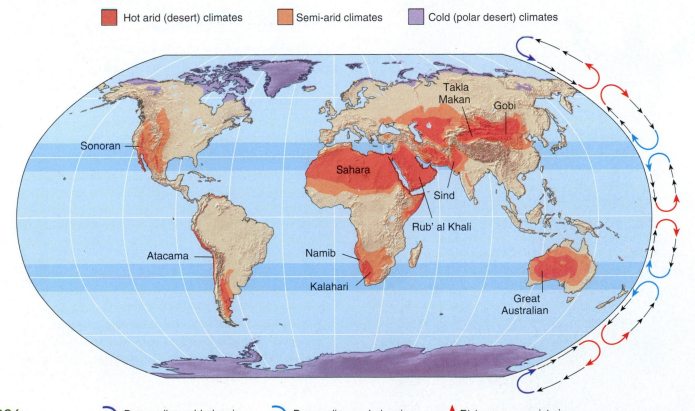

■ Hot arid (desert) climates   ■ Semi-arid climates   ■ Cold (polar desert) climates

Takla Makan · Gobi · Sonoran · Sahara · Sind · Rub' al Khali · Atacama · Namib · Kalahari · Great Australian

↓ Descending cold, dry air   ↓ Descending cool, dry air   ↑ Rising warm, moist air

Arid and semi-arid lands, collectively known as **drylands**, are characterized by a lack of available water. This is not just measured in terms of low precipitation but also the balance between precipitation and evaporation. If evaporation (i.e., water losses from the system) exceeds precipitation (i.e., water inputs into the system) on a regular basis, then water availability will be low.

Deserts can be separated into five categories, of which we have already mentioned two. The most extensive, the **subtropical deserts**, are associated with the two belts of low rainfall near the 30°N and 30°S latitudes. These include the Sahara, Kalahari, and Great Australian deserts. **Polar deserts** receive as little precipitation as subtropical deserts, but because the precipitation comes in the form of snow that never gets a chance to melt, these deserts gradually build up a thick ice sheet. The other three categories are less related to global air circulation patterns and more related to local geography. They are **continental interior deserts**, **rainshadow deserts**, and **coastal deserts** (**Figure 13.2**). Later in the chapter, we will take a closer look at the unique environment of the polar desert. For now, let's focus on the geological processes that characterize the world's hot deserts.

Hot deserts • Figure 13.2

a. The Sahara is the greatest of the world's subtropical deserts. Here, a camel caravan crosses the desert in Libya. ▼

b. Mongolian nomads transport their belongings through the Altai Mountains, which border on the Gobi desert. This is a typical inland continental desert, which (even though it lies outside the subtropical latitudes) receives little rain because it is so far from the ocean. ▼

c. A rainshadow desert forms when a mountain range creates a barrier to the flow of moist air, causing a zone of low precipitation to form on the downwind side of the range. Mountains do not completely block air, but they do effectively remove most of its moisture. Death Valley, seen here, lies just east of the Sierra Nevada, the tallest mountain range in the continental United States. ▼

d. Baja, California, the long, narrow peninsula in Mexico just south of its border with the western United States, consists mostly of coastal desert. Coastal deserts occur locally along the western margins of continents, where cold, upwelling seawater cools and stabilizes maritime air flowing onshore, decreasing its ability to form precipitation. ▼

Wind Erosion

As discussed in Chapter 7, wind is an important agent of erosion, transport, and deposition. Processes related to wind—**eolian** processes—are particularly effective as agents of erosion in arid and semi-arid regions.

Wind-blown sediment

Sediment carried by the wind tends to be finer than that moved by water or ice. Because air is far less dense than water, air cannot move as large a particle as water flowing at the same velocity. Typically, the largest particles that can be lifted in the airstream are grains of sand (**Figure 13.3**).

Although smaller in size, the particles in wind-blown sediment are similar to those in water-borne sediment in the way they travel. The largest grains are transported through **surface creep**. As wind speed increases, smaller grains may be bumped or lifted into the air, where they experience **saltation**. Finer dust-sized particles may be carried aloft to heights of a kilometer or so, where they can travel along in **suspension** as long as the wind keeps blowing. Such particles can be carried all the way across the ocean.

Mechanisms of wind erosion

Flowing air erodes the land surface in two ways. The first, **abrasion**, results from the impact of wind-driven grains of sand. Abraded rock acquires distinctive, curved shapes and a surface polish. A bedrock surface or stone that has been abraded and shaped by wind-blown sediment is a **ventifact** ("wind artifact"). When preserved in sedimentary strata, ventifacts can tell geologists the direction of prevailing winds in the past. Other landforms characteristic of desert regions, such as steep-sided but flat-topped **buttes**, also result, at least in part, from wind erosion of bedrock (**Figure 13.4**).

> **surface creep**
> Sediment transport in which the wind causes particles to roll along the ground.
>
> **saltation** Sediment transport in which particles move forward in a series of short jumps along arc-shaped paths.

> **suspension**
> Sediment transport in which the wind carries very fine particles over long distances and periods of time.
>
> **abrasion** Wind erosion in which airborne particles chip small fragments off rocks that protrude above the surface.

Movement of sediment by wind
• Figure 13.3

a. Under conditions of moderate wind speed, sand grains larger than 500 micrometers (0.5 mm) in diameter move by surface creep, while smaller grains (70–500 µm) move by saltation (see Figure **b**). Still finer particles (20–70 µm) are carried aloft in turbulent eddies.

b. Fast wind speeds cause movement of sand grains by saltation. Impacted grains bounce into the air and are carried along by the wind as gravity pulls them back to the land surface where they impact other particles, repeating the process.

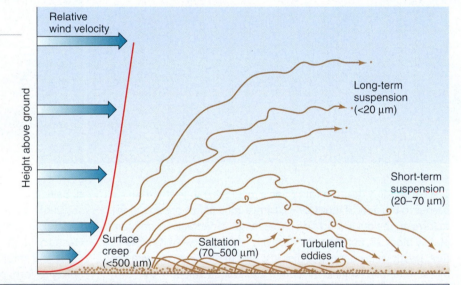

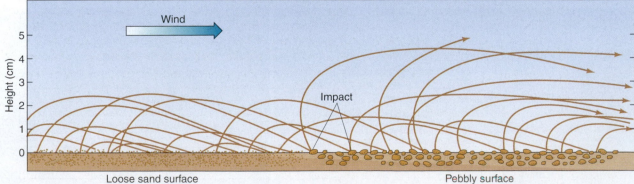

Erosion by wind • Figure 13.4

✓ THE PLANNER

The two main mechanisms of wind erosion are abrasion and deflation.

Abrasion

a. Wind-blown sand peppers the upwind side of an exposed rock, eventually abrading it to a smooth, inclined surface. ▼

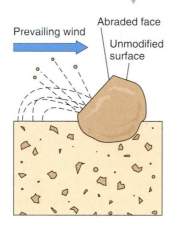

Prevailing wind

Abraded face

Unmodified surface

b. Ventifacts, each with at least one smooth, abraded surface facing upwind, litter the ground near Lake Vida in Victoria Valley, Antarctica. ▼

From deflation to desert pavement

c. These drawings and accompanying photos show how the progressive removal of sand from sediment with different-sized particles can lead to the formation of desert pavement. ▼

Deflation

Deflation

No further deflation

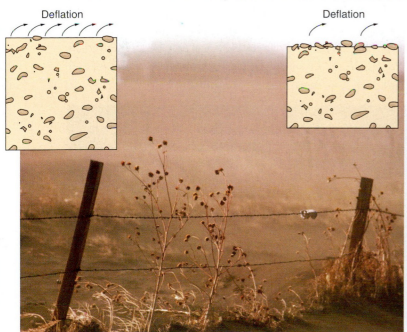

In the first photo, deflation is in progress in a plowed field in eastern Colorado.

The second photo shows desert pavement on the floor of Searles Valley in California. One hypothesis for its formation is that the gravel is too coarse for the wind to move, and the pavement is formed by deflation of fine-grained material. A second hypothesis is that a layer of gravel on the surface traps wind-transported dust.

Think Critically

What clues could you use to determine whether a rock is weathered by wind or by water?

The second important wind erosional process is called **deflation** (see Figure 13.4). Deflation on a large scale takes place only where there is little or no vegetation and where

loose particles are fine enough to be picked up by the wind. It is especially severe in deserts but can occur elsewhere during times of drought, when no vegetation or moisture is present to hold soil particles together. Continued deflation sometimes leads to the development of **desert pavement**; most of the fine particles are removed, leaving a continuous pavement-like covering of coarse particles.

Desert Landforms

Like water, wind as an agent of erosion removes material, transports it as sediment, and then deposits it elsewhere. This means that eolian landforms can be either erosional or depositional, or they can be a combination of both. One distinctive desert landform arises from wind erosion when a resistant stratum is underlain by a softer, more easily eroded stratum that is removed and carried away (**Figure 13.5**).

Any sediment that the wind removes from one place must eventually be deposited somewhere else—sometimes very far away. For example, distinctive reddish dust particles from the Sahara Desert have been identified in the soils of Caribbean islands, in the ice of Alpine glaciers, in deep-sea sediment, and in the tropical rainforests of Brazil. However, the most distinctive and characteristic eolian deposits found in deserts themselves are dunes.

Dunes Although little is known about how **dunes** begin, it is likely that they start where some minor surface irregularity or obstacle distorts the flow of air. Upon encountering an impediment, the wind sweeps over and around it but leaves a pocket

of slower-moving air immediately downwind. In this pocket of low wind velocity, sand grains moving with the wind drop out and begin to form a mound. The mound in turn influences the flow of air over and around it and may continue to grow into a dune.

A typical dune is asymmetrical, with a gentle windward slope (i.e., the side facing toward the wind) and a steep leeward face (i.e., the side facing away from the wind). Pushed by the wind, sand moves by surface creep and saltation up the gentle windward slope (**Figure 13.6**). When it reaches the top, the sand cascades down the steep leeward slope, also called the **slip face**. The slip face is always on the leeward side, so we can tell which way the wind was blowing from the asymmetrical form of the dune. Crisscrossed strata within the dune, called **cross beds**, are former slip faces. (These can be preserved as cross bedding in sedimentary rock derived from former sand dunes, as discussed in Chapter 8.)

Desert landform • Figure 13.5

Wind erosion created the "Finger of God," a pillar-like butte of sandstone in the Kalahari Desert of South Africa. The erosion-resistant sandstone rests precariously on a pyramid of easily eroded shale. In North America, rock pillars like these are known as hoodoos. Note the sandstone mesa, the table-like landform in the background, which was formed in a similar way.

How sand dunes form • Figure 13.6

a. This cross section through a sand dune shows the typical gentle windward slope (facing into the wind) and the steep slip face (facing away from the wind). The solid lines inside the dune show old slip faces.

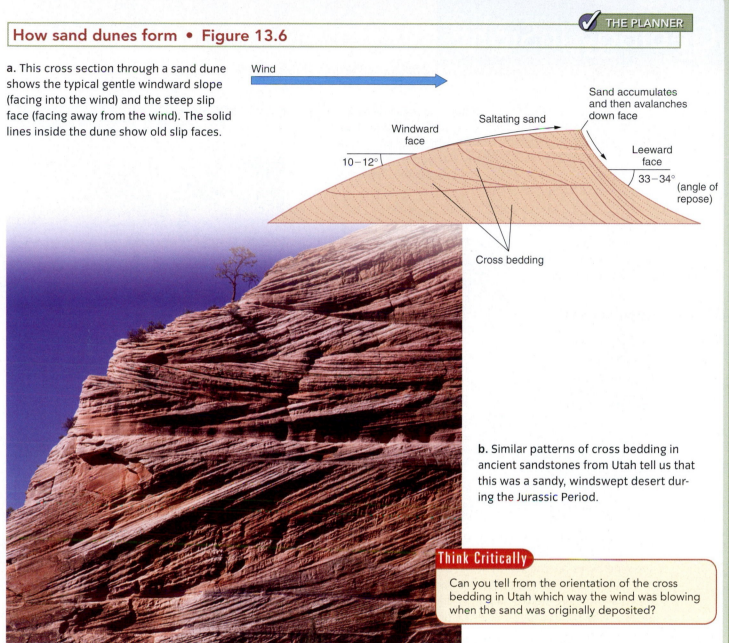

b. Similar patterns of cross bedding in ancient sandstones from Utah tell us that this was a sandy, windswept desert during the Jurassic Period.

Think Critically

Can you tell from the orientation of the cross bedding in Utah which way the wind was blowing when the sand was originally deposited?

The sliding sand on a slip face comes to rest at the **angle of repose**, the steepest angle at which loose particles will come to rest. The angle of repose varies for different materials, depending on factors such as the size and angularity of the particles. For dry, medium-sized sand particles, it is about 33−34°; the angle is generally steeper for coarser materials such as gravel and gentler for fine materials such as silt.

On Earth, three factors control the shape and behavior of a dune: the wind conditions, the amount of vegetation cover, and the characteristics and quantity of sand available. These three factors are shown schematically in the triangle in **Figure 13.7** (on the next page) along with photos of common types of dunes. Clearly, if there is no sand, or if there is plenty of vegetative cover to anchor sediment in place, dunes will not form. Some types of dunes tend to remain stationary. Others may migrate over long distances, and they can cause severe degradation and loss of agricultural productivity when they invade non-desert lands. This is part of the process of **desertification**.

Dunes are known to form on at least one other planet in our solar system: Mars. There, where there is no vegetative cover, the kinds of dunes observed are barchan and transverse, two of the kinds of dunes controlled principally by sand supply.

Geology InSight

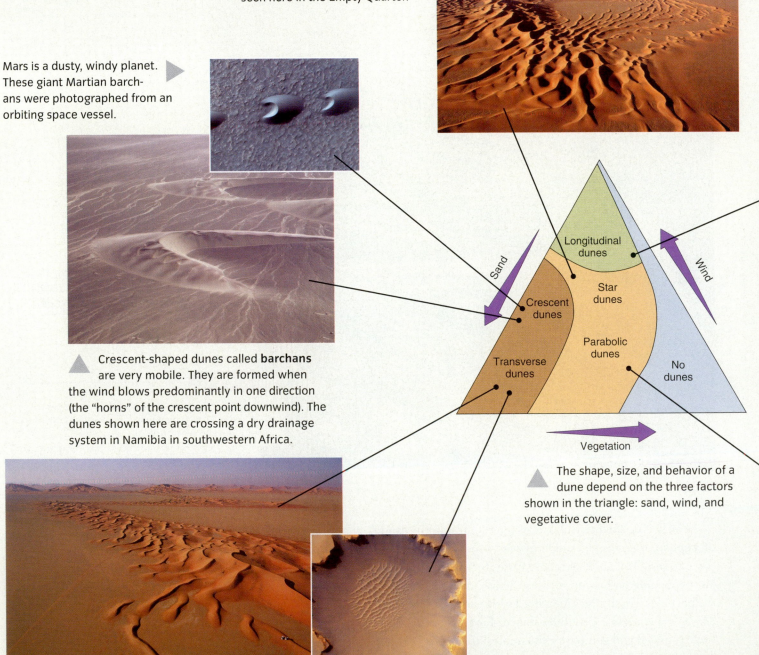

If the wind regularly blows in several different directions, it piles the sand up into stationary **star dunes**, such as the ones seen here in the Empty Quarter.

Mars is a dusty, windy planet. These giant Martian barchans were photographed from an orbiting space vessel.

Crescent-shaped dunes called **barchans** are very mobile. They are formed when the wind blows predominantly in one direction (the "horns" of the crescent point downwind). The dunes shown here are crossing a dry drainage system in Namibia in southwestern Africa.

When there is a copious sand supply, barchan dunes can merge and form **transverse dunes,** such as these dunes in the Empty Quarter of Saudi Arabia. They are oriented perpendicular to the prevailing winds.

A field of transverse dunes in Victoria Crater, Mars. This crater was explored by *Opportunity,* one of two robotic vehicles on Mars. The edge of the crater has been extensively modified by mass wasting and wind erosion.

The shape, size, and behavior of a dune depend on the three factors shown in the triangle: sand, wind, and vegetative cover.

Sand

Wind

Vegetation

Longitudinal dunes

Star dunes

Crescent dunes

Parabolic dunes

Transverse dunes

No dunes

Longitudinal dunes, like these dunes in Death Valley, run parallel to the prevailing winds. They form in deserts with meager sand supply. They can also form in areas with bidirectional winds, which push first on one side of the dune and then on the other.

Coastal regions, where the moist wind off the ocean allows vegetation to grow, form a typical environment for **parabolic dunes**. These are oriented in the opposite direction from barchan dunes: The arms, stabilized by vegetation, point upwind.

Deserts and Drylands **393**

a. After a storm, a flash flood thunders down this arroyo on the Navajo reservation in Arizona. As the floodwater subsides, sediment is deposited across the alluvial floor of the canyon.

b. Large alluvial fan extending out from the Altun Mountains on the southern border of the Taklimakan Desert, China.

Arroyos and other water-related desert landforms Contrary to popular belief (and many Hollywood movies), most deserts do not consist of endless expanses of sand dunes. Only one-third of the Arabian Peninsula, the sandiest of all dry regions, and only one-ninth of the Sahara Desert are covered with sand dunes. The remaining land area is either crossed by systems of stream valleys or covered by alluvial fans and alluvial plains. Running water, therefore, is important in the erosion of deserts, just as it is in rainy regions. However, instead of acting slowly and steadily, running water in deserts acts suddenly and in short bursts.

Rainfall in a desert region typically occurs during intense downpours that occur during a brief rainy season. The rapid runoff erodes steep-sided canyons, called **arroyos** into the landscape (**Figure 13.8a**). These are likely to be dry most of the year but are subject to flash floods during the wet season. When arroyos draining an upland region meet the flat desert floor below, they lose their ability to transport sediment and drop their load of sand and gravel in an alluvial fan (**Figure 13.8b**). If canyons are closely spaced along the base of a mountain range, the alluvial fans will sometimes coalesce into a broad alluvial apron called a **bajada**.

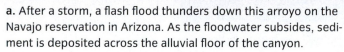

CONCEPT CHECK STOP

1. **Why** is a desert more likely to develop in the subtropics than in the tropics?

2. **What** is the role of wind in shaping desert landforms?

3. **What** structure of sedimentary layers is characteristic of sand dunes?

4. **What** three variables control the formation of the five main types of sand dunes?

Desertification and Land Degradation

LEARNING OBJECTIVES

1. **Clarify** the difference between desertification and land degradation, and explain how human activity can cause desertification.

2. **Outline** the major signs of desertification.

3. **Suggest** two things that can be done to slow or reverse desertification and land degradation.

Deserts are constantly changing—migrating, growing, or shrinking. **Desertification** of non-desert lands can result from natural environmental processes (mainly climatic change), or from human activities, or from a combination of the two. To distinguish between natural and anthropogenic desertification, the United Nations Environment Program and others who work in dryland management prefer to use the terms **land degradation** and **accelerated desertification** in reference to changes that result partially or primarily from human activity. Not all land degradation leads to desertification; for example, chemical contamination of soil is also a form of land degradation.

> **desertification**
> Invasion of desert conditions into non-desert areas.
>
> **land degradation**
> Land damage or loss of productivity caused by human activity, which may lead to the advance of desert conditions into non-desert areas.

Changing Deserts

In the region south of the Sahara lies a drought-prone belt of dry grassland known as the Sahel. There the annual rainfall is normally only 10 to 30 centimeters, most of it falling during a single brief rainy season. In the early 1970s, the Sahel experienced the worst drought of the 20th century. For several years in a row, the annual rains failed to appear, causing the adjacent desert to spread southward by as much as 150 kilometers. The drought extended from the Atlantic Ocean to the Indian Ocean, and it affected a population of at least 20 million.

Naturally occurring processes involving fluctuations in both precipitation and temperature have caused the Sahara Desert to advance and retreat many times over the past 10,000 years. However, the results of the drought were intensified by the fact that between about 1935 and 1970, the human population of the region had doubled, and the number of domestic livestock also increased dramatically. This resulted in severe overgrazing, and the grass cover was devastated by the drought. Millions of people suffered from thirst and starvation. The overgrazing is continuing today, leaving the Sahel at risk for another devastating famine if the dry weather should return.

The major signs of desertification include lower water tables, higher levels of salt in water and topsoil, reduction in surface water supplies, unusually high rates of soil erosion, and destruction of vegetation. Areas that are the most susceptible to desertification, whether by natural or human causes, are shown in the *Case Study* on the next page. Most are semi-arid fringe lands adjacent to the world's great deserts. In many of these places, humans have lived successfully in a semi-arid environment for centuries; what has changed is an explosion in the size of the population and the adoption of agricultural practices that may not be suited to that part of the world.

Mitigating Land Degradation

One of the best-known examples of accelerated desertification occurred in the United States and southern parts of Canada during the mid-1930s, when huge dust storms swept across the Great Plains and drove many farm families off their land. John Steinbeck, in his award-winning novel *The Grapes of Wrath*, described this resettlement, the largest forced migration in the country's history. The Southern Plains came to be called the "Dust Bowl," and historians refer to that period as the Dust Bowl years.

CASE STUDY

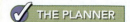

The Sahel and the Dust Bowl

Climate change and specifically drought are not the only causes of desertification. Human-induced stresses by overgrazing or poor farming practices are equally important.

Land degradation and resulting desertification can happen anywhere that inappropriate land uses cause stress that exacerbates the effects of natural climatic change. This happened in the North American Prairies during the 1930s Dust Bowl. Shown here, a massive dust storm closed in on Stratford, Texas, on April 18, 1935. Within a few minutes, the town would be enveloped in pitch darkness, and it would be impossible to see even the house in the foreground of this photo. The dust resulted from prairie lands being laid open to the elements by plowing. ▼

In the Sahel region of Niger, a herd of goats grazes on pasture at the edge of the desert. As the goats consume the remaining grass and bushes, the dunes of the desert will inevitably advance. Advances and retreats of the Sahara Desert in this region, caused by natural climatic changes, have been exacerbated by overly stressful use of the land. ▼

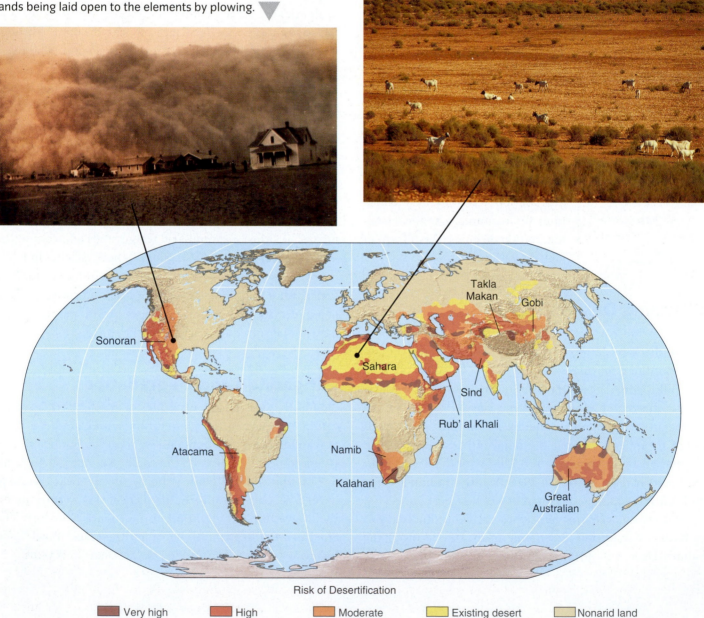

Risk of Desertification

■ Very high ■ High ■ Moderate ■ Existing desert ■ Nonarid land

This map identifies some of the world's great deserts, and regions that are most at risk of desertification today.

wheat

sorghum

corn

Stabilizing soil • Figure 13.9

One of the biggest changes in agriculture in the Great Plains since the 1930s has been the extensive use of groundwater for irrigation. These fields in Kansas use central-pivot irrigation, which minimizes evaporative loss of water and gives the fields a distinctive circular shape. In June, when this satellite photo was taken, wheat fields are bright yellow. Corn fields, dark green, are growing vigorously, and sorghum, light green, is just starting to come up. Irrigation helps to stabilize the topsoil. Farmers also use contour plowing, windbreaks, and other methods to protect the soil, with the hope of preventing another Dust Bowl. The protection afforded by irrigation comes at a cost, though: It is slowly depleting the High Plains Aquifer, which underlies much of the Midwest.

The Dust Bowl had both natural and human causes. Like the Sahel famine, it was triggered by a multiple-year drought. The effects of the drought were exacerbated by decades of poor land-use practices. The grasses that originally grew on the prairies had protected the rich topsoil from wind erosion. However, settlers gradually replaced those tall grasses with plowed fields and seasonal grain crops, which left the ground bare and vulnerable for part of the year.

Today, improved farming and irrigation practices have greatly reduced the risk of similar catastrophes (**Figure 13.9**). In both the United States and Canada, the Dust Bowl experience led to the establishment of soil management agencies at both the federal and state/provincial levels. Some programs and incentives for farmers to adopt better soil management practices

that were started just after the Dust Bowl continue even today.

How can desertification be halted or even reversed? The answer lies largely in understanding the geological principles involved and in the application of measures designed to reestablish a natural balance in the affected areas. Soil management techniques are widely known and readily available; they simply

need to be applied more aggressively. These techniques include crop rotation, windbreaks, contour plowing and terracing of steep slopes, low-till and no-till farming, and reforestation of vulnerable lands. Geologists and soil scientists can identify and map soils that are unsuitable for agriculture. Land use planners can eliminate the

Where Geologists CLICK

Desertification

http://www.fao.org/desertification

For a more extensive discussion of desertification and land degradation, go to the Web site of the United Nations Food and Agricultural Organization. At this site, you can search through a very large collection of photos showing desertification, soil erosion, and some of the efforts to manage them, categorized by region and theme.

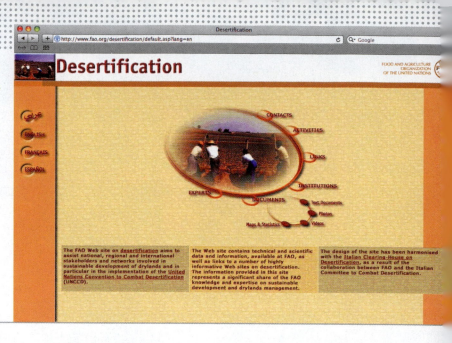

incentives to exploit arid and semi-arid lands beyond their capacity. The preservation of productive lands is essential to maintaining the world's food production capacity at the level needed for an increasing population. You can learn much more about how scientists, farmers, and other land managers are working to protect the world's productive lands from erosion and desertification by visiting the United Nations Food and Agricultural Organization Web site (see *Where Geologists Click*).

CONCEPT CHECK 🛑 STOP

1. **What** are the main causes of desertification and land degradation?

2. **What** are the main signs or symptoms of desertification?

3. **What** are some ways that farmers and others can address the problem of land degradation?

Glaciers and Ice Sheets

LEARNING OBJECTIVES

1. **Distinguish** between several different kinds of glaciers and ice formations.

2. **Explain** how temperate and polar glaciers differ.

3. **Describe** how ice in a glacier changes form, accumulates, ablates, and moves.

4. **Identify** several kinds of landforms created by glacial erosion and deposition.

Desertification can be an expression of natural climatic change, but another climate battle is played out in the vast deserts of the polar ice sheets. The expansion and shrinking of glaciers and ice sheets, both in the polar regions and in more temperate alpine settings, are expressions of the complex interplay between temperature and precipitation in the global climate system, which we will investigate in greater detail in Chapter 14.

The existence of glaciers and ice sheets is linked to interactions among several parts of the Earth system: tectonic forces that produce high, mountainous areas; the ocean, a source of moisture; and the atmosphere, which delivers moisture to the land in the form of snow. We now turn our attention to the great polar deserts and other parts of the **cryosphere**.

> **cryosphere** The perennially frozen part of the hydrosphere.

Components of the Cryosphere

Annual precipitation (snowfall) is generally very low in polar regions because the air is too cold to hold much moisture. The small amount of snow that does fall doesn't usually melt because summer temperatures stay very low. In areas where more snow falls each winter than melts during the following summer, the covering of snow gradually grows

glacier A semi-permanent or perennially frozen body of ice, consisting largely of recrystallized snow, which moves under the pull of gravity.

thicker. As the snow accumulates, its increasing weight causes the snow at the bottom to compact into a solid mass of ice. When the accumulating snow and ice become so thick that the pull of gravity causes the frozen mass to move, a **glacier** is born.

Ice caps that occur at high latitudes are worth special mention. An **ice sheet** is the largest type of glacier on Earth, a continent-sized mass of ice that covers all or nearly all the land within its margins. At present, ice sheets are found only in Greenland and Antarctica, though they have been much more extensive in the past. They contain 95% of the world's glacier ice (and 70% of the world's fresh water). They are so thick and heavy that some of the land underneath Antarctica has actually been pushed below sea level (**Figure 13.10**).

Ice shelves are thick sheets of floating ice hundreds of meters thick that adjoin glaciers on land. They are constantly replenished by land-based glaciers and also lose ice when large pieces called **icebergs** break off from them.

Sea ice, by contrast, is a form of ice cover that never touches land at all, but forms by the direct freezing of seawater. Antarctica is surrounded by sea ice, which in wintertime roughly doubles the apparent area of the continent. Most of the Arctic Ocean is covered by sea ice year-round.

Glaciers are cold because they consist primarily of ice and snow. However, scientists have found by drilling holes through glaciers that interior temperatures are not all the same. Some glaciers are warmer than others, and a difference in temperature influences the behavior and movement of the ice. In one kind of glacier, the ice is near its melting point throughout the interior. These glaciers, called **temperate glaciers**, form in low and middle latitudes. Meltwater and ice can exist together at equilibrium in temperate glaciers. At high latitudes and high altitudes, where the mean annual temperature is below freezing, the temperature in the glacier remains low, and little or no seasonal melting occurs. Such a cold glacier is commonly called a **polar glacier**. The five main types of glaciers are illustrated in **Figure 13.11**, on the next page.

Welcome to Antarctica • Figure 13.10

The East Antarctic Ice Sheet covers most of the continent of Antarctica, and the West Antarctic Ice Sheet overlies a volcanic island arc and the surrounding seafloor. In this satellite image, you can also see four ice shelves that occupy large bays. Glaciers that flow down from the mainland feed these shelves.

a. A cirque glacier, such as this one in Montana's Glacier National Park, occupies a bowl-shaped depression on a mountainside and typically serves as the source for a valley glacier.

b. Valley glaciers flow down valleys and are fed either from cirque glaciers or ice caps. This valley glacier is in Alaska's Wrangell Saint Elias National Park.

c. An **ice cap** covers a mountaintop (or low-lying land in the polar regions) completely and usually displays a radial-outward flow pattern. In this aerial photograph, the Greenland Ice Cap surrounds the Nunatak Mountains.

d. When a glacial valley is partly filled by an arm of the sea, the valley is called a **fjord**, and the glacier is a **fjord glacier**. Such glaciers often give rise to icebergs that break off and float away.

e. When a glacier flows all the way out of the mountains and onto the surrounding lowlands, it is called a **piedmont glacier**. The Columbia Glacier in Alaska starts as a valley glacier and then spreads out as a piedmont glacier.

Glaciers and Ice Sheets **401**

AMAZING PLACES

Glacier Bay National Park and Wilderness, Alaska

At the upper end of the Alaska panhandle is Glacier Bay National Park and Wilderness, a wonderland of snow-capped peaks, fjords, and numerous glaciers reaching down to the sea. In 1741, when Russian explorer Vitus Bering passed the place where today boats enter the bay from the Gulf of Alaska, he was blocked from entry by a wall of ice 3 km wide and more than 1 km thick (**Figure a**). Behind the blockage lay a huge ice mass known as the Great Pacific Glacier. Bering's visit was near the end of what today is known as the Little Ice Age (see Chapter 14). Returning home, Bering was wrecked and died on the Commander Islands at the western end of the Aleutians.

A century and a half later, in 1888, naturalist and conservationist John Muir visited the location. He found that he could to sail into the bay because the Great Pacific Glacier had retreated about 70 km. The ice retreat has continued since then, and today the head of the bay is over 100 km from the place where Bering's entry was blocked (**Figure b**). As the great glacier has retreated, 20 separate glaciers that once fed the giant now reach down to the sea, and the land exposed by melting ice is being covered by vegetation.

Ever-Changing Glaciers

Although we have defined a glacier as a semi-permanent or perennially frozen body of ice, glaciers are constantly changing in several ways. For example, the snow that falls on the surface of glaciers gradually changes into ice. Glaciers also shrink and grow in response to seasonal changes in temperature and precipitation. The ice in a glacier also moves, slowly but surely, under the influence of gravity. This movement is usually slow, but in some circumstances can be surprisingly rapid. Changes in climatic conditions also cause the margins of glaciers to advance or retreat. A dramatic example of glacial change is profiled in *Amazing Places*. Let's take a closer look at some of the processes of change in glaciers.

How glaciers form Newly fallen snow is very porous and easily penetrated by air. The presence of air in the pore spaces allows the delicate points of each snowflake to **sublimate** (change from solid directly to vapor, without melting). The resulting water vapor crystallizes in tiny spaces in the snowflakes, eventually filling them. In this way, the ice crystals in the snow pack slowly become smaller, rounder, more compact, and denser, until the pore spaces between them disappear (**Figure 13.12**). Snow that survives for a year or more becomes more compact as it is buried by successive snowfalls. As the years go by, the snow gradually becomes denser and denser, until it is no longer penetrable by air and becomes **glacier ice.** This process may take decades in the case of temperate glaciers to millennia in the case of polar ice sheets.

From snow to ice • Figure 13.12

As a new snowflake is slowly converted into a granule of glacier ice, it loses its delicate points through evaporation and recrystallization and becomes much more compact.

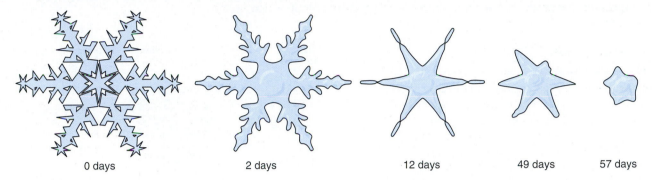

0 days   2 days   12 days   49 days   57 days

Further changes take place as the ice is buried deeper and deeper. As snowfall adds to the glacier's thickness, the increasing pressure causes the small grains of glacier ice to grow. This increase in size is similar to what happens when fine-grained rock recrystallizes as a result of metamorphism (see Chapter 10). Ice is, in fact, a mineral (see Chapter 2), and therefore glacier ice is technically a rock. However, the properties of this rock are very different from those of any other naturally occurring rock because of its very low melting temperature and its unusually low density. Ice floats in water, because it is only nine-tenths as dense as water.

How glaciers grow and shrink The mass of a glacier constantly changes as the weather varies from season to season and with local and global climatic changes over time. In a way, a glacier is like a checking account. Instead of being measured in terms of money, the balance of a glacier's account is measured in terms of the amount of snow deposited, mainly through snowfall in the winter, and the amount of snow (and ice) withdrawn, mainly through melting during the summer. The additions are collectively called **accumulation**, and the losses are collectively called **ablation** (**Figure 13.13**). The total added to the account at the end of a year—the difference between accumulation

A glacier has a budget • Figure 13.13

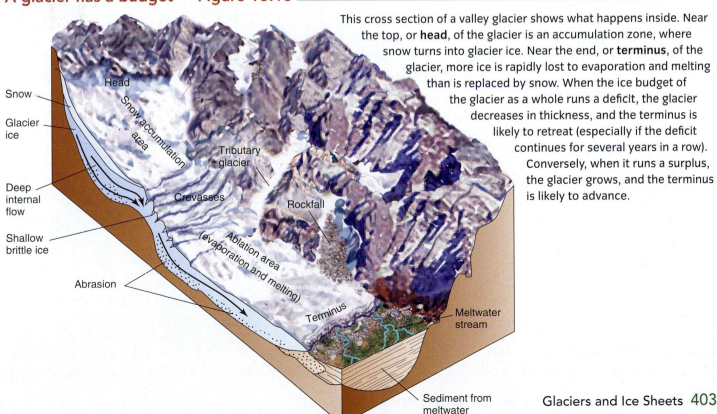

This cross section of a valley glacier shows what happens inside. Near the top, or **head**, of the glacier is an accumulation zone, where snow turns into glacier ice. Near the end, or **terminus**, of the glacier, more ice is rapidly lost to evaporation and melting than is replaced by snow. When the ice budget of the glacier as a whole runs a deficit, the glacier decreases in thickness, and the terminus is likely to retreat (especially if the deficit continues for several years in a row). Conversely, when it runs a surplus, the glacier grows, and the terminus is likely to advance.

Snow

Glacier ice

Head

Snow accumulation area

Deep internal flow

Tributary glacier

Crevasses

Rockfall

Shallow brittle ice

Ablation area (evaporation and melting)

Abrasion

Terminus

Meltwater stream

Sediment from meltwater

How ice deforms and flows internally • Figure 13.14

Two processes that produce changes in a mass of ice are recrystallization and internal flow.

Recrystallization

a. A deep ice core drilled at Russia's Vostok Station penetrates through the East Antarctic Ice Sheet to a depth of 2083 m. Microscopic examination of samples taken from different depths in the core show a progressive increase in the size of ice crystals, the result of slow recrystallization as the thickness and weight of overlying ice slowly increases with time.

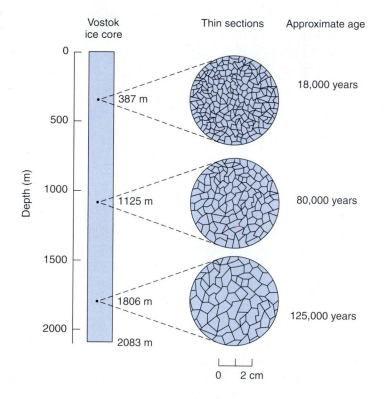

Internal Flow

b. Through a combination of recrystallization and slippage along crystal lattice planes, randomly oriented ice crystals are reorganized by stress so that their internal crystal planes are parallel.

c. The ice creeps (slips) along its internal crystal lattice planes, in a process very similar to playing cards in a deck of cards sliding past one another.

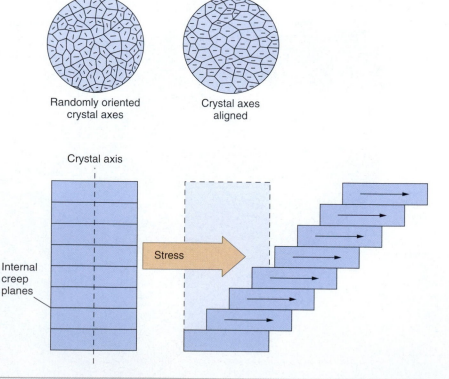

and ablation—is a measure of the glacier's **mass balance**. The account may have a surplus (a positive balance) or a deficit (a negative balance), or it may hold the same amount at the end of the year as it did at the beginning.

How glaciers move

According to our definition, a glacier moves because of the pull of gravity. How can we detect this movement? One method is to carefully measure the position of a boulder on its surface relative to a fixed point beyond the glacier's edge. If you measure the boulder's position again a year later, you will find that it has moved "downstream," usually by several meters. Actually, it is the ice that has moved, carrying the boulder along.

Measurements of velocity show that the ice in the central part of the glacier moves faster than the ice at the sides, and the uppermost layer moves faster than the lower layers. This is similar to what happens to water flowing in a stream (see Chapter 7), though it occurs much more slowly; in most glaciers, flow velocities range from a few centimeters to a few meters a day. It may take hundreds of years for an ice crystal that fell as a snowflake at the head of a glacier to reach the terminus and melt. The glacier ice moves in two basic ways: by internal flow, and by basal sliding across the underlying rock or sediment.

Deformation and internal flow

As the weight of overlying snow and ice in a glacier increases, individual ice crystals are subjected to higher and higher stress. Under this stress, ice crystals deep within the glacier undergo very slow movements, called **creep**, along internal crystal planes (see **Figure 13.14**). As the compacted, frozen mass moves, the crystal axes of the individual ice crystals are forced into the same orientation and end up with their internal crystal planes oriented in the same direction.

In contrast to the deep parts of a glacier, where ice flows by internal creep, the surface portion has relatively little weight on it and is brittle. When a glacier passes over a change in slope, such as a cliff, the surface ice cracks as tension pulls it apart. When the crack opens up, it forms a **crevasse**, a deep, gaping fissure in the upper surface of a glacier (**Figure 13.15**). Thus, ice moves in a glacier through a combination of ductile deformation at depth and brittle deformation at the surface—a pattern not dissimilar to the pattern seen in deformation of rock (see Chapter 9).

Crevasses: Brittle deformation • Figure 13.15

Deep fissures in the ice, called crevasses, open up as a result of stresses in the brittle surface layer of a glacier. The glacier flows in a direction perpendicular to a crevasse.

Glacial flow

Basal sliding Sometimes ice at the bottom of a glacier slides across its bed (the rock or sediment on which the glacier rests). This is called **basal sliding**. In temperate glaciers, meltwater at the base can act as a lubricant. Basal sliding may account for up to 90% of total observed movement in such a glacier, with the remaining 10% being internal flow. In contrast, polar glaciers are so cold that they are frozen to their bed; they seldom move by basal sliding, and all movement is by internal flow.

On infrequent occasions, a glacier may seem to go wild when ice in one part of the glacier begins to move rapidly downslope, producing a chaos of crevasses and broken pinnacles. Rates of movement have been observed that are up to 100 times those of ordinary glaciers. These episodes are called **surges**, and their causes are not fully understood (**Figure 13.16**). Geologists hypothesize that, in many cases, a buildup of water pressure at the base of the glacier reduces friction and permits very rapid basal sliding. The surge stops when the water finds an exit.

The "galloping glacier" • Figure 13.16

a. According to one hypothesis, a glacial surge gets started when water at the base of the glacier gets blocked from flowing out. The buildup of pressure lubricates the base and allows the glacier to flow very rapidly, until the water finds a way out again. ▶

b. In 1986, Hubbard Glacier in Alaska surged across the mouth of Russell Fjord, damming it and creating a fresh-water lake. This picture was taken soon after the dam broke and reopened the fjord. ▽

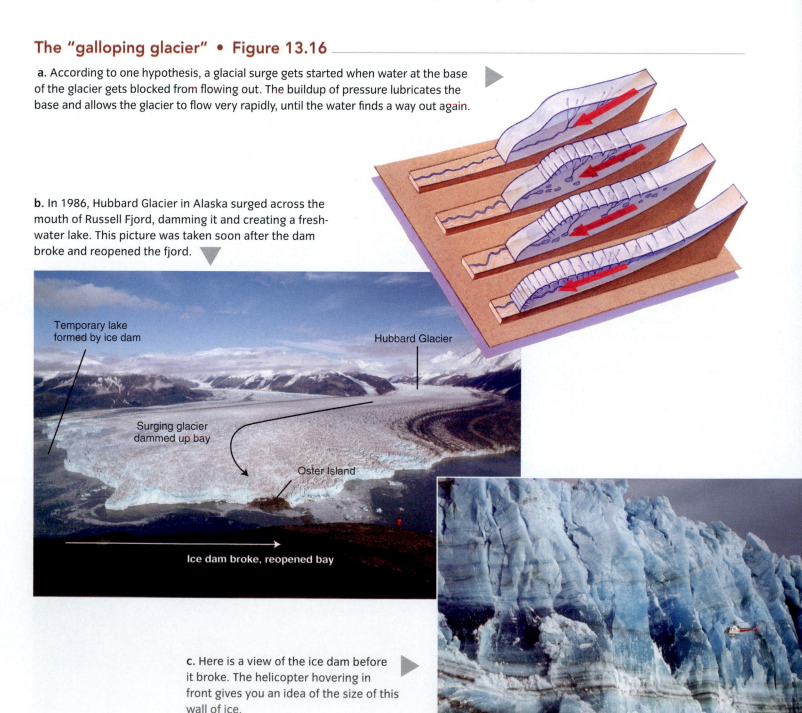

Temporary lake formed by ice dam

Hubbard Glacier

Surging glacier dammed up bay

Oster Island

Ice dam broke, reopened bay

c. Here is a view of the ice dam before it broke. The helicopter hovering in front gives you an idea of the size of this wall of ice. ▶

The Glacial Landscape

As glaciers move, they change the landscape by eroding and scraping away material, as well as by transporting and depositing material at their ends and along their margins. In changing the surface of the land over which it moves, a glacier acts like a file, a plow, and a sled. As a file, it rasps away firm rock. As a plow, it scrapes up weathered rock and soil and plucks out blocks of bedrock—glacial erosion. As a sled, it carries away the load of sediment acquired by plowing and filing, along with rock debris that falls onto it from adjacent slopes—glacial deposition. **Figure 13.17** illustrates some of the types of terrain that result from these processes.

Glacial erosion

The base of a glacier is studded with rock fragments of various sizes that are carried along with the moving ice. When basal sliding occurs, small fragments of rock embedded in the basal ice scrape away at the underlying bedrock and produce long, nearly parallel scratches called **glacial striations**. Larger particles gouge out deeper **glacial grooves** (**Figure 13.17a** on the next page). Because glacial striations and grooves are aligned parallel to the direction of ice flow, they help geologists reconstruct the flow paths of former glaciers.

Mountain glaciers produce a variety of distinctive landforms. Bowl-shaped **cirques** are found at a glacier's head. Two cirques on opposite sides of a mountain can meet to form a sharp-crested ridge called an **arête**. Cirques developing on all sides of a mountain may carve its peak into a prominent **horn** (like Mt. Everest, shown in **Figure 13.17b**).

When glacier ice moves downward from a cirque, it scours a valley channel with a distinctive U-shaped cross section and a floor that usually lies well below the level of tributary valleys (**Figure 13.17c**). Continental ice sheets can gouge the bedrock to form lakes; some examples of very large glacially formed lakes are the Great Lakes, Lake Winnipeg, and Great Bear Lake. Large glaciers and ice sheets are more effective agents of erosion and carve deeper valleys and lakes than small tributary glaciers. At the intersection of a smaller glacier and a larger glacier, there is usually an abrupt change in elevation of the valley floor due to the different depths of erosion.

Glacial deposition

Like streams, glaciers carry a load of sediment particles of various sizes. Unlike a stream, however, a glacier can carry part of its load at its sides and even on its surface. A glacier can carry very large rocks and small fragments side by side. When deposited by a glacier, the mixed rocky fragments (called glacial **till**) are not sorted, rounded, or stratified the way stream deposits usually are. In most cases, the boulders and rock fragments in a till are different from the underlying bedrock (**Figure 13.17d**).

> **till** A heterogeneous mixture of crushed rock, sand, pebbles, cobbles, and boulders deposited by a glacier.

The boulders, rock fragments, and other sediment carried by a glacier may be deposited along its margins or at its terminus. These form ridges called **moraines**. Specifically, **lateral moraines** form along the edges, a **terminal moraine** forms at the terminus (**Figure 13.17e**), and a **recessional moraine** forms as a glacier melts and recedes. If two glaciers converge, they may trap lateral moraines between them, forming a ridge of material that rides along the middle of the ice stream, called a **medial moraine** (**Figure 13.17f**). Geologists have used the locations of glacial moraines in the United States and Canada to determine how far the glacier ice cover extended over North America during the last ice age.

> **moraine** A ridge or pile of debris that has been, or is being, transported by a glacier.

The sinuous deposit shown in **Figure 13.17g** may look perplexing at first: It seems like an upside-down stream bed embossed upon the landscape. What could create such a feature? We have already mentioned that the bottoms of some temperate glaciers contain meltwater. This water may form a stream that tunnels through the glacier. (Such streams can sometimes be seen emerging from the terminus of an active glacier.) Like any other stream, it deposits sediment. If the glacier subsequently retreats, that sediment is left behind in a raised bed called an **esker**, like the one shown in the photo.

The retreat of a glacier can leave behind a terrain full of pits and pockmarks from abandoned blocks of ice embedded in the glacial debris. These subsequently melt, and the depressions left behind are called **kettles**. Many kettles fill with water to form kettle ponds and lakes. One famous example of a small kettle pond is Walden Pond, immortalized by the writer Henry David Thoreau.

Glacial erosion

a. These glacial grooves in Ohio were etched into limestone by the Wisconsin Glacier during the most recent ice age, about 35,000 years ago.

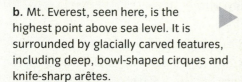

b. Mt. Everest, seen here, is the highest point above sea level. It is surrounded by glacially carved features, including deep, bowl-shaped cirques and knife-sharp arêtes.

c. The gorgeous Lauterbrunnen Valley in Switzerland has the classic U-shape of a glacial valley. The glacier that formed it no longer exists.

Glacial deposition

◄ **d.** Glacial till can sometimes include very large boulders, such as these boulders in Yellowstone National Park. When they are different from the bedrock, such boulders are called **erratics**.

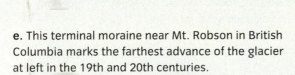

e. This terminal moraine near Mt. Robson in British Columbia marks the farthest advance of the glacier at left in the 19th and 20th centuries. ►

▲ **f.** The dark stripes running down the center of Kaskawulsh Glacier in the Yukon are part of a medial moraine.

▲ **g.** The curving ridge of sand and gravel in this photo is an esker, in Kettle Moraine State Park in Wisconsin.

WHAT A GEOLOGIST SEES

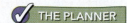

Periglacial Landforms

a. Looking at this curious patterned ground in Beacon Valley, Antarctica, a geologist would suspect that the pattern might somehow be connected with freezing and thawing processes in the soil.

c. The geologist would reason that the process of melting and freezing must have continued for a long time to develop such an extensive area of patterned ground. In summer, the crack opens or partially melts, allowing more water to enter. In winter, the ice freezes again. The ice wedge continues to grow as the melting and refreezing cycle repeats hundreds of times. By excavation, the geologist would discover that such wedges can grow as wide as 3 meters and as deep as 30 m.

b. Close examination reveals that each of the patterns is surrounded by fractures filled by wedges of ice. The geologist realizes that the ice wedges must have formed when water seeped into open cracks in the ground and subsequently froze.

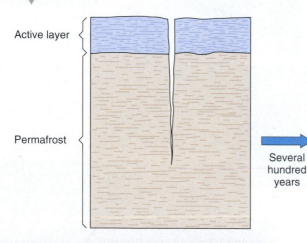

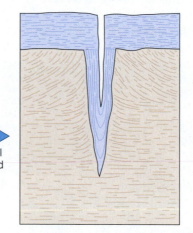

Active layer

Permafrost

Several hundred years

d. In 2008, a small robot called *Phoenix* landed in a northern region of Mars where the terrain has a curious pattern that geologists thought might be patterned ground like that in Antarctica. To test their hypothesis, they activated a chemical analyzer on *Phoenix* that confirmed the presence of ice in the wedges by baking a soil sample just enough to melt the ice.

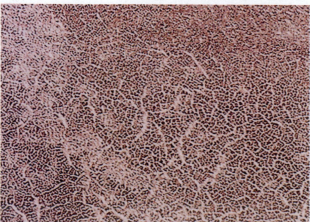

Think Critically

Can you think of any other natural environments where you have seen a pattern of hexagonal polygons similar to this one?

Periglacial landforms Periglacial areas, which lie near or adjacent to glaciers, also have distinctive landforms that result from intense frost action and a large annual range in temperatures. The most common type of landscape in present-day periglacial regions is **tundra**, a treeless landscape with long winters, very short summers, poorly developed soils, and low, scrubby vegetation. Tundra is usually characterized by a layer of **permafrost**. During the short summer, the ice in permafrost melts only in a thin layer near the surface, called the **active layer**. The freeze–thaw cycle produces characteristic geological formations called **ice wedges** and **patterned ground** (see *What a Geologist Sees*).

> **permafrost** Ground that is perennially below the freezing point of water.

CONCEPT CHECK

1. **How** does a cirque glacier differ from a valley glacier?

2. **What** processes and characteristics distinguish temperate glaciers from polar glaciers?

3. **What** has to happen for a glacier to advance or retreat?

4. **What** is a moraine? What is significant about a terminal moraine?

 THE PLANNER

Summary

1 Deserts and Drylands 386

- The term **desert** refers to arid lands where annual rainfall is less than 250 mm. Five types of deserts have been identified: subtropical, continental interior, rainshadow, coastal, and polar. Subtropical and polar deserts result from the global wind patterns that create dry high-pressure air masses around 30°N and 30°S and at the poles, as shown in the map. The other three kinds of deserts result from more local geological conditions.

The world's deserts • Figure 13.1

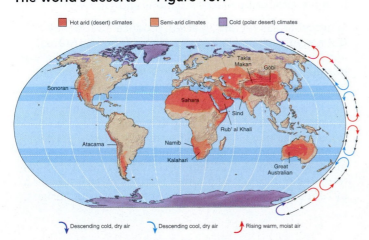

Hot arid (desert) climates   Semi-arid climates   Cold (polar desert) climates

Sonoran
Takla Makan
Gobi
Sahara
Sind
Rub' al Khali
Atacama
Namib
Kalahari
Great Australian

Descending cold, dry air   Descending cool, dry air   Rising warm, moist air

- Eolian (wind) erosion is particularly effective in arid and semi-arid regions. Wind moves particles through **surface creep**, **saltation**, and **suspension**. Flowing air erodes the land surface through the processes of **abrasion** and **deflation**.

- **Dunes** are hills or ridges of sand deposited by winds. They are asymmetrical, with a gentle slope facing the wind and a steeper slip face on the leeward side. Common types of dunes are barchan, transverse, star, parabolic, and longitudinal. The type of dune that will form in a given place depends on the amount of sand, the wind conditions, and the amount of vegetation.

- Contrary to the popular image of a desert, the majority of desert lands are not covered by sand. Water erosion is an important geologic process in deserts. Flash floods carve deep canyons called arroyos and create depositional landforms such as alluvial fans.

2 Desertification and Land Degradation 395

- **Desertification** involves the invasion of desert conditions into nondesert lands, which can result in accelerated soil erosion, decline in water tables, and loss of agricultural productivity. Desertification can be caused by natural environmental changes or by human activities, and can lead to famines affecting millions of people.

- When deserts advance as a result of poor land-use practices, it is referred to as **land degradation** or accelerated desertification.

- Appropriate land use, the application of simple soil management approaches, and carefully designed irrigation see photo, can prevent or mitigate many of the negative consequences of desertification.

Stabilizing soil • Figure 13.9

3 Glaciers and Ice Sheets 398

- The perennially frozen part of the hydrosphere is called the **cryosphere**. It includes **glaciers** on land as well as sea ice. Glaciers come in several varieties and can form either in polar regions or at high altitudes in temperate regions. Most of the world's fresh water is locked in vast ice sheets in Antarctica and the North Pole region.

- Glacier ice is formed by the compaction of grains of snow and recrystallization of small ice crystals into larger ones—processes that are similar to lithification and metamorphism in ordinary rock. The mass of a glacier can change from season to season and year to year through accumulation and ablation as shown in the diagram. Accumulation (by new snowfall) predominates at the head of the glacier, and ablation (by melting) predominates at the terminus.

A glacier has a budget • Figure 13.13

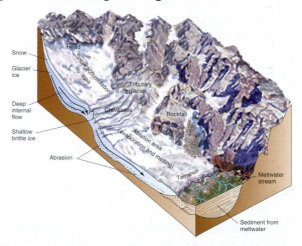

- Just like ordinary rock, glacial ice can move by either ductile or brittle deformation. The internal movement is ductile and aligns the ice crystals in the direction of flow. Brittle deformation occurs when a glacier goes over a change in slope, opening up fissures called crevasses. At the bottom of a glacier, where basal sliding occurs, small fragments of rock embedded in the ice scrape away at the underlying bedrock, producing glacial striations and grooves.

- In mountainous regions, glaciers produce a variety of distinctive erosional landforms, such as cirques, arêtes, and U-shaped valleys. Common periglacial features include **permafrost** and patterned ground, which are formed by the repeated freezing and thawing of groundwater.

- Glacial deposits often consist of unsorted **till**. A **moraine** is a ridge or pile of debris being carried along by a glacier or deposited along its edge or terminus. When a glacier retreats, the moraine is left behind. Such deposits are useful for identifying the previous extent of glaciers that have retreated or disappeared. Other features often left behind by retreating glaciers are kettle lakes and elevated eskers.

Key Terms

- abrasion 388
- cryosphere 398
- deflation 390
- desert 386
- desertification 395
- dune 390
- glacier 399
- land degradation 395
- moraine 407
- permafrost 411
- saltation 388
- surface creep 388
- suspension 388
- till 407

Critical and Creative Thinking Questions

1. We have discussed the connection between large atmospheric circulation patterns and the occurrence of subtropical deserts. What climatic and atmospheric processes contribute to the formation of rainshadow deserts, coastal deserts, and continental interior deserts?

2. Investigate the current status of drought and land degradation in the Sahel or elsewhere. Can you find any information about soil erosion control techniques or other methods that are being used to combat desertification?

3. Investigate the history of soil conservation programs and organizations that emerged from the experiences of the Dust Bowl years. How have these organizations changed over time?

4. Do some research on the most recent ice age. Do you live in an area that was formerly covered by ice? If so, how thick was the ice? Is there any evidence in the landforms around you to indicate that the area was formerly glaciated?

5. Many glaciers in locations around the world have been retreating at a rapid rate for the past few decades. Why? Is the retreat caused by natural or anthropogenic processes? We will discuss this in greater detail in Chapter 14. Choose one glacier and investigate its current status; find out whether the glacier is advancing or retreating, and how quickly.

What is happening in this picture?

This house was built on top of permafrost in Dawson City, Yukon Territory, Canada.

Think Critically

1. How might warming temperatures affect the permafrost?
2. Why would this cause the surface to buckle and subside?

Self-Test

(Check your answers in Appendix D.)

1. _____ and _____ deserts result from the global wind patterns that create dry high-pressure air masses around 30°N and 30°S and at the poles.
   a. Subtropical, continental interior
   b. Subtropical, polar
   c. Subtropical, rainshadow
   d. Continental interior, rainshadow
   e. Continental interior, polar

2. Deflation on a large scale takes place only _____.
   a. where there is little or no vegetation
   b. where loose particles are fine enough to be picked up by the wind
   c. in desert environments
   d. Both a and b are correct.
   e. Both b and c are correct.

3. A typical sand dune _____.
   a. is asymmetrical
   b. has a gentle windward slope
   c. has a steep leeward face
   d. All of the above are correct.

4. The diagram below depicts dune formation as a function of wind, sand supply, and vegetation cover. Label the diagram with the following terms:

   longitudinal dunes   transverse dunes
   barchan dunes   parabolic dunes
   star dunes

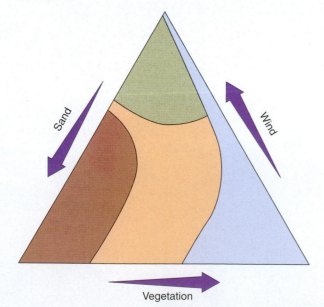

5. Which one of the following is *not* one of the major indicators of desertification?
   a. lower water tables and a reduction of surface waters
   b. increased surface temperatures in summer months
   c. higher levels of salt in water and topsoil
   d. unusually high rates of soil erosion
   e. destruction of vegetation

6. Arroyos and bajadas are _____.
   a. common products of wind erosion
   b. two possible end results of subsidence in desert environments
   c. desert landforms created by running water
   d. signs of human impact on the desert landscape

7. This diagram shows various glaciers at Earth's surface. Label the illustration with the appropriate terms for the individual glaciers A through E:

   cirque glacier   fjord glacier
   piedmont glacier   ice cap
   valley glacier

8. In _____, meltwater and ice can exist together in equilibrium.
   a. temperate glaciers
   b. polar glaciers
   c. ice caps

9. In _____, little or no seasonal melting occurs.
   a. temperate glaciers
   b. polar glaciers
   c. ice caps
   d. Both b and c are correct.

10. Ice in a glacier typically undergoes both brittle and ductile deformation.
   a. true
   b. false

11. Glacier ice moves under the influence of gravity through _____.
   a. basal sliding
   b. internal flow
   c. Both a and b occur in all glaciers.
   d. Either a or b can occur, depending on the characteristics of the glacier.

12. Which of the following is formed by glacial erosion?
   a. striations
   b. permafrost
   c. eskers
   d. till

13. Which of the following is formed by the deposition of glacial sediment?
   a. grooves
   b. arêtes
   c. moraines
   d. crevasses

14. This photograph shows a glacial landscape. What is the name for the curving ridge of sand and gravel that dominates the picture?
   a. erratic
   b. esker
   c. kettle
   d. terminal moraine

15. The term **periglacial** refers to land that _____.
   a. was glaciated in a former time
   b. is currently lying underneath glacier ice
   c. is near or adjacent to a glacier
   d. is at risk of becoming glaciated

THE PLANNER

Review the Chapter Planner on the chapter opener and check off your completed work.

Earth's Climates: Past, Present, Future

14

The farms of Kufra Oasis in the Sahara Desert depend on irrigation by groundwater from the Nubian Aquifer. Radar imaging from space (inset) has allowed scientists to see what lies beneath desert sand, revealing ancient buried river systems in the Sahara Desert. The image in the inset photo was acquired using Spaceborne Imaging Radar-C (SIR-C) on the space shuttle Endeavour in 1994, part of NASA's Mission to Planet Earth. The image shows the Kufra region of southeast Libya. The stream channels make up the **paleodrainage** system of Wadi Kufra. In the past, when the climate was much wetter, these channels carried running water.

Today the area is extremely arid, with only about a millimeter of rainfall annually. The ancient river channels are now dry valleys, called *wadis*, deeply buried by wind-blown sand. The two channels come together at Kufra Oasis, marked by a group of circular agricultural fields at the righthand edge of the image. The most productive wells at Kufra are located where the ancient river valleys converge, indicating that the buried channels serve as conduits for groundwater recharge.

The Sahara Desert has advanced and retreated numerous times in both historic and prehistoric times. These ancient channels show that the desert once had a much wetter and more hospitable climate.

CHAPTER OUTLINE

CHAPTER PLANNER ✔

- ❑ Study the picture and read the opening story.
- ❑ Scan the Learning Objectives in each section:
  p. 418 ❑   p. 420 ❑   p. 431 ❑   p. 442 ❑
- ❑ Read the text and study all visuals.
  Answer any questions.

Analyze key features

- ❑ Process Diagram, p. 422 ❑
- ❑ Case Study, p. 421 ❑   p. 436 ❑
- ❑ Geology InSight, p. 434 ❑
- ❑ What a Geologist Sees, p. 438 ❑
- ❑ Amazing Places, p. 428 ❑
- ❑ Stop: Answer the Concept Checks before you go on:
  p. 420 ❑   p. 431 ❑   p. 441 ❑   p. 449 ❑

End of chapter

- ❑ Review the Summary and Key Terms.
- ❑ Answer the Critical and Creative Thinking Questions.
- ❑ Answer What is happening in this picture?
- ❑ Complete the Self-Test and check your answers.

The Climate System

LEARNING OBJECTIVES

1. **Describe** the main components of Earth's climate system.

2. **Explain** why the climate system is so complex.

3. **Identify** the six major climate categories.

Climatically extreme areas such as deserts, glaciers, and polar regions (see Chapter 13) have taken on special importance in recent years because they are particularly sensitive to changes in **climate** and thus are likely to be early indicators of **global warming**. Global warming is a scientifically complex issue, and it is also politically and emotionally complex. It is important to emphasize the difference between what we know is happening and what we think is happening—a distinction that is often lost in the doomsday accounts one sees on television, online, and in the press. This is a scientific approach, and it is the approach that we will take in this chapter.

In this chapter we investigate the natural causes of climate change, and we examine the evidence demonstrating that climates have changed throughout Earth history. We look at some of the tools and techniques that scientists use to study climates of the past. We also consider how human activities are affecting the climate system. We start our study of climate by looking at the components of the climate system and the interacting processes and feedbacks that characterize it.

climate The average weather conditions of a location or region over time.

global warming Present-day warming of the world's climate that most scientists believe is likely to continue and is at least partly caused by human activities.

Components of the Climate System

Earth's climate system is complex and is driven by interactions among all of the planet's major subsystems (**Figure 14.1**): atmosphere, hydrosphere (mainly the ocean), cryosphere, geosphere, and biosphere. Most recently, these have been further influenced by processes in the **anthroposphere**, the realm of human activity. The components of the climate system interact so closely that a change in one of them causes changes in the others.

Earth's climate system • Figure 14.1

The climate system has five major interacting components: geosphere, atmosphere, hydrosphere, cryosphere, and biosphere.

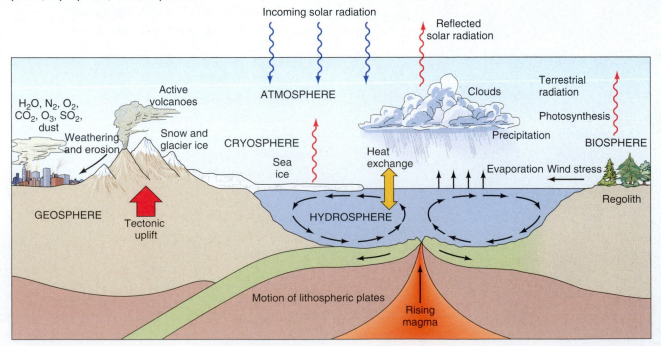

Global climate zones • Figure 14.2

In a simplified version of the Köppen climate classification system, there are six basic climate zones: tropical, dry, temperate-humid, cold-humid, polar, and highland. These zones are strongly related to the global pattern of atmospheric circulation and the distribution of major ecosystems.

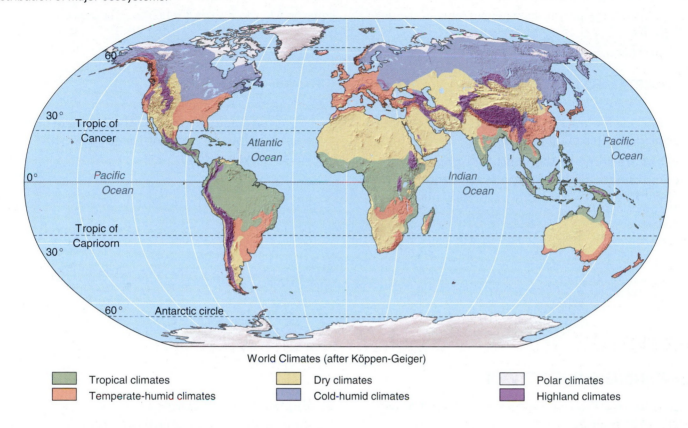

World Climates (after Köppen-Geiger)

- Tropical climates
- Temperate-humid climates
- Dry climates
- Cold-humid climates
- Polar climates
- Highland climates

All of the components of the climate system, except the geosphere, are driven by solar energy. Some incoming solar radiation is reflected back to space by clouds, atmospheric pollutants, ice, snow, and other reflective surfaces. The remainder of the incoming solar energy is absorbed, stored, and transferred through various reservoirs in the Earth system. It is subsequently re-emitted at longer wavelengths, from the three great reservoirs—atmosphere, ocean, and land. The radiative balance between incoming and outgoing energy, moderated by the **greenhouse effect** (see Chapter 12), is crucial in maintaining a stable global climate.

Vegetation is particularly important in the climate system because it influences the composition of air by acting as a reservoir for carbon dioxide (a greenhouse gas) and by affecting humidity and local cloud cover. It also helps determine the reflectivity of the land surface. Its absence can increase wind erosion, which can influence climate by affecting how much dust is in the atmosphere.

The ocean serves as a great reservoir of thermal energy that helps moderate climate because of its ability to absorb energy and retain it in the form of heat. As you learned in Chapter 12, the ocean also is extremely important in controlling the composition of the atmosphere because seawater contains a large amount of dissolved carbon dioxide. If the balance between oceanic and atmospheric carbon dioxide reservoirs were to change by even a small amount, the radiation balance of the atmosphere would be affected, bringing about a change in world climates.

Present-Day Climates

Scientists commonly use variations of the **Köppen system** to describe the present-day distribution of climate zones (**Figure 14.2**). This climate classification system, originally developed by Austrian climatologist Vladimir Köppen, is based on the distribution of native vegetation types. It combines measurements of average temperature and precipitation to define six major climate categories. The principal categories can be refined by modifiers that indicate variations of precipitation, temperature, and seasonality.

The most notable feature of a map showing the world distribution of climate zones is the close link between climate zones and the prevailing wind currents in the global atmospheric circulation system, shown in Figure 12.22. For example, the world's dry climates and deserts are mostly concentrated along the 30°N and 30°S latitudes, where the descending air masses are dry. The wet tropical climates and rainforests of Africa and South America lie near the equator, where we expect lots of warm, moist air.

CONCEPT CHECK STOP

1. **What** are the main components that interact to control Earth's climate?

2. **What** are some of the ways that different components of the climate system interact with each other?

3. **How** can scientists modify the six main climate classifications to account for regional and seasonal variations?

Natural Causes of Climate Change

LEARNING OBJECTIVES

1. **Summarize** the main external factors that cause climate change.

2. **Summarize** the main internal factors that cause climate change.

3. **Explain** how feedbacks contribute to complexity in the climate system.

s discussed in the chapter opener, we know that deserts and glaciers expand and retreat as a result of natural causes, and have done so throughout Earth history (see the *Case Study*). Several natural mechanisms cause changes in global climate, on both short and long time scales. Some of these mechanisms are external to the Earth system, and others are internal. It is important that we understand these natural mechanisms of change, before we go on to consider the human input into climate change.

External Influences on Climate

The Sun, the main source of energy driving Earth's climate system, varies in its energy output on both long and short time scales. Fluctuations in the energy output of the Sun should result in overall cooling of the planet when the output is low and warming when the output is high. One problem with this hypothesis is that there is no direct way to measure what the solar output actually was at times in the distant past, when Earth experienced **glaciations** or **interglaciations**. Tests must therefore be based on computer modeling of solar output, and the results of such tests have been contradictory. Solar variability must play a fundamental role, but its influence is tempered by many other factors (both external and internal) that contribute to climate change.

> **glaciation** (or **glacial period** or **ice age**) A relatively cold period, when Earth's ice cover greatly exceeded its present extent.
>
> **interglaciation** (or **interglacial period**) A relatively warm period, when Earth's ice cover and climate resembled those of the present day.

CASE STUDY

Wetter Times in Wadi Kufra

At Kufra Oasis (the cluster of circular fields in the chapter-opening inset photo), there is an abundance of fresh water of very high quality, stored deep underground in the Nubian Aquifer. Geologists have studied the isotopic composition of the water to determine its source and its age (i.e., how long it has been underground). They have also analyzed woody organic matter from ancient soils (**paleosols**) along the now-buried river channels; radiocarbon dating of these materials provides ages for the ancient soil surfaces.

Studies have shown that present-day groundwater recharge in this extremely arid environment is virtually nonexistent; all water that falls as precipitation evaporates before it is able to infiltrate. However, there have been at least three significant recharge events of long duration (i.e., hundreds of years) during the Holocene, the geologic epoch that began 11,700 years ago. These humid periods, called **pluvials**, were characterized by much higher rainfall (on the order of 300 to 400 mm/yr). The last of the pluvials ended about 3500 years ago, and the Sahara has been both drier and hotter since then.

The pluvial periods provided enough rainfall to support surface water bodies and comparatively lush vegetation. Evidence for wetter periods has been bolstered by discoveries of the remains of large mammals, such as elephants and hippos, which cannot survive in the current arid conditions. The pluvials also supported early humans in the region. Archaeologists have used SIR-C images (like the chapter-opening photo) as "road maps" to help locate human artifacts buried under the sand and to better understand the history of early people and climatic conditions in the region. The nicknames "Wet Sahara" and "Green Sahara" are sometimes used to describe the Neolithic Period (around 9500 years ago) in this region, when inhabitants of the Sahara maintained settlements of significant size and carried out farming and cattle herding.

A scientist holds up a belly plate from a soft-shelled turtle from the Tenerian culture, dating from the "wet period" of the Sahara. The presence of this animal's remains indicates a much wetter climate at that time.

Cave paintings like this one from the Adrar Akakus region of modern-day Libya, some as old as 12,000 years, depict palm trees, animals, and activities indicative of a much greener and wetter climate where there is now only desert.

The 11-year-old girl in this burial at Gobero, Niger is wearing a bracelet on her upper arm that was carved from the tusk of a hippo 4800 years ago. Today hippos only live in much wetter parts of Africa, farther to the south. Pluvial periods in the Sahara supported human populations.

Think Critically

The southern boundary of the Sahara Desert has both advanced and retreated in recent decades, although overall it has advanced. This desertification is partly a result of natural climatic variations (mainly rainfall) and partly a result of human land mismanagement. How might a scientist begin to figure out which of these influences is the most important in causing desertification?

Natural Causes of Climate Change   421

How orbital changes influence climate • Figure 14.3

Three kinds of orbital change combine to affect Earth's climate. These factors interact to control how much sunlight reaches Earth at any given time.

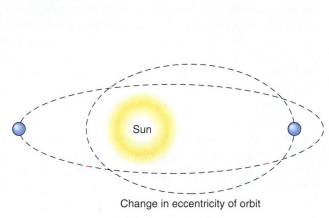

Change in eccentricity of orbit

Axial tilt change

Wobble (Precession)

a. Earth's orbit becomes more and less elongate, varying on a period of about 100,000 years.

b. Earth's axis changes its tilt, varying on a period of about 41,000 years.

c. Earth's axis wobbles in a circle, with a period of about 26,000 years.

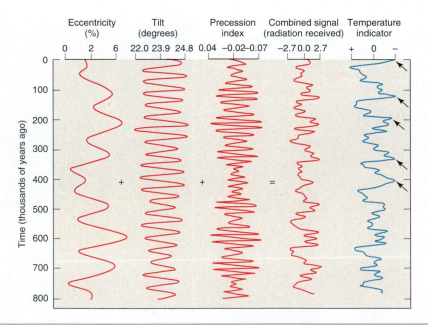

d. These curves show variations in eccentricity, tilt, and precession during the past 800,000 years. Summing these factors produces a combined signal that shows the amount of radiation received on Earth at any given latitude, over time (see the curve labeled "combined signal"). The magnitude and frequency of changes in the combined signal closely matches those of the curve at the far right, which is an indicator of temperature. This supports the hypothesis that Earth's orbital changes influence the timing of glacial–interglacial cycles (arrows).

Variations in Earth's orbit also affect the climate system (**Figure 14.3**), mainly by influencing how much solar radiation reaches Earth's surface, where, and at what times of year. Changes in the **eccentricity** (departure from circularity) of Earth's orbit, variations in the degree of **tilt** of the planet's axis of rotation, and **precession** (wobbling) of the axis are all important influences that vary on different time cycles but with great predictability. Sometimes these factors cancel each other out. At other times, they align in such a way that incoming solar radiation is minimized or maximized. The times and magnitudes of these variations, called **Milankovitch cycles**, can be calculated to a high degree of certainty, and they correspond fairly closely to the timing of past glacial and interglacial cycles (**Figure 14.3d**).

Milankovitch cycles The combined influences of astronomical-orbital factors that produce changes in Earth's climate.

Internal Influences on Climate

External factors such as the Sun's output and Earth's orbital variations set the stage for our climate system, but there are many internal factors that moderate these influences.

Atmospheric filtering

The first obvious candidate for an internal influence on climate is the atmosphere. Solar radiation must pass through the atmosphere before it reaches Earth's surface, and it passes through again—this time as terrestrial radiation—on its way back to outer space. As you learned in Chapter 12, gases in the atmosphere interact with both solar and terrestrial radiation as it passes through, reflecting, refracting, and scattering some of it and absorbing some of it in particular wavelengths. Radiatively active gases, mainly in the troposphere, selectively absorb infrared radiation, causing a layer of warmed air to accumulate close to the surface (see Figure 12.18); this is the greenhouse effect, which warms the surface of the planet sufficiently to support life as we know it. The most important of the naturally occurring greenhouse gases, by far, is water vapor. Other important contributors include carbon dioxide (CO_2), methane (CH_4), and nitrous oxide (NO_2), all of which are minor constituents of the atmosphere but very important in terms of their contribution to Earth's radiation balance.

The greenhouse gas composition of the atmosphere is controlled by many natural processes, including seasonal and long-term variations in photosynthesis, variations in forest cover and soil moisture, decay of organic matter, and burial (or exposure) of seafloor sediment. Another important factor is volcanic gas, of which carbon dioxide is a major component. Vast lava outpourings have occurred in the geologic past, and geologists have hypothesized that these could have raised the carbon dioxide content of the atmosphere to as much as 20 times the current level. If this hypothesis is correct, then the geosphere and the atmosphere are linked dynamically in their contribution to the greenhouse effect and climate change.

Reflectivity

Another factor that controls how much solar radiation reaches Earth's surface is the **albedo** of the atmosphere. **Aerosols**—extremely fine, suspended particles of dust, volcanic ash, ice, pollutants, and other substances—can cause dramatic changes in atmospheric albedo. Large explosive volcanic eruptions can eject huge quantities of fine ash and sulfur into the atmosphere, creating a veil of aerosols that encircles the globe (**Figure 14.4**). The aerosols scatter incoming solar

> **albedo** The reflectivity of a surface, as a percentage of the total reflected radiation.

Volcanoes and climate • Figure 14.4

Major volcanic eruptions can cause global cooling.

a. Mt. Pinatubo, a stratovolcano in the Philippines, erupted violently in 1991, producing a sulfur-rich aerosol haze that encircled the globe. The color scale on this diagram shows atmospheric sulfur dioxide in parts per billion, in 1991; the warmer colors indicate higher levels of sulfur.

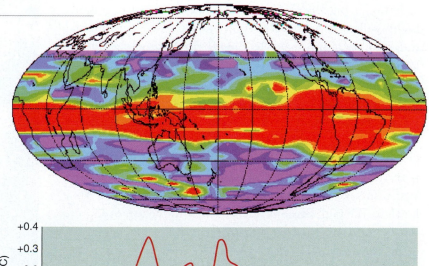

b. Sulfate aerosols from the eruption of Mt. Pinatubo caused a global decrease in temperature of at least 0.4°C—and more in the northern hemisphere.

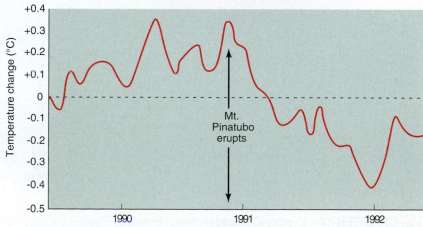

radiation, resulting in cooling of Earth's surface. Volcanic dust settles out rather quickly, generally within a few months to a year; however, tiny droplets of sulfate, produced by the interaction of volcanic sulfur gases and other atmospheric components, can remain in the upper atmosphere for years. These aerosols scatter the Sun's rays and increase the reflectivity of clouds; therefore, explosive, sulfur-rich volcanic eruptions are strongly connected with global cooling.

Clouds influence both the albedo and the heat-retaining capacity of the atmosphere (**Figure 14.5**). Different clouds contribute in different ways. The reflectivity of different cloud types varies dramatically, from about 10% to about 90%. This variation results from differences in thickness, as well as the size and abundance of water droplets and other particles (aerosols) that make up the cloud. Wispy, high-altitude clouds warm the surface by trapping warm air underneath them. Thick, low-altitude clouds trap some heat, but they cool the surface by increasing the reflectivity of the atmosphere.

The albedos of land, water, and ice surfaces also play significant roles in the climate system (**Figure 14.5c**). Ice and snow are high-reflectivity surfaces; forests, rock, soil, and water are generally low-reflectivity surfaces. Whenever the world enters a glacial age, large areas of land and water are progressively covered by snow and ice. Instead of radiation being absorbed and heating the land surface, the highly reflective surfaces of snow and ice scatter incoming radiation back into space, leading to further cooling of the lower atmosphere.

Changes in ocean circulation

As discussed in Chapter 12, the global circulation of the world ocean plays an important role in the climate system. The thermohaline circulation system (see Figure 12.5) links the atmosphere with the deep ocean, by way of the ocean's surface layer. Warm surface water moving northward into the North Atlantic in the Gulf Stream releases heat to the atmosphere by evaporation, maintaining the relatively mild climate of northwestern Europe. As a result of evaporation, the remaining water becomes cooler and more saline—and therefore denser. Cold, saline water sinks to produce North Atlantic Deep Water in the northern hemisphere and Antarctic Bottom Water in the southern hemisphere, kick-starting the deep-water parts of the thermohaline circulation. The locations where this sinking occurs are surprisingly limited because the conditions that produce the necessary cold, dense water are very specific.

Clouds, albedo, and climate • Figure 14.5

a. High-altitude, wispy clouds (as seen here in a satellite image, off the coast of North Carolina) have low albedos; they allow the Sun's shortwave radiation to pass through (red arrows). However, they absorb outgoing longwave radiation (yellow arrows), re-radiating some of it back to the surface and thus contributing to greenhouse warming.

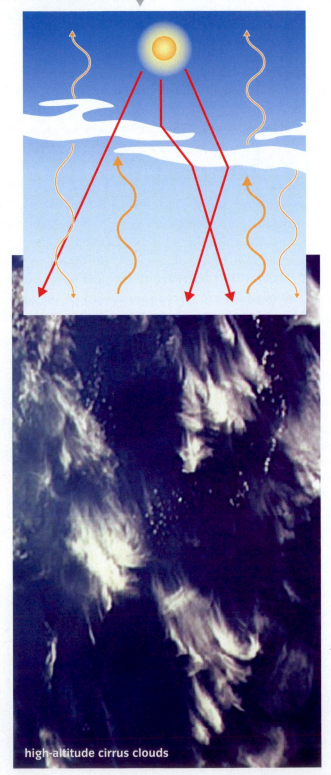

high-altitude cirrus clouds

b. Low-altitude stratocumulus clouds (as seen here in a satellite image, off the coast of West Africa) are thicker and less transparent than high-altitude clouds, so they have higher albedos and reflect a lot of the incoming solar energy (red arrows) back to outer space. They also trap outgoing longwave energy (yellow arrows) and send it back to the surface, which contributes to greenhouse warming. However, the effectiveness of their high reflectivity means that the net effect is cooling. ▼

c. Land, water, forests, snow, ice, and human-made surfaces all have characteristic reflectivities. ▼

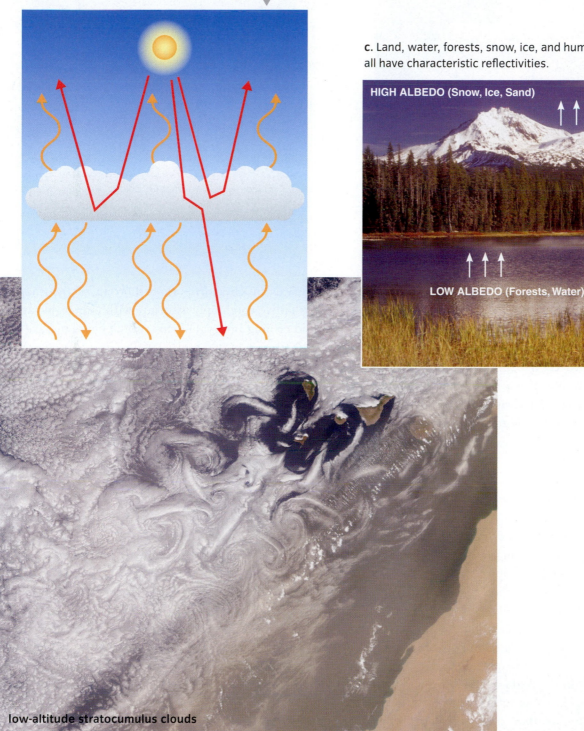

HIGH ALBEDO (Snow, Ice, Sand)

LOW ALBEDO (Forests, Water)

low-altitude stratocumulus clouds

Younger Dryas cold snap • Figure 14.6

a. Under full-glacial conditions, plants that are currently limited to polar and high-altitude regions could move into forests in northwestern Europe. Among these plants is *Dryas octopetala*, shown here. A large amount of *Dryas* pollen was found in deposits dating to the cold period now known as the Younger Dryas event.

b. Oxygen isotopes are an indicator of temperature, as you will learn shortly. Measurements of oxygen isotopes in sediment from a Swiss lake (left) and an ice core from the Greenland ice sheet (right) show that the onset and the end of the Younger Dryas event were both rapid. At the end of the event, the climate over Greenland warmed by about 7°C in only 40 years.

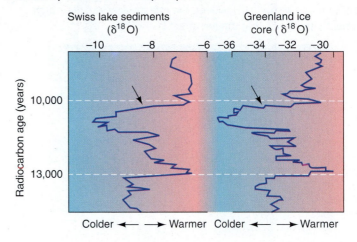

Consider what would happen if this system closed down. If the water of the North Atlantic were to become warmer or fresher, for example, such that North American Deep Water failed to form, the thermohaline circulation could be disrupted, bypassing the northern Atlantic altogether. This would effectively shut down the warm Gulf Stream Current, plunging the coastal areas of North America and western Europe into a deep freeze. You might recognize this as the scenario in the 2004 science fiction movie *The Day After Tomorrow*.

Once the Gulf Stream portion of the thermohaline circulation is cut off, the expanding sea-ice cover in the North Atlantic and extensive ice sheets on the adjacent continents would cause the climate to become increasingly cold. Thus, changes in the thermohaline circulation system may help explain why Earth's climate system appears to fluctuate between two relatively stable modes—one in which the ocean conveyor system is operational (interglaciation) and one in which it has shut down (glaciation).

In *The Day After Tomorrow*, the change from interglacial to glacial happened over the course of a few days, an impossible time scale in the real world. However, over the past decade or so, considerable interest has been focused on fluctuations of climate that start and end abruptly, last only a few hundred to a thousand years, and recur

at intervals much shorter than the tens of thousands of years that we commonly associate with major glaciations. A rapid cooling interval of this type appears to have happened between 11,000 and 10,000 years ago, when Earth was in the process of emerging from the last glaciation. This cold episode lasted about 1300 years, and scientists refer to it as the **Younger Dryas** event (**Figure 14.6**). The Younger Dryas was not unique; it now appears that the climate fluctuated rapidly in and out of glacial conditions during much of the last ice age.

Plate tectonics The global oceanic circulation is directly influenced by the geographic distribution of continents and ocean basins, which in turn influence climate. For example, the Middle Cretaceous Period (about 100 million years ago) had one of the warmest climates in Earth's history, with an average global temperature at least 6°C warmer than today. Sea level was 100 to 200 m higher, and warm, shallow seas covered many continental coastal areas. The continents were clustered together, with an enormous world ocean covering more than half of the globe. This dramatically affected the thermohaline circulation and, in combination with a high level of atmospheric carbon dioxide (possibly from extremely vigorous volcanic emissions), profoundly influenced the global climate.

Closing the ocean gateway • Figure 14.7

a. Surface waters flowed freely from the Pacific into the Atlantic 10 million years ago via an oceanic gateway known as the Central American Seaway. ▼

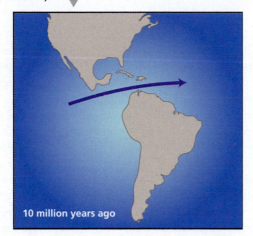

10 million years ago

b. About 5 million years ago, the Pacific and Caribbean plates began to converge. The resulting rise of the Isthmus of Panama restricted water exchange between the Atlantic and Pacific basins. The warm Gulf Stream current began to flow northward along the eastern coast of North America. ▼

5 million years ago

c. Today, the Isthmus of Panama completely blocks the flow of water between the Atlantic and Pacific in this region, and the strong Gulf Stream is responsible for the mild climate in the North Atlantic. ▼

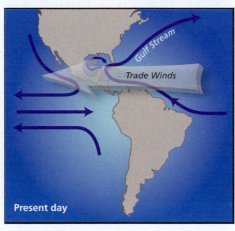

Gulf Stream

Trade Winds

Present day

Another example of geographic and tectonic influences on climate is the Isthmus of Panama, a narrow strip of land that joins North America and South America (**Figure 14.7**). The land of the isthmus was pushed above sea level by about 3 million years ago, during the Pliocene Epoch, as a result of the collision of the Pacific and Caribbean plates. Before then, water circulated freely through an oceanic "gateway" between the Pacific and Atlantic basins. The temperature of the ocean overall seems to have been much warmer than it is presently, which likely contributed to a rise in global surface temperature that culminated about 55 million years ago in an event termed the **Paleocene-Eocene Thermal Maximum**. Plant and animal remains typical of redwood and cypress forests and dating from that time have been found in the cold, treeless tundra of the High Arctic, providing evidence of temperate or even tropical climates in these northern locations (see *Amazing Places* on the next page). The gradual rising of the land bridge between North and South America eventually blocked the circulation of ocean water between the basins. Warm water flowing from the Caribbean (on the Atlantic side), instead of passing through to the Pacific, was diverted, flowing to the north alongside North America. This warm current eventually became the Gulf Stream.

So ice leads to more ice, and bare ground leads to more bare ground. Glaciations last for many thousands of years, but they may start and stop rapidly. Clouds can cool the Earth's surface, but they also can warm the surface. Volcanic eruptions can cause global cooling by ejecting volcanic dust and sulfur aerosols, but they can also cause global warming by adding greenhouse gases to the atmosphere. All of this may seem confusing and contradictory, but this kind of complexity is quite typical of the climate system. When one part of a delicately balanced system is changed, the effects and readjustments—both complementary and contradictory—are propagated throughout the system. The challenge in understanding the climate system is to determine which response will dominate when changes occur. To get a better handle on this, we need to consider the role of feedbacks in the climate system.

AMAZING PLACES

Fossil Forests of the High Arctic

Axel Heiberg Island and Ellesmere Island, in Nunavut's Qikiqtaaluk Region in the far north of Canada, are cold, windswept, and treeless. In the 1980s, a team of scientists began to study an amazing collection of fossilized remains on these islands. Actually, they are more like mummified remains, since entire plants, wood (Figure a), leaf litter, seeds, and cones have been preserved without the mineral replacement that technically characterizes fossilization (see Chapter 15).

The preserved remains include the stumps and root systems of tall trees (Figure b), including redwood, cypress, oak, pine, and sycamore—trees that could not survive in the current tundra conditions. Also present are the remains of ferns and flowering plants, as well as tortoises, snakes, and even crocodiles (Figure c), hippos, and tapirs. The remains reveal that Axel Heiberg, Ellesmere, and other locations in Canada's High Arctic were covered by many generations of lush forests and wetlands during the warm Eocene Period, about 45 million years ago.

The Eocene climate in this location was probably similar to that of present-day temperate rainforests of the Pacific Northwest—and perhaps even as warm as the wetland forests of Florida (Figure d). At this high latitude, there is constant daylight in the summer and 24-hour darkness in the winter—an interesting survival challenge for the plants and animals that inhabited these regions during the Eocene Period.

One might be tempted to think that plate tectonics simply rearranged the land masses, moving Axel Heiberg and Ellesmere from earlier, more southern latitudes to their present-day polar locations; however, paleomagnetic studies show that their locations have been relatively stable for the past 45 million years. This indicates that the polar paleoclimate was radically different from what it is today. Evidence also suggests that global ocean temperatures were much warmer than they are at present.

The government of the Canadian territory of Nunavut is considering establishing a park to protect these extraordinary fossil forests. The proposed name for the park is Napaaqtulik, which means "where there are trees" in the Inuktitut language.

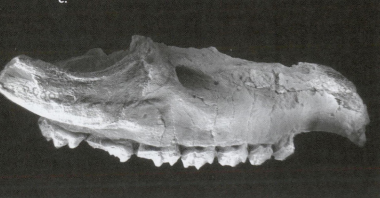

Feedbacks

The recognition of abrupt climatic shifts shows how much scientists still have to learn about the global climate system. What makes Earth's climate system even more challenging is that the interacting components are linked by **feedbacks**, both positive and negative.

> **feedback** A cycle in which the output from a process becomes an input into the same process.

A **negative feedback** occurs when a system is stabilizing or self-limiting (**Figure 14.8a**). In this case, the system's response to a change is in the opposite direction from the initial input. For example, if the global surface temperature warms, evaporation would be expected to increase. This would lead to more water vapor in the atmosphere and thus to more cloud cover. If these were low clouds that raise the albedo of the atmosphere and block incoming solar radiation, the surface temperature would cool down again. The response of the system is in the opposite direction to the initial change, and the process is self-limiting.

In contrast, a **positive feedback** is self-perpetuating and self-reinforcing—what we might refer to as a "vicious cycle." The system's response to a change, in this case, is in the same direction as the initial input (**Figure 14.8b**). For example, if the global surface temperature were to warm, evaporation—and thus water vapor in the atmosphere—would be expected to increase. Water vapor is a highly effective greenhouse gas, so an increase would lead to further warming, more evaporation, more water vapor, and still more warming. This is thought to be one of the most significant positive feedbacks in the climate system.

There are many positive feedbacks in Earth's climate system. It is important to understand that while positive feedbacks are self-reinforcing, they don't always act to reinforce warming. For example, an ice-covered surface increases albedo, reflecting solar radiation and further cooling the surface; this is a positive feedback. If some ice melts, revealing an underlying surface of bare rock or water, the darker material will absorb more solar radiation, warming the surface and leading to further melting of ice; this is also a positive feedback.

Positive and negative feedbacks • Figure 14.8

Feedbacks are common in the climate system.

a. This diagram shows a negative feedback cycle. Warming leads to increased evaporation and cloud cover. If these are low-altitude clouds that reflect incoming solar radiation, the surface will cool (compare this to Figure 14.5).

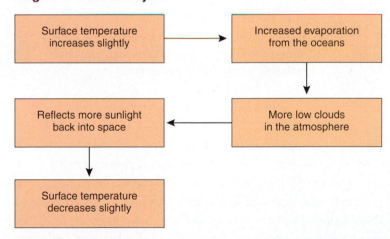

Negative feedback cycle

b. This diagram shows a positive feedback cycle. Warming leads to increased evaporation and water vapor in the atmosphere. Water vapor, an effective greenhouse gas, causes additional warming. The net warming or cooling effect also depends on what types of clouds form as a result of the additional evaporation; compare to Figure 14.5.

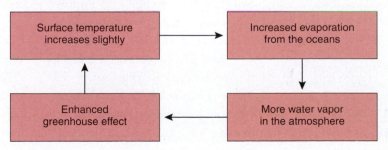

Positive feedback cycle

Global carbon cycle and climate • Figure 14.9

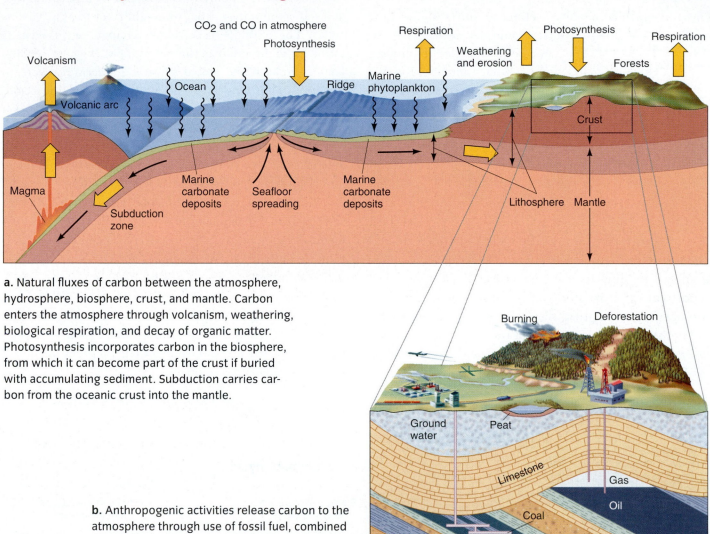

a. Natural fluxes of carbon between the atmosphere, hydrosphere, biosphere, crust, and mantle. Carbon enters the atmosphere through volcanism, weathering, biological respiration, and decay of organic matter. Photosynthesis incorporates carbon in the biosphere, from which it can become part of the crust if buried with accumulating sediment. Subduction carries carbon from the oceanic crust into the mantle.

b. Anthropogenic activities release carbon to the atmosphere through use of fossil fuel, combined with deforestation and cultivation.

Feedbacks make Earth's climatic processes extremely challenging to disentangle. For example, if there were an increase in evaporation, which type of cloud would mainly be formed—the warming kind (high altitude) or the cooling kind (low altitude)? Or would both responses occur, with one overpowering the other? These are the types of research issues that climatologists are attempting to re-

> **carbon cycle** The set of processes by which carbon cycles from reservoir to reservoir through the global environment.

solve as they struggle to improve our understanding of the processes that control Earth's climate.

The global **carbon cycle** has many feedbacks that affect climate (**Figure 14.9**). For example, seafloor spreading regulates atmospheric CO_2 by controlling the rate of volcanism and,

therefore, the production of volcanic CO_2. During periods of intense tectonic activity, volcanic emissions contribute large quantities of CO_2 to the atmosphere; during periods of tectonic quiescence, volcanic emissions are less. The metamorphism of carbon-rich sediment and limestone carried downward in subduction zones is an additional and possibly larger source of atmospheric CO_2 that is controlled by the tectonic cycle.

Carbon dioxide is also removed from the atmosphere by the weathering of surface rock. As you learned in Chapter 7, rainwater combines with CO_2 to form carbonic acid (H_2CO_3), which causes chemical weathering of silicate rock. The weathering products are carried to the ocean by streams, where marine organisms use the carbonate and silica to build their shells or skeletons. When they die, their remains accumulate on the seafloor and are buried, stored as sediment, and perhaps even converted by lithification into limestone. Weathering of rock, therefore, is a negative feedback in the climate system; it removes carbon from the atmosphere and stores it for a long time. This **sequestration** of carbon helps move the global climate system toward an equilibrium condition.

> **sequestration** Long-term storage of a material, in isolation from the atmosphere.

One major source of uncertainty in the carbon cycle lies in the response of terrestrial ecosystems to a warmer global climate. Increased CO_2 in the atmosphere leads to enhanced growth of vegetation, which removes CO_2 from the atmosphere through photosynthesis—a negative feedback. However, warmer conditions can lead to an increase in soil respiration, resulting in increased movement of CO_2 and CH_4 to the atmosphere—a positive feedback. In balance, it appears that a warmer climate could decrease the ability of the land to act as a storage **reservoir** and a **sink** for carbon, therefore contributing to a positive feedback.

> **reservoir** A place in the Earth system where a material is stored for a period of time.
>
> **sink** A reservoir that takes in more of a given material than it releases.

The ocean, too, is crucial as a sink for atmospheric carbon dioxide. If the atmospheric concentration of CO_2 increases, the uptake of CO_2 by ocean water should also increase. However, warmer conditions decrease the ability of the ocean to absorb CO_2. An additional complexity is that as the ocean takes in CO_2 the water becomes more acidic, as discussed in Chapter 13. A more acidic ocean would be less capable of absorbing atmospheric CO_2.

In the deep ocean, methane (CH_4) is stored, frozen into molecules of ice in seafloor sediment. An enormous volume of methane—perhaps more than 6 trillion tonnes—is trapped in this form, called **gas hydrates**. If ocean water warms, even by a small amount, the ice that traps this vast quantity of methane could melt, allowing it to bubble out. This could lead to a significant positive feedback because methane is a potent greenhouse gas (many times more powerful than carbon dioxide). Its release into the atmosphere would cause warming, potentially leading to further warming of ocean water—another positive feedback. In fact, such a feedback process is thought to have been a factor in causing the Paleocene-Eocene Thermal Maximum.

CONCEPT CHECK STOP

1. **What** do astronomical cycles have to do with Earth's climate?

2. **How** can plate tectonics affect the global climate?

3. **What** is the difference between a negative feedback and a positive feedback, and what are some of the important feedbacks in the climate system?

The Record of Past Climate Change

LEARNING OBJECTIVES

1. **Summarize** the evidence of past climate change.

2. **Explain** how geologists learn about past climates.

3. **Describe** the trends in Earth's climate over the past few million years, the past few thousand years, and the past few hundred years.

B efore we can understand the human role in climate change, present or future, we must first look back in time and find out how Earth's climate has changed over geologic history. The evidence of past temperature fluctuations and glaciations comes from multiple sources; no single source is definitive.

Evidence of Climate Change

Last winter may have been colder than the winter before, or last summer may have been wetter than the previous summer, but such observations do not mean that the climate is changing. The identification of real climatic change must be based on a shift in average conditions over a span of many years. Several years of abnormal weather may not mean that a change is occurring, but a **trend** that persists for more than a decade may signal a shift to a new climate regime.

> **trend** A long-term or underlying pattern in a time series of data.

Climate change, past to present • Figure 14.10

This graph shows estimates of global temperatures over (**a**) the past 60 million years, (**b**) the past 10 million years, (**c**) the past 150,000 years, and (**d**) the past 130 years. a, b, and c are based on data from seafloor sediment; D is based on instrumental weather records.

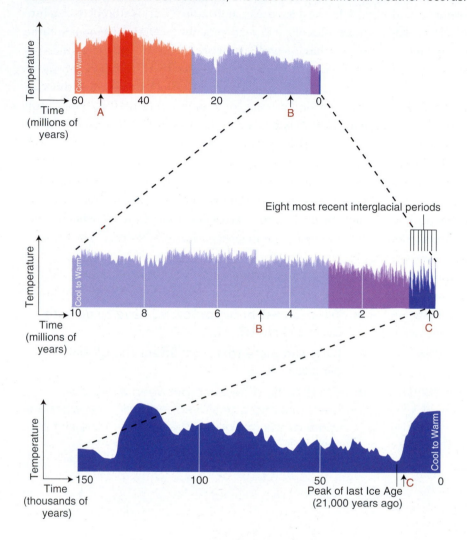

a. Global surface temperatures, 60 million years ago to present Earth's surface was largely free of ice. Sea levels were higher and seawater circulated freely between the Atlantic and Pacific.

b. Global surface temperatures, 10 million years ago to present As plate motions moved the major landmasses near their present locations, temperatures fell and glaciers appeared at the poles. In the past 800,000 years (the dark blue band in the temperature graph), the climate has fluctuated eight times between ice ages and warm interglacial periods.

c. Global surface temperatures, 150,000 years ago to present At the peak of the last Ice Age, 21,000 years ago, glaciers blanketed most of North America.

d. Global surface temperatures, 130 years ago to present This graph shows global mean surface temperature anomalies. An "anomaly" is a departure from a long-term baseline; in this case, the baseline is the average surface temperature for the 20th century. A positive anomaly indicates that the observed temperature was warmer than the average; a negative anomaly indicates that the observed temperature was cooler than the average. Earth is now warmer than it has been at any time in the past 100,000 years, and it is roughly the same temperature as it was in the last interglacial period, 120,000 years ago.

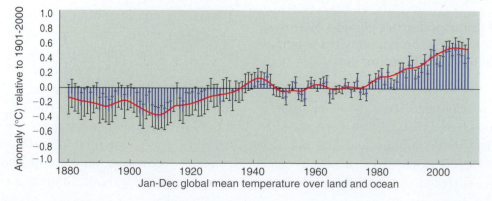

Today, changes in the atmospheric conditions that define weather and climate are carefully tracked and measured on an almost continuous basis at many locations around the world. However, such instrumental tracking has been recorded systematically for only the past century. Let's take a look at the records of past climate change (summarized in **Figure 14.10**), and then consider how scientists have managed to extend those records even farther back, into deep prehistoric time.

Figure 14.10 highlights three very important points. First, change is the norm, not the exception, in our climate system. No matter how much global warming occurs in the next century, it is unlikely to show up as more than a barely perceptible blip on any of these graphs. Second, the overall trend in temperature in the Cenozoic Era has been downward. The current global warming episode will do nothing to change that. Third, superimposed on the overall trend of cooling is a series of glaciations, each lasting about 100,000 years, followed by warm interglacial periods. We are currently near the peak of a warm interglacial cycle. In 50,000 years or so, Earth will probably experience another ice age.

Historical records of climate
In some places, instrumental records of climate have been maintained for a century or more, which is long enough to see if average conditions have really shifted. One of the longest continuous climate records comes from Great St. Bernard Hospice in the Alps. The Augustinian friars have recorded temperature in that location since the 1820s and snowfall since the 1850s. The temperature pattern demonstrated by the instrumental record from the Alps is representative of other parts of the northern hemisphere, where average temperatures experienced a fluctuating rise after the 1880s to reach a peak in the 1940s (Figure 14.10d). Thereafter, average temperatures declined slightly until the 1970s, when they again began to rise. In the 1990s and again in the 2000s, they reached the highest values yet recorded.

The geologic record of climate
Historical instrumental records represent only a tiny fraction of the evidence of climate change; the majority comes from the geologic record. Scientists have long puzzled over geologic features that seem out of place in their present climate. In the chapter opener and the *Case Study* about Wadi Kufra, you learned that ancient buried river channels have revealed a much more humid Sahara in earlier times, and in *Amazing Places* you learned about fossilized tropical forests in the High Arctic. On the other hand, fossilized plant remains in the temperate north-central United States show that this region formerly resembled Arctic landscapes like those now seen in far northern Canada. All of these and other lines of evidence from **paleoclimatology** demonstrate that ancient climates in many localities differed dramatically from present-day climates and that, throughout Earth history, change has been the norm in the climate system rather than the exception.

> **paleoclimatology**
> The scientific study of ancient climates.

Climate proxy records
The fossil record provides a broad overview of climatic changes over vast spans of time, but **climate proxy records** can provide evidence of year-to-year or season-to-season variability in specific locations. Although lacking the precision of instrumental data, climate proxy records can add up to a detailed picture of local, regional, and global climate trends.

> **climate proxy record** Records of natural events that are influenced by, and closely mimic, climate.

One approach to climate proxies is to look at historical records of events and processes that were controlled by climate. These records go as far back as people have kept track of climate-controlled events. For example, just about everything related to farming, fishing, and harvesting is controlled by climate. These activities are important for human well-being, so people have been keeping track of them for a long time—more than 1000 years, in some cases. Three examples of climate proxies in human historical records are shown in **Figure 14.11** (on the next page). Other useful records include variations in the height of the Nile River at Cairo, the number of weeks of sea ice off the coast of Iceland (recorded by Icelandic fishers since about 1200 C.E.), the quality of wine harvests in Germany, the date of blooming of cherry trees in Japan, and variations in wheat prices in Europe.

Climate proxies give scientists information about how the climate has changed over time. These records can be kept by humans or by nature, but a crucial aspect is the ability to attach dates to the phenomena.

Human records of climate

Some human records of climate proxies span more than 1000 years.

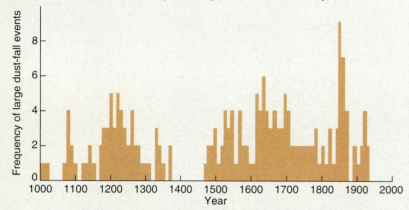

a. The frequency of major dust-fall events in China is useful as a proxy because a dusty atmosphere is indicative of a cold, glacier-dominated climate.

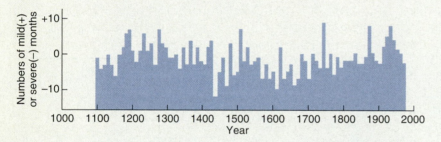

b. The severity of winters in England is based on the number of mild or severe months experienced.

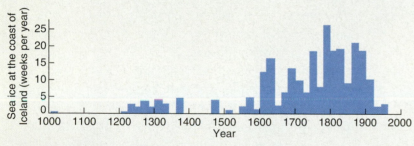

c. The number of weeks per year during which sea ice reached the coast of Iceland is a record that fishers have kept for almost 1000 years.

Natural records of climate

d. Glacier ice forms from snow, which is deposited in annual layers, as seen here in Glacier Bay National Park, Alaska.

e. The width of annual tree rings varies with growing conditions, including sunlight, temperature, and precipitation.

f. The growth of some corals produces structures similar to the annual rings of trees. Analyzing the chemical composition of the coral can provide a temperature record.

g. Seafloor sediment contains fossils of tiny sea organisms, **foraminifera**, which once lived in surface waters. These microfossils contain abundant information about the chemistry and temperature of the ocean.

h. Fossil pollen can be used to reconstruct past vegetation and climate. This is a scanning electron microscope photograph of a grain of *Drymiswinterii* pollen (42 micrometers), from a tree that is native to temperate rainforests.

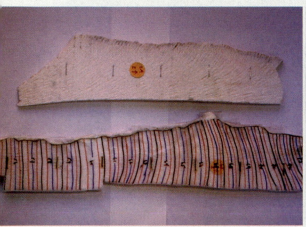

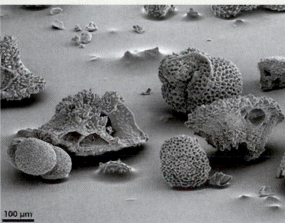

100 μm

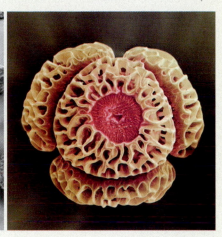

i. Instrumental measurements of temperature in the northern hemisphere from the middle of the 19th century onward are shown by the black curve. The other curves are based on reconstructions of temperature from climate proxies: tree rings (green), glacier lengths (dark blue), and ice cores (light blue). The remaining curves (yellow, red, and purple) were reconstructed using a combination of different proxies. All curves show that temperatures during the past few decades of the 20th century were higher than during any comparable period in the past 1000 years.

SURFACE TEMPERATURE RECONSTRUCTIONS FOR THE LAST 2,000 YEARS

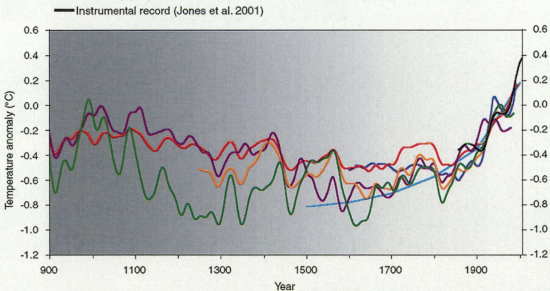

- Borehole temperatures (Huang et al. 2000)
- Multiproxy (Mann and Jones 2003a)
- Multiproxy (Hegerl et al. 2006)
- Instrumental record (Jones et al. 2001)
- Glacier lengths (Oerlemans 2005b)
- Multiproxy (Moberg et al. 2005a)
- Tree rings (Esper et al. 2002a)

CASE **STUDY**

Using Isotopes to Measure Past Climates

Arguably the most important paleoclimate data comes from the analysis of isotopes in water molecules from ice cores (**Figure a**). Isotopes (Chapter 2) are naturally occurring variations of elements, which differ just slightly from one another in mass but not in other chemical characteristics. For example, oxygen has three naturally occurring isotopes—^{16}O, ^{17}O, and ^{18}O. Of these ^{18}O is the heaviest; it behaves chemically just like the others, but it is preferentially concentrated during processes influenced by mass. Many natural processes that separate isotopes, including evaporation and precipitation, are temperature dependent—that is, they are influenced by variations in temperature.

Sampling and analysis of the isotopic composition of any material that was affected by a temperature-dependent process and that was in equilibrium with its surroundings can reveal the past temperature history of that material. A number of isotopes are useful in paleoclimate analysis, but water is so ubiquitous and so closely tied up with climatic processes that oxygen isotopes (from H_2O) are the most widely used for this purpose. Groundwater, sediment, shells, bones, and other Earth materials also contain oxygen, preserve the distinctive isotopic signature of their surroundings, occur in layers that can be dated, and are influenced by temperature; they, too, can be used as climate proxies.

Some isotopic variations in glacier ice from ice cores provide information about fluctuations in the air temperature near the glacier surface (**Figures b** and **c**). Other variations in ice cores and in isotopic studies of marine sediment reveal changes in ice volume on a global scale. To get a truly global perspective, scientists need to combine multiple records from various areas. Traditionally, Antarctica and Greenland have been the principal sources for ice core data, but more attention has been focused lately on retrieving the records from high-altitude temperate glaciers, many of which are disappearing.

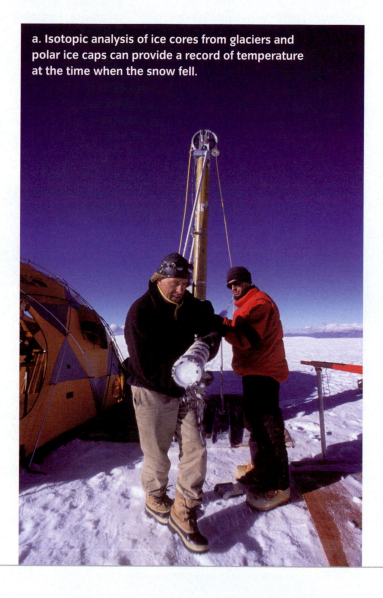

a. Isotopic analysis of ice cores from glaciers and polar ice caps can provide a record of temperature at the time when the snow fell.

Even the longest human historical records go back only 1000 years or so. To extend our use of climate proxies farther into the past, we must rely on nature's own record of climate, preserved by processes such as the deposition of annual or seasonal growth rings. To use this approach, there must be a mechanism for determining the date of the natural record; without this, we would end up with a record of changes in climate, but no way to determine when the changes actually occurred. Let's look at a few of the most useful natural climate proxies.

Ice core data Probably the most important source of information about paleoclimates is preserved in ice cores from the polar ice sheets (mainly Greenland and Antarctica) and lower-latitude alpine glaciers. Ice cores provide nearly continuous records of weather conditions, some extending to hundreds of thousands of years before the present. Measurements of the isotopes of oxygen (^{18}O and ^{16}O) in glacier ice enables scientists to calculate the air temperature when the snow that later was transformed into ice accumulated at the glacier surface.

b. Ice core data from Antarctica show a reconstruction of temperature, in terms of the difference (Δ, the Greek letter delta, which stands for "change") from present-day temperature, going back 420,000 years. This temperature reconstruction is based on analysis of oxygen isotopes from the ice.

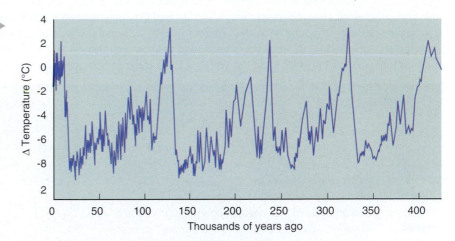

c. The temperature reconstruction shown here, also from Antarctica, is based on the analysis of deuterium (an isotope of hydrogen) in ice cores from Antarctica. Peaks on the graph indicate warm interglacial periods. This record has now been extended to 740,000 years before the present.

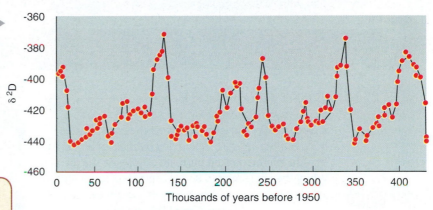

Think Critically

One of the most interesting recent challenges in paleoclimatology has been to explain slight mismatches in timing (called **asynchronicity**) between major climatic events recorded in ice core records from Antarctic and Greenland ice cores. Can you think of reasons a "global" change in climate might be recorded slightly earlier at one pole than at the other?

Other isotopic measurements reveal how much of Earth's global water budget was locked up in glaciers at various times in the past (see the *Case Study*).

An important feature of ice core analysis is that glacier ice is laid down in annual layers (as shown in Figure 14.11d). By counting the annual layers, scientists can count back in time to attach dates to the temperature determinations. Without the layering, ice core data would be of much less use, since there would be no precise time scale attached to the temperature changes. Scientists have been able to correlate ice core data from several different locations around the world, contributing to our understanding of paleoclimates on a global scale.

Another useful aspect of ice core analysis is that small bubbles of air are trapped in the glacier ice during its transformation from snow to ice. These provide samples of the ambient air at the time the snow fell, and they can be extracted and chemically analyzed to determine the composition of the atmosphere at that time. This makes an extremely powerful data set; we can not only decipher

WHAT A GEOLOGIST SEES

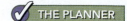

Ice Cores

The geologist will see that at times when the temperature was low, CH_4 and CO_2 were also low, and at times when temperature was high, CH_4 and CO_2 were also high. However, this is a chicken-and-egg question: Was the temperature low because the greenhouse gas concentrations were low, or were the greenhouse gas concentrations low because the temperature was low? Scientists do not yet have the answer to this question.

a. This geologist is holding an ice core that was extracted from a depth of 3,000 meters in the Antarctic Ice Cap. The ice is estimated the be one million years old. The core will be cut into thin slices which will be examined under a microscope.

b. This is a slice prepared for microscopic study, from the core of Antarctic ice. As the snow in a glacier recrystallizes, ice fills up the pore spaces between the crystals. Some tiny bubbles of air fail to escape and are locked permanently inside the ice. When the ice is melted under controlled conditions in a laboratory, the chemical composition of the "fossil air" can be measured.

Think Critically

How could a geologist determine the age of the "fossil air" trapped inside an ice core?

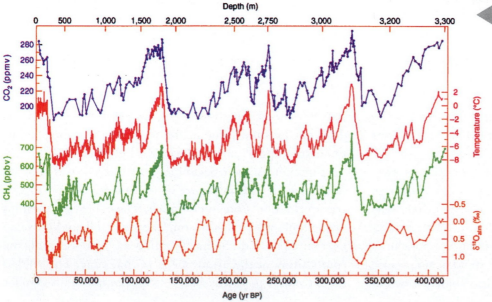

c. A geologist studying the core will be particularly interested in the concentrations of carbon dioxide and methane in the air bubbles. Samples from Antarctica and Greenland have shown that the atmosphere contained far less carbon dioxide and methane during glacial ages than during interglacials. Calculations suggest that the low levels of these two important greenhouse gases during glacial times can account for nearly half of the ice age temperature lowering. In contrast, the rapid increase of atmospheric carbon dioxide in our time is unprecedented in the ice core record and implies that an unusual warming is occurring.

The curve labeled $\delta^{18}O$ refers to the oxygen isotopic composition of the ice, which provides a proxy for temperature.

what the climate was like in the past but also learn the chemistry of the atmosphere at that time (see *What a Geologist Sees*).

Annual growth rings

Many organisms deposit annual growth rings and are therefore potentially useful as climate proxies. For example, a tree living in middle latitudes typically adds a growth ring each year (as you saw in Figure 14.11e), the width and density of which reflects the local climate. Many species live for hundreds of years; a few, like the Giant Sequoia and Bristlecone pine of the California mountains, live for thousands of years. Living corals and similar organisms, such as coralline red algae, also deposit growth rings (as shown in Figure 14.11f). These organisms equilibrate to the water in which they grow, so their chemical composition changes in response to the composition and temperature of the water. From the annual growth rings, scientists can obtain a record of changes in water temperature spanning thousands of years.

Sediment, microfossils, and pollen

Deep-sea and lake-bottom sediment cores provide some of the best indirect evidence of past climatic changes. Seafloor sediment contains abundant fossils of tiny sea creatures called foraminifera, which once lived in surface waters. These microfossils (seen in Figure 14.11g) equilibrate with the water around them, preserving a chemical record of past climatic changes. Plant pollen fossils like those in Figure 14.11h also contribute greatly to our knowledge of past climatic conditions. A sample of bog or lake sediment typically yields a vast number of pollen grains that can be identified by type, counted, and analyzed statistically. At any given level in a sediment core, the pollen grains reveal the assemblage of plants that flourished near the site when the enclosing sediment layers were deposited. The precipitation and temperature at the site of a similar modern plant assemblage can be used to estimate the climate represented by the fossil assemblage.

Earth's Past Climates

By using climate proxies and other evidence of past climates preserved in the geologic record, scientists have established a detailed chronology of climatic changes over Earth's long history. Let's summarize these changes, starting with the most recent millennium and moving backward in time.

Climate of the past millennium

An episode of mild climate during the Middle Ages, called the **Medieval Warm Period**, gave way about 800 years ago to a colder period, when temperatures in western Europe averaged 1 to 2°C lower. Scientists refer to this as the **Little Ice Age**, and it is discernible as a tiny dip in the temperature reconstruction in Figure 14.10c. In western Europe, the Little Ice Age was characterized by unusually snowy winters; cool, wet summers; and expansion of sea ice in the North Atlantic. By the early 17th century, advancing glaciers were overrunning farms in the Alps, Iceland, and Scandinavia. During the worst years of that century, sea ice completely surrounded Iceland, causing fisheries to fail. Erratic weather led to crop failures, rising grain prices, and famines, which resulted in large-scale emigrations of Europeans to North America. Thus, many Canadians and Americans owe their present nationality to the Little Ice Age climate.

Little Ice Age conditions persisted until the middle of the 19th century, when a general warming trend caused mountain glaciers to retreat and the North Atlantic sea ice to retreat northward. Minor fluctuations of climate have continued to take place since then, but the overall trend of warming brought conditions that were increasingly favorable for crop production at a time when the human population was expanding rapidly and entering the industrial age.

The last glaciation

The last time Earth's climate was dramatically different from what it is now was during the last glacial period. The last glaciation, which started about 70,000 years ago and ended about 10,000 years ago, was the most recent of a long succession of glaciations in the Pleistocene Epoch. To reconstruct the climate of this latest ice age, scientists have relied on analyses of periglacial features, glacier ice, and sediment that contain isotopic and fossil evidence of paleoclimate.

In the popular imagination, glacial ages were times when temperatures were very cold, perhaps rivaling those in Antarctica today. Although such extreme cold did exist in some regions, in other places the average temperatures at the culmination of the last glaciation were not very different from what they are now. In the tropics, temperatures

North America in the last glaciation • Figure 14.12

This map shows some aspects of the geography of North America about 20,000 years ago, during the last glaciation. Coastlines lie farther seaward due to a drop in sea level of about 120 m. Sea-surface temperatures are based on analysis of microfossils in deep-sea cores. Circled numbers show the estimated temperature lowering, relative to present temperatures, at selected sites, based on climate-proxy evidence.

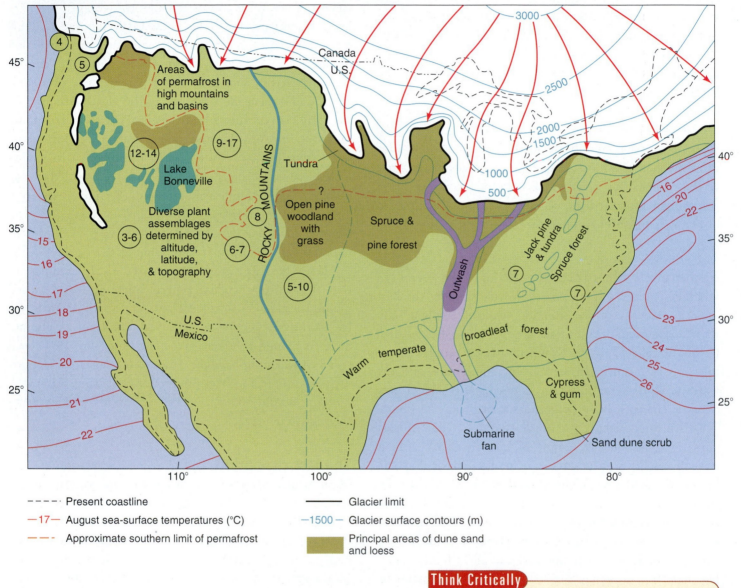

- - - - Present coastline

—17— August sea-surface temperatures (°C)

- - - Approximate southern limit of permafrost

——— Glacier limit

—1500— Glacier surface contours (m)

Principal areas of dune sand and loess

Think Critically

Why would temperature reductions during the last glaciation have been greater in continental interiors compared to coastal regions?

were about the same as they are today. In midlatitude coastal regions, temperatures on land were generally reduced by about 5 to 8°C, whereas in continental interiors, reductions of 10 to 15°C occurred (**Figure 14.12**).

During the last glaciation, the climate of the northern middle and high latitudes became so cold that a vast ice sheet formed over central and eastern Canada and expanded southward toward the United States and westward toward the Rocky Mountains. Other great ice sheets formed over the mountains of western Canada, northern Europe, and northwestern Asia. As ocean water evaporated and was deposited as snow on the growing ice sheets, world sea level fell to about 120 m below the present level. This changed the shapes of coastlines and continents (as shown in Figure 14.12). The great ice sheets of Greenland and Antarctica spread across the adjacent, exposed continental shelves. Large glacier systems formed in the Alps, Andes, Himalayas, and Rockies, as well as on isolated peaks and mountain ranges scattered through all latitudes. Some scientists postulate that an ice shelf may have completely covered the Arctic Ocean, extending into the North Atlantic.

At the height of the glacial age, the middle latitudes were both windier and dustier than they are today. Glacial-aged aeolian deposits of fine sediment called loess are found south of the ice limit in the midwestern United States, where cold winds flowing south from the ice mass picked up fine sediment from the floodplains of glacial meltwater streams and carried it further south. The same happened in central Asia and western Europe. In each of these regions, successive layers of loess are separated by soils that formed during the interglacial periods.

In many arid and semi-arid regions, including the Sahara, the Middle East, southern Australia, and the American Southwest, the shift to cooler glacial-age climates resulted in the enlargement of existing lakes and rivers or the creation of new ones. Some of this may have been caused by increased precipitation, but evidence in some locations actually points to reduced precipitation during glacial times. An alternative explanation is that lower temperatures led to reduced evaporation.

The glacial ages witnessed major changes not only in the cryosphere, hydrosphere, and atmosphere but also in the biosphere. Pollen studies show that in glacial times, the vegetation distribution was quite different from what we see today. About 20,000 years ago, a belt of tundra existed immediately south of the glacier margin, and today's grasslands of the Great Plains were mostly open pine woodlands. The changes accompanying the advance and retreat of the ice sheets were dynamic and complicated; both plant and animal species were displaced

in various directions, forming ecological communities that are unknown on the present landscape.

Pleistocene and older glacial ages

As recently as a few decades ago, it was thought that Earth had experienced only 4 glacial ages. This traditional view was discarded when studies of deep-sea sediment disclosed evidence of a long succession of glaciations during the Pleistocene. Paleomagnetic dating (see Chapter 3) of deep-sea cores shows that during the past 800,000 years, there have been 8 such episodes. All the available evidence shows that in the Pleistocene Epoch as a whole (the past 1.8 million years), about 30 glacial ages are recorded.

Ancient glaciations, identified mainly by rock of glacial origin and associated polished and striated rock surfaces, are known from farther back in Earth's history, as well. The earliest glaciation dates to about 2.4 billion years ago, in the early Proterozoic, and evidence of other glacial episodes has been found in rock of late Proterozoic, early Paleozoic, and late Paleozoic age (see the geologic time scale in Figure 3.8). During the late Paleozoic (Carboniferous and Permian) glaciation, 50 or more glacial advances and retreats are believed to have occurred. The geologic record is fragmentary and not always easy to interpret, but evidence suggests that Earth's land areas must have had a very different relationship to one another during the late Paleozoic glaciation than they do today. In the Mesozoic Era, glaciation of similar magnitude apparently did not occur, consistent with geologic evidence that points to a long interval of mild temperatures both on land and in the ocean.

CONCEPT CHECK

1. **What** are the most important sources of evidence about changes in Earth's climate over time?

2. **How** do scientists use ice cores to learn about past climates?

3. **How** many major glaciations occurred during the Pleistocene Epoch?

Predicting the Future

LEARNING OBJECTIVES

1. **Examine** the evidence for anthropogenic climate change.

2. **Describe** the possible effects of continued global warming in the next century.

3. **Summarize** what is known about our changing climate with significant certainty and with less certainty, as well as some of the major uncertainties in predicting future climate change.

Our climate has changed dramatically in the past and will continue to change. However, the causes of change are complicated, and this makes it difficult to predict how the climate will change and at what rate. Furthermore, in the past two centuries, a new component has been introduced: an industrialized human population that is now capable of significantly affecting the climate not only locally but globally.

That the world's climates can change measurably within a human lifetime (whether by natural or human causes) is a relatively new realization. With this realization has come increasing concern about the impacts of such changes on nature and on society. Scientists continue to grapple with the challenges of measuring the changes, separating short-term fluctuations from long-term trends, and disentangling natural influences from human influences.

Because the stakes in predicting future climate change are so high—for the environment, the economy, and society as a whole—it is important to distinguish between what we know with a high degree of certainty, what we think we know but with somewhat less certainty, and what we recognize as true uncertainties.

What We Know

Earth's atmosphere has changed as a result of human activity, particularly since the Industrial Revolution; this fact is not controversial. These **anthropogenic** changes to atmospheric chemistry will have an impact on the climate system. Some of the changes have resulted from the emission of gases that are of wholly synthetic origin, such as chlorofluorocarbons that contribute to the breakdown of ozone in the stratosphere. Other changes have resulted from human activities that cause the mobilization of

naturally occurring compounds, such as sulfur and carbon compounds from the burning of fossil fuels. Among these, the greenhouse gases have received increasing scientific and public attention over the past two decades because their atmospheric concentrations are rising. While it may be troubling, in itself, that human action has had a measurable effect on the chemistry of the atmosphere, our real concern in this context is the impact of these changes on the stability of Earth's climate system.

When considering the human contribution to the greenhouse effect, also known as the **anthropogenic greenhouse effect** (or **accelerated,** or **enhanced, greenhouse effect**), we are primarily concerned with carbon dioxide (CO_2) because of the magnitude of our emissions and methane (CH_4) because of its extreme efficiency as a greenhouse gas.

> **anthropogenic greenhouse effect** The portion of greenhouse warming that results from human activities rather than natural processes.

The first indisputable evidence of an anthropogenic effect on atmospheric chemistry with the potential to affect climate was the steady rise in carbon dioxide levels observed over the past 50 years (**Figure 14.13**). Atmospheric carbon dioxide levels are higher now and rising much more rapidly than at any other time in the past 100,000 years. Samples of ancient air from ice cores show that the preindustrial concentration of CO_2 was about 280 parts per million by volume, a typical value for an interglacial age. The subsequent rapid increase to 387 ppm during the past 200 years is unprecedented in the ice core record and implies that something very unusual is taking place.

What is the cause of this increase? Humans pump roughly 8 billion tons of carbon into the atmosphere each year, most of it as carbon dioxide from the burning of fossil fuels—more than enough to account for the

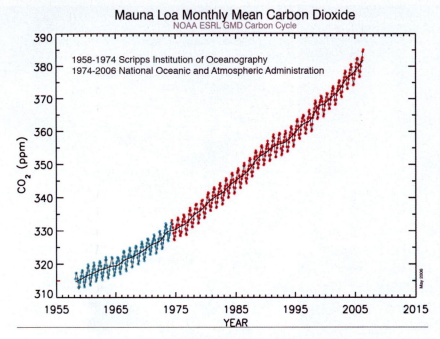

Mauna Loa Monthly Mean Carbon Dioxide
NOAA ESRL GMD Carbon Cycle

1958-1974 Scripps Institution of Oceanography
1974-2006 National Oceanic and Atmospheric Administration

a. The concentration of carbon dioxide in the atmosphere (shown here in ppm) has been measured since 1958 at the Mauna Loa Observatory in Hawaii. The "zig-zag" pattern is caused by annual fluctuations related to seasonal variations in the biological uptake of CO_2 as a result of photosynthesis. The long-term trend shows a persistent increase in this important greenhouse gas.

b. The concentration of methane (CH_4) in the atmosphere, shown here in parts per billion (ppb), has also increased dramatically since the 1800s.

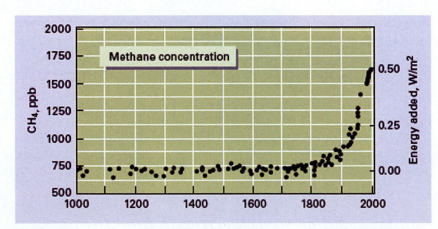

increase. Additional contributing factors are widespread deforestation, with its attendant burning and decay of cleared vegetation, and the use of wood for fuel in many underdeveloped countries that have rapidly growing populations.

Methane gas (CH_4) absorbs infrared radiation 25 times more effectively than CO_2, making methane an important greenhouse gas despite its relatively low atmospheric concentration. The concentration of atmospheric methane has increased by about 150% since the Industrial Revolution (from 700 ppb to about 1745 ppb) (**Figure 14.13b**). The increase essentially parallels the rise in the human population. This is not surprising, for much of the methane now entering the atmosphere is generated either by biological activity related to rice cultivation or as a by-product of the digestive processes of domestic livestock, especially cattle. The global livestock population has increased greatly in the past two centuries, and the total acreage under rice cultivation has increased more than 40% since 1950.

Logic and past climate records suggest that an increase in greenhouse gases should be accompanied by an increase in temperature. The evidence for this increase in temperature is becoming more conclusive each year. Worldwide

a. An 1870 post card shows the position of Rhone Glacier. The photo of the glacier today was taken in 2006.

b. The pack ice in the Beaufort Sea, off the North Slope of Alaska, now breaks up weeks earlier in the spring than it once did.

◀ **c.** On the West Antarctic Peninsula, the rise in temperatures has led to an influx of gentoo penguins, which prefer warmer subarctic temperatures, and a sharp decline in the numbers of Adélie penguins (shown here).

temperature records show an increase of as much as 0.8°C in the past century, and the first decade of the 21st century is now recognized as the hottest on record (in terms of difference from the 20th century average). Whatever its cause (or causes), the temperature increase has begun to have real, observable effects. These include the retreat of glaciers, calving of large icebergs from ice shelves, and shrinking of some animal habitats and expansion of others (**Figure 14.14**).

What We Think We Know

In recent years, climatologists have developed **general circulation models** that have successfully simulated the general character of present-day climates and have greatly improved weather forecasting. These successes, and the availability of ever-more-powerful computing capability, encourage scientists to use such models to obtain a general picture of future climate change.

> **general circulation model**
> A computer model of the climate system, linking processes in the atmosphere, hydrosphere, biosphere, and geosphere.

General circulation models allow us to incorporate different assumptions about anthropogenic factors that are driving climate change. The models differ in various details, but they all predict that if greenhouse gas emissions stay where they are today—an extremely optimistic assumption—we can expect a global temperature increase of 0.5°C to 1.5°C over a 50-year period. However, if fossil fuel consumption continues to grow at an increasing rate, atmospheric carbon dioxide will likely double by 2100, if not sooner. If this proves to be the case, climate models predict that average global temperatures will rise between 1.5° and 4.5°C. The rise will not be uniform all over the globe, and it will likely be greatest in polar regions (**Figure 14.15**).

Despite the uncertainties inherent in climate modeling and data analysis, there is now a strong scientific consensus that (1) human activities have led to increasing atmospheric concentrations of carbon dioxide and other trace gases that have enhanced the greenhouse effect; (2) global mean surface air temperature has increased by up to 0.8°C during the past 100 to 150 years, an increase that appears to be a direct result of the anthropogenic greenhouse effect; and (3) during the next century, global average temperature will likely continue to increase.

The most thoroughly reviewed and widely accepted synthesis of scientific information concerning climate change is a series of reports issued by the **Intergovernmental Panel on Climate Change (IPCC)**, established in 1988 by the United Nations Environment Programme (UNEP) and the World

> **Intergovernmental Panel on Climate Change (IPCC)** An international, interdisciplinary panel of scientists and other experts, established to keep the world community up to date on the science of the global climate system.

Temperature rise if carbon dioxide doubles • Figure 14.15

This map shows projected surface temperature changes for the late 21st century (2090–2099), relative to the average temperature in the period 1980–1999. Note that the changes will not be uniform throughout the Earth system, with temperature increases particularly high in the Arctic. (*Source*: IPCC Climate Change, 2007.)

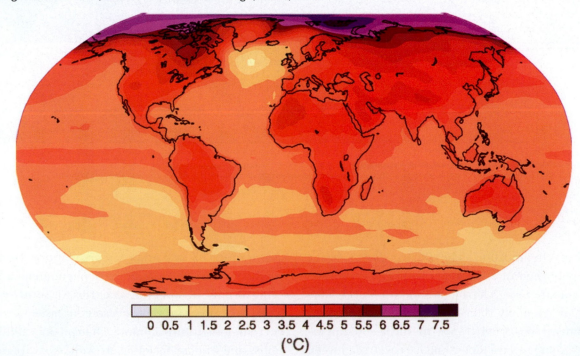

0 0.5 1 1.5 2 2.5 3 3.5 4 4.5 5 5.5 6 6.5 7 7.5

(°C)

Projected effects of warming • Figure 14.16

This figure provides examples of impacts associated with projected global warming. Examples of global impacts are projected for climate changes (and sea level and atmospheric CO_2, where relevant) associated with different amounts of increase in global average surface temperature in the 21st century. The black lines link impacts; broken-line arrows indicate impacts continuing with increasing temperature. The starting point of each line of text at the left indicates the approximate level of warming associated with the onset of a given impact. Adaptation to climate change and large-scale intervention in the climate system are not included in these estimations. Confidence levels for all statements are high. (*Source*: Figure SPM-7 from IPCC *Fourth Assessment Report, Climate Change 2007: Synthesis Report*. www.ipcc.ch/publications_and_data/ar4/syr/en/spms3.html.)

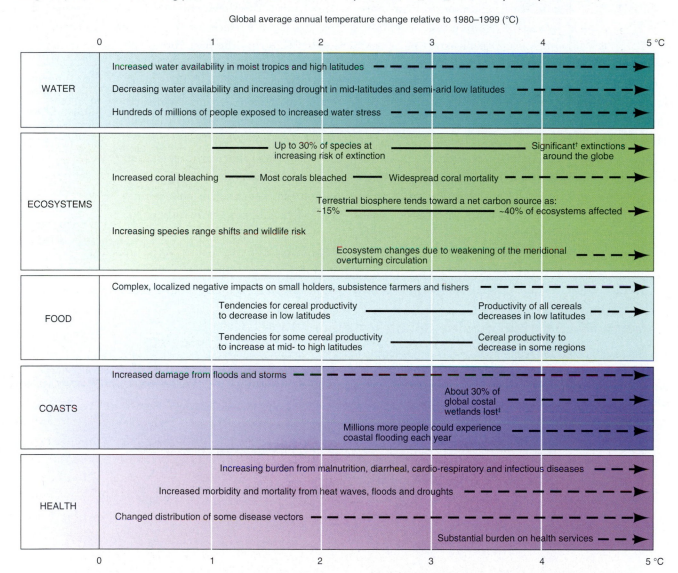

Global average annual temperature change relative to 1980–1999 (°C)

† Significant is defined here as more than 40%.   ‡ Based on average rate of sea level rise of 4.2 mm/year from 2000 to 2080.

Meteorological Organization. The IPCC's *Fourth Assessment Report*, published in 2007, represents the consensus of scientific climate research from around the world. This report summarizes many thousands of scientific studies, and it documents observed trends in surface temperature, precipitation patterns, snow and ice cover, sea levels, storm intensity, and other factors (**Figure 14.16**). It also predicts future changes in these phenomena after considering a range of potential scenarios for future greenhouse gas emissions. The report also addresses the impacts of current and future climate change on wildlife, ecosystems, and human societies, as well as strategies that we

Where Geologists CLICK

The Intergovernmental Panel on Climate Change

> **www.ipcc.ch**

The *Fourth Assessment Report* of the IPCC is available online at www.ipcc.ch. (Click on AR4 Climate Change 2007.) The *Fifth Assessment Report*, due to be completed in 2013 or 2014, will be available in phases at the same site. These are extremely important documents that you should read for yourself.

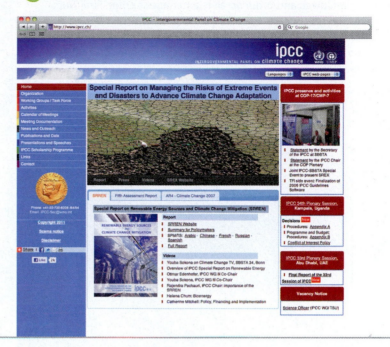

might pursue in response to climate change. You can—and should—read this very important document yourself, and other reports produced by the IPCC; they are all available online (see *Where Geologists Click*).

As with all other scientific endeavors, the IPCC must deal in uncertainties; its authors have therefore assigned statistical probabilities to all conclusions and predictions. In addition, estimates regarding impacts of change on human societies are conservative because scientific conclusions had to be approved by representatives of the world's national governments, some of which are reluctant to move away from a fossil fuel–based economy. The IPCC report concludes that average surface temperatures on Earth increased by an estimated 0.74°C in the century from 1906 to 2005, with most of this increase occurring in the later decades of that century. In the future, we can expect average surface temperatures on Earth to rise by at least 0.2°C per decade for the next two decades.

And what of the human role in climate change? The IPCC has strengthened its wording with each new report, and the *Fourth Assessment Report* in 2007 concluded that human actions are very likely the cause of global warming. "Very likely" in this case indicates a probability of 90% or greater, and "global warming" refers to the observed increase of 0.74°C in average surface temperatures over the past 100 years. As of 2011, no major internationally or nationally recognized scientific organization publicly maintains a position that differs from this conclusion, although some organizations still remain noncommittal.

The Big Uncertainties

It is of obvious importance to human society to be able to apply the scientific understanding of Earth's climate system to the prediction of future climate changes. However, there are a number of basic uncertainties: Will concentrations of greenhouse gases continue to increase, and, if so, how rapidly? How will the ocean, a major reservoir of heat and a fundamental element in the climate system, respond to climatic changes? Will the thermohaline circulation shut down, as it has done in the past, and what further effects would that have on climate? How will changes affect ice sheets and cloud cover, and how will those in turn affect the planet's albedo? How will soils respond to warming? Will species and ecosystems be able to adapt? How will climatic changes and their impacts differ from region to region? When feedbacks with multiple responses occur, which response will dominate?

Many of the linkages in the climate system are still poorly understood and therefore difficult to represent accurately in general circulation models. For instance, computer models do not yet adequately portray the dynamics of ocean circulation or cloud formation, two of the most important elements of the climate system. Because of such

Potential effects of global warming • Figure 14.17

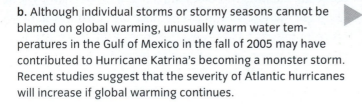

a. The Maldives and other small-island nations could find themselves underwater if sea levels rise. The Maldives, in the Indian Ocean, are built on coral reefs that grow over time but could not possibly grow fast enough to keep up with a rapid change in sea level.

b. Although individual storms or stormy seasons cannot be blamed on global warming, unusually warm water temperatures in the Gulf of Mexico in the fall of 2005 may have contributed to Hurricane Katrina's becoming a monster storm. Recent studies suggest that the severity of Atlantic hurricanes will increase if global warming continues.

uncertainties and complexities, scientists are reluctant to make firm forecasts, and they tend to be cautious in their predictions. They qualify their conclusions with adjectives such as "possible," "probable," and "uncertain." To the scientists who carry out this research, these terms have specific, clearly defined meanings, and they represent an appropriate level of acknowledgment of the scientific uncertainty involved in the study of Earth's climate system.

This understandably cautious approach has complicated the public discourse on climate change; it emphasizes the gap between what we know about the Earth system and what we would like to know, and it points to the many challenges that face both the scientists who study global change and the policymakers who must manage it.

The indirect effects of warming are particularly difficult to quantify. Likely impacts include the rising of sea level due to the melting of land ice; increased intensity of severe hurricanes and typhoons, which draw energy and moisture from warm seawater; increased frequency and severity of forest fires; spread of diseases that thrive in hotter climates, such as malaria; and significant changes in animal and plant populations due to death or migration. Some of the possible effects of global warming are illustrated in **Figure 14.17**.

c. In much of the western United States, the freshwater supply is dependent on the melting of the winter snow pack in the Rocky Mountains. A smaller snow pack could cause streams like this one, at the foot of the Sierra Nevada, to dry up earlier in the year.

d. Increased frequency and severity of forest fires is a likely impact of a hotter, drier climate. This fire is in Shasta-Trinity National Forest, California, 2009.

e. The effect of global warming on crops is difficult to assess. An atmosphere with a higher carbon dioxide concentration could have a fertilizing effect on crops. An extended growing season in areas such as Siberia could also promote agricultural productivity. On the other hand, a hotter, drier climate could be devastating for crops in areas such as this drought ridden field in Delaware.

An increase in global surface air temperature by a few degrees may not sound like much; surely, we can put up with this rather insignificant change. However, the difference in average global temperature between the present and the coldest part of the last ice age was only about 5°C, so a temperature change of even a degree or two could have significant global repercussions. In extracting and burning Earth's immense supply of fossil fuels, people have unwittingly begun a great "geochemical experiment" that is having a significant impact on our planet and its inhabitants. It is extremely important, not only scientifically, but socially, economically, and politically, that we seek and find the answers to questions about future climate change.

CONCEPT CHECK   🛑 STOP

1. **How** does the current global warming trend compare with previous climatic fluctuations?

2. **What** effects are likely to happen over the century if, as scientists now believe, Earth is getting warmer and humans are at least partially responsible for the changes?

3. **What** does it mean when scientists use terms like "probable" and "likely" to describe the potential effects of climate change?

Summary

1 The Climate System 418

- Earth's **climate** system is complex as shown in the diagram. It is driven by interactions among the atmosphere, hydrosphere (mainly the ocean), cryosphere, geosphere, biosphere, and, most recently, the anthroposphere. The components of the climate system interact so closely that a change in one of them causes changes in the others. The main source of energy that drives the climate system is the Sun.

Earth's climate system • Figure 14.1

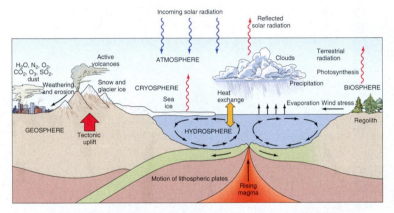

- Climate classification systems such as the Köppen system use average temperature and precipitation measurements and vegetation types to define major climate zones. Climate zones are closely related to global atmospheric circulation patterns.

2 Natural Causes of Climate Change 420

- Climate change is nothing new; it is a fundamental characteristic of the climate system.

- External influences on climate include the Sun's radiative output and cyclical variations in Earth's orbital characteristics as shown in the diagram. These **Milankovitch cycles** are closely related to the timing of glacial–interglacial cycles.

How orbital changes influence climate • Figure 14.3

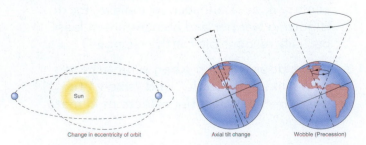

Change in eccentricity of orbit   Axial tilt change   Wobble (Precession)

- Internal influences on climate include the filtering capacity and composition of the atmosphere, particularly with regard to greenhouse gas concentrations; the **albedo** or reflectivity of the atmosphere and other Earth surfaces; volcanic emissions; cloud cover; variations in oceanic circulation; and plate tectonics.

- **Feedbacks** are characteristic of the climate system and add to its complexity. **Positive feedbacks** are self-reinforcing; **negative feedbacks** are self-limiting. Many important feedbacks in the climate system are related to process in the global **carbon cycle**.

3 The Record of Past Climate Change 431

- Geologists study **paleoclimate** in a variety of ways, including analysis of fossils, sediment, and ice cores. Ocean sediment contains microscopic fossils, and changes in the proportion of warm-water and cold-water plants reflect the changes in world temperatures. Ice cores preserve trapped bubbles of "fossil air" that help scientists determine the concentration of greenhouse gases in the past.

- **Climate proxies** are processes that are controlled or influenced by and that mimic climatic variations. Climate proxies can be tracked in human records or in the geologic record, but for a proxy to be useful, it must be possible to attach a date to the record. Isotopic analyses of ice cores provide the most useful and extensive climate proxy record.

- Over the past 1.8 million years, Earth has experienced repeated **glaciations** and warm **interglacial** periods. We are now in an interglacial period (as shown in the graph). The last glaciation, or ice age, peaked about 21,000 years ago. The current trend of global warming may create a more pronounced interglacial period but is unlikely to significantly delay the next ice age, which can be expected in 50,000 years or so.

Climate change, past to present • Figure 14.10

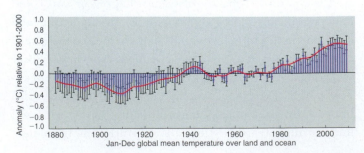

Jan-Dec global mean temperature over land and ocean

4 Predicting the Future   442

- Greenhouse gases have always affected the climate, but the possibility of an **anthropogenic greenhouse effect**, caused by greenhouse gas emissions from the burning of fossil fuel, has arisen only in the past century as shown in the graph.

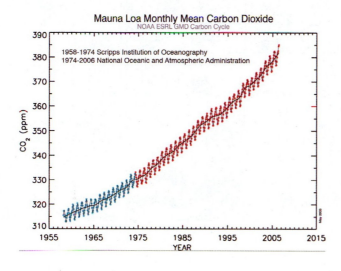

- The current expectation of many scientists is that **global warming** will continue through the 21st century, and that its extent will depend on human actions to limit the production of greenhouse gases. The **Intergovernmental Panel on Climate Change** keeps the international community up to date on developments in climate science.

- The effects of global warming are so far most pronounced in the polar regions, where sea ice is thinning, glaciers are retreating, and ice caps are melting. Worldwide effects in the future may include a rise in sea levels, increases in the severity of weather systems, and changes in animal habitat that will cause some species to die and force others to migrate to new territory.

Key Terms

- albedo 423
- anthropogenic greenhouse effect 442
- carbon cycle 430
- climate 418
- climate proxy record 433
- feedback 429

- general circulation model 445
- glaciation 420
- global warming 418
- interglaciation 420
- Intergovernmental Panel on Climate Change (IPCC) 445

- Milankovitch cycles 422
- paleoclimatology 433
- reservoir 431
- sequestration 431
- sink 431
- trend 431

Critical and Creative Thinking Questions

1. How did sea-surface temperatures at the peak of the last glaciation differ from those of the present? Why do you think some regions of the ocean have shown more change than others? What influence would these changes have had on atmospheric circulation and weather?

2. How can isotopic analyses of deep-sea sediment reveal changes in global ice volumes?

3. At the height of the most recent Ice Age, vegetation in North America south of the ice front must have been very different from the vegetation today. Do some research and find out what is known of vegetation changes in your area over the past 20,000 years.

4. Choose two human climate proxy records and two natural climate proxy records. Describe in detail how the phenomena are controlled by or mimic climate, what type of climate information can be derived from them, and how long the records have been kept.

5. Find out if your city, state, province, or country has set goals for the reduction of carbon dioxide emissions to limit its contribution to global warming. What steps have been taken to meet these goals?

What is happening in this picture?

In this "Calvin and Hobbes" comic strip, Calvin is telling his pet tiger, Hobbes, about his theory that the days are getting colder because the Sun is going out. Let's call this Hypothesis 1. His father, on the other hand, says the days are getting colder because Earth is getting farther from the Sun—Hypothesis 2. Another reasonable hypothesis is that Earth's axis is tilted, which makes the sunlight less direct in winter—Hypothesis 3.

Think Critically

1. What scientific tests could you perform to determine why—or, indeed, if—the weather is getting colder?
2. Which explanation do you think is right—Calvin's, his dad's, or Hypothesis 3?

Self-Test

(Check your answers in Appendix D.)

1. Which one of the following is an example of climate change?
   a. an extremely hot summer
   b. three very cold winters in a row
   c. a very intense hurricane season
   d. two years of above-normal precipitation
   e. All of these indicate that the climate is changing.
   f. None of these is definitively indicative of climate change.

2. The Köppen climate classification system is based on _____.
   a. variations in oceanic circulation
   b. temperature, precipitation, and vegetation patterns
   c. global atmospheric circulation patterns
   d. paleoclimate records
   e. All of the above answers are correct.

3. The orbital variations that combine to produce Milankovitch cycles are _____.
   a. seasons, day–night cycles, and tides
   b. radiative balance, solar output, and albedo
   c. tilt, eccentricity, and precession
   d. glaciations, interglaciations, and feedbacks

4. Which one of these surfaces would have the highest albedo?
   a. forest   c. ice cap   e. rock
   b. ocean   d. soil

5. This graph of Earth's climate record shows that temperatures have _____ overall, in the past 2 million years.
   a. fluctuated greatly
   b. remained constant
   c. been slowly decreasing
   d. been slowly increasing

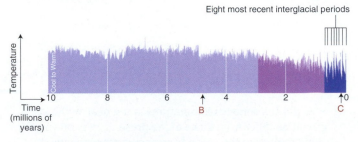

6. Geologic studies have revealed a number of natural agents that can act together to cause climate change. Which of the following is not considered a possible cause of climate change?
   a. changes in the composition of the atmosphere
   b. tectonism
   c. interruptions of ocean circulation
   d. changes in solar output
   e. All of the above are considered to be natural causes of climate change.

7. Which of the following would definitely not be useful as a climate proxy for the study of paleoclimate?

   a. growth rings in corals

   b. layered ice from polar regions and glaciers

   c. ocean sea-surface temperature

   d. tree rings

   e. Any of these would likely be useful as a climate proxy.

8. A climate proxy must be influenced by or mimic some aspect of climate and then preserve a record of that climatic influence. What is the other crucial characteristic of a useful climate proxy?

   a. It must be global, rather than local, in its extent.

   b. It must be something that can be duplicated in a laboratory.

   c. It must have occurred in the recent past.

   d. It must be able to have a date attached to it.

9. Imagine this scenario: The climate warms by 1°C. Evaporation of ocean water increases. As a result, cloud cover increases. The clouds that form are mainly stratus clouds, which block incoming solar radiation, cooling the surface. Is this scenario an example of a positive feedback or a negative feedback?

   a. negative feedback

   b. positive feedback

   c. This scenario shows both positive and negative feedback.

   d. This scenario shows neither a positive nor a negative feedback.

10. Geologists use a number of techniques to study paleoclimates, including data collected from _____.

    a. fossil pollen grains from peat bogs and lakes

    b. seafloor sediment

    c. paleosols

    d. ice cores

    e. All of the above are techniques geologists use to study paleoclimate.

11. What can scientists learn from air bubbles trapped in polar ice, like the ones shown here?

    a. greenhouse gas composition of ancient air

    b. temperature changes over the past few thousand years

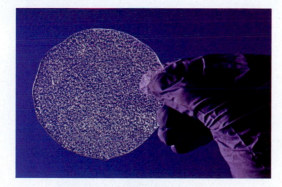

c. variations in snowfall at the poles

d. date of ice core climate proxy

12. When did the last major glaciation end?

    a. about 1 million years ago   b. about 10,000 years ago

    c. about 1000 years ago   d. about 150 years ago

13. What is the cause of the zig-zag pattern seen in the curve on this graph?

    a. human contributions to atmospheric carbon dioxide caused by the burning of fossil fuels

    b. variations in atmospheric carbon dioxide that result from seasonal variations in photosynthesis

    c. regional and global climate change

    d. changes in sea-surface temperature and oceanic thermo-haline circulation

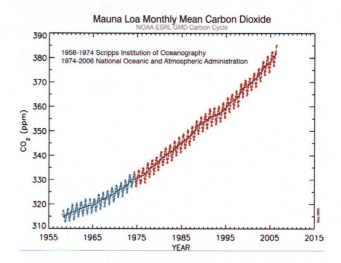

14. Anthropogenic carbon dioxide in the atmosphere results from the burning of fossil fuels, in addition to some other important sources. How far back does the recent record of atmospheric carbon dioxide measurements extend?

    a. about 1000 years   c. about 50 years

    b. about 100 years   d. about 10 years

15. What would be the effects of continued global warming over the next century?

    a. retreat of glaciers

    b. calving of large icebergs from ice shelves

    c. shrinking of some animal habitats

    d. expansion of some animal habitats

    e. All of the above are possible effects of global warming.

THE PLANNER ✓

Review the Chapter Planner on the chapter opener and check off your completed work.

A Brief History of Life on Earth

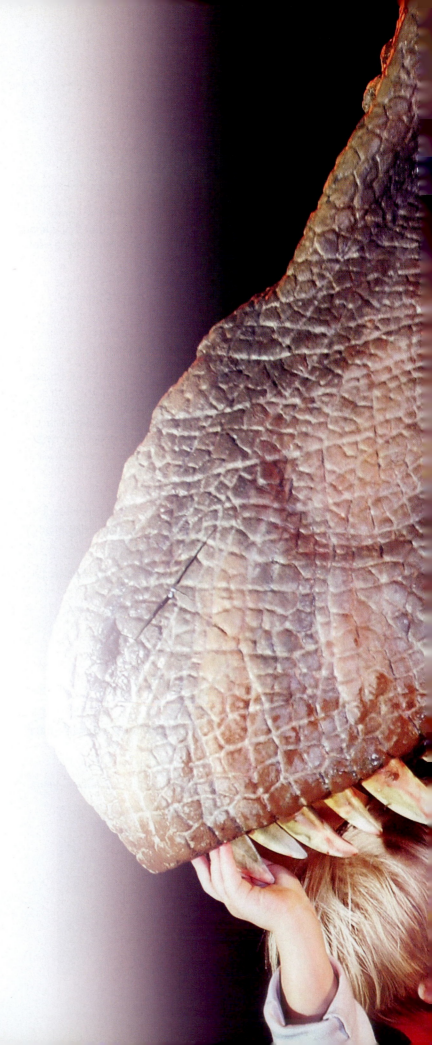

The history and diversity of life, from this ferocious *Tyrannosaurus rex* (at the Natural History Museum in London, England) to the humblest microbe, are very much a part of the planet's geologic history. Throughout this book are numerous examples of interactions between the biosphere and other parts of the Earth system. Plants and microorganisms accelerate the weathering of rock and formation of soil. Marine organisms sink to the bottom of the sea, where they form sediment that eventually turns into limestone. Land plants accumulate, undergoing lithification to form coal. Biologic processes, including the transpiration of plants and respiration of animals, regulate the air that we breathe.

This interaction also works in the opposite direction: The Earth system has fundamentally affected the course of life on this planet. Earth's atmosphere and hydrosphere provide a habitat in which life can develop and prosper. Life is shaped by the need to survive in particular environments—the scalding waters around a midocean rift, the arid land of a desert, the humid climate of a tropical rainforest—all of which result from natural processes.

The study of geology is thus inseparable from the study of life on Earth. In this chapter, we trace the story of life and its interactions with the atmosphere, hydrosphere, and geosphere.

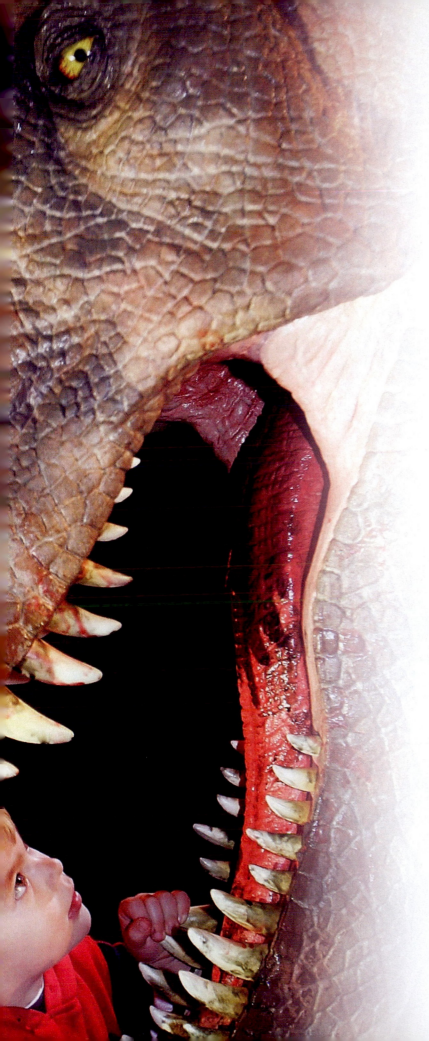

CHAPTER OUTLINE

CHAPTER PLANNER ✓

☐ Study the picture and read the opening story.
☐ Scan the Learning Objectives in each section:
   p. 456 ☐   p. 459 ☐   p. 464 ☐   p. 470 ☐
☐ Read the text and study all visuals.
   Answer any questions.

Analyze key features

☐ Geology InSight, p. 461 ☐   p. 470 ☐   p. 482 ☐   p. 485 ☐
☐ Process Diagram, p. 466 ☐
☐ What a Geologist Sees, p. 472 ☐
☐ Amazing Places, p. 473 ☐
☐ Stop: Answer the Concept Checks before you go on:
   p. 459 ☐   p. 464 ☐   p. 469 ☐   p. 485 ☐

End of chapter

☐ Review the Summary and Key Terms.
☐ Answer the Critical and Creative Thinking Questions.
☐ Answer What is happening in this picture?
☐ Complete the Self-Test and check your answers.

Ever-Changing Earth

LEARNING OBJECTIVES

1. **Describe** how Earth's environment has changed over the past 4 billion years.

2. **Explain** how photosynthesis adds oxygen to the atmosphere.

3. **Describe** how the oxygen cycle affects life on Earth.

4. **Summarize** the probable origin of the ocean.

T hroughout this book, we have seen evidence that Earth is a place of constant change and that negative feedbacks seem to moderate the changes so that Earth continues to be habitable. Perhaps the most dramatic changes are the rearrangements of the continents and oceans resulting from plate tectonics. Refer to Figure 1.19 and review the continental rearrangements that have occurred over the past 500 years. Abundant evidence demonstrates that plate motions have been operating for at least 2 billion years and probably even longer, but the exact locations of continents and oceans during the Precambrian is still uncertain and the focus of much research.

In this chapter, we present evidence of other changes within Earth's life zone. **Figure 15.1** presents a summary of these changes—changes that have supported life and were, to a certain extent, driven by life itself.

The changing Earth • Figure 15.1

From left to right, this diagram illustrates some of the major events in the history of Earth's surface environment during its first 4.6 billion years.

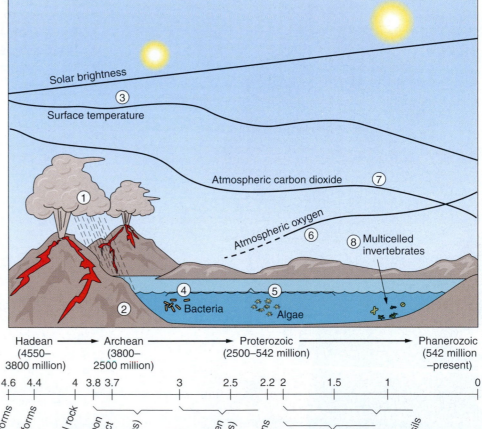

a. This pea plant, like all other green, leafy plants, produces oxygen through the process of photosynthesis.

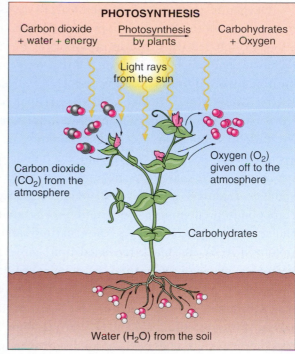

PHOTOSYNTHESIS

Carbon dioxide + water + energy	Photosynthesis by plants	Carbohydrates + Oxygen

Light rays from the sun

Carbon dioxide (CO_2) from the atmosphere

Oxygen (O_2) given off to the atmosphere

Carbohydrates

Water (H_2O) from the soil

b. The photosynthetic reaction combines carbon dioxide (CO_2) and water (H_2O) to make carbohydrates (molecules containing C, H, and O), which the plant needs to grow. The reaction produces oxygen molecules (O_2). The plant releases the oxygen into the atmosphere through pores in its leaves. The reaction does not happen spontaneously but requires energy from sunlight; that is why it is called "photosynthesis" (*photo-* means "light").

Photosynthesis • Figure 15.2

Compared to the Sun, which is representative of the raw materials from which Earth and the other planets of our solar system formed (see Chapter 1), Earth contains less of some volatile elements, such as nitrogen, argon, hydrogen, and helium. As discussed in Chapter 12, these elements were lost when the envelope of gas, or **primary atmosphere**, that surrounded early Earth was stripped away by the solar wind, by meteorite impacts, or both. Little by little, Earth generated a new, **secondary atmosphere** by volcanic outgassing of volatile materials from its interior.

Volcanic outgassing continues to be the main process by which volatile materials are released from Earth—although it is now going on at a much slower rate. The main chemical constituent of volcanic gas (as much as 97% by volume) is water vapor, with varying amounts of nitrogen, carbon dioxide, and other constituents. In fact, the total volume of volcanic gas released over the past 4 billion years or so accounts for the present composition of the atmosphere, with one extremely important exception: oxygen. As you can see in Figure 15.1, Earth had virtually no oxygen in its atmosphere more than 4 billion years ago, but the atmosphere is now approximately 21% oxygen.

Traces of oxygen were probably generated in the early atmosphere through the breakdown of water molecules into oxygen and hydrogen by ultraviolet light (a process called **photodissociation**). Although this is an important process, it doesn't even come close to accounting for the present high levels of oxygen in the atmosphere. Almost all of the free oxygen now in the atmosphere originated through **photosynthesis** (**Figure 15.2**).

photosynthesis
A chemical reaction whereby plants use light energy to induce carbon dioxide to react with water, producing carbohydrates and oxygen.

Oxygen is a very reactive chemical. Early in Earth's history, most of the free oxygen produced by photosynthesis was combined with iron in ocean water to form iron oxide–bearing minerals. The control that oxygen exerted on iron dissolved in the ocean is discussed in Chapter 8's *What a Geologist Sees*. Evidence of the gradual transition from oxygen-poor to oxygen-rich ocean water is preserved in seafloor sedimentary rock. The minerals in seafloor sedimentary rock more than about 2.5 billion years old contain reduced (oxygen-poor) iron compounds; in rock younger than 1.8 billion years, oxidized (oxygen-rich) compounds predominate. Sediment precipitated during the transition period contains alternating bands of red (oxidized iron) and black (reduced iron) minerals. This sediment, originally deposited as chemical sediment on the seafloor, eventually became the rock known as banded iron formation. Because ocean water is in constant contact with the atmosphere, and the two systems function together in a state of dynamic equilibrium, the transition from an oxygen-poor to an oxygen-rich atmosphere also must have occurred during this period.

Along with the buildup of molecular oxygen (O_2) came an eventual increase in ozone (O_3) levels in the atmosphere. Ozone filters out harmful ultraviolet radiation, so its accumulation in the stratosphere eventually made it possible for life to flourish in shallow water—and finally on land. This critical stage in the evolution of the atmosphere was reached between 1100 and 542 million years ago. Interestingly, the fossil record shows an explosive diversification of life-forms at the same time (the beginning of the Phanerozoic Eon).

Oxygen has continued to play a key role in the evolution and form of life. Over the past 200 million years, the concentration of oxygen has risen from 10% to as much as 25% of the atmosphere, before settling (probably not permanently) at its current value of 21% (**Figure 15.3**). This increase has benefited humans and all other mammals because mammals are voracious oxygen consumers. Not only do we require oxygen to fuel our high-energy, warm-blooded metabolism, our unique reproductive system demands even more. An expectant mother's used (venous) blood must still have enough oxygen in it to diffuse through the placenta into her unborn child's bloodstream. It would be very difficult for any mammal species to survive in an atmosphere of only 10% oxygen.

Scientists cannot yet be certain why atmospheric oxygen levels increased, but they have a hypothesis, and it illustrates the interactions between all parts of the Earth system. It begins with photosynthesis, but photosynthesis is only one part of the oxygen cycle. The cycle is completed by decomposition, in which organic carbon combines

The oxygen content of the atmosphere • Figure 15.3

Over the past 200 million years, oxygen levels in the atmosphere have increased markedly. The rise of mammals, aided by the demise of the dinosaurs 65 million years ago, may also have resulted partly from the plentiful oxygen supply.

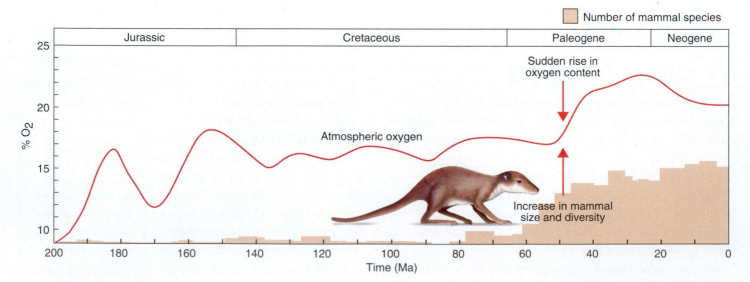

with oxygen and forms carbon dioxide. But if organic matter is buried as sediment before it fully decomposes, its carbon is no longer available to react with the free oxygen in the atmosphere. In that case, there should be a net accumulation of carbon in sediment and a buildup of oxygen in the atmosphere.

What could cause carbon to be buried on such a massive scale? This is where plate tectonics enters the picture. As you learned in Chapter 4, the most important tectonic event of the past 200 million years was the breakup of Pangaea at about the boundary between the Triassic and Jurassic periods. This breakup started the opening of the Atlantic Ocean, which created broad new continental margins. Vast amounts of sediment accumulated on the new margins, accelerating the burial rate of organic matter and isolating carbon from the atmosphere. The lithosphere, hydrosphere, and biosphere have therefore all interacted to produce the oxygen-rich atmosphere we breathe today.

The role of life in the chemical evolution of the atmosphere and hydrosphere does not stop with the oxygen cycle. Nitrogen, carbon, phosphorus, and sulfur also circulate through the Earth system, and their flows are also strongly affected by living organisms.

The global cycling of water has also played a fundamental role in the evolution of this planet. Water is the single most important ingredient for the development and maintenance of life. Earlier in this chapter, it was stated that as much as 97% of volcanic gas is water vapor. As water vapor cools and condenses, it precipitates, so the condensation of ancient volcanic gas is thought to be the origin of much of the water in the ocean. Another hypothesis maintains that some water also arrived early in Earth's history, through the impact of icy comets.

When did water, wherever it came from, condense to form the first ocean? Remember from Chapter 3 that the oldest Earth object found to date is a 4.4-billion-year-old mineral grain from sandstone in Australia. That mineral is zircon ($ZrSiO_4$), and careful study of the isotopes of oxygen in the grain indicates that it was influenced by water. The hypothesis is, therefore, that the ocean must be at least 4.4 billion years old. So far, scientists have been unable to suggest how large the primeval ocean was or what its composition was. The most ancient evidence of salts dissolved in the ocean comes from sedimentary rock formed about 3 billion years ago, which contains fossil imprints of crystals of halite ($NaCl$) deposited by evaporation.

CONCEPT CHECK 🛑 STOP

1. **Where** did Earth's primary and secondary atmospheres come from?

2. **Why** was there little or no oxygen in the secondary atmosphere early in Earth's history?

3. **What** are the sources of atmospheric oxygen, and what processes remove oxygen from the atmosphere?

4. **When** did ocean water first accumulate on Earth, and what is the source of its saltiness?

Early Life

LEARNING OBJECTIVES

1. **Identify** the minimum requirements for life.
2. **Describe** the system that cells use for replication.
3. **Summarize** the differences between aerobic and anaerobic metabolism.
4. **Explain** the difference between prokaryotic and eukaryotic organisms.

S cientists have not found conclusive evidence for life on Earth prior to the beginning of the Archean Eon, about 3.8 billion years ago. In rock of that age, isotopic signals from graphite in metamorphosed shale suggest that the carbon in the graphite was once part of living cells. The environment of Earth before 3.8 billion years ago was absolutely hostile to life, with a noxious atmosphere, constant volcanic eruptions, and a very high rate of bombardment by colossal meteorites. Yet a remarkably short time later, 3.55 billion years ago, entire colonies of bacterial life-forms had developed, as shown by layered structures called **stromatolites** (**Figure 15.4**) that are fossilized in rock of that age. How microscopic life emerged and spread so quickly on a previously barren and inhospitable planet is still one of the greatest unsolved mysteries of science.

Life in Three Not-So-Easy Steps

What is a living organism? There is no absolute agreement on this question, but it seems clear that any living organism must have at least two abilities. First, it must have a means of replication, creating a more or less accurate copy of itself. Second, it must have a metabolism, a means of extracting energy and material sustenance from its environment.

In addition, all life on Earth (though perhaps not life on other planets, if it exists) has certain other properties in common. All organisms on Earth use the same set of 20 carbon-based molecules, called **amino acids**, as "building blocks." Furthermore, all known living organisms are made of one or more **cells**, enclosed by a membrane. The cell membrane serves a dual purpose: maintaining a relatively constant chemical environment (or **homeostasis**) within the cell and allowing materials and energy to pass in and out as needed. (Cells are open systems!)

Carbon is essential to life on Earth because carbon atoms can **polymerize**, or form very long chains and complex molecules (such as proteins and enzymes), which enable a cell to function. The first step in the emergence of life was likely the creation of these chemicals out of inorganic ingredients. However, amino acids and biopolymers do not form by themselves. At a minimum, they require an energy source. In the 1950s, Stanley Miller (**Figure 15.5**) showed that it is possible to assemble amino acids out of a mixture of inorganic ingredients by passing an electric spark (e.g., from lightning) through them. However, it is still unclear whether the earliest amino acids really formed this way.

An ancient life-form • Figure 15.4

a. These odd-looking bumps in Shark's Bay, Western Australia, are present-day stromatolites, which are formed in warm, shallow seas by photosynthetic bacteria.

b. This 2.2-billion-year-old fossilized stromatolite from Michigan shows a pattern of growth rings that is identical to a cross-section of a modern-day stromatolite.

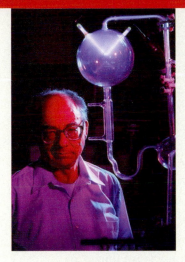

The first not-so-easy step: Making organic molecules

Miller's experiment used a "primordial soup" of methane, ammonia, and hydrogen, in the central flask shown here, to create amino acids. Unfortunately, scientists now believe that he used the wrong ingredients: The primitive atmosphere that developed from Earth's outgassing probably consisted mainly of carbon dioxide, water vapor, and nitrogen. No one has come up with a convincing mechanism for producing organic molecules out of these ingredients in conditions that resemble those of early Earth.

The second not-so-easy step: Replication

The two strands of the twisted molecule of DNA are held together by organic molecules called nucleotides. Nucleotides come in complementary pairs (shown as orange and brown, pink and green). When the two strands are separated, either one can act as a sort of photographic template to recreate the other. This allows for the duplication of genetic information, which is needed for an organism to grow or reproduce.

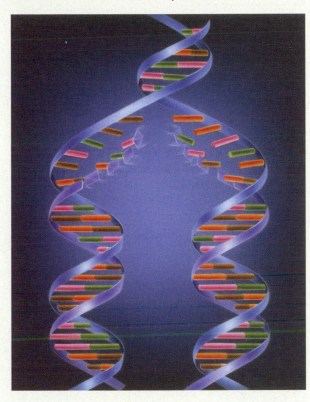

The third not-so-easy step: Metabolism

Seafloor hydrothermal vents give us an idea of how life might have emerged in the very hot environment of early Earth.

"Black smokers," such as this vent west of Vancouver Island, Canada, release superheated water and a variety of minerals (Figure a). Sunlight never penetrates to these depths, so organisms must rely on chemosynthesis rather than photosynthesis to obtain energy. Deep-sea bacteria like these (blue spheres on surface of a tube worm), extract energy from the chemicals produced by the vent and form the base of a food chain that includes tubeworms, crabs, and shrimp (Figures b and c).

a.
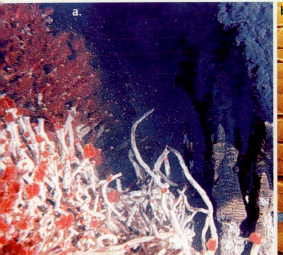

b.
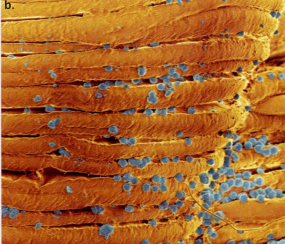

c.

Replication in modern life relies on a truly remarkable mechanism. Every cell nucleus contains a molecule called **DNA** (**deoxyribonucleic acid**). DNA is shaped like a twisted ladder (see Figure 15.5), in which each rung consists of two complementary organic molecules called **nucleotides**, which fit together like a lock and key.

DNA Deoxyribonucleic acid; a double-chain biopolymer that contains all the genetic information needed for an organism to grow and reproduce.

DNA is far too complex to have been the first self-replicating molecule. Most experts believe that a simpler, single-stranded molecule called **RNA** (**ribonucleic acid**) must have appeared first, and they have dubbed the hypothetical pre-DNA version of life the "RNA world." But even RNA is too complex and fragile to assemble by itself; proteins are required. How would these proteins work together without DNA or RNA to direct them? This version of a chicken-and-egg paradox has yet to be resolved.

Although present-day organisms use one method of replication exclusively (DNA), several different kinds of metabolism have evolved. Plants obtain energy via photosynthesis. Animals obtain energy from the food they eat, and oxygen is needed to release the energy; this is **aerobic metabolism**. In contrast, many kinds of bacteria have **anaerobic metabolisms**, for example, obtaining energy through the nonoxygenated breakdown of food by the process of **fermentation**. Other one-celled organisms can extract energy directly from hydrogen, sulfur, or salt via **chemosynthesis**. This allows them to survive in high-temperature or high-salinity environments that scientists once considered uninhabitable (see Figure 15.5). Such organisms are often called **extremophiles** because they live in conditions that are the extreme end of the chemical and thermal circumstances in which organic molecules can remain chemically stable.

Archean and Proterozoic Life

prokaryote A single-celled organism with no distinct nucleus—that is, no membrane separates its DNA from the rest of the cell.

Although some chemical signs of life date as far back as the beginning of the Archean Eon, the most ancient fossils known are the remains of microscopic **prokaryotes**. Prokaryotes are still present on Earth today: All bacteria are prokaryotes, and the extremophiles shown

in (**Figure 15.5b**) are part of a separate **domain** of prokaryotes called **Archaea**. Prokaryotes are unicellular (although they may form colonies, they can survive on their own), and they have rudimentary nuclei. Bacteria are extremely well adapted to their own specific environments, and far more numerous than anything else on Earth. Your own body contains 10 times as many bacterial cells as human cells!

domain The broadest taxonomic category of living organisms; biologists today recognize three domains: bacteria, *Archaea*, and eukaryotes.

For the first 2 billion years of life, the fossil record is rather sparse because the only forms of life were microscopic and had no hard parts that could be preserved. However, we do have one important piece of indirect evidence of ancient life. Stromatolites, as you saw in Figure 15.4, are mound-like structures that consist of many thin layers of calcium carbonate. Similar structures can be found today; they are formed in seawater by the action of photosynthetic organisms called **cyanobacteria**. Fossilized stromatolites have been found in rock up to 3.55 billion years old, providing evidence that photosynthesis and the production of oxygen is an ancient process on Earth.

At some time in the Proterozoic Eon, a new kind of lifeform appeared: **eukaryotes** (from the Greek words meaning "true nucleus"). In addition to a nucleus and membrane, a eukaryote (**Figure 15.6**) contains a number of smaller structures called organelles, some of which may originally have been prokaryotic bacteria that were engulfed by the larger eukaryotes and came to live in symbiosis with them.

eukaryote An organism composed of eukaryotic cells—that is, cells that have a well-defined nucleus and organelles.

Eukaryotes appeared at least 1.4 billion years ago, and perhaps as long ago as 2.7 billion years. Like most of the other dates in the early history of life, these dates are uncertain because the first eukaryotes were microscopic in size and lacked hard parts and therefore are not well preserved as fossils. It is no accident that they emerged after the transition to an oxygenated atmosphere; they had aerobic metabolisms and thus were better equipped to make use of free oxygen than prokaryotic bacteria had been.

Prokaryotes and eukaryotes • Figure 15.6

Both cells have been stained to enhance their visibility under a microscope.

a. In prokaryotic cells, such as this bacterium, the **nucleus** is poorly defined and not contained by a **membrane**.

b. In eukaryotic cells, such as this cell from a plant root, the **nucleus** and several other structures called **organelles** are clearly defined.

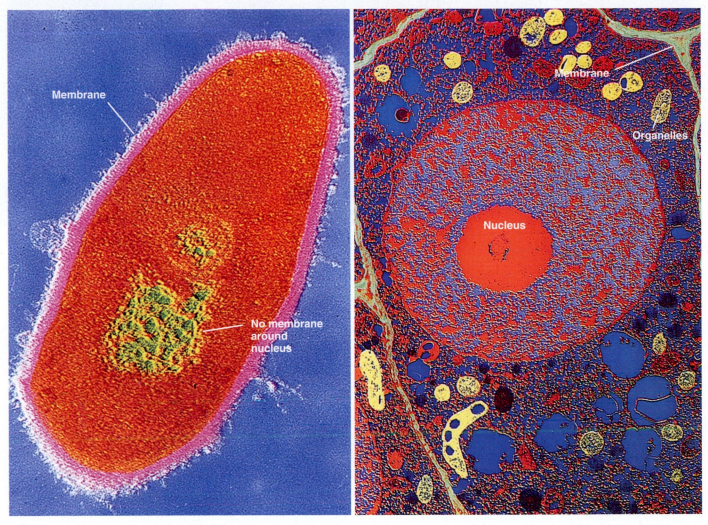

The use of oxygen as a fuel had profound ramifications. Aerobic metabolisms are more efficient than anaerobic metabolisms, allowing more energy to be extracted from each molecule of food. Eukaryotic cells grew larger and became more complex than prokaryotic cells. Eukaryotes were also more tolerant of crowding than anaerobic bacteria had been. Although the first eukaryotes were still single-celled organisms, similar to modern slime molds, the stage was set for multicellular life.

The earliest fossils of multicellular eukaryotic organisms appeared at the end of the Proterozoic Eon, in rock about 630 million years old. These fossils, which have now been found in a number of locations, are called the **Edia-cara fauna**, after the site where they were first discovered, the Ediacara Hills of South Australia. The Ediacara fauna

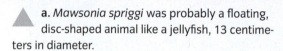

a. *Mawsonia spriggi* was probably a floating, disc-shaped animal like a jellyfish, 13 centimeters in diameter.

b. *Dickinsonia costata* was a worm-like creature, 7.5 cm in diameter.

The first multicellular animals • Figure 15.7

These strangely shaped fossils, specimens of the Ediacara fauna, are the most ancient evidence of multicellular animals.

lived in quiet marine bays. They were jelly-like animals with no hard parts (**Figure 15.7**). These organisms represent a huge jump in complexity from the first unicellular eukaryotes, which appeared at least 800 million years earlier. Scientists still do not know much about what happened during those 800 million years because fossil evidence is sparse and difficult to interpret.

CONCEPT CHECK   STOP

1. **What** are the basic building blocks and fundamental processes that must be in place for life to exist?

2. **How** are replication and metabolism carried out by terrestrial life-forms?

3. **Why** did eukaryotic organisms become more complex than prokaryotes?

4. **Are** there still prokaryotic organisms today, and if so, what are they like?

Evolution and the Fossil Record

LEARNING OBJECTIVES

1. **Define** the theory of evolution.
2. **Describe** how natural selection provides a means for evolution to occur.
3. **Explain** the role of genetics and environmental change in evolution.
4. **Describe** several ways in which fossils form.

A t the beginning of the Phanerozoic Eon, about 542 million years ago, the fossil record suddenly shows an explosion in the diversity of species. We have explored some of the possible reasons for this explosion: the oxygenated atmosphere, the protective layer of ozone, and the emergence of eukaryotes and then multicellular organisms. At this point, it seems appropriate to pause and describe how new species arise and how they are preserved as fossils.

Evolution and Natural Selection

On December 27, 1831, Charles Darwin set sail from England as an unpaid naturalist and "gentleman's companion" for the captain of the *H.M.S. Beagle* (**Figure 15.8**). Trained as a clergyman, Darwin at the time of his departure believed in the biblical account of creation and the fixity of species. By the time he returned in 1836, his views had changed considerably, though he still saw the hand of God at work.

Darwin kept scrupulous notes and made paintings and sketches of the plants, animals, and fossils he saw on the voyage. In 1859, long after his return to England, Darwin published his observations and ideas in a book called *On the Origin of Species by Means of Natural Selection*. He waited this long to publish his findings because he was concerned about the uproar it might—and did—cause. From the time of Wil-

evolution The theory that life on Earth has developed gradually, from simple organisms to more complex organisms.

natural selection The process by which individuals that are well adapted to their environment have a survival advantage and pass on their favorable characteristics to their offspring.

species A population of genetically and/or morphologically similar individuals that can interbreed and produce fertile offspring.

liam "Strata" Smith, who formulated the Principle of Faunal and Floral Succession (Chapter 3), it had been widely recognized that life has changed through time. By the time of Darwin, the concept of **evolution** had already reached the status of a theory.

In his book, Darwin outlined a hypothesis of **natural selection** to explain how evolution had occurred. Darwin proposed that new **species** develop from existing ones by a gradual process of change through inheritable characteristics. All present-day organisms are descendants of different kinds of organisms that existed in the past, whose populations slowly changed through adaptation to changing environmental conditions. Darwin was not the first to suggest evolution as an explanation for the variety and distribution of species on Earth, but he was the first to provide a thorough discussion of

Charles Darwin: Decipherer of evolution's clues • Figure 15.8

Though he was a retiring and unpretentious man, Charles Darwin wrote one of the most controversial and influential books in scientific history, *On the Origin of Species*. Originally printed in a small edition of 1250 copies, his book has been through at least 400 printings and translated into at least 29 languages.

Darwin's finches • Figure 15.9

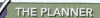

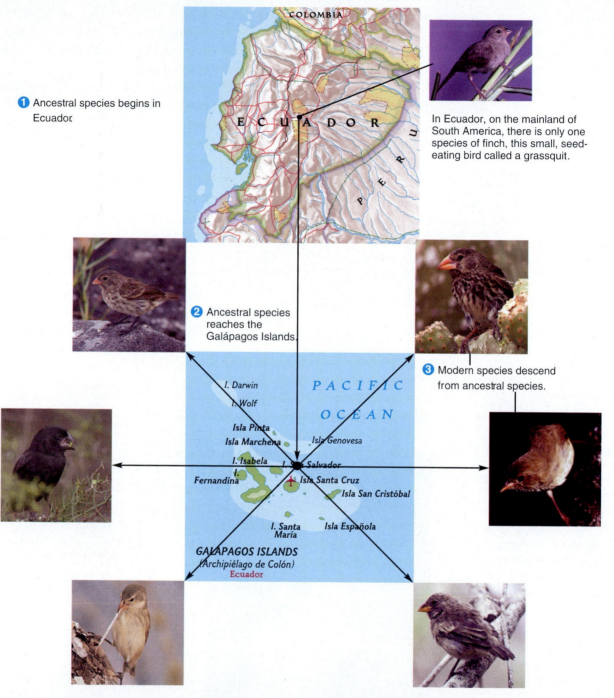

❶ Ancestral species begins in Ecuador.

COLOMBIA

ECUADOR

PERU

In Ecuador, on the mainland of South America, there is only one species of finch, this small, seed-eating bird called a grassquit.

❷ Ancestral species reaches the Galápagos Islands.

❸ Modern species descend from ancestral species.

PACIFIC OCEAN

I. Darwin
I. Wolf
Isla Pinta
Isla Marchena
Isla Genovesa
I. Isabela
I. San Salvador
I. Fernandina
Isla Santa Cruz
Isla San Cristóbal
I. Santa María
Isla Española

GALÁPAGOS ISLANDS
(Archipiélago de Colón)
Ecuador

On the Galápagos Islands, Darwin counted 13 species of apparently related birds, but with different beak shapes and different diets. What could account for such a dramatic difference in diversity between the islands and the mainland? Darwin reasoned that long ago, finches from the mainland had colonized the volcanic islands and had subsequently changed as a result of having to adapt to their new environments. Different beak shapes had developed as the birds adapted to different diets. It was a logical idea, but it contradicted the widely believed theological doctrine that no new species had appeared on Earth since creation.

Think Critically

Darwin reasoned that the original grassquits that colonized the islands had become stranded and therefore isolated from the population in Ecuador. How might they have become stranded?

the available evidence, gathered during his voyage and over the following years (**Figure 15.9**).

Because the word **theory** is so often misunderstood (see Chapter 1), it is important to emphasize that there is no doubt among legitimate biologists that evolution does occur, has occurred throughout the history of life on this planet, and has repeatedly been shown to occur in laboratory situations. The question is how it occurs. The late paleontologist Stephen Jay Gould said that Darwin's two great accomplishments were establishing the fact of evolution and proposing the theory of natural selection to explain it. There is an analogous situation in physics: It is a fact (established by Galileo) that gravitation exists, but there are different theories (proposed by Newton and Einstein) of how it works.

Darwin was motivated to publish *On the Origin of Species* when another young naturalist, Alfred Russel Wallace, independently hit on the same idea of natural selection. According to this theory, any generation of a species will have a broad range of genetic characteristics. Individuals who are better adapted to their surroundings will have more survival success and more reproductive success. In later generations, their descendants will be more numerous than those of less-well-adapted individuals. Over time, the entire population will evolve, as natural selection favors better-adapted individuals.

Natural selection also allows for relatively rapid changes. In modern laboratories, scientists can observe evolution in progress, through the expression of specific genetic traits in organisms that reproduce rapidly, such as fruit flies. In nature, if a useful new trait emerges that provides a competitive advantage for an individual and its descendants—for example, a different beak shape, which might help a finch procure seeds—then after several generations, most individuals will have that trait. Note that the organism does not develop this trait in response to an environmental need or constraint. The trait occurs by chance, but its occurrence enhances the survival and therefore the reproductive success of the individual.

Since Darwin's time, the science of genetics has greatly strengthened the support for natural selection as the mechanism for evolution. Darwin did not know how organisms transmit their traits to later generations; now we know that heritable traits are encoded in **genes** on an organism's DNA. The random variation that is essential to Darwin's theory is explained by **mutations** in those genes—accidental substitutions of one nucleotide for another, or deletions or transpositions.

There is still some scientific debate over the relative importance of gradual change versus rapid change in evolution. Darwin favored extremely slow, gradual change, or **gradualism**, whereas Gould, among others, advocated **punctuated equilibrium**, in which species persist for a very long time with few changes and undergo occasional periods of very rapid change. In recent years, scientists have begun to recognize the importance of occasional catastrophic events. The best-documented example is the meteorite impact that is thought to have wiped out the dinosaurs. These animals died out, in all likelihood, not because they were less-well-adapted to their environment than other species but because of a chance event that they had never experienced before and had no opportunity to adapt to.

To explore the topics of evolution and natural selection in greater detail, see *Where Geologists Click*.

Where Geologists CLICK

Evolution 101

http://evolution.berkeley.edu/evolibrary/article/evo_01

For a beautifully detailed but very clear and understandable explanation of all aspects of evolution and natural selection, visit the University of California at Berkeley's "Evolution 101" Web site. The sections The History of Life: Looking at the Patterns and Mechanisms: The Processes of Evolution are particularly relevant to this chapter.

How Fossils Form

The best evidence for evolution lies in the vast numbers of **fossils** chronicling the succession of species that no longer exist on Earth. Although fossilization can occur in many ways, it is always a relatively rare event. If a dead animal or plant is exposed to air, running water, scavengers, or bacteria, it will decompose or be eaten, and its parts will be scattered or destroyed. Hard parts such as bones, shells, and teeth are less easily destroyed and thus more likely to be preserved than soft or delicate parts such as skin, hair, feathers, or leaves. For an organism to be preserved as a fossil, it must be quickly covered up by a layer of protective material—usually sand or mud but sometimes tree sap, ice, tar, or volcanic ash.

fossil Remains of an organism from a past age, embedded and preserved in rock.

Sometimes a deceased organism is preserved with little or no alteration. For example, insects many millions of years old have been trapped in tree sap, which hardens into amber (**Figure 15.10a**). This seals off the specimen from the elements so completely that parts of its original organic matter can still be recovered. Ice and tar are also excellent preservatives. In dry climates, natural mummification can occur, in which the soft parts desiccate and harden before they have a chance to decompose.

WILEY PLUS NATIONAL GEOGRAPHIC Video

More often, however, fossils reflect the original shape of an organism but do not contain the original

Examples of fossil formation • Figure 15.10

a. Even the delicate legs and wings of this ancient mosquito are preserved in its casing of amber, which is fossilized tree resin. This mosquito is more than 24 million years old.

b. Over many millennia, this log in Petrified Forest National Park, Arizona, became mineralized. Although it looks uncannily like fresh wood, preserved even down to the cellular level, it is completely made of stone.

Not all fossils are bones • Figure 15.11

a. This 2-meter-long fossilized dinosaur (*Oviraptor*) was found curled protectively around a nest containing at least 20 eggs. This is strong evidence that dinosaurs cared for their young.

b. Over 65 million years ago, hadrosaurs in what is now Argentina left their tracks in soft, red mud, which turned to rock. The formation of the Andes Mountains tilted a formerly horizontal mud flat to such an extent that it is now a vertical rock wall.

materials. Bones and other hard parts are replaced by minerals carried in solution by groundwater (a process called **mineralization**). This is the process that creates **petrified wood** (Figure 15.10b). The remains of plants are sometimes preserved by carbonization, which occurs when volatile material in the plant evaporates, leaving behind a thin film of carbon.

Some fossils do not contain any actual parts of the organism. For example, the organism may leave behind an imprint, or a **mold,** in the soft sediment that covered it, as with the Ediacara fauna. Other kinds of indirect evidence of previous animal life include eggshells and **trace fossils** (**Figure 15.11**). Finally, prehistoric animals left behind fe-

> **trace fossil**
> Fossilized evidence of an organism's life processes, such as tracks, footprints, and burrows.

cal droppings, which are called **coprolites** when preserved and fossilized. In spite of their unappealing origin, such fossils can provide useful clues about animals' characteristics, habits, and diets.

CONCEPT CHECK   🛑 STOP

1. **What** is the difference between evolution and natural selection?

2. **How** did finches lead Charles Darwin to the idea of natural selection as a mechanism for evolution?

3. **Why** does natural selection rely on random variations in the genetic code?

4. **Why** do the hard parts of organisms appear more often as fossils than do soft parts?

Life in the Phanerozoic Eon

LEARNING OBJECTIVES

1. **Describe** the dramatic changes in Earth's biota during the Cambrian Period.

2. **Identify** the requirements for a living organism to survive on land.

3. **Describe** how plants, amphibians, and reptiles met the requirements for surviving on land.

4. **Outline** the most recent evolutionary steps in the development of humans.

5. **Define** mass extinction and identify two theories of what causes such events.

Geology InSight | Timeline of life • Figure 15.12

The geologic timescale provides a context for our understanding of the origin and evolution of life on Earth.

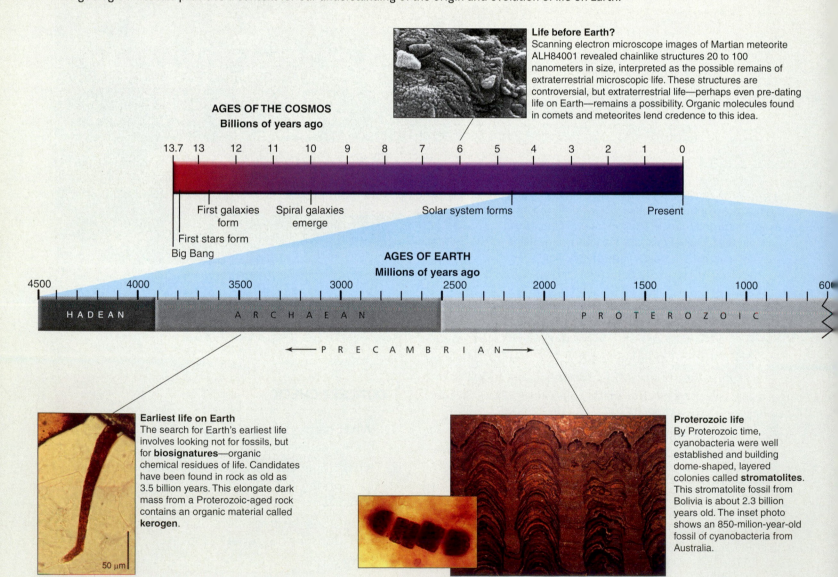

Life before Earth?
Scanning electron microscope images of Martian meteorite ALH84001 revealed chainlike structures 20 to 100 nanometers in size, interpreted as the possible remains of extraterrestrial microscopic life. These structures are controversial, but extraterrestrial life—perhaps even pre-dating life on Earth—remains a possibility. Organic molecules found in comets and meteorites lend credence to this idea.

AGES OF THE COSMOS
Billions of years ago

13.7 13 · 12 · 11 · 10 · 9 · 8 · 7 · 6 · 5 · 4 · 3 · 2 · 1 · 0

First galaxies form
Spiral galaxies emerge
Solar system forms
Present
First stars form
Big Bang

AGES OF EARTH
Millions of years ago

4500 · 4000 · 3500 · 3000 · 2500 · 2000 · 1500 · 1000 · 600

HADEAN · ARCHAEAN · PROTEROZOIC

← PRECAMBRIAN →

Earliest life on Earth
The search for Earth's earliest life involves looking not for fossils, but for **biosignatures**—organic chemical residues of life. Candidates have been found in rock as old as 3.5 billion years. This elongate dark mass from a Proterozoic-aged rock contains an organic material called **kerogen**.

50 µm

Proterozoic life
By Proterozoic time, cyanobacteria were well established and building dome-shaped, layered colonies called **stromatolites**. This stromatolite fossil from Bolivia is about 2.3 billion years old. The inset photo shows an 850-million-year-old fossil of cyanobacteria from Australia.

During most of the history of life on Earth—3 billion years of it—the only living organisms were of microscopic size. But about 630 million years ago, with the appearance of larger multicellular animals, the Ediacara fauna, life began to diversify very rapidly into a wide variety of sizes and body types. We will take a whirlwind tour through this last phase of development of life, which continues to the present day (**Figure 15.12**).

The Phanerozoic Eon is named from the Greek *phaneros*, meaning "visible," because it is that part of Earth's history in which evidence of life is abundantly visible. It was no accident that early geologists placed the boundaries of the Cambrian Period where they did; they recognized the appearance of macro-fossils as a significant turning point for life on Earth. The Phanerozoic Eon is divided into three eras (see Figure 3.8): the Paleozoic Era (old life), the Mesozoic Era (middle life), and the Cenozoic Era (recent life).

THE PLANNER

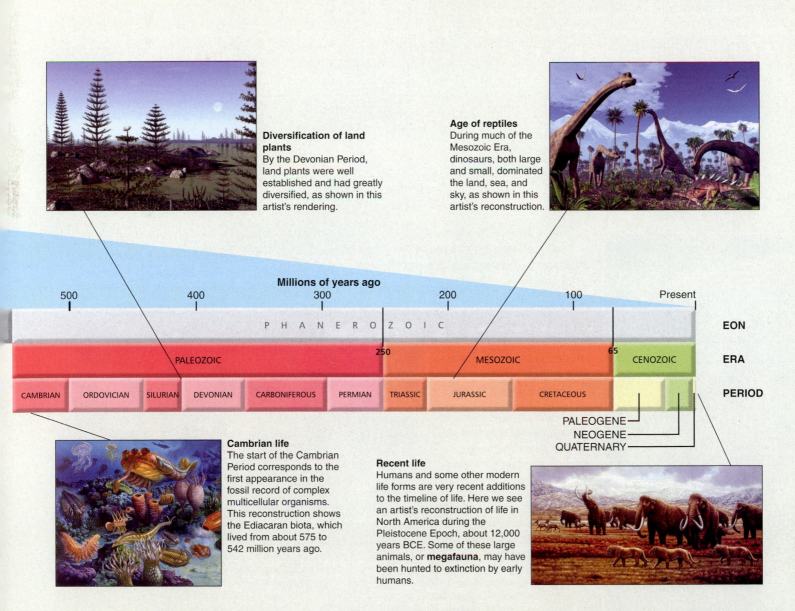

Diversification of land plants
By the Devonian Period, land plants were well established and had greatly diversified, as shown in this artist's rendering.

Age of reptiles
During much of the Mesozoic Era, dinosaurs, both large and small, dominated the land, sea, and sky, as shown in this artist's reconstruction.

Millions of years ago

500   400   300   200   100   Present

PHANEROZOIC — EON

250   65
PALEOZOIC   MESOZOIC   CENOZOIC — ERA

CAMBRIAN | ORDOVICIAN | SILURIAN | DEVONIAN | CARBONIFEROUS | PERMIAN | TRIASSIC | JURASSIC | CRETACEOUS — PERIOD

PALEOGENE
NEOGENE
QUATERNARY

Cambrian life
The start of the Cambrian Period corresponds to the first appearance in the fossil record of complex multicellular organisms. This reconstruction shows the Ediacaran biota, which lived from about 575 to 542 million years ago.

Recent life
Humans and some other modern life forms are very recent additions to the timeline of life. Here we see an artist's reconstruction of life in North America during the Pleistocene Epoch, about 12,000 years BCE. Some of these large animals, or **megafauna**, may have been hunted to extinction by early humans.

The Paleozoic Era

The Paleozoic Era started 542 million years ago, with the Cambrian Period. It was a time of incredible diversification of life (see *What a Geologist Sees*). Why was this so? One hypothesis is that sexual reproduction, which developed with the eukaryotes, afforded a more rapid way for new "experimental" forms of life to emerge. Another hypothesis is that there was finally enough oxygen in the atmosphere to support the metabolism of larger organisms. The presence of oxygen led to the development of an ozone layer, which would have shielded Cambrian life-forms from harmful ultraviolet radiation. The rising oxygen content of the atmosphere may have influenced the biochemistry of calcium phosphate and calcium carbonate in seawater, allowing animals to grow skeletons and/or shells. There are other hypotheses, too; one is that predation started, and it led to the preservation of creatures that had developed protective hard parts; another is that extreme climate changes at the end of the Proterozoic Eon led to accelerated evolutionary responses of surviving organisms.

Whatever the reasons, a great many changes began to occur about 542 million years ago, in what has been called the **Cambrian explosion**, or **Cambrian radiation**. Compact animals were evolving to replace the soft-bodied, jelly-like organisms of Ediacara times. These included trilobites, mollusks (clams and sea snails), and echinoderms (sea urchins). All the new animals were equipped with gills, filters for feeding, efficient guts, a circulatory system, and other characteristics that have continued to serve animals well up to the present.

The Cambrian Period also ushered in the development of skeletons, both internal and external. Skeletons gave many organisms a selective advantage, protecting them against predators, against drying out, against being injured in turbulent water, and so on. It is not surprising that hard-shelled organisms, such as trilobites, are better preserved in the fossil record. However, a wide variety of soft-bodied creatures have been found in the Burgess Shale (see *Amazing Places*), and even richer deposits in Chengjiang, China. From these deposits,

WHAT A GEOLOGIST SEES

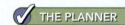
THE PLANNER

A Sample of Cambrian Life

a. This geologist is holding a fossil of a trilobite, one of the first animals to develop a hard covering, presumably to defend against predators. Trilobites were extremely numerous and long-lasting as a species; they are ubiquitous among Cambrian fossils. This 34-cm-long specimen of *Olenellus getzi* was found in Lancaster County, Pennsylvania, and is Lower Cambrian in age.

b. This specimen of *Waptia fidensis*, a soft-bodied arthropod, is preserved in the Burgess Shale of British Columbia. This species did not have a hard shell as the trilobites did, and therefore fossils of it are much rarer. As a geologist, what processes do you think needed to occur for the soft parts of this specimen to be preserved in such detail?

Think Critically

Despite their efficient defense against predation, trilobites eventually became extinct. Arthropods, however, are still with us, and they are a hugely diverse family that includes organisms from lobsters and crabs to millipedes and spiders. Some scientists estimate that arthropods account for more than 80% of living animal species. Why have arthropods been so successful overall?

AMAZING PLACES

The Burgess Shale

In 1909, the American paleontologist Charles Walcott discovered the world's preeminent site for Cambrian fossils, a rock formation called the Burgess Shale, about 90 km from Banff, British Columbia. It is a treasure trove for specimens such as this trilobite (**Figure a**), as well as more exotic organisms such as the five-eyed *Opabinia* (**Figure b**) and others. It was only in the 1970s that paleontologists realized that many of the organisms found there are unrelated to any living animals. They are not only extinct species but members of extinct phyla. Exploring the Burgess Shale is like exploring the evolutionary paths that life could have taken but didn't.

Many of the animals preserved in the Burgess Shale were soft bodied. Such creatures fossilize only under very special circumstances. This is how scientists reasoned that it happened: The Burgess fauna lived in shallow, oxygen-rich waters atop an algal reef with steep sides (Figure b). Periodically, mudslides would sweep some unlucky animals off the side of the reef and deposit their bodies at the base (**Figure c**). At that time, the deep waters were still so oxygen poor that no microorganisms existed at that depth to decompose them. The fine silt particles from the mudslide encased the bodies and hardened into shale, preserving a highly detailed imprint of the animals; the *Waptia fidensis* fossil in *What a Geologist Sees* is an example.

a.

b.

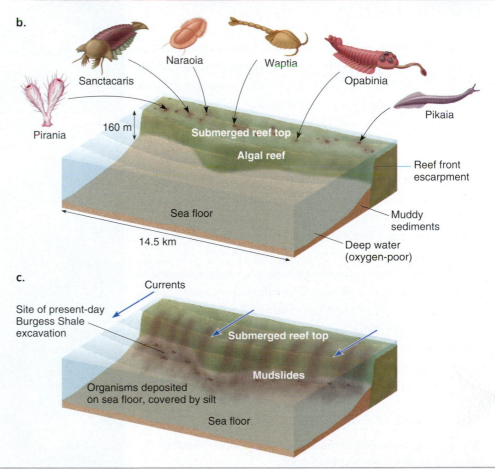

Sanctacaris · Naraoia · Waptia · Opabinia · Pirania · Pikaia

160 m · Submerged reef top · Algal reef · Reef front escarpment · Sea floor · 14.5 km · Muddy sediments · Deep water (oxygen-poor)

c.

Currents · Site of present-day Burgess Shale excavation · Submerged reef top · Mudslides · Organisms deposited on sea floor, covered by silt · Sea floor

kingdom The second-broadest taxonomic category. There are six recognized kingdoms, including animals and plants.

we now know that every **phylum** in the animal **kingdom** was present in the Cambrian Period, as were a number of phyla that no longer exist. In this sense, the Cambrian marine environment had the greatest diversity of life-forms in Earth's history.

From Sea to Land

The great proliferation of life in the Cambrian explosion was confined to the sea. In order to spread to land, it was essential for living organisms to meet certain requirements. The first requirement is some means of *structural support*. Whereas aquatic organisms are buoyed by water, land-based organisms must contend with gravity. But because water remains critical to all the chemical processes of life, any land-based organism must maintain an *internal aquatic environment*. Living in an environment surrounded by air, rather than water, the organism must develop some way of *exchanging gases with air* instead of with water. Finally, all sexually reproducing organisms require a moist environment for the reproductive system. The first organisms to overcome these four hurdles were plants.

Plants Evidence suggests that land plants evolved from green algae more than 600 million years ago (**Figure 15.13**). Eventually, during the Silurian Period, **vascular plants** evolved. These plants have structural support from stems and limbs (requirement 1), and they have a set of vessels through which water and dissolved elements are transferred from the roots to the leaves (requirement 2). Gas exchange (requirement 3) occurs by diffusion through adjustable openings in the leaves called **stomata**. When carbon dioxide pressure inside the leaf is high, the stomata open; when it is low, they close. The stomata also close when the plant is short of water, thereby protecting it from drying out.

The means of reproduction in plants (requirement 4) has passed through several evolutionary stages. The earliest plants were seedless (**Figure 15.14**), and some seedless plants still exist, such as mosses and ferns. These can reproduce either sexually or asexually, but for

The earliest life on land? • Figure 15.13

The first plants to colonize the land may have looked like the green algae shown in this photograph.

The evolution of plant life • Figure 15.14

a. This fossil fern, preserved in shale, is about 300 million years old.

b. Compare the fossil fern to a modern fern. This photograph shows the spore-producing organs, the dark spots on the underside of the frond.

c. Naked-seed plants developed from the seedless plants late in the Devonian Period. These are leaves of modern and fossilized ginkgos.

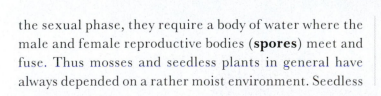

d. Flowers and fruits represented an advantageous adaptation for plants because they serve as an incentive for insects to do the work of distributing pollen. This 15-million-year-old fossil found in Idaho is shown next to its modern equivalent, the sweet gum fruit.

the sexual phase, they require a body of water where the male and female reproductive bodies (**spores**) meet and fuse. Thus mosses and seedless plants in general have always depended on a rather moist environment. Seedless plants reached their peak in the Carboniferous Period, and their fossils created the extensive deposits of coal (carbon-rich sedimentary rock) that gave the period its name.

In the middle Devonian Period, some plants began to evolve their own moist environment to facilitate sexual reproduction. The first plants to do this were **gymnosperms**, including ginkgos and conifers. The female cell of a gymnosperm is attached to the vascular system and therefore has a supply of moisture. The male cell is carried in a pollen grain with a waxy coating. When the two fuse, a **seed** results. The seed contains moisture and a food supply that sustains the growth of the young plant until it can support itself. With the evolution of seeds, vascular plants were able to spread beyond swampy lowlands to other habitats.

> **gymnosperm** A naked-seed plant.

Arthropods

Many of the creatures in Cambrian seas belonged to the phylum *Arthropoda*, named for their jointed legs. Modern arthropods include crabs, spiders, centipedes, and insects—the most diverse phylum on Earth. They were the first creatures to make the transition from sea to land.

With a few exceptions, early arthropods were small and light. They were covered with a hard shell of **chitin** (a fingernail-like material). Thus, they were well adapted to life on land, in regard to the need for structural support and water conservation. The first arthropods on land were probably centipedes and millipedes, in the Silurian Period. By the Mississippian Period, insects were abundant and included dragonflies with a wingspan of up to 60 cm (**Figure 15.15**). But for all their success as land creatures, the arthropods have very primitive respiratory and vascular systems. For example, insects breathe through tiny tubes that penetrate their outer coating. This mode of respiration severely limits the size of an insect and explains why almost all insects are small.

The "blood" of arthropods is simply body fluid that bathes the internal organs; it does not circulate in blood vessels. The fluid is generally kept in motion by a sluggish "heart" that is little more than a contracting tube. However, "primitive" does not mean poorly adapted. The arthropods have branched into more than a million species, and they are nearly indestructible. Who has ever heard of a cockroach having a heart attack?

Ancient insects • Figure 15.15

A single wing of the giant dragonfly *Megatypus schucherti* is 16 cm long. The largest dragonfly today is only 15 cm in total width. This specimen, Lower Permian in age, is from Dickinson County, Kansas.

Pioneer fish? • Figure 15.16

a. When a living coelacanth was caught in the Indian Ocean in 1938, it created a sensation because not only its species but its entire order, the crossopterygians, had been believed to be extinct. The first fish to haul themselves onto land may have been relatives of the coelacanth. Today's coelacanths, like this one photographed near the Comoro Islands, are exclusively deep-sea creatures.

b. Another candidate for the first fish to make the transition to land is the lungfish. Unlike the coelacanth, today's lungfish still survive for short periods on land without water.

Fishes and amphibians The phylum of greatest interest to most of us, because humans belong to it, is the **chordates**. These are animals that have at least a primitive version of a spinal cord (called a **notochord**). Like all the other animal phyla, chordates can be found as fossils of the Cambrian Period, although they are relatively inconspicuous. The earliest so far discovered is *Haikouichthys*, a jawless fish akin to a hagfish or a lamprey that lived 525 million years ago—only 17 million years after the beginning of the Cambrian explosion.

Jawed fish arrived next. With jaws, fish could lead a predatory lifestyle and grow to much larger sizes; the jawless fish had been limited to filtering food out of the water or dredging it from the seafloor. Among the first large jawed fish were sharks and ray-finned fish. The earliest known intact shark fossil is 409 million years old.

The first fish to venture onto land may have been a member of an obscure order called *Crossopterygii*, or lobe-finned fish (**Figure 15.16**). These fish had several features that could have enabled them to make the transition to land. Their lobe-like fins contained all the elements of a quadruped limb. They had internal nostrils, characteristic of air-breathing animals. As fish, they had already developed a vascular system that was adequate for life on land.

Amphibians, the first terrestrial chordates, originated in the Devonian Period. Amphibians have never become wholly independent of aquatic environments because they have not developed an effective method for conserving water. To this day, they retain permeable skins. They have also never really met the reproductive requirement for life on land. In most amphibian species, the female lays her eggs in water, and the male fertilizes them there after a courtship ritual. The young (e.g., tadpoles) are fish-like when first hatched. Just like the seedless plants, amphibians have kept one foot (figuratively speaking) on land and one foot in the water.

The Mesozoic Era

The Paleozoic Era closed with the extinction of trilobites and a great many other marine creatures. An estimated 96% of all living species disappeared at the end of the Permian Period, in the greatest mass extinction in Earth's history. We will return to the question of mass extinctions at the end of this chapter, but the cause of this extinction remains an enigma that has yet to be solved. The era that followed the Paleozoic, the Mesozoic Era, commenced with all the major landmasses on Earth joined together in the supercontinent Pangaea (see Figure 1.19). Then, in the middle of the era, the supercontinent began to split apart in a process that continues today.

Flowering plants Whatever the cause of the great extinction at the end of the Permian, it did not seem to greatly affect plant life on land. Gymnosperms, which had appeared in the Devonian Period, dominated the plant world during the Triassic and Jurassic periods. Gymnosperms, however, have an important liability: The male cell carrier, the pollen, is spread through the air. This is extremely inefficient; what chance does a pollen grain in the air have of finding a female cell? Eventually, at the beginning of the Cretaceous Period, **angiosperms**—flowering plants—found a more efficient solution.

> **angiosperm** A flowering, or seed-enclosed, plant.

For a small incentive, such as nectar or a share of the pollen, insects deliver pollen directly from one flower to another or from one part of a plant to another. After pollination, a plant develops a seed in much the same way as gymnosperms. In many cases, birds and other animals help distribute the seeds by eating a plant's seed-bearing fruits and distributing the seeds in their feces. Flowering plants still dominate land plants today.

Reptiles, birds, and mammals The Carboniferous and Permian periods were times when amphibians were abundant. Most families died out in the end-of-Permian extinction. One branch of the amphibians, however, evolved into reptiles—the first fully terrestrial animals.

Reptiles were freed from the water by evolving an egg that contained amniotic fluid for the young to grow in and by developing a watertight skin. These two evolutionary advances enabled them to occupy many terrestrial niches that the amphibians had not been able to exploit because of their need to live near water. The amniotic egg led to an explosion in reptile diversity, much as the evolution of the jaw had done for fish. Moving out of the Mississippian and Pennsylvanian swamps, some reptiles colonized the land, some moved back into the water, and a few took to the air. Not only did reptiles greatly increase in diversity during the Jurassic Period, they also grew to tremendous size. The dinosaurs were the largest land animals that have ever lived, possibly ranging up to 100 metric tons in weight and 35 m in length.

Birds first appeared near the end of the Jurassic Period, and they are now considered to be direct descendants of the dinosaurs. An early bird, *Archaeopteryx* (**Figure 15.17**), would have been classified as a dinosaur if not for the discovery that it had feathers. Even before *Archaeopteryx*, vertebrates had made the transition to the air in the form of pterosaurs—flying reptiles with long wings and tails. Most paleontologists do not consider pterosaurs to be true dinosaurs, although they were very closely related. The detail of the transition from reptiles to birds remains one of the most highly controversial topics in paleontology today.

Mammals are descended from a class of "mammal-like reptiles" that existed as long ago as the Permian Period. The transition from reptiles to mammals is well understood, though perhaps not quite as well understood as the transition from fish to amphibians. It is difficult to pick out a single mammalian adaptation that is comparable in survival value to the jaws of fish, the eggs of reptiles, or the feathers of birds. During the Cretaceous Period, mammals certainly did not outcompete the dinosaurs; they survived by being small and inconspicuous.

The Cenozoic Era

Evidence suggests that an accidental catastrophe—a giant meteorite impact—was at least partially responsible for the environmental changes that wiped out the great reptiles and 70% of all other species at the end of the Mesozoic Era. The departure of dinosaurs gave mammals a chance to grow larger and to diversify. In the Cenozoic Era, mammals benefited from the atmosphere's high oxygen level, which is conducive to fast metabolism. Unlike reptiles, whose brain sizes have not grown (relative to their body size), mammals have continued to evolve toward larger brain size throughout the Cenozoic Era. This may be one key to the success of mammals because it has enabled

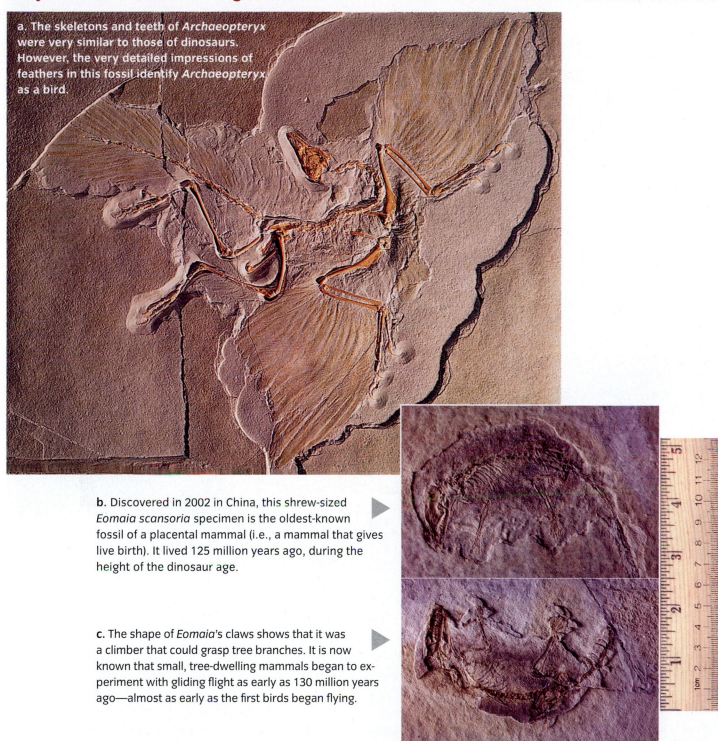

a. The skeletons and teeth of *Archaeopteryx* were very similar to those of dinosaurs. However, the very detailed impressions of feathers in this fossil identify *Archaeopteryx* as a bird.

b. Discovered in 2002 in China, this shrew-sized *Eomaia scansoria* specimen is the oldest-known fossil of a placental mammal (i.e., a mammal that gives live birth). It lived 125 million years ago, during the height of the dinosaur age.

c. The shape of *Eomaia*'s claws shows that it was a climber that could grasp tree branches. It is now known that small, tree-dwelling mammals began to experiment with gliding flight as early as 130 million years ago—almost as early as the first birds began flying.

them to diversify their lifestyles to a much greater extent than reptiles ever could.

The Cenozoic was also a developmental time for land plants. The last frontier for plants—the dry steppes, savannas, and prairies—was not colonized until the Paleogene Period, when grasses evolved. This process involved the assistance of animals, in particular the great grazing herds that lived on all continents except Antarctica.

a. This drawing by Michael Rothman depicts a mother and child of the species *Australopithecus afarensis*. These human-like individuals lived together in small groups, formed lasting bonds with mates, and looked after their children through infancy.

b. This 3.3-million-year-old fossil is the skull of an *A. afarensis* baby, who probably looked much like the child in the drawing.

c. From footprints like these, preserved in soft volcanic mud, scientists know that australopithecines walked upright on two feet. This 70-m trail includes the footprints of two adults and possibly a child, stepping in the footprints of one of the adults. To the right are footprints of an extinct three-toed horse.

The human family From a human point of view, the most remarkable thing that has happened during the Cenozoic Era is the emergence of our own species. Charles Darwin was often accused of believing that humans are descended from the apes. In fact, the family of humans, *Hominidae*, and the family of apes, *Pongidae*, are both descended from an earlier common ancestor that was neither human nor ape.

The emergence of humans is one of the most complex and controversial fields in paleontology, in part because of gaps in the fossil record and the lack of transitional forms. But it is clear that **hominids**—human-like organisms—are a very recent evolutionary development. The first hominid that was clearly **bipedal** (i.e., routinely walked upright) was *Australopithecus*, of which the famous fossil "Lucy" is an example (**Figure 15.18**). These hominids were only about 1.2 m in height but had a brain capacity larger than that of chimpanzees. Their fossils range from about 3.9 million to 3.0 million years in age. From the shape of its pelvis and from footprints left in soft volcanic mud (see Figure 15.18), we know that *Australopithecus* walked upright, though its skull looked more apelike than human. Lucy's descendents never spread beyond Africa, and they disappeared altogether about 1.1 million years ago.

Genus *Homo* *Homo erectus*, one of the first species of our own genus (*Homo*), was more widely traveled than *Australopithecus*. Fossils of *Homo erectus*, dating back about 1.8 million years, have been found in Africa, Europe, China, and Java. An earlier species called *Homo habilis* is believed to have mastered the use of stone tools as far back as 2.4 million years ago, and toolmaking is a distinguishing feature of the genus *Homo*. The earliest known (thus far) species of the genus *Homo*, called *Homo gautengensis*, was discovered in 2010; this species may represent a link to the earlier Australopithecines.

Homo erectus disappeared 300,000 years ago and was replaced by *Homo neanderthalensis* ("Neanderthal man") no later than 230,000 years ago. Unfortunately, the poor fossil record between 400,000 and 100,000 years ago has made it difficult for scientists to determine how this transition occurred. We hypothesize from burial sites that Neanderthals might have practiced some form of religion. Because of similarities in teeth and brain size (slightly larger than our own), some experts have argued that Neanderthal was part of our own species; however, recent DNA studies suggest that *Homo sapiens* is not a direct descendant of Neanderthals. The Neanderthals disappeared about 30,000 years ago and were replaced rather suddenly by the biologi-

cally modern people, the first indisputable members of our own species, *Homo sapiens*.

Did *Homo sapiens* evolve from the Neanderthals, or were they a distinct species? What happened during the 5000-year period when both kinds of humans were alive and overlapped geographically in Europe? Did the modern *Homo sapiens* kill the Neanderthals? These and many other questions await answers as paleontologists continue to look for clues in the fossil record.

Mass Extinctions

Species extinctions have happened throughout the history of life on this planet; in fact, by far the majority of species that have ever lived are now extinct. Scientists refer to these as **background extinctions**. At several places in the fossil record of the Phanerozoic Eon, however, paleontologists have found abrupt and profound changes in the fossil assemblage—not just in one location but worldwide; and not just affecting one species, but vast numbers of species in geologically short periods of time. These dramatic changes record **mass extinctions**. For geologists, mass extinctions are a convenience because they delineate so many of the time periods in the geologic column, but they are also a conundrum: What could possibly cause most of the species on Earth to become extinct in a short period of time?

> **mass extinction**
> A catastrophic episode in which a large fraction of living species become extinct within a geologically short time.

The most famous mass extinction occurred 65 million years ago, at the end of the Cretaceous Period (the boundary between the Mesozoic and Cenozoic eras). It is sometimes called the **K-T extinction** (K stands for Cretaceous, and T stands for Tertiary, an old term for the period immediately following the Cretaceous). Paleontologists estimate that around 70% of all species, including the dinosaurs, died out during this extinction. Many of the species that survived must also have come perilously close to extinction.

In the past quarter-century, scientific opinion has coalesced around the hypothesis that the end-of-Cretaceous extinction was at least partly caused by environmental changes that resulted from the impact of a large meteorite with Earth (**Figure 15.19** on the next page). When it was proposed by Walter and Luis Alvarez in 1980, the meteorite-impact theory was highly controversial, but many lines of evidence now support it. We can be reasonably certain now that the events described in Figure 15.19 did happen, although we cannot be absolutely certain that this event killed all the dinosaurs.

a. About 65 million years ago, a meteorite as large as Mt. Everest struck Earth at a spot just offshore from Mexico's Yucatan Peninsula. It blasted a crater 180 kilometers across. Although the crater is now covered by younger sedimentary rock, a satellite image reveals a faint circular depression around its edge and drilling has confirmed that the rocks were shattered by a great impact. The crater is named Chicxulub (pronounced "tjik-ju-lub") after a Mayan town near the center of impact. ▼

b. The blast wave killed plants and animals all over the western hemisphere. As broiling-hot debris from the impact rained down, forests ignited, creating continent-wide forest fires. ▼

IMPACT

40 MINUTES

ONE WEEK

◀ **c.** Soot from the fires started by the impact debris may have remained in the atmosphere for months, or even years, blocking the sunlight and halting photosynthesis all over the world.

ONE YEAR

◄ **d.** A year after the impact, algae and ferns may have begun to grow again, but the forests were still bare.

e. A thin layer of clay, shown here from the Raton Basin in Colorado, can be found all over the world in rock units marking the end of the Cretaceous period. Here the whitish layer of clay is about 2 cm thick but it becomes much thicker closer to the impact site. Above the clay is a thinner layer that is enriched in iridium and contains shocked quartz grains. Iridium is an element that is rare on Earth but more common in meteorites. Evidence strongly suggests that both layers were formed by a meteorite impact. Dinosaur fossils can be found below the clay layer but not above.

▼

Shocked quartz fragments

Clay

Impact debris

f. In strata just above the impact, fern spores are much more prevalent than angiosperm pollen—a sign of a regenerating plant community.

▼

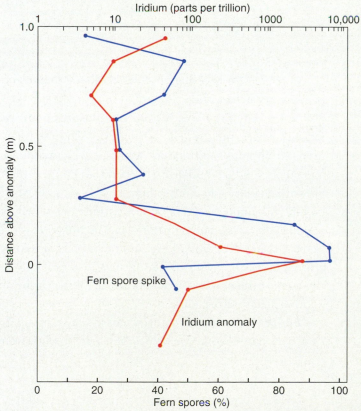

Iridium (parts per trillion)

Distance above anomaly (m)

Fern spore spike

Iridium anomaly

Fern spores (%)

Life in the Phanerozoic Eon **483**

Volcanism and mass extinctions • Figure 15.20

Several of the world's largest flood basalt deposits formed at roughly the same time as mass extinctions occurred. Scientists are still debating whether, and how, an enormous lava flow in one region could cause mass extinctions all over the world. (Ma = millions of years before present.) (*Source*: Courtesy Dr. Richard Ernst.)

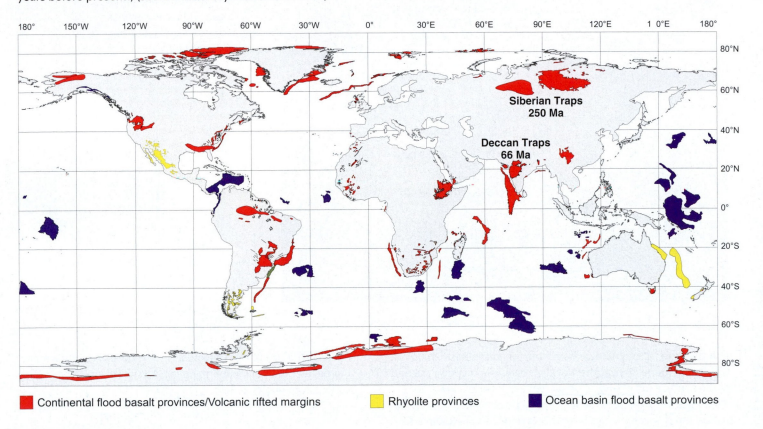

One of the remaining problems with the meteorite impact theory is that the great end-of-Cretaceous extinction was not unique, nor was it even the most dramatic of all extinctions. As mentioned earlier in this chapter, the most devastating mass extinction occurred 245 million years ago, at the end of the Permian Period, when an estimated 95% of all species died out. There have been at least 5 and possibly as many as 12 mass extinctions during the past 250 million years. Some of them can be linked to massive meteorite impacts, but others cannot. Several other known craters on Earth rival the size of the Chicxulub crater, yet the ages of the craters do not correlate well with the dates of other mass extinctions.

On the other hand, some scientists have pointed out that a strong correlation exists between the eruption of flood basalts (see Figure 6.2i) and mass extinctions. The end-of-Cretaceous boundary lies very close in time to the eruption of the Deccan Traps in India, and the Permian extinction occurred around the time of the largest known flood basalts, the Siberian Traps (**Figure 15.20**).

Perhaps massive outbreaks of volcanism, producing worldwide climate change, are the "normal" agent of mass extinctions, and the end-of-Cretaceous meteorite was merely an interloper that happened to strike Earth. The study of mass extinctions and the role that impacts may have played in them continues to be an area of active research. It is especially appropriate today because the human footprint on planet Earth is so heavy (**Figure 15.21**) that the current rate of species extinctions now rivals the rate of extinctions during the previous die-offs.

Geology InSight

✓ THE PLANNER

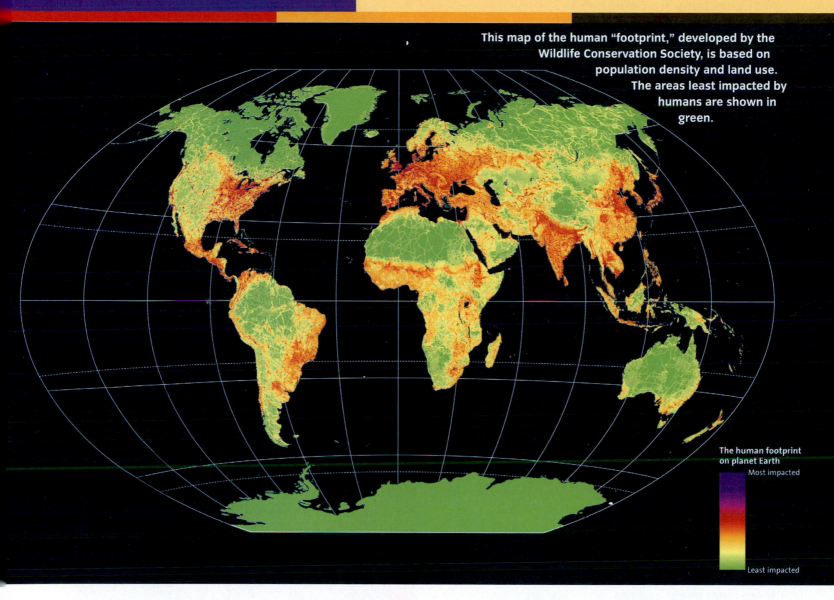

This map of the human "footprint," developed by the Wildlife Conservation Society, is based on population density and land use. The areas least impacted by humans are shown in green.

The human footprint on planet Earth

Most impacted

Least impacted

CONCEPT CHECK   STOP

1. **What** are some of the possible reasons for the explosion of animal diversity in the Cambrian Period?

2. **How** have plants adapted to life on land, particularly in their reproductive systems?

3. **What** are the major milestones in the evolution of chordates?

4. **Why** is it inaccurate to say that humans descended from apes?

5. **How** do we know that mass extinctions have happened in the past, and what evidence suggests that a meteorite impact killed the dinosaurs?

Summary

1 Ever-Changing Earth 456

- The history of life is intertwined with that of the atmosphere and hydrosphere. Earth's early atmosphere consisted primarily of water vapor, carbon dioxide, and nitrogen, the products of outgassing from volcanoes.

- Oxygen gradually accumulated over more than a billion years, primarily through the process of **photosynthesis** as shown in the diagram. Photosynthesis, a reaction by which plants convert carbon dioxide and water into carbohydrates, is by far the most important source of oxygen in our atmosphere.

Photosynthesis • Figure 15.2

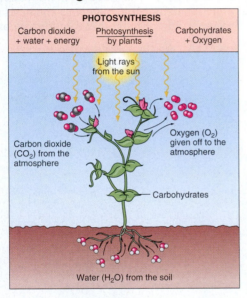

- The buildup in oxygen was accompanied by a buildup of ozone in the upper atmosphere. Earth's land surfaces were probably uninhabitable until the ozone layer began blocking harmful ultraviolet radiation from the Sun.

- Living plants produce oxygen. When dead organisms decay, the bacteria that decompose them consume oxygen and release carbon dioxide. This cycle maintains the balance of oxygen in the atmosphere and hydrosphere. Nevertheless, atmospheric oxygen levels have fluctuated dramatically over geologic time. One reason this can occur is the burial of organic matter before it has completely decomposed. Plate tectonics can accelerate the rate of burial of organic matter and thereby affect the composition of the atmosphere.

2 Early Life 459

- The minimum requirements for life are metabolism and a means of replication. All life on Earth is based on a small set of carbon-based building blocks called amino acids. Also, all living organisms use **DNA**, a biopolymer, to copy genetic information. However, different organisms have evolved different kinds of metabolism, and this is one reason that life is able to flourish in a great variety of environments.

- The fundamental unit of life is the **cell**, which has two fundamental varieties: **prokaryotes** and **eukaryotes**. Eukaryotic cells enclose their DNA within a membrane, whereas prokaryotic cells lack a well-defined nucleus.

- Biologists have classified all living organisms into three **domains**—bacteria, Archaea, and eukaryotes. The first two domains are prokaryotes, which are exclusively single-celled organisms. Prokaryotes evolved long before eukaryotes. Eukaryotes evolved at least 1.4 billion years ago and can be either one celled or multicellular.

- The most ancient fossils are about 3.5 billion years old, although chemical evidence of the presence of life is preserved in rock considerably older than this. The best pieces of evidence for early life are fossilized stromatolites, thick mats of one-celled organisms that can still be found in some places today (see photo). Most other fossils of early life are microscopic, and consequently they are hard to find and difficult to interpret. The earliest known fossils of large multicellular organisms date back 635 million years.

An ancient life-form • Figure 15.4

3 Evolution and the Fossil Record 464

- As a result of **evolution**, all present-day organisms are descendants of different kinds of organisms that existed in the past. The mechanism by which evolution occurs is **natural selection**, in which well-adapted individuals tend to have greater survival and more breeding success and thus pass along their traits to later generations.

- New **species** may emerge slowly, as the environment changes gradually over time, or quickly, when a beneficial mutation appears or when a population moves to a new habitat. It is not clear which process is more common.

- Organisms can be preserved as **fossils** in many. different ways. The body may be kept essentially intact (see photo), or its shape may be preserved while minerals replace its contents. Some organisms leave behind evidence, such as molds of footprints or **trace fossils**, without leaving behind any body parts.

Examples of fossil formation • Figure 15.10

4 Life in the Phanerozoic Eon 470

- The Cambrian Period, at the beginning of the Phanerozoic Eon, was a time of explosive diversification of life-forms. Several factors may have contributed to the Cambrian explosion: the emergence of multicellular life (just before the Cambrian Period), the protective ozone layer, the rise in oxygen levels, and the ability to form shells.

- All of the extant phyla in the animal **kingdom**, as well as several extinct phyla, emerged during the Cambrian Period. All of the Cambrian fauna were aquatic, and some had hard internal or external skeletons. In order to colonize land,

plants and animals had to develop a means of structural support, an internal aquatic environment, and a way of exchanging gases with the atmosphere. For sexual reproduction, a moist environment was also essential.

- The earliest land plants were seedless plants, such as ferns. The next major event in the plant kingdom was the emergence of **gymnosperms**, or naked-seed plants, which flourished during the age of dinosaurs. **Angiosperms**, or flowering plants, developed last, with a much more efficient reproductive system, facilitated by insect pollination.

- The earliest land animals were arthropods (a phylum that includes insects). Fish first ventured onto land during the Devonian Period and eventually gave rise to the amphibians. These, however, were limited in their geographic range because they still required water to spawn.

- Reptiles evolved a watertight skin and eggs that could be incubated outside water. This freed them from dependence on watery environments. Both mammals and birds descended from reptiles, by different evolutionary pathways.

- Mammals have existed since the Jurassic Period. It was apparently the disappearance of the dinosaurs, along with a rise in atmospheric oxygen, that gave mammals the opportunity to diversify and increase in size. Hominids are a very recent family of mammals, which appeared within the last 3.9 million years. Anatomically modern humans appeared only 30,000 years ago. Scientists are not certain whether Neanderthals, which died out around the same time, were a distinct species.

- Several geologic time periods are delineated by **mass extinctions**. The causes of mass extinctions are not thoroughly understood, but most scientists now believe that the end-of-Cretaceous extinction, in which the dinosaurs died out, resulted from a meteorite impact as shown below. Other mass extinctions may have been caused by volcanism.

The day the dinosaurs died? • Figure 15.19

Key Terms

- angiosperm 478
- cell 460
- DNA 462
- domain 462
- eukaryote 462

- evolution 465
- fossil 468
- gymnosperm 476
- kingdom 474
- mass extinction 481

- natural selection 465
- photosynthesis 457
- prokaryote 462
- species 465
- trace fossil 469

Critical and Creative Thinking Questions

1. Recall from Chapter 1 (see Table 1.1) that Earth and Venus are so similar in size and overall composition that they are almost "twins." Why did these two planets evolve so differently? Why is Earth's atmosphere rich in oxygen and poor in carbon dioxide, whereas the reverse is true on Venus? What would happen to Earth's oceans if Earth were a little bit closer to the Sun?

Goldilocks planets: Too much, too little, just right   Table 1.1

In spite of their similar origins, Venus, Earth, and Mars have profound geologic differences that have made Earth the only one that is hospitable for life.

	Venus	**Earth**	**Mars**
	 Visible Light / Cloud-penetrating radar	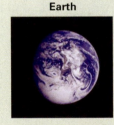	
Atmosphere	White in photos, 97% carbon dioxide; temperature averages a blistering 480°C (hot enough to melt lead)	78% nitrogen, 21% oxygen; average temperature 14.6°C	Thin and insufficient to retain much heat; average temperature 63°C; temperatures usually too low to melt water ice (0°C)
Hydrosphere	Exists only as vapor in the atmosphere due to high temperatures	Contains water as solid, liquid, and vapor	Water cannot exist in liquid form on the surface due to low temperatures and pressure
Biosphere	None	Only known biosphere	None known

2. One popular criticism of the theory of evolution by natural selection was the paucity of transitional form. However, Archaeopteryx was such an organism. In what ways did it resemble its dinosaur predecessors? In what ways was it like or unlike modern birds?

3. Another common criticism of Darwin's theory was its alleged inconsistency with the biblical account of creation. Yet Darwin himself was a Christian, and many scientists since Darwin have had no difficulty reconciling the evidence for evolution with their religious beliefs. Investigate how they have done so. Do you personally feel that evolution by natural selection conflicts with your religious beliefs?

4. What do you think might have happened to mammals if the end-of-Cretaceous extinction had not wiped out the dinosaurs?

What is happening in this picture?

In this fossilized bird and frog, found in the Messel Oil Shale near Darmstadt, Germany, the animals' soft tissues, such as feathers and skin, are exceedingly well preserved. (Note the faint imprint of the frog's skin around its bones, for example.)

Think Critically

Why do you think the soft body parts were preserved in this case, even though they are seldom found in most fossils?

Self-Test

(Check your answers in Appendix D)

1. Which one of the following statements best describes changes in Earth's environment over the past 4 billion years?

   a. Solar brightness and oxygen generally increased, while surface temperature and atmospheric CO_2 decreased.

   b. Solar brightness and oxygen generally decreased, while surface temperature and atmospheric CO_2 increased.

   c. Solar brightness and surface temperature generally increased, while oxygen and atmospheric CO_2 generally decreased.

   d. Solar brightness and surface temperature generally decreased, while oxygen and atmospheric CO_2 generally increased.

2. Photosynthesis is fundamentally important to life on Earth. On this diagram, label the inputs and outputs of the photosynthetic process, using the following terms:

   carbon dioxide, CO_2 light rays
   water, H_2O carbohydrates
   oxygen, O_2

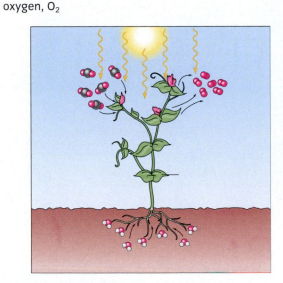

3. Although scientists cannot be certain why atmospheric oxygen levels increased, they hypothesize that atmospheric oxygen has increased over time _____.

   a. because of a continued increase in the number of photosynthetic organisms

   b. because of increased rates of seafloor spreading

   c. because organic matter is buried before it decomposes, reducing the formation of carbon dioxide (CO_2)

   d. All of the above statements are correct.

4. The minimum requirements for life are _____.

   a. photosynthesis and a means of replication

   b. mobility and a means of replication

   c. metabolism and a means of replication

   d. mobility and metabolism

5. Two cells are shown in these photographs. One is from a prokaryotic organism and the other from a eukaryotic organism. Label structures in each of the cells with the following terms and then label each cell either *prokaryotic* or *eukaryotic*.

   organelles   cell membrane   nucleus

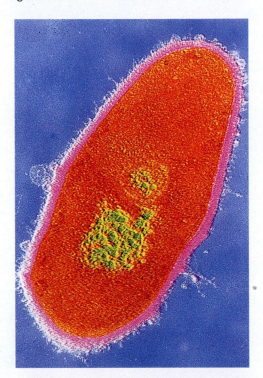

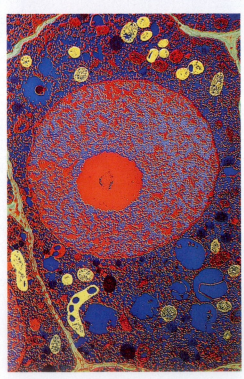

6. _____ is an example of anaerobic metabolism.

   a. Chemosynthesis

   b. Photosynthesis

   c. Fermentation

   d. Both a and b are correct.

   e. Both a and c are correct.

7. Natural selection is the process by which _____.

   a. individuals migrate to environments better suited to survival of the species

   b. individuals adapt to their environment over time

   c. well-adapted individuals pass on their survival advantages to their offspring

   d. All of the above are correct.

8. As a mechanism for evolution, natural selection requires the passing on of specific traits from one generation to the next. These specific traits may reflect _____ passed on through an organism's _____.

   a. genetic mutations; metabolism

   b. genetic mutations; DNA

   c. genetic mutations; RNA

   d. metabolism; DNA

   e. metabolism; RNA

9. How are organisms preserved as fossils within the rock record?

   a. through the preservation of the organism in substances such as sap, tar, or ice

   b. through minerals replacing the contents of the dead organism (mineralization)

   c. by the leaving of impressions, molds, or footprints

   d. through mummification, in especially dry environments

   e. All of the above are methods through which organisms could be preserved as fossils.

10. Which of the following hypotheses do scientists think best explains the explosion of life-forms represented by the Cambrian fossil record?

    a. Sexual reproduction began in the early Cambrian, leading to a greater diversity of living organisms.

    b. An oxygen-rich atmosphere allowed for the metabolism of larger organisms.

    c. Predation began in the early Cambrian, leading to the development of many more organisms with shells and skeletons that might be better preserved in the fossil record.

    d. Extreme climatic changes at the end of the Proterozoic Eon led to accelerated evolutionary responses in surviving organisms.

    e. All of the above hypotheses are considered valid by scientists today.

11. Which of the following is not a requirement for a living organism to survive on land?

    a. structural support

    b. a means of locomotion

    c. an internal aquatic environment

    d. a mechanism of exchanging gases with the air

    e. if sexually reproducing, a moist environment for the reproductive system

12. Vascular plants exchange gas with the air by _____.

    a. diffusion through adjustable openings in the leaves called *stomata*

    b. decay of leaves

    c. photosynthesis

    d. All of the above are correct.

13. Reptiles were freed from the water by _____.

    a. evolving an egg that contained amniotic fluid for the young to grow in

    b. developing a watertight skin

    c. developing a warm-blooded metabolism

    d. Both a and b are correct.

    e. Both b and c are correct.

14. Of the five species from the family Hominidae listed here, which one is likely the earliest species from the genus to which *Homo sapiens* belongs?

    a. *Australopithecus afarensis*

    b. *Homo erectus*

    c. *Homo neanderthalensis*

    d. *Homo gautengensis*

    e. *Homo habilis*

15. Mass extinctions on Earth may be correlated with _____.

    a. massive eruptions of flood basalts

    b. the impact of massive meteorites

    c. widespread glaciation

    d. Both a and b are correct.

    e. Both b and c are correct.

THE PLANNER

Review your Chapter Planner on the chapter opener and check off your completed work.

Understanding Earth's Resources

In the 1980s, these gold miners in Brazil worked claims the size of one of the square pedestals in the photograph. Miners carried the ore out on their backs. Injury and death were common.

Why are people willing to risk life and limb for gold? Gold is rare, and scarcity drives up its value. It lasts because it doesn't tarnish, corrode, or dissolve in common liquids. Its value ensures that owners take very good care of it. Gold is the ultimate recyclable metal. The gold from Cleopatra's bracelets may now reside in a modern wedding band.

Because gold is so durable and so assiduously guarded, we can estimate how much has been mined in human history: about 180,000 tons. If all of this gold were stacked on a football field, it would make a pile only 1.75 meters high—not much of a return on 17,000 years of work!

Gold is one of many resources we extract from Earth; others include copper, iron, tin, and zinc. They are less romantic than gold but much more essential to our daily lives. It may seem as though our supply of these resources is limitless, but ore comes in finite amounts. As time goes by, the limits will become more and more apparent.

CHAPTER PLANNER ✓

- ❏ Study the picture and read the opening story.
- ❏ Scan the Learning Objectives in each section:
  p. 494 ❏   p. 498 ❏   p. 507 ❏   p. 511 ❏
- ❏ Read the text and study all visuals.
  Answer any questions.

Analyze key features

- ❏ Amazing Places, p. 496 ❏
- ❏ Geology InSight, p. 504 ❏   p. 512 ❏
- ❏ Process Diagram, p. 499 ❏
- ❏ Case Study, p. 510 ❏
- ❏ What a Geologist Sees, p. 517 ❏
- ❏ Stop: Answer the Concept Checks before you go on:
  p. 498 ❏   p. 506 ❏   p. 511 ❏   p. 520 ❏

End of chapter

- ❏ Review the Summary and Key Terms.
- ❏ Answer the Critical and Creative Thinking Questions.
- ❏ Answer What is happening in this picture?
- ❏ Complete the Self-Test and check your answers.

Renewable and Nonrenewable Resources

LEARNING OBJECTIVES

1. **Describe** how the loss of resources has affected past civilizations.

2. **Explain** the difference between renewable and nonrenewable resources.

3. **Identify** two kinds of nonrenewable resources that are crucial to modern society.

From the time our ancestors first picked up conveniently shaped stones and used them for hunting and skinning wild animals, to the present day, humans have relied on a bountiful Earth to supply a seemingly endless flow of useful materials. Today it is hard to imagine life without myriad such resources making life easier and more convenient. But, as you will learn, history suggests that there may be limits to the bounty.

Natural Resources and Ancient History

Silent and inscrutable after 10 centuries, the giant stone heads of Easter Island (**Figure 16.1**) are the principal remains of a civilization that once flourished on this re-

Easter Island megaliths • Figure 16.1

The Polynesians who erected these stone heads called *moai* on Easter Island were unwittingly sowing the seeds of their own culture's demise. They stripped the island of its palm trees, at least in part because they needed the trunks to roll the megaliths into place.

mote outpost in the South Pacific. The Rapa Nui people transported hundreds of the megaliths, called *moai*, over hilly terrain, from the quarry where they were carved to their final locations. Archaeologists believe that the inhabitants used tree trunks and ropes to roll the stone figures across the land.

Just as interesting as the questions of how and why the Rapa Nui transported the *moai* is the reason they stopped. Scientists have found that palm trees grew on Easter Island until around 1500. Apparently the islanders cut down every last tree, leaving the ground bare and vulnerable to rapid erosion. In so doing, the Rapa Nui lost not only the ability to transport megaliths but possibly also the resources necessary to sustain their civilization, which was decimated by famine and warfare over the next century.

The decline of past civilizations is an endlessly fascinating and endlessly controversial topic. No civilization's decline can be attributed to a single cause. Nevertheless, a common thread runs through the stories of many societies that flourished and then collapsed. That thread is the depletion of **natural resources**. The Easter Islanders failed to conserve their trees, a biologic resource, and soil, an Earth resource.

> **natural resource**
> A useful material that is obtained from the lithosphere, atmosphere, hydrosphere, or biosphere.

Another critical Earth resource is water. In the area that is now present-day Libya, the Garamantian culture arose in the Sahara Desert around the same time as the Roman Empire, and it prospered for about 1000 years, despite the nearly complete lack of rainfall. The Garamantes survived by tapping into immense underground aquifers and transporting the groundwater through a channel of aqueducts called *foggara* (**Figure 16.2**). Unfortunately, around 500 CE, they apparently reached the end of their technological ability to retrieve groundwater. The *foggara* had been depleted, and their cities became ghost towns.

Like the Rapa Nui and the Garamantes, every other human society depends on natural resources. Besides the most important resources—soil, water, and air—Earth's resources include building materials, metals, fertilizers, fuel, and gemstones. Biological resources include crops, wild plants, and animals. Some of these resources are

Foggara

a. An ancient city in the Sahara Desert, Garama was the center of the Garamantian culture from 500 BCE to 500 CE, and as many as 10,000 people may have lived here.

b. The large population was supported by a vast underground network of *foggara*, or aqueducts, whose openings to the surface can still be seen.

Ghost town in the Sahara • Figure 16.2

renewable. For example, even though we may consume a food crop each season, a new crop grows during the follow-

> **renewable resource** A resource that can be replenished or regenerated on the scale of a human lifetime.

ing season. A layer of soil lost to erosion will eventually be regenerated through the physical, chemical, and biologic processes of soil formation. Groundwater drawn from wells may eventually be replenished by rainwater. But what does "eventually" mean? Some resources take a very long time to regenerate—longer than humans are willing or able to wait. For example, the aquifers under the Sahara Desert formed tens of thousands of years ago, when the cli-

> **nonrenewable resource** A resource that cannot be replenished or regenerated on the scale of a human lifetime.

mate was moister than it is today, and they will not be replenished until the climate changes again, perhaps tens of thousands of years from now. For practical purposes, we consider these resources to be **nonrenewable**.

Resources such as coal, oil, copper, iron, gold, and fertilizers are mined from mineral deposits. Mineral deposits are known to be forming today, but the rate of formation is exceedingly slow. For example, it may take 600,000 years for a large copper deposit to be formed. From a human point of view, all mineral resources are one-crop resources, and Earth's supply of those "crops" is fixed.

Humans are the first and only species to routinely use nonrenewable resources. It does not seem to be in our nature to stop. However, history suggests that we should use such resources very judiciously and monitor how much we have left. That is what the Rapa Nui and the Garamantes failed to do. After the cultures that produced the *moai* and the *foggara* died out, their descendants had to adapt to the new conditions. The Easter Islanders developed new rituals that allocated their limited food resources. The inhabitants of the Sahara developed a nomadic lifestyle that was not dependent on the extraction of deeply buried groundwater. Early European settlers in North America began smelting and processing iron in the early 1600s, almost as soon as

AMAZING PLACES

Saugus Iron Works

Just a few kilometers northeast of the center of Boston, Massachusetts, is a truly amazing place. It is a National Historical Site maintained by the National Park Service called the Saugus Iron Works, the first integrated iron works in North America. The Massachusetts Bay Colony grew rapidly during the 1630s, as immigrants, mainly Puritans, arrived. The English Civil War (1642–1651) slowed trade across the Atlantic Ocean, and the colonists soon found it necessary to start smelting and processing iron in order to have a supply of the iron articles they needed.

In the swamps at the mouth of the Saugus River, the colonists found a supply of bog iron ore. This soft, spongy material is hydrous iron oxides (limonite) precipitated from iron-bearing waters by oxidation. Using water power **(a)** from the Saugus River to pump bellows **(b)**, they started smelting iron in the first blast furnace in the New World in 1646. For fuel they used charcoal produced from the hardwood forests of the region. Two skilled metallurgists were brought from Europe, and they soon had a forge operating so that the iron from the blast furnace could be forged into useful tools. Other facilities followed, a rolling mill for making sheets and rods, a shearing device for cutting sheets, and a drop-hammer weighing 500 pounds for pounding hot iron into desired shapes. The entire operation was powered by seven large water wheels.

The Saugus works lasted for 22 years, closing in 1668, when the local bog-iron ore ran out. While it lasted, the Saugus operation was equal to the best of such integrated iron works in Europe, and it was a fertile training ground for skilled iron workers who moved on, opening iron works elsewhere, and founding America's iron and steel industry.

A lot is known about the Saugus Iron Works because there are contemporary written records and the site was intensively investigated during archaeological excavations between 1948 and 1953. What you see today are reconstructions based on the research and the written record. The entire complex is thought to be the most detailed and accurate 17th-century iron works anywhere in the world.

a.

b.

they arrived in the New World (see *Amazing Places*). Like our ancestors, our descendants, too, will be fundamentally dependent on the availability of critical Earth resources. They may have to make serious changes in their lifestyles as critical resources, such as oil, become even more scarce.

Resources and Modern Society

In our world of the 21st century, there are now 7.0 billion human inhabitants, and the population is growing larger by millions each year. Each of us uses—directly or indirectly—a very large amount of material derived from nonrenewable **mineral resources**. Nearly every chemical element is used in one way or another, and these elements are extracted from more than 200 different minerals. Without them, we could not build planes, cars, televisions, or computers. We could not distribute electric power or build tractors to till fields and produce food. The metals needed to build machines for manufacturing, transportation, and communications are nonrenewable resources extracted from the ground. Without these resources, industry would collapse, and living standards would deteriorate dramatically.

We are equally dependent on **energy resources**. Imagine what life would be like if we had to rely entirely on human muscle power. If a healthy adult rides an exercise bike that drives an electric generator hooked to a lightbulb, the best an average person can do in a nonstop eight-hour day of pedaling is to keep a 75-watt bulb burning. In North America, the same amount of electricity can be purchased from a power company for about 10 cents. Viewed in this way, we can see that human muscle power is puny. Over many thousands of years, our ancestors found ways to supplement muscle power. At first, they did this by domesticating beasts of burden such as horses, oxen, camels, elephants, and llamas. They later learned to make sails to use wind power, dams to use water power, and engines to convert the heat energy of wood, oil, and coal into mechanical power.

Today we use supplementary energy in every part of our lives, from food production and transportation to housing and recreation. North Americans are among the world's biggest energy consumers (**Figure 16.3**). Whereas soil and water were the most critical nonrenewable

How much energy do we use? • Figure 16.3

An average American uses energy, directly or indirectly, at a rate equivalent to burning more than 150 75-watt lightbulbs every minute of the day, every day of the year. (Canadians, with their cold winters, use even more energy.)

Here a highway sign admonishes Los Angeles commuters, "Don't be fuelish—be carpoolish."

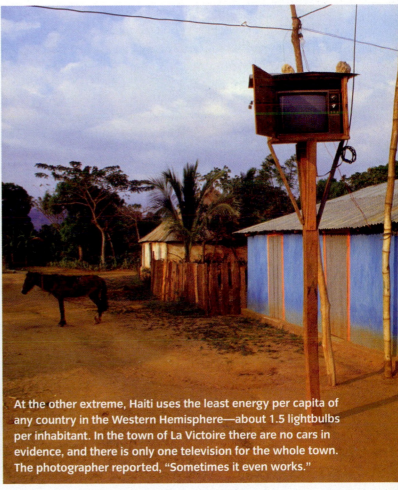

At the other extreme, Haiti uses the least energy per capita of any country in the Western Hemisphere—about 1.5 lightbulbs per inhabitant. In the town of La Victoire there are no cars in evidence, and there is only one television for the whole town. The photographer reported, "Sometimes it even works."

resources for earlier civilizations, our society's weakest link may be its excessive dependence on nonrenewable energy sources. How and whether we can reduce this dependence is both a political problem and a social problem. Geology plays an important role in three ways: identifying new supplies of traditional resources, allowing us to estimate how much we have left, and evaluating the consequences of using new and unconventional resources.

CONCEPT CHECK   STOP

1. **How** can the loss of resources affect the stability of a society?
2. **Why** is the distinction between renewable and nonrenewable resources so important?
3. **What** are some common renewable resources? What are some common nonrenewable resources?

Fossil Fuels

LEARNING OBJECTIVES

1. **Identify** the four principal fossil fuels.
2. **Identify** the sources of coal and petroleum.
3. **Explain** how coal and petroleum form from these sources.
4. **Describe** the types of petroleum and their uses.

People make use of renewable energy sources, such as wood, wind, solar and water power, and nonrenewable energy sources. Everywhere in the world, even in the least-developed countries, nonrenewable sources supply at least half of the energy used (**Figure 16.4**), and the main nonrenewable source of energy is **fossil fuels**.

Fossil fuels are animal or plant remains that have been buried and trapped in sediment or sedimentary rock and have undergone chemical and physical changes during and after burial. All fossil fuels are composed primarily of **hydrocarbon** (hydrogen + carbon)

fossil fuel
Combustible organic matter that is trapped in sediment or sedimentary rock.

Energy sources and uses
• Figure 16.4

In the United States, fossil fuels (oil, natural gas, and coal) account for 83% of the energy used. The large amount of lost energy—approximately half, as shown in the upper-right arrow—arises both from inefficiencies in energy use and from the fundamental physical limits on the efficiency of any heat engine.

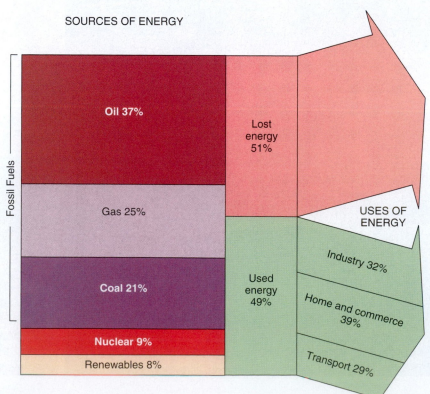

SOURCES OF ENERGY

Fossil Fuels

Oil 37%

Gas 25%

Coal 21%

Nuclear 9%

Renewables 8%

Lost energy 51%

Used energy 49%

USES OF ENERGY

Industry 32%

Home and commerce 39%

Transport 29%

Coalification • Figure 16.5

The conversion of plant matter to coal, or **coalification**, happens over a period of millions of years, as layers of peat are buried and compressed by overlying sediments.

1 Swamps are thick with the organic remains of vegetation. As the organic matter decomposes, it is buried by more vegetation and sediment. The plant matter is converted into peat.

2 As the thickness of overlying sediment increases over time, causing higher pressures and temperatures on the organic layer, water and other volatile components are expelled.

3 By the time a layer of peat has been converted into coal, its thickness has been reduced by 90 percent, most of the volatile components are gone, and carbon (the heat source) has been greatly concentrated.

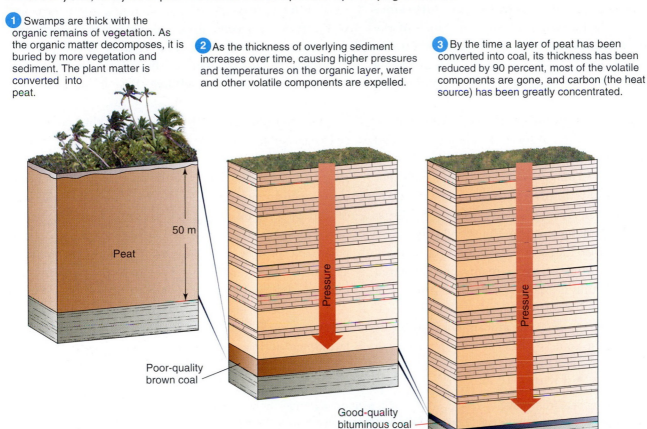

50 m

Peat

Pressure

Poor-quality brown coal

Good-quality bituminous coal

Increasing thickness of overlying strata through time

Think Critically

If coal and diamond are made from the same element, carbon, why won't you ever find a diamond in a coal deposit?

compounds. Because fossil fuels are derived from the altered remains of plants or animals, their energy content is derived originally from the Sun, via photosynthesis. The principal fossil fuels are peat, coal, oil, and natural gas. The kind of sediment, the specific type of organic matter, and the postburial changes that occur determine which type of fossil fuel is formed.

Peat and Coal

Organic matter that accumulates on land comes from trees, bushes, and grasses. These land plants are rich in organic compounds that tend to remain solid after burial. In water-saturated places such as swamps and bogs, the remains accumulate to form **peat**.

Peat is the initial stage in the formation of **coal** (**Figure 16.5**). As layers of peat are compressed and heated by being buried under more and more sediment, water and gaseous compounds such as carbon dioxide (CO_2)

peat A biogenic sediment formed from the accumulation and compaction of plant remains from bogs and swamps, with a carbon content of about 25%.

coal A combustible rock (50 to 95% carbon), formed by the compression, heating, and lithification of peat.

and methane (CH_4) escape. The loss of these constituents leaves a higher proportion of carbon in the residue, and carbon is what we want for burning—the higher the carbon content, the greater the **rank** of the coal. The lowest rank of coal is called **lignite**, and **anthracite** is the name for the highest rank of coal. Much of the world's coal is **bituminous coal**, intermediate between lignite and anthracite. Anthracite is so changed from its original form that it is considered a metamorphic rock.

Coal occurs in strata (miners call them seams) along with other types of sedimentary rock, mainly shale and sandstone. Anthracite occurs with slate—low-grade metamorphic rock. Coal seams tend to occur in groups. For example, 60 seams of bituminous coal have been found in western Pennsylvania. This abundant coal powered Pittsburgh's steel mills and made it into the "Steel City."

Peat (**Figure 16.6**) and coal have been forming more or less continuously since vascular plants first appeared on Earth about 450 million years ago. However, the size of peat swamps has varied greatly throughout Earth's history, and therefore the amount of coal formed has also varied. By far the largest amounts of peat swamp formation occurred on Pangaea during the warm Carboniferous and Permian periods, 360 to 245 million years ago. The great coal seams of Europe and eastern North America formed from peat deposited at that time, when the swamp plants were different from today and included giant ferns and scale trees. A second great period of peat formation peaked during the Cretaceous Period, 144 to 66.4 million years ago. During this period, the plants of peat swamps were flowering plants, much like those found in swamps today.

The luxuriant growth needed to form thick and extensive coal seams is most likely to occur in a tropical or semi-tropical climate. We can therefore deduce that either the global climate was warmer in the past or the lands where these seams occur were once in the tropics. Probably both conditions were involved. Coal deposits that were formed in warm low-latitude environments but are now in frigid polar lands provide compelling evidence for plate tectonics. The many places where coal is found in the United States today provide an excellent example (**Figure 16.7**). Today peat formation is occurring in wetlands such as the Okefenokee Swamp in Florida and Georgia and the Great Dismal Swamp in Virginia and North Carolina (**Figure 16.7b**).

For peat's sake • Figure 16.6

A peat cutter harvests peat from a bog in Ireland. When dried, peat provides fuel for heat and cooking. It is higher in energy content than firewood but lower than coal because it is in the process of changing from plant matter to coal. If the peat cutter could wait a few million years, he could harvest much higher-energy coal.

Coal seams and coal-forming environments • Figure 16.7

a. Although the Appalachian Mountains of western Pennsylvania and West Virginia are synonymous with coal mining, much of the industry has moved west. The largest coal mine in the western hemisphere, the Black Thunder mine, is located in Gillette, Wyoming.

b. Okefenokee Swamp in Florida and Georgia is a modern day example of a peat swamp. Given a few million years and the right conditions, this peat swamp will one day turn into a coal seam.

c. All of the coal deposits shown on this map were once swamps, resembling the Okefenokee Swamp.

Bighorn basin

Northern Great Plains region

Green River basin

Powder River basin

Uinta basin

Denver basin

Illinois basin

Black Mesa field

Raton Mesa field

Western interior basin

Appalachian basin

San Juan basin

■ Anthracite ■ Bituminous coal ■ Lignite

The first commercial oil well • Figure 16.8

On August 27, 1859, Edwin Drake's drill "struck oil" in a deposit 21 m underground in Titusville, Pennsylvania. This shed at Drake's Well Museum is a reconstruction of the world's first commercially productive oil well.

wells were being dug by hand near Oil Springs, Ontario. In 1859, the first commercial oil well was drilled in the United States (**Figure 16.8**), and a modern industry was born.

The modern use of **natural gas** started with an accidental discovery at Fredonia, New York, in 1821, when an ordinary water well began to produce bubbles of a mysterious gas. The gas was accidentally ignited and produced a spectacular flame. Wooden pipes were installed to carry the gas to a nearby hotel, where it was used in gaslights. Though rarely used today, gaslights preceded the electric lightbulb, and many people considered them to be superior to early lightbulbs.

> **natural gas** The gaseous form of petroleum.

Today, petroleum products are used for a wide variety of purposes in addition to their main use as fuel for heating and operating vehicles and machines. Different components of petroleum are used in fertilizers, lubricants, asphalt, and the most ubiquitous material in contemporary life, plastic (**Figure 16.9**).

Deposits of petroleum are nearly always found in association with sedimentary rock that formed in a marine environment. When marine microorganisms die, their remains settle to the bottom and collect in the fine seafloor mud, where they start to decay. The decay process quickly uses up any oxygen that is present. The remaining organic material is thereby preserved without decaying and is covered with more layers of mud and decaying organisms. The increasing heat and pressure associated with burial initiate a series of complex physical and chemical changes, called **maturation**, which break down the solid organic material and turn it into liquid and gaseous hydrocarbons.

The temperature and depth under which maturation takes place determine the breakdown products

Petroleum

In the ocean, microscopic phytoplankton (tiny floating plants) and bacteria are the principal sources of organic matter in sediment. However, the chemicals found in these organisms differ from those in land plants. When buried and compressed under the right conditions, they are slowly transformed into **petroleum**.

> **petroleum** Naturally occurring gaseous, liquid, and semisolid substances that consist chiefly of hydrocarbon compounds.
>
> **oil** The liquid form of petroleum.

Oil first came into widespread use in about 1847, when a merchant in Pittsburgh started bottling and selling oil from natural seeps. Five years later, a Canadian chemist discovered that heating and distilling oil yields kerosene, a liquid that could be used in lamps. Kerosene lamps quickly replaced whale oil lamps and greatly reduced the use of candles in household lighting. Soon oil

One raw material, many uses • Figure 16.9

Crude oil consists mostly of hydrocarbons—molecules containing carbon and hydrogen but no oxygen. The molecules come in many different sizes. At an oil refinery, the crude oil is distilled into heavier and lighter components. Each type of hydrocarbon has different uses. The ones with fewer carbon atoms generally have a lower boiling point and are more useful as fuels. (The "octane" in gasoline, for instance, has eight carbon atoms.)

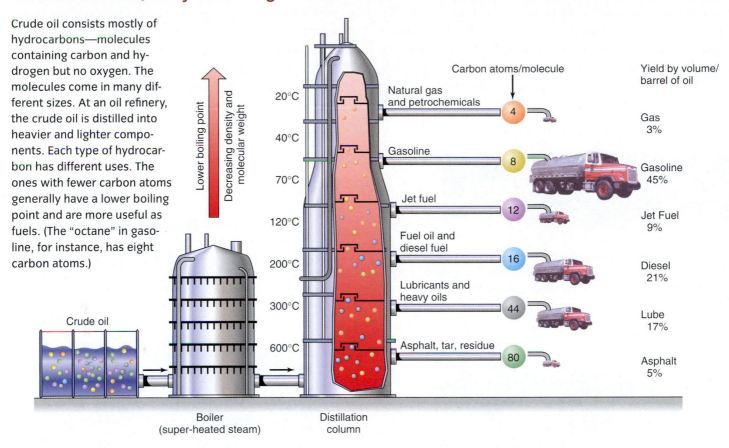

(**Figure 16.10**). As maturation is proceeding, the sediment itself is turned into rock, such as shale or limestone. A rock in which organic material has been converted into oil and natural gas is called a **source rock**. Oil and gas occupy more volume than solid organic matter, so the conversion process creates internal stresses that cause small fractures to form in the source rock, and the fractures allow oil and gas to escape into adjacent, more porous rock.

Oil and gas are light, and they slowly migrate toward the surface. If nothing happens to stop this slow percolation, they will eventually reach the surface and evaporate into the atmosphere. It is estimated that 99.9% of the petroleum in the world escapes this way.

The petroleum window • Figure 16.10

This diagram shows the regions of depth and temperature in which organic matter is converted to oil and gas and at which trapping occurs.

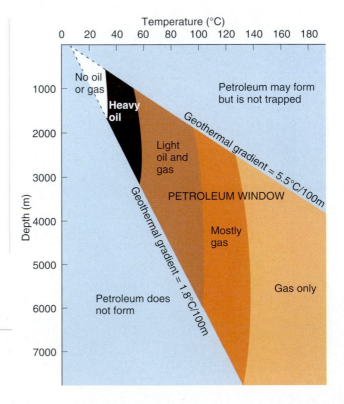

On occasion, however, the migrating oil and gas encounter an obstacle—a layer of impermeable rock that prevents it from reaching the surface. Such a formation is called a **cap rock**, and it most commonly consists of shale. A geologic situation that includes a source rock to contribute organic material, a porous **reservoir rock** in which the oil accumulates, and a cap rock to stop the migration is referred to as a petroleum trap (or a hydrocarbon trap). Although this may seem like a rather rare combination of circumstances, several types of petroleum traps are known (**Figure 16.11**). Recognizing and finding these formations is the job of a petroleum geologist.

Geology InSight   Traps for oil and gas • Figure 16.11   ✓ THE PLANNER

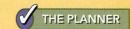

This figure illustrates six geologic circumstances in which oil can become trapped. Each kind of trap requires a source rock, a reservoir rock to contain the oil or gas, and a cap rock. Note that oil does not form an underground pool; it is trapped in the pores of the reservoir rock. Oil is usually found together with natural gas and water. The gas lies on top because it is the least dense, and the water lies underneath because it is densest.

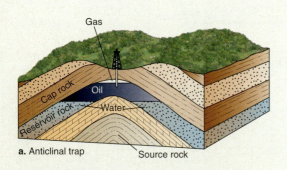

a. Anticlinal trap

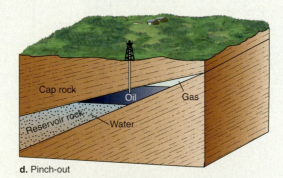

d. Pinch-out

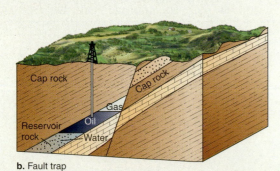

b. Fault trap

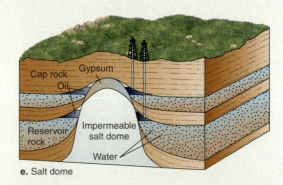

e. Salt dome

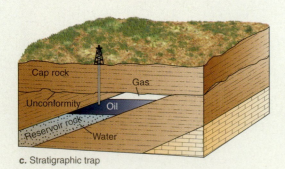

c. Stratigraphic trap

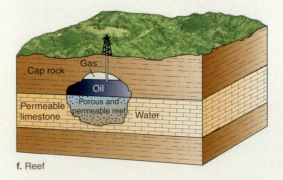

f. Reef

Nontraditional Sources of Oil: Tar Sands and Oil Shales

Oil that is too viscous (thick) to be readily pumped is called heavy oil, or **tar**. **Tar sands** are an increasingly important source of oil. The world's largest-known deposit is in Alberta, Canada (**Figure 16.12a**); geologists estimate that it contains as much as 600 billion barrels. (The world's annual oil production is about 30 billion barrels.) Sandstones in this region are cemented together by tar. To soften and extract the tar, workers cook the sandstone with hot water and steam inside a rotating drum. Once extracted, big organic molecules in the tar must be refined and reduced to the smaller molecules in gasoline and other useful petroleum products.

> **tar sand** A sediment or sedimentary rock in which the pores are filled by dense, viscous, asphalt-like oil.

Another source of petroleum is a waxlike organic substance called **kerogen**, from very fine-grained sedimentary rock such as shale. If burial temperatures are not high enough to form oil and natural gas, kerogen is formed instead. If the kerogen is mined and heated, it breaks down and forms oil and gas. To be worth extracting, kerogen in **oil shales** must yield more energy than the amount needed to mine and heat it. The world's largest deposit of oil shale is found in Colorado, Wyoming, and Utah (**Figure 16.12b**). The U.S. Geological Survey estimates that these deposits, if mined and processed, could yield about 2000 billion barrels of oil.

> **oil shale** A fine-grained sedimentary rock with a high content of kerogen.

Nontraditional sources of oil • Figure 16.12

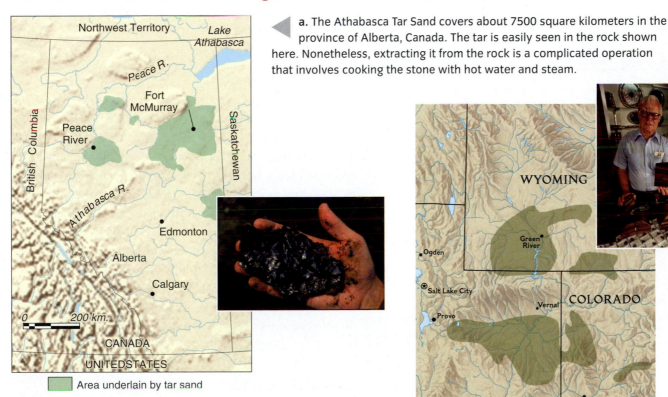

a. The Athabasca Tar Sand covers about 7500 square kilometers in the province of Alberta, Canada. The tar is easily seen in the rock shown here. Nonetheless, extracting it from the rock is a complicated operation that involves cooking the stone with hot water and steam.

Area underlain by tar sand

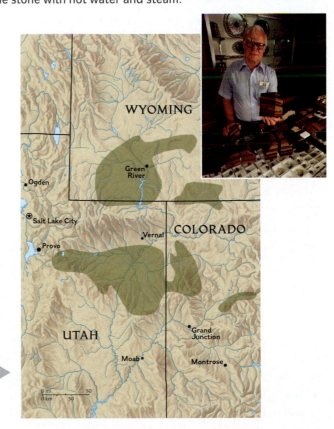

b. Another hydrocarbon-rich rock is oil shale, found in huge amounts in the Green River formation in Colorado, Wyoming, and Utah. The shale bookends held by this gem dealer contain about half a pint of oil.

Nontraditional Sources of Gas: Coal Beds and Shale

Most gas is produced from fields that also produce oil, the amount of each commodity depending on the depth and temperature reached by the source bed. Note in Figure 16.10 that when temperature and depth are great enough, gas is the only product formed in a source bed. It has been known for nearly a century that gas is present in source beds, but conventional vertical drilling did not yield a profitable return. In recent years, gas has been recovered by a technology called horizontal drilling, developed for drilling from deep-water platforms, combined with a technique of creating fractures in the source bed using a technique called hydrofracturing. Gas-rich source beds are usually deep extensions of source beds that produced oil at shallower depths, so **shale gas**, as the product has come to be called, is generally found in areas that already have a history of petroleum production.

Coal mines have long been understood to be dangerous because the coal contains methane gas that can be accidentally ignited and cause an explosion. Coal may

appear to be a tight solid, but it actually contains myriad of micropores, and methane is present in all the pores. In recent years it has been found possible to drill into coal seams to create small fractures into which the gas can migrate and be recovered.

It is estimated that about 25% of the gas produced in the United States now comes from coal beds and shale.

Relying on Fossil Fuels

Because most of us grew up in an oil-powered economy, it is easy to forget just how recent our dependence on oil is. **Figure 16.13** shows the history of energy consumption and the changing character of the world's energy "mix." Until 1860 or so, the world ran mainly on energy provided by wood; from then until World War II, coal was the dominant energy source. The huge increase in demand for energy after World War II has mostly been filled by oil and natural gas. The ability to discover new fossil fuel deposits to meet these increasing demands was greatly improved by the advent of new technologies for exploration, including seismic studies, satellite imagery, computer modeling, and digital recording methods.

Despite our improved ability to locate and retrieve fossil fuels, we cannot ignore the fact that they are nonrenewable resources. Most projections indicate that world production

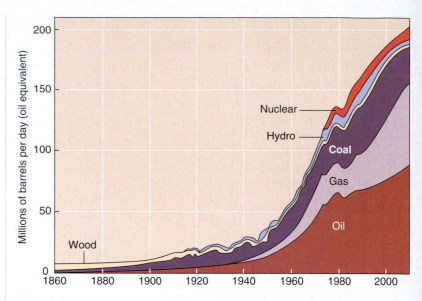

World energy use • Figure 16.13

The history of world energy consumption shows oil's rise to prominence in the current energy mix. The brief decline in energy use in the late 1970s and early 1980s was a result of the "energy crises" of 1973 and 1979, when countries in the Middle East cut production and industrialized countries began to take conservation more seriously.

of oil will peak and begin to decrease before the year 2020. What will happen then? One possibility is that, as in the 1970s, increased efficiency and conservation measures will reduce the demand for energy. There will presumably be an increased contribution from so-called nontraditional sources of oil and gas, including tar sands and oil shales. Coal is still abundant in the United States, but it is unlikely to replace oil because it creates more severe air pollution. Some researchers are working on new technologies for cleaner-burning coal. However, to fulfill the worldwide demand for energy in a reliable, affordable, and environmentally acceptable manner, it will be necessary to place greater emphasis on other sources of energy besides fossil fuels.

CONCEPT CHECK   STOP

1. **What** is the origin of the energy in a fossil fuel?
2. **What** kinds of organisms are the source material of petroleum?
3. **What** kinds of environments lead to the formation of coal and petroleum?
4. **Name** three different kinds of petroleum.

Renewable Energy Resources

LEARNING OBJECTIVES

1. **Identify** the major alteratives to fossil fuels.
2. **Explain** why biomass, wind, and wave energy are indirect forms of solar energy.

3. **Explain** why solar energy is only a minor part of our energy mix.
4. **Describe** how nuclear power works and identify its advantages and disadvantages.

Each of the world's 7.0 billion people uses energy at an average rate of approximately 2370 watts, for a total rate of 16.6×10^{12} watts. However, this is only a minute fraction of the amount of energy that enters the Earth system, primarily from the Sun (**Figure 16.14**). It is clear, then, that we are not about to run out of energy in an absolute sense. The question is whether we are clever enough to learn how to use alternative sources of energy rather than rely so heavily on nonrenewable fossil fuels.

The potential alternatives to fossil fuels come from solar, biomass (fuel wood and animal waste), wind, wave, tidal, hydroelectric, nuclear, and geothermal sources. Most of these are renewable. Nuclear power is technically nonrenewable, but the amount of energy available is so vast that it is, for all practical purposes, inexhaustible. (The real issues for nuclear power are safety and waste disposal.) Sources that derive their energy directly or indirectly from the Sun will not be exhausted as long as the Sun continues to shine. Tidal energy, too, will continue as long as Earth rotates and we have a Moon.

Earth's energy budget • Figure 16.14

Earth's primary energy sources are the Sun, geothermal energy, and the tides. The incoming energy from the Sun—173,000 terawatts—dwarfs the other two sources. Most of our other energy sources are, in effect, converted solar energy. For instance, the energy in fossil fuels comes from photosynthesis by plants, and hydro power comes from the energy stored in evaporated water that has been precipitated on land.

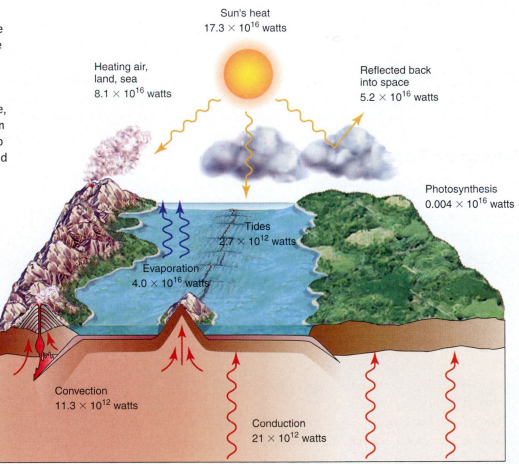

Sun's heat
17.3×10^{16} watts

Heating air, land, sea
8.1×10^{16} watts

Reflected back into space
5.2×10^{16} watts

Photosynthesis
0.004×10^{16} watts

Tides
2.7×10^{12} watts

Evaporation
4.0×10^{16} watts

Convection
11.3×10^{12} watts

Conduction
21×10^{12} watts

Solar energy • Figure 16.15

The solar panels here were installed on the roof of San Francisco's Moscone Convention Center in 2004. With 5400 solar panels generating 825,000 kilowatt-hours of electricity per year—about enough electricity to power 100 homes—it is the largest city-owned solar installation in the United States.

Power from Sun, Wind, and Water

Solar energy reaches Earth from the Sun at a rate more than 10,000 times greater than the sum of all human energy demands. The hyper-abundance of solar energy makes it a very logical candidate for the power source of the future. However, the reality is that the solar future is still far away. Solar energy is best used for direct heating at or below the boiling point of water, as in greenhouses and solar water heaters. But it will make a significant dent in our energy budget only if we can convert it directly into electricity. Unfortunately, the best available solar (or **photovoltaic**) cells (**Figure 16.15**) are still too costly and too inefficient for most uses. Photovoltaic technology is constantly improving, however, and the cost of energy generated in this manner is decreasing.

Plants are very efficient converters of solar energy. Our oil-based economy runs on the energy stored by plants millions of years ago. However, living plant matter also contains stored solar energy—and, unlike fossil fuels, it is renewable. **Biomass energy** derived from fuel wood was the dominant source of energy until the end of the 19th century, when it was displaced by coal. Biomass fuels are still widely used throughout the world, particularly in less developed countries, where the cost of fossil fuel is very high in relation to income.

> **biomass energy** Any form of energy that is derived more-or-less directly from plant life, including fuel wood, peat, animal dung, and agricultural wastes.

Wind is another indirect form of solar energy because the heating of land and oceans creates the giant convection cells that drive the winds (see Chapter 12). For thousands of years, wind has been used as a source of power for ships and windmills. Today, huge windmill "farms" are being erected in particularly windy places (**Figure 16.16**). In Denmark, about 3000 wind turbines supply electricity throughout the breezy coastal country. Although windmills are still expensive, it seems likely that their cost will soon be competitive with the costs of coal-burning electric power plants. One

Wind energy • Figure 16.16

The windmills at this wind farm at Tehachapi Pass, California, generate "clean" (pollution-free) electricity by harnessing the energy of the wind. Each fan's rotary motion turns an electric generator. Wind farms can operate only where steady surface winds prevail year-round; consequently, wind energy is likely to remain a niche energy source.

Hydropower • Figure 16.17

The Hoover Dam on the Colorado River in Arizona and Nevada looms in the background of this photograph. Behind it lies Lake Mead, an artificial lake created by the construction of the dam. Water from the lake drains through the hydroelectric power station (foreground). The ultimate sources of hydroelectric energy are the Sun, which powers the processes of the hydrologic cycle, and gravity, which causes the water behind the dam to fall and release its stored potential energy.

challenge for wind and solar energy is intermittency: the fact that the wind is not always blowing and the sun is not always shining when customers need the energy. Engineers need to design efficient and economical ways to store the energy when it is not needed and release it when it is needed, or wind and solar will be limited to being components in a power system backstopped by a coal, oil, or nuclear power plant.

Several methods have been proposed to harness the power of waves, tides, or sea currents. However, **hydroelectric energy** is at present the only form of water-derived power that meets a significant portion of

> **hydroelectric energy** Electricity generated by running water.

the world's energy needs. In order to convert the power of flowing water into electricity, it is necessary to build dams (**Figure 16.17**). The flowing water is used to run turbines, which in turn convert the energy into electricity. The total recoverable energy from the water flowing

in all of the world's streams is estimated to be equivalent to the energy obtained by burning 15 billion barrels of oil per year. Thus, even if all the potential hydropower in the world were developed, it could not meet all of today's energy needs. Another problem is that reservoirs eventually fill up with silt, so even though the source is inexhaustible, dams and reservoirs have limited lifetimes.

An energy source that will undoubtedly play a larger role in the future is hydrogen, which can be used to power **fuel cells**. Like batteries, fuel cells depend on chemical energy. But unlike batteries, they are easily replenished with new fuel. The only waste product produced by a hydrogen fuel cell is water (H_2O), so it is very clean compared with fossil fuels. However, the downside to fuel cells is that there is no place to "mine" hydrogen. It can be extracted by separating water into hydrogen and oxygen, but the separation process uses up just as much energy as it produces. Hence hydrogen cannot really be thought of as a fuel in its own right but as a way of storing energy that is produced in some other way (perhaps with solar or nuclear power). Even so, there is a great deal of excitement these days about the "hydrogen economy," and future developments will be worth watching.

Nuclear and Geothermal Power

Of the alternative power sources mentioned earlier, **nuclear energy** is probably the only one with the potential to completely replace fossil fuel energy. However, it faces serious drawbacks that have so far prevented it from reaching that potential.

Present-day nuclear reactors exploit an energy-producing process called radioactive decay (see Chapter 3). Two isotopes of uranium, a radioactive element, occur naturally in certain minerals and can be mined and used as fuel for nuclear reactors. The natural radioactivity of uranium-235 is augmented by purifying and concentrating it and by bombarding it with neutrons to stimulate the process of fission. Under precisely controlled conditions, this gives rise to a **chain reaction** that sustains itself as long as there is enough uranium fuel present (see the *Case Study* on the next page). An uncontrolled chain reaction would cause a nuclear explosion. But if the reaction proceeds under control, a great deal of useful energy can be generated. The fissioning of just 1 gram of uranium-235 produces as much heat as the burning of 2 million grams (or 13.7 barrels) of oil. This vast difference in energy content means that nuclear energy is essentially limitless, even though it comes from a nonrenewable resource.

CASE STUDY

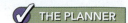

Thinking Critically About Nuclear Power

Approximately 17% of the world's electricity is derived from nuclear power plants. In several European countries, the production of nuclear power is rising rapidly because these countries have very limited fossil fuel supplies. In the United States, two new reactors were approved in 2012, but prior to that the last new nuclear power plant was ordered in 1978.

The industry has been dogged by concerns about safety, especially after a 1986 accident in a reactor in Chernobyl, Ukraine, killed about 50 people and released radioactivity over a wide area, and after a 2011 earthquake and a tsunami crippled a Japanese reactor and released a lot of radioactivity. Another unresolved problem is the disposal of radioactive waste.

◀ **A tremendous source of energy?**

In a chain reaction, a neutron strikes a uranium nucleus and causes it to split into smaller nuclei. The process releases two more neutrons, which can go on to collide with more uranium nuclei. Each time a uranium nucleus splits, it releases heat. An atomic power plant uses that heat to generate steam, which turns turbines that generate electricity.

Smaller nucleus

Other particles and energy

Neutron

^{235}U ^{235}U ^{235}U

Smaller nucleus

. . . Or a tremendous source of toxic waste? ▲

Nuclear reactors produce radioactive waste that must be isolated from human contact for thousands of years. Here, a technician uses a radiation detector at a waste storage site in France where reprocessed waste from 10 reactors is stored for five years before final disposal.

◀ In the United States, nuclear waste is presently stored at 125 temporary sites in 39 states. No permanent storage site has been built yet. In 2002, the U.S. government finally designated Yucca Mountain in Nevada, shown here, as a permanent storage site. However, the selection remains highly controversial.

Another possibility is **geothermal energy**, which comes from Earth's interior, produced by the radioactive decay of uranium, thorium, and other radioactive isotopes, as well as the slow escape of heat from Earth's core. As explained in Chapters 4 and 6, this is the energy source that powers plate tectonics and volcanic eruptions. It has been utilized for heat and power generation in locations where it is easily tapped, including New Zealand, Italy, Iceland, and the United States.

For the most part, the geothermal energy used by humans comes from **hydrothermal reservoirs**, underground systems of hot water or steam that circulate in fractured or porous rock. To be used efficiently, hydrothermal reservoirs should be 200°C or hotter, and this temperature must be reached within 3 km of the surface. Therefore, most of the world's hydrothermal reservoirs are close to the margins of tectonic plates, where hot rock or magma can be found close to the surface. The Geysers geothermal installation in northern California—the largest producer of geothermal power in the world—is a hydrothermal reservoir system.

CONCEPT CHECK STOP

1. **What** are the two major sources of non-fossil–fuel energy being used today?
2. **What** is the source of energy in wind and waves?
3. **What** are the limitations to solar energy?
4. **What** are the environmental drawbacks to nuclear power?

Mineral Resources

LEARNING OBJECTIVES

1. **Define** economic geology.
2. **Explain** what distinguishes ores from other mineral deposits.
3. **Identify** six ways in which minerals become concentrated.
4. **Describe** the connection between plate tectonics and the location of mineral deposits.

The number and diversity of minerals and rocks that provide materials used by humans is so great as to defy attempts to organize them conveniently. Nearly every kind of rock and mineral can be used for something (**Figure 16.18**); however, we only mine about 200 different minerals and about a dozen types of rock. Mineral resources that are most valuable also tend to be rare. **Metallic minerals** are mined specifically for the metals that can be extracted from them. Examples are zinc, a valuable plating metal, extracted from the mineral sphalerite (ZnS), and lead, used in automobile batteries, extracted from the mineral galena (PbS). **Nonmetallic minerals** are mined for their properties as minerals, not for the metals they contain. Examples are salt, gypsum, and clay.

The branch of geology that is concerned with discovering new supplies of useful minerals is **economic geology**. Economic geologists seek mineral deposits from which the desired minerals can be recovered least expensively. To distinguish between profitable and unprofitable deposits, we use the word **ore**. The difference depends on the extraction cost of the mineral and

> **ore** A deposit from which one or more minerals can be extracted profitably.

Some mineral products used in everyday life • Figure 16.18

Sand, gravel, stone, brick (clay), cement, steel, tar (asphalt)

Iron and steel, copper, lead, tin, cement, asbestos, glass, tile (clay), plastic

Mineral pigments (e.g., iron, zinc, titanium), and fillers (e.g., talc, mica, asbestos)

Iron, copper, many rare metals

Grown with mineral fertilizers; processed, packaged, and delivered by machines made of metal

how much people are prepared to pay for it. A formerly unprofitable deposit may become a profitable ore if extraction technology improves, new uses are identified for the mineral, or the mineral becomes scarce enough that the price increases.

Finding and Assessing Mineral Resources

Mineral resources are nonrenewable, and can be exhausted through mining. Economically exploitable minerals are localized in distinct areas within Earth's crust, as a result of plate tectonic processes (**Figure 16.19**). The

Geology InSight   World map of mineral resources • **Figure 16.19**

This map hints at the complexity of the distribution of mineral resources around the world. Even a mineral-rich nation like the United States must depend on tiny Jamaica for its aluminum.

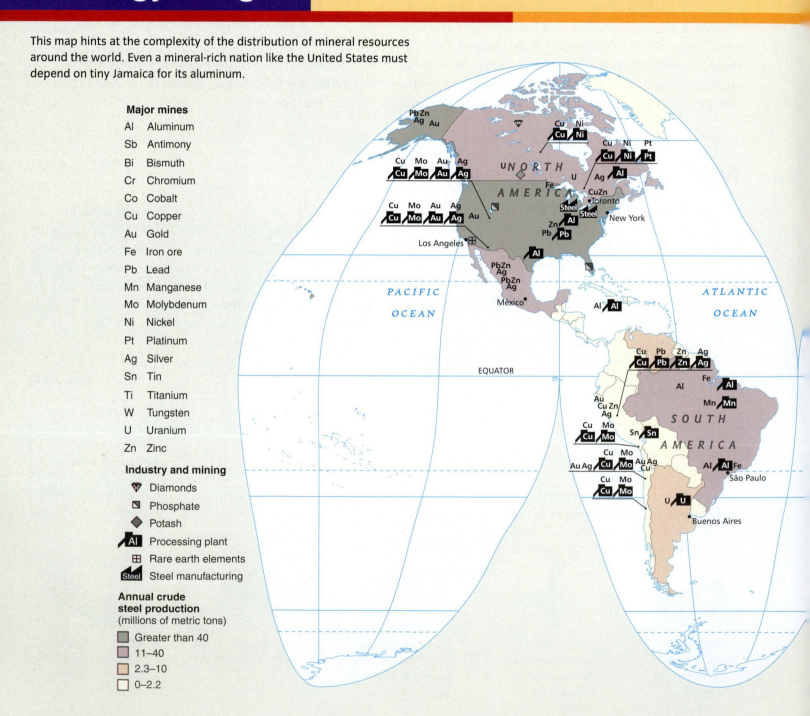

Major mines

Al	Aluminum
Sb	Antimony
Bi	Bismuth
Cr	Chromium
Co	Cobalt
Cu	Copper
Au	Gold
Fe	Iron ore
Pb	Lead
Mn	Manganese
Mo	Molybdenum
Ni	Nickel
Pt	Platinum
Ag	Silver
Sn	Tin
Ti	Titanium
W	Tungsten
U	Uranium
Zn	Zinc

Industry and mining

- Diamonds
- Phosphate
- Potash
- Al Processing plant
- Rare earth elements
- Steel Steel manufacturing

Annual crude steel production
(millions of metric tons)

- Greater than 40
- 11–40
- 2.3–10
- 0–2.2

uneven distribution of exploitable deposits means that no nation is self-sufficient in mineral supplies. For example, the United States has little or no aluminum, manganese, nickel, or chromium and must import most of what is used.

Because minerals are nonrenewable, countries that can meet their needs for a given mineral today may not be able to in the future. An example is England, one of the first industrialized nations. A century ago, England was a great mining nation, producing and exporting such

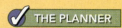

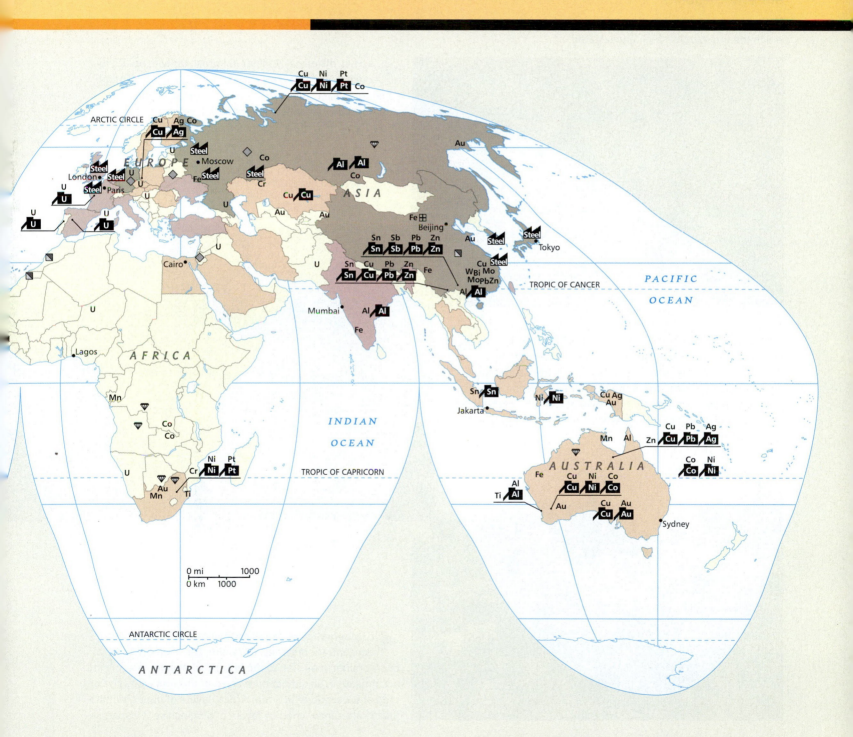

materials as tin, copper, tungsten, lead, and iron. Today, the known deposits of those minerals have been exhausted. The pattern followed by England—intensive mining followed by depletion of the resource, declining production and exports, and increasing dependence on imports—has been repeated often enough that geologists can estimate the remaining effective lifetime of a given mine or mineral resource. Such projections are necessarily tentative; it is especially difficult to anticipate whether new deposits will be discovered.

Over the past few decades, there has been a shift of mineral exploration and production away from industrialized nations and toward the less developed parts of the world. This shift will undoubtedly continue; as

Mineral deposits • Figure 16.20

There are many ways mineral deposits can form. They are classified by the processes by which the ore minerals are concentrated.

a. Hydrothermal A miner in Potosi, Bolivia, points to a rich vein containing chalcopyrite (a copper mineral), sphalerite (a zinc mineral), and galena (a lead mineral). The vein formed when volcanically heated water forced its way into a fracture in the surrounding andesite.

b. Metamorphic Hot fluids can alter rock by contact metamorphism, or metasomatism, often concentrating the minerals into distinct zones. This rock is from a metamorphic mineral deposit at the Tempiute Mine in Arizona. The ore minerals present are sphalerite (brown, lower left), pyrite (gold colored), and scheelite (pale grayish brown, lower right), a tungsten mineral.

c. Magmatic Fractional crystallization in magma can concentrate minerals. Chromite, the main ore mineral of chromium, may sink to the bottom of basaltic magma and accumulate in an almost pure layer. In this unusually fine outcrop at Dwars River in South Africa, layers of pure chromite (black) are sandwiched by layers of plagioclase feldspar.

you can see in Figure 16.19, there are many nations of the world whose mineral resources are relatively unexploited. This doesn't mean that there are no more mineral deposits to be found in industrialized countries, but geologists will have to develop new exploration techniques. When new deposits are found their exploitation will have to be done with minimal environmental degradation.

How Mineral Deposits Are Formed

Minerals can become concentrated into locally enriched deposits as a result of six general types of processes, which we illustrate in **Figure 16.20**.

Hydrothermal deposits are formed when minerals precipitate from hot water solutions. These solutions frequently deposit their mineral loads in cracks in the rock, creating mineral veins like the one in **Figure 16.20a**.

d. Sedimentary Evaporite deposits, such as this salt pan in Death Valley, California, form when lake water or seawater evaporates, leaving behind its dissolved mineral. Household products such as baking soda and borax come from lake deposits. Death Valley itself was briefly mined for borax from 1883 to 1888.

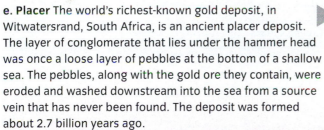

e. Placer The world's richest-known gold deposit, in Witwatersrand, South Africa, is an ancient placer deposit. The layer of conglomerate that lies under the hammer head was once a loose layer of pebbles at the bottom of a shallow sea. The pebbles, along with the gold ore they contain, were eroded and washed downstream into the sea from a source vein that has never been found. The deposit was formed about 2.7 billion years ago.

f. Residual Chemical weathering of rock can concentrate insoluble minerals by removing the more soluble ones. In this bauxite, a residual ore sample from Queensland, Australia, rounded masses of aluminum hydroxide (gibbsite) are imbedded in a matrix of iron and aluminum hydroxides.

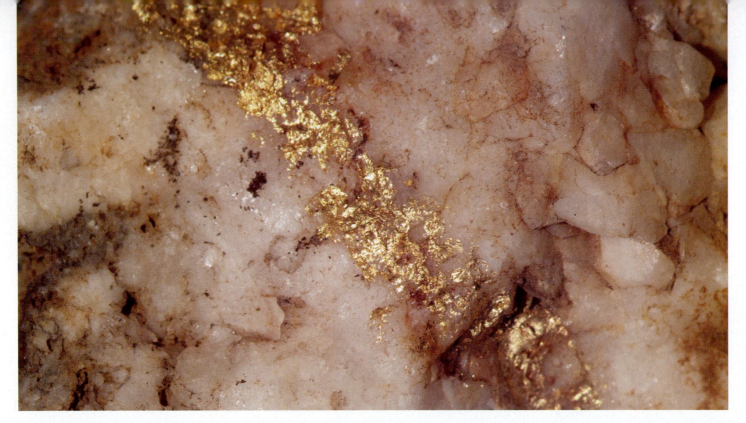

Hydrothermal gold deposit • Figure 16.21

Delicate sheets of native gold were deposited in the center of a quartz vein, Burgin Hill Mine, California.

Such mineral deposits usually contain many different ores. For example, 90% of the silver produced in the United States is a by-product of copper mined from hydrothermal deposits. In many cases, the ore mineral is not visible to the naked eye. Remember that the size of a crystal depends on the amount of time it has to grow. Near the surface, hydrothermal solutions cool very rapidly and thus do not have time to grow large crystals (but see **Figure 16.21** for a beautiful exception).

In recent years, geologists have gotten a chance to see hydrothermal deposits forming right before their eyes in three distinct locations: a desert valley, a shallow sea, and a midoceanic rift zone (see *What a Geologist Sees*).

Many of the processes discussed elsewhere in this book play important roles in concentrating minerals and creating ore deposits. For instance, contact metamorphism and regional metamorphism (see Chapter 10) give rise to **metamorphic mineral deposits** of the kind shown in Figure 16.20b. Fractional crystallization (see Chapter 6) produces **magmatic mineral deposits** (Figure 16.19c and **Figure 16.22**). Commercially valuable ores that arise in magmatic deposits include chromium, which is used to make steel tougher, and titanium, a corrosion-resistant metal that is lighter and stronger than steel.

Magmatic mineral deposits • Figure 16.22

Chromite, the main ore mineral of chromium, crystallizes from a basaltic magma. Because they are denser than the magma, chromite crystals sink to the bottom and accumulate in an almost pure layer.

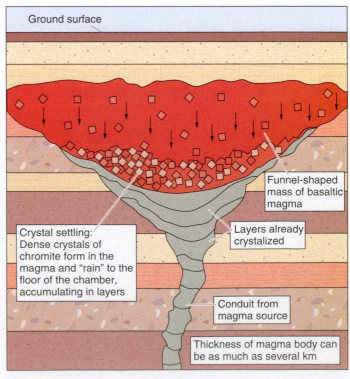

Ground surface

Funnel-shaped mass of basaltic magma

Crystal settling: Dense crystals of chromite form in the magma and "rain" to the floor of the chamber, accumulating in layers

Layers already crystalized

Conduit from magma source

Thickness of magma body can be as much as several km

WHAT A GEOLOGIST SEES

Mineral Deposits in Progress

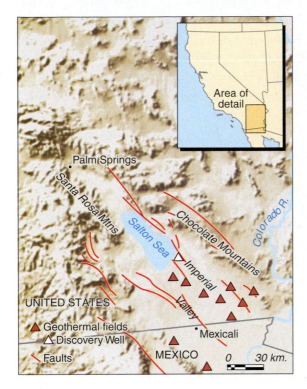

Area of detail

Palm Springs

Santa Rosa Mtns

Salton Sea

Chocolate Mountains

Colorado R.

Imperial

Valley

UNITED STATES

▲ Geothermal fields
△ Discovery Well
Mexicali
╱ Faults
MEXICO   0   30 km.

a. For many years, geologists could only speculate about how hydrothermal mineral deposits might have formed. Then three discoveries changed everyone's ideas about mineral deposits. First, in 1962, oil prospectors in the Imperial Valley of southern California unexpectedly tapped a superheated (320°C) brine at a depth of 1.5 km.

Geologists noticed that as the brine flowed upward, it cooled and deposited minerals it had been carrying in solution. Over a period of three months, more than 1.5 tons of copper and half a ton of silver were deposited in this manner. Geologists quickly realized that the same process could happen naturally under the right circumstances.

b. Two years later, oceanographers found a series of hot, dense brine pools at the bottom of the Red Sea. Like the Imperial Valley brine, this water is trapped in a graben, formed in this case over the spreading center between the African and Arabian plates. Even more surprisingly, when geologists took samples from the bottom of the brine pools, they found ore minerals such as chalcopyrite, galena, and sphalerite. In other words, they found a sedimentary mineral deposit in the process of formation.

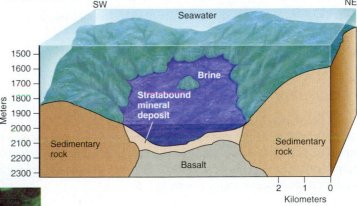

SW   NE

Seawater

Meters
1500
1600
1700   Brine
1800   Stratabound
1900   mineral
2000   deposit
2100   Sedimentary   Sedimentary
2200   rock   rock
2300   Basalt

2   1   0
Kilometers

Water temperature—320°C
Black color from suspended iron sulfide

c. Finally, in 1978, geologists in deep-diving submarines discovered 320°C hot springs emerging from the ocean floor at a midocean ridge in the Pacific. Around the hot springs lay a blanket of sulfide minerals. Again, they had found a modern hydrothermal ore deposit forming before their eyes.

Think Critically

In what other modern geologic environments are ore deposits currently being formed?

"Chimney" is made of chalcopyrite and pyrite deposited by hydrothermal solution.

Placer deposits always contain minerals that originated somewhere else and were transported away from their original site (sometimes not very far) by water. A careful look at Chapter 11 will show you where to look for placer deposits: They accumulate in places such as sandbars, where the water slows down and loses its carrying power (**Figure 16.23**). Placer deposits are associated with minerals that have a high specific gravity, such as gold and platinum. Our last category, **residual deposits**, are the minerals that are left behind by chemical weathering of a host rock (see Chapter 7). In summary, nearly everything you have learned about the Earth system in this book has some relevance to economic geology.

Plate Tectonic Controls on Ore Deposits

To a certain extent, the distribution of mineral resources can be linked to specific tectonic environments. This permits economic geologists to identify locations where certain types of mineral deposits are most likely to be found (**Figure 16.24**). However, the correlation is not nearly as precise or as reliable as we would like, which is why we still need exploration geologists to go out to the sites.

Many kinds of mineral deposits occur in groups, forming what exploration geologists call **metallogenic provinces**. These are regions of the crust within which mineral deposits occur in unusually large numbers. Metallogenic provinces form as a result of either climatic (as in the formation of bauxite deposits in the tropics) or plate tectonic processes. For example, both magmatic and hydrothermal deposits typically form near present or past plate boundaries. This is hardly surprising, as the deposits are related directly or indirectly to igneous activity, and most igneous activity is related to plate tectonics.

Likewise, some energy resources are concentrated in particular tectonic environments. Geothermal energy is closely related to active plate boundaries and other locations where heat flow is high. Fossil fuels occur in environments where organic-rich sediments accumulated long ago—terrestrial swamp environments for peat and coal and marine sedimentary basins for oil and natural gas.

Placer deposits • Figure 16.23

Dense ore minerals, such as gold, platinum, and diamond, tend to sink and accumulate at the bottom of sediment layers deposited by moving water, such as streams or longshore currents. Such concentrations are called **placers**. More than half of the gold recovered in human history has come from placers. This diagram shows three typical geologic locations where placer concentrations can be found.

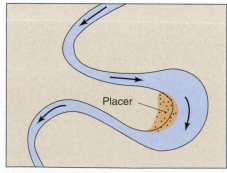

Inside meander loops

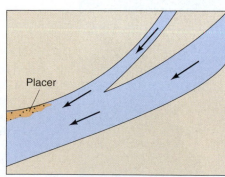

Downstream from a tributary

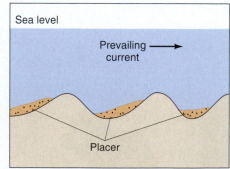

Behind undulations on ocean floor

Plate tectonics and minerals • Figure 16.24

Many materials are concentrated in particular plate tectonic settings. For instance, tungsten is found in metamorphic rock adjacent to granite intrusions. Manganese is found on the deep seafloor in great abundance in "manganese nodules," which may be formed partly through chemical precipitation and partly by biochemical reactions.

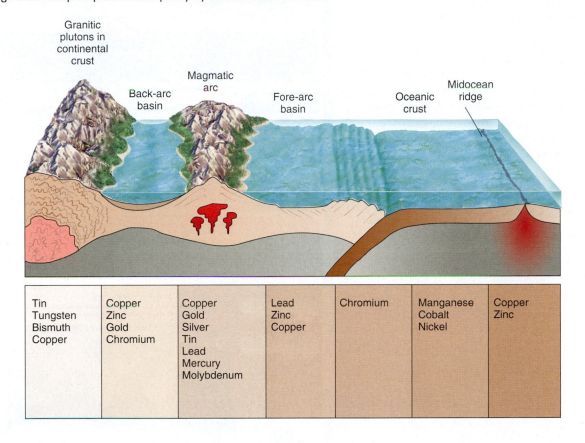

Tin Tungsten Bismuth Copper	Copper Zinc Gold Chromium	Copper Gold Silver Tin Lead Mercury Molybdenum	Lead Zinc Copper	Chromium	Manganese Cobalt Nickel	Copper Zinc

Will We Run Out?

Are there enough mineral and energy resources available on this planet to raise the living standards of all people to the levels they desire? The answer to this question is not clear. Many geologists are concerned that shortages of nonrenewable resources will eventually hamper development. Unfortunately, there is no sure way to know exactly how much of any mineral or energy resource remains to be found. At present, it appears that shortages of energy resources are more likely to affect us in the short run than are shortages of metallic minerals.

Throughout the history of energy and mineral exploration, geologists have continually refined and improved the techniques used to locate new deposits. Much ingenuity has been expended in bringing mineral and energy resource production to its present state. By a combination of new discoveries by geologists, new extraction and processing technologies, and careful use and conservation practices, it is likely that Earth's mineral and energy resources can be extended far into the 21st century. However, in the deep future—by the end of the 21st century and beyond—our descendants may face some very hard choices, as the Easter Islanders did in the 1500s. At the very least, they will have to manage Earth's resources very differently from how we and our predecessors have managed them.

Where Geologists CLICK

Natural Resources Canada

http://gdr.nrcan.gc.ca/index_e.php

The Natural Resources Canada and the Canadian Geological Survey provide very useful databases, online reports, and interactive map resources with extensive information about mineral resources. The Mineral Deposits Web Map Server option will allow you to make your own maps of minerals deposits, with various map layers, using a database query tool to download extensive data about ore deposits in a spatial format.

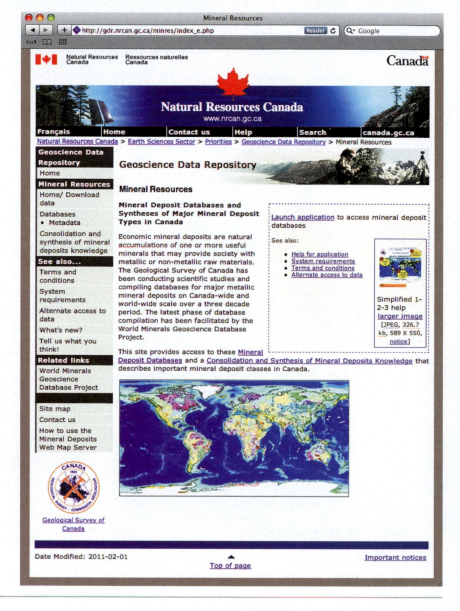

Extensive information is available online concerning the status of mineral resources and exploration in Canada and the United States (see *Where Geologists Click*).

CONCEPT CHECK

1. **What** do economic geologists do?

2. **Why** are some mineral deposits not considered to be ores?

3. **How** do placer deposits form?

4. **What** is a natural tectonic setting for magmatic mineral deposits?

Summary

1 Renewable and Nonrenewable Resources 494

- **Natural resources** can be either **renewable** or **nonrenewable**. Biologic resources are mostly renewable, whereas Earth's resources, particularly minerals and fossil fuels, are mostly nonrenewable. A resource such as soil or groundwater is considered nonrenewable if it cannot be replenished on the scale of a human lifetime (see photo).

How much energy do we use? • Figure 16.3

- Past civilizations have gone through major upheavals when they ran out of a nonrenewable resource. The most likely shortages to face our civilization in the near future are shortages of energy resources.

2 Fossil Fuels 498

- Currently, most of the world's energy needs are met by **fossil fuels**, which are nonrenewable resources. These resources form over the course of millions of years through the burial and chemical alteration of animal and plant remains.

- The major types of fossil fuel are **peat**, **coal**, **petroleum**, and **natural gas**. Peat and coal are formed through the compression and heating of land plants in swampy or boggy conditions, with higher grades of coal being formed by deeper burial as shown in the diagram. Although peat swamps continue to exist today, the most prolific periods for the production of peat were the Carboniferous and Permian periods, when warm and moist conditions prevailed over most of Europe and North America.

Traps for oil and gas • Figure 16.11

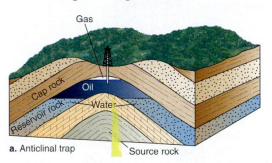

a. Anticlinal trap

- Petroleum and natural gas are formed by the compression and heating of the remains of marine organisms. Petroleum consists of a variety of hydrocarbon molecules, which contain hydrogen and carbon but no oxygen. The molecules with more carbon atoms tend to be heavier, less volatile, and more useful as lubricants, while the molecules with fewer carbon atoms (including natural gas) are lighter and more useful as fuel.

- The **oil** component of petroleum can be refined into gasoline or kerosene, and it is also the raw material for plastics. Nonconventional sources of oil include **tar sand**, which is already economically important in Canada, and **oil shale**. These require more intensive processing to yield their oil, but they may become economically important in the future.

3 Renewable Energy Resources 507

- By far the greatest source of energy available at Earth's surface comes from solar radiation. However, humans are so far able to capture only a minuscule fraction of that energy. Direct solar energy remains too expensive for electric generation in most places, although it is well suited for heating.

- Wind power and **biomass energy** can be considered indirect versions of solar energy. They are renewable, but at present they provide only a small fraction of our energy. Biomass (primarily wood) was the world's leading source of energy until the late 1800s, when it was replaced by coal.

- **Hydroelectric energy**, derived from running water, is a proven energy source in North America. It is not capable of significant further growth. Hydroelectric energy is generated when water flows over a dam and drives the turbines of a power station.

- Nuclear energy, provided by the radioactive decay of uranium as shown in the diagram, is virtually limitless and probably the only resource that could in theory replace fossil fuels. In Japan and certain European countries, it is already a major and growing source of energy. However, in North America it faces a serious waste-disposal problem and public fears about its safety.

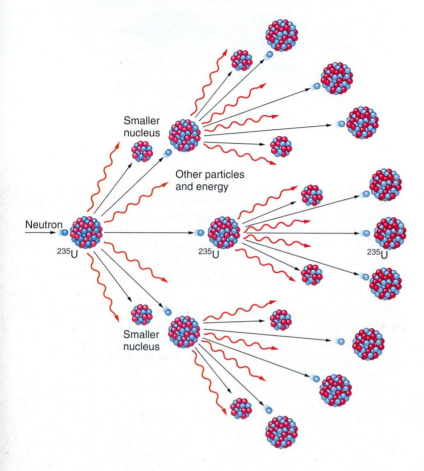

- Geothermal energy, powered by the heat radiating from magma chambers and recently solidified volcanic rock, is plentiful in certain locations near tectonic plate boundaries.

4 Mineral Resources 511

- Economic geologists search for new mineral deposits and assess how long they are likely to last. Not all mineral deposits can be mined profitably; those that can are called **ores**. About 200 minerals are mined in the world today.

- Mineral deposits form when natural processes concentrate certain types of material in one location. Hydrothermal deposits occur when superheated water cools rapidly and dissolved minerals precipitate out of it. Metamorphic deposits occur as a result of contact metamorphism, and magmatic deposits can result from crystallization. Sedimentary deposits include the salts left behind by evaporating water from a lake or shallow sea. Placer deposits form when mineral-bearing rock is eroded and transported by water as shown in the diagram. They are usually found where the water slows down and drops its sediment load. Finally, residual deposits occur in the rock that is left behind after chemical weathering. Examples of ores that are formed by these six processes are (respectively) copper, zinc, chromium, gypsum, gold, and aluminum.

Placer deposits • Figure 16.23

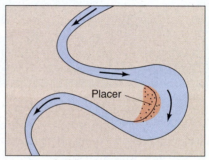

Inside meander loops

- The location of several of these processes is controlled by plate tectonics, and an understanding of an area's geologic history can guide economic geologists to promising locations. For example, magmatic mineral deposits are most likely to be found near present or former plate boundaries. Oil and gas are typically found in former marine sedimentary basins.

- New discoveries and new technologies can prolong to some extent our supply of nonrenewable resources. However, in the long run, society must confront the challenge of managing Earth's finite resources.

Key Terms

- biomass energy 508
- coal 499
- fossil fuel 498
- hydroelectric energy 509
- natural gas 502

- natural resource 494
- nonrenewable resource 495
- oil 502
- oil shale 505
- ore 511

- peat 499
- petroleum 502
- renewable resource 495
- tar sand 505

Critical and Creative Thinking Questions

1. Oil production in the United States satisfies only half of the country's needs; the rest is imported. If imports were cut off, what changes would you expect to occur in your lifestyle?

2. Investigate how geophysicist M. King Hubbert predicted in 1956 that annual U.S. oil production would peak in the early 1970s (the actual peak occurred in 1971). Also, investigate the debate over whether the same predictive technique can be applied to world oil supplies. When do you think world oil production will reach its peak? Or has it done so already?

3. Many hydrothermal mineral deposits of copper, gold, silver, and other metals have been found in the countries bordering the Pacific Ocean. Can you offer an explanation for this remarkable concentration? If you were part of a team of exploration geologists looking for large copper deposits, where would you focus your search?

4. Given that we are now dependent on nonrenewable resources of energy and minerals and that the world's population continues to increase, how do you think human societies will adjust in the future? Do we have a resource problem or a population problem (or both)?

5. Some people think that sustainable development is not a useful concept because it may be impossible to implement—or even to define—in the case of nonrenewable resources. Others think that it is an extremely important concept, if only because it makes us think about the needs of future generations in planning resource management. What do you think?

What is happening in this picture?

An abandoned salt mine in Germany, which lies 1 km underground, has become a repository for the casks of spent nuclear fuel shown in this picture.

Think Critically

What are some of the characteristics that might make this a good place to store nuclear waste?

Self-Test

(Check your answers in Appendix D.)

1. Which of the following statements is incorrect?
   a. The depletion of natural resources apparently led to the collapse of some ancient civilizations.
   b. Nonrenewable resources are never replenished.
   c. Renewable resources must be managed so they are not used at a rate that is greater than the rate of renewal or replenishment of the resource.
   d. Groundwater is, in principle, a renewable resource, but once depleted, it may take a very long time to be replenished.

2. _____ is an example of a nonrenewable resource.
   a. Petroleum
   b. Wind power
   c. Tidal power
   d. Solar power

3. _____ is an example of a renewable resource.
   a. Hydroelectric power
   b. Nuclear power
   c. Coal
   d. Natural gas

4. This diagram shows the proportion of energy resources used in the United States today. The sources of energy have been listed for you. Label the right side of the graph to show lost and used energy. What are the three main categories for the used energy? What is the main form of the lost energy?

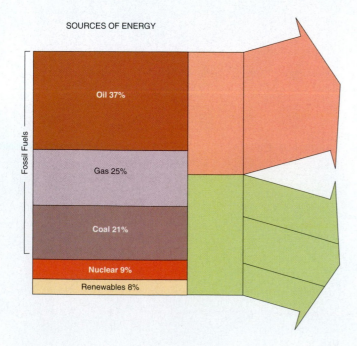

SOURCES OF ENERGY

Fossil Fuels

Oil 37%

Gas 25%

Coal 21%

Nuclear 9%

Renewables 8%

5. _____ forms through the compression heating and lithification of plant-rich sediment.
   a. Petroleum
   b. Coal
   c. Biomass energy
   d. All of the above answers are correct.

6. _____ forms from the decomposition of ancient plankton buried in marine sediment.
   a. Petroleum
   b. Coal
   c. Biomass energy
   d. All of the above answers are correct

7. The diagram below is a graph showing world energy consumption over time. Complete the graph by labeling the fields showing the proper distribution of types of energy consumed.

   oil   wood
   nuclear   hydroelectric
   coal   gas

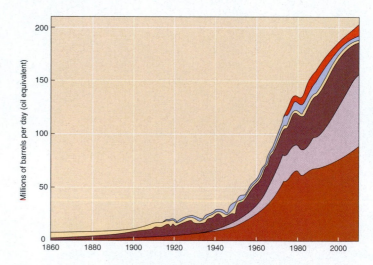

8. _____ is technically a nonrenewable resource, but the amount of energy available is so vast that it is, for all practical purposes, inexhaustible.
   a. Geothermal energy
   b. Nuclear power
   c. Biomass energy
   d. Solar power
   e. Wind energy

9. Which of the following renewable resources derives its energy from the Sun?

   a. wind energy

   b. wave energy

   c. biomass energy

   d. All of the above resources derive their energy from the Sun.

10. Why doesn't solar energy play a more significant role in electricity production?

    a. With even the most current solar photovoltaic technologies, it is still relatively costly to produce electricity.

    b. There is not enough energy in sunlight to help with our current levels of consumption.

    c. Current photovoltaic technologies do not allow for electricity production at high enough voltages.

    d. It can be challenging and costly to install solar technologies on existing buildings.

11. Which of the following statements about nuclear power is (are) correct?

    a. Nuclear power is one of the few energy sources with the potential to completely replace fossil fuels.

    b. Nuclear reactors generate waste that must be isolated from human contact for thousands of years.

    c. Approximately 17% of the world's electricity is currently generated by nuclear power.

    d. Nuclear power is generated through a controlled reaction of fission decay.

    e. All of the above statements are correct.

12. _____ is the branch of geology that is concerned with discovering new supplies of useful minerals.

    a. Geophysics

    b. Environmental geology

    c. Physical geology

    d. Economic geology

    e. Mineral geology

13. What is the difference between an ore and a mineral deposit?

    a. There is no difference.

    b. Ore deposits are mineral deposits that contain metals.

    c. Ore deposits are mineral deposits that can be profitably mined.

    d. Ore deposits are mineral deposits that have formed under high temperature.

14. _____ deposits are formed when minerals precipitate out of hot water solutions.

    a. Hydrothermal

    b. Metamorphic

    c. Placer

    d. Sedimentary

    e. Magmatic

15. This diagram shows a cross section through Earth's crust across a number of tectonic settings. Draw an arrow from each of the metallic resources listed below to the location where it can be found.

    copper   gold

    manganese   chromium

    lead

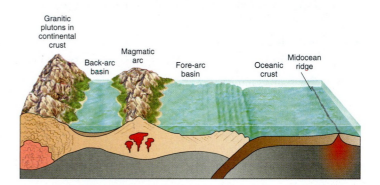

THE PLANNER ✓

Review the Chapter Planner on the chapter opener and check off your completed work.

Commonly Used Units of Measure

Length

Metric Measure

1 kilometer (km)	= 1000 meters (m)
1 meter (m)	= 100 centimeters (cm)
1 centimeter (cm)	= 10 millimeters (mm)
1 millimeter (mm)	= 1000 micrometers (μm) (formerly called microns)
1 micrometer (μm)	= 0.001 millimeter (mm)
1 angstrom (Å)	= 10^{-8} centimeters (cm)

Nonmetric Measure

1 mile (mi)	= 5280 feet (ft) = 1760 yards (yd)
1 yard (yd)	= 3 feet (ft)
1 fathom (fath)	= 6 feet (ft)

Conversions

1 kilometer (km)	= 0.6214 mile (mi)
1 meter (m)	= 1.094 yards (yd)
	= 3.281 feet (ft)
1 centimeter (cm)	= 0.3937 inch (in)
1 millimeter (mm)	= 0.0394 inch (in)
1 mile (mi)	= 1.609 kilometers (km)
1 yard (yd)	= 0.9144 meter (m)
1 foot (ft)	= 0.3048 meter (m)
1 inch (in)	= 2.54 centimeters (cm)
1 inch (in)	= 25.4 millimeters (mm)
1 fathom (fath)	= 1.8288 meters (m)

Area

Metric Measure

1 square kilometer (km^2)	= 1,000,000 square meters (m^2)
	= 100 hectares (ha)
1 square meter (m^2)	= 10,000 square centimeters (cm^2)
1 hectare (ha)	= 10,000 square meters (m^2)

Nonmetric Measure

1 square mile (mi^2)	= 640 acres (ac)
1 acre (ac)	= 4840 square yards (yd^2)
1 square foot (ft^2)	= 144 square inches (in^2)

Conversions

1 square kilometer (km^2)	= 0.386 square mile (mi^2)
1 hectare (ha)	= 2.471 acres (ac)
1 square meter (m^2)	= 1.196 square yards (yd^2)
	= 10.764 square feet (ft^2)
1 square centimeter (cm^2)	= 0.155 square inch (in^2)
1 square mile (mi^2)	= 2.59 square kilometers (km^2)
1 acre (ac)	= 0.4047 hectare (ha)
1 square yard (yd^2)	= 0.836 square meter (m^2)
1 square foot (ft^2)	= 0.0929 square meter (m^2)
1 square inch (in^2)	= 6.4516 square centimeter (cm^2)

Volume

Metric Measure

1 cubic meter (m^3)	= 1,000,000 cubic centimeters (cm^3)
1 liter (l)	= 1000 milliliters (ml)
	= 0.001 cubic meter (m^3)
1 centiliter (cl)	= 10 milliliters (ml)
1 milliliter (ml)	= 1 cubic centimeter (cm^3)

Nonmetric Measure

1 cubic yard (yd^3)	= 27 cubic feet (ft^3)
1 cubic foot (ft^3)	= 1728 cubic inches (in^3)
1 barrel (oil) (bbl)	= 42 gallons (U.S.) (gal)

Conversions

1 cubic kilometer (km^3)	= 0.24 cubic miles (mi^3)
1 cubic meter (m^3)	= 264.2 gallons (U.S.) (gal)
	= 35.314 cubic feet (ft^3)
1 liter (l)	= 1.057 quarts (U.S.) (qt)
	= 33.815 ounces (U.S. fluid) (fl. oz.)
1 cubic centimeter (cm^3)	= 0.0610 cubic inch (in^3)
1 cubic mile (mi^3)	= 4.168 cubic kilometers (km^3)
1 acre-foot (ac-ft)	= 1233.46 cubic meters (m^3)
1 cubic yard (yd^3)	= 0.7646 cubic meter (m^3)
1 cubic foot (ft^3)	= 0.0283 cubic meter (m^3)
1 cubic inch (in^3)	= 16.39 cubic centimeters (cm^3)
1 gallon (gal)	= 3.784 liters (l)

Mass

Metric Measure

1000 kilograms (kg)	= 1 metric ton (also called a tonne) (m.t)
1 kilogram (kg)	= 1000 grams (g)

Nonmetric Measure

1 short ton (sh.t)	= 2000 pounds (lb)
1 long ton (l.t)	= 2240 pounds (lb)
1 pound (avoirdupois) (lb)	= 16 ounces (avoirdupois) (oz) = 7000 grains (gr)
1 ounce (avoirdupois) (oz)	= 437.5 grains (gr)
1 pound (Troy) (Tr. lb)	= 12 ounces (Troy) (Tr. oz)
1 ounce (Troy) (Tr. oz)	= 20 pennyweight (dwt)

Conversions

1 metric ton (m.t)	= 2205 pounds (avoirdupois) (lb)
1 kilogram (kg)	= 2.205 pounds (avoirdupois) (lb)
1 gram (g)	= 0.03527 ounce (avoirdupois) (oz) = 0.03215 ounce (Troy) (Tr. oz) = 15,432 grains (gr)
1 pound (lb)	= 0.4536 kilogram (kg)
1 ounce (avoirdupois) (oz)	= 28.35 grams (g)
1 ounce (avoirdupois) (oz)	= 1.097 ounces (Troy) (Tr. oz)

Pressure

Metric Measure

1 pascal (Pa)	= 1 newton/square meter (N/m^2)
1 kilogram-force square centimeter (kg/cm^2 or kgf/cm^2)	= 1 technical atmosphere (at) = 98,067 Pa = 0.98067 bar
1 bar	= 10^5 pascals (Pa) = 1.02 kilogram-force/square centimeter (kgf/cm^2 or at)

Nonmetric Measure

1 atmosphere (atm)	= 1.01325 bar = 14.696 lb/in^2 (psi)
1 pound per square inch (lb/in^2 or psi)	= 68.046 × 10^{-3} atmospheres (atm)

Conversions

1 kilogram-force/ square centimeter (kgf/cm^2 or at)	= 0.96784 atmosphere (atm) = 14.2233 pounds/square inch (lb/in^2 or psi) = 0.98067 bar = 98,067 Pa
1 bar	= 0.98692 atmosphere (atm) = 10^5 pascals (Pa) = 1.02 kilogram-force/square centimeter (kgf/cm^2)
1 Pa = 10^{-5} bar	= 10.197 × 10^{-6} kgf/cm^2 (or at) = 9.8692 × 10^{-6} atm = 145.04 × 10^{-6} lb/in^2 (or psi)
1 atm	= 101,325 Pa 5 1.01325 bar

Temperature

Metric Measure

0 degrees Celcius (°C)	= freezing point of water at sea level
100 degrees Celcius (°C)	= boiling point of water at sea level
0 degrees Kelvin (K)	= −2273.15°C = absolute zero
1°C	= 1 K (temperature increments)
273.15 K	= 0.0° C

Nonmetric Measure

Fahrenheit (°F)	= (K · 9/5) − 459.67
Fahrenheit (°F)	= (°C · 9/5) + 32

Conversions

degrees Kelvin (K)	= °C + 273.1
degrees Celcius (°C)	= K − 273.15
degrees Fahrenheit (°F)	= (°C · 9/5) + 32
degrees Celcius (°C)	= (°F − 32) · 5/9

The periodic table lists the known **chemical elements**, the basic units of matter. The elements in the table are arranged left-to-right in rows in order of their **atomic number**, the number of protons in the nucleus. Each horizontal row, numbered from 1 to 7, is a **period**. All elements in a given period have the same number of electron shells as their period number. For example, each atom of hydrogen or helium has one electron shell, while each atom of potassium or calcium has four electron shells. The elements in each column, or **group**, share chemical properties. For example,

the elements in column IA are very chemically reactive, whereas the elements in column VIIIA have full electron shells and thus are chemically inert.

Scientists now recognize up to 118 different elements; 92 occur naturally on Earth, and the rest (with the exception of element 117) have been produced synthetically using particle accelerators. Elements are designated by **chemical symbols**, which are the first one or two letters of the element's name in English, Latin, or another language.

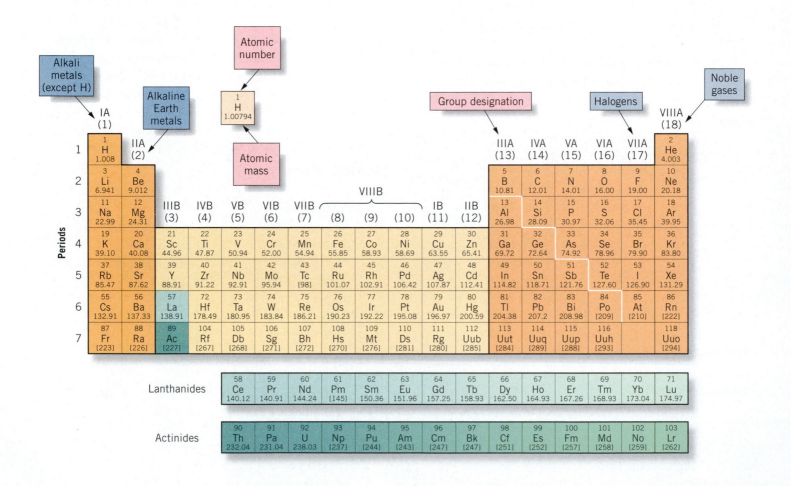

Appendix C Tables of the Properties of Selected Common Minerals

Properties of the common minerals with metallic luster Table C.1

Mineral	Chemical Composition	Form and Habit	Cleavage	Hardness / Specific Gravity	Other Properties	Most Distinctive Properties
Chalcopyrite	$CuFeS_2$	Massive or granular.	None. Uneven fracture.	3.5–4 / 4.2	Golden yellow to brassy yellow. Dark green to black streak.	Streak. Hardness distinguishes from pyrite.
Copper	Cu	Massive, twisted leaves and wires.	None. Can be cut with a knife.	2.5–3 / 9	Copper color but commonly stained green.	Color, specific gravity, malleable.
Galena	PbS	Cubic crystals, coarse or fine-grained granular masses.	Perfect in three directions at right angles.	2.5 / 7.6	Lead-gray color. Gray to gray-black streak.	Cleavage and streak.
Hematite	Fe_2O_3	Massive, granular, micaceous.	Uneven fracture.	5–6 / 5	Reddish-brown, gray to black. Reddish-brown streak.	Streak, hardness.
Limonite (*Goethite* is most common.)	A complex mixture of minerals, mainly hydrous iron oxides.	Massive, coatings, botryoidal crusts, earthy masses.	None.	1–5.5 / 3.5–4	Yellow, brown, black, yellowish-brown streak.	Streak.
Magnetite	Fe_3O_4	Massive, granular. Crystals have octahedral shape.	None. Uneven fracture.	5.5–6.5 / 5	Black. Black streak. Strongly attracted to a magnet.	Streak, magnetism.
Pyrite ("Fool's gold")	FeS_2	Cubic crystals with striated faces. Massive.	None. Uneven fracture.	6–6.5 / 5.2	Pale brass-yellow, darker if tarnisned. Greenish-black streak.	Streak. Hardness distinguishes from chalcopyrite. Not malleable, which distinguishes from gold.
Sphalerite	ZnS	Fine to coarse granular masses. Tetrahedron shaped crystals.	Perfect in six directions.	3.5–4 / 4	Yellowish-brown to black. White to yellowish-brown streak. Resinous luster.	Cleavage, hardness, luster.

Properties of rock-forming minerals with nonmetallic luster   Table C.2

Mineral	Composition	Chemical Form and Habit	Cleavage	Hardness	Specific Gravity	Other Properties	Most Distinctive Properties
Amphiboles. (A complex family of minerals, *Hornblende* is most common.)	$X_2Y_5SigO_{22}(OH)_2$ where X = Ca, Na; Y = Mg, Fe, Al.	Long, six-sided crystals; also fibers and irregular grains.	Two; intersecting at 56° and 124°.	5–6	2.9–3.8	Common in metamorphic and igneous rocks. *Hornblende* is dark green to black; *actinolite*, green; *tremolite*, white.	Cleavage, habit.
Apatite	$Ca_5(PO_4)_3$ (F, OH, Cl)	Granular masses. Perfect six-sided crystals.	Poor. One direction.	5	3.2	Green, brown, blue, or white. Common in many kinds of rocks in small amounts.	Hardness, form.
Aragonite	$CaCO_3$	Massive, or slender, needle-like crystals.	Poor. Two directions.	3.5	2.9	Colorless or white. Effervesces with dilute HCl.	Effervescence with acid. Poor cleavage distinguishes from calcite.
Calcite	$CaCO_3$	Tapering crystals and granular masses.	Three perfect; at oblique angles to give a rhomb-shaped fragment.	3	2.7	Colorless or white. Effervesces with dilute HCl.	Cleavage, effervescence with acid.
Dolomite	$CaMg(CO_3)_2$	Crystals with rhomb-shaped faces. Granular masses.	Perfect in three directions as in calcite.	3.5	2.8	White or gray. Does not effervesce in cold, dilute HCl unless powdered. Pearly luster.	Cleavage. Lack of effervescence with acid.
Feldspars: Potassium feldspar (*orthoclase* is a common variety)	$KAlSi_3O_8$	Prism-shaped crystals, granular masses.	Two perfect, at right angles.	6	2.6	Common mineral. Pink, white, or gray in color.	Color, cleavage.
Garnets	$X_3Y_2(SiO_4)_3$; X = Ca, Mg, Fe, Mn; Y = Al, Fe, Ti, Cr.	Perfect crystals with 12 or 24 sides. Granular masses.	None. Uneven fracture.	6.5–7.5	3.5–4.3	Common in metamorphic rocks. Red, brown, yellowish-green, black.	Crystals, hardness, no cleavage.

(continued)

Table C.2 (continued)

Mineral	Chemical Composition	Form and Habit	Cleavage	Hardness / Specific Gravity		Other Properties	Most Distinctive Properties
Graphite	C	Scaly masses.	One, perfect. Forms slippery flakes.	1–2	2.2	Metamorphic rocks. Black with metallic to dull luster.	Cleavage, color. Marks paper.
Gypsum	$CaSO_4.2H_2O$	Elongate or tabular crystals. Fibrous and earthy masses.	One, perfect. Flakes bend but are not elastic.	2	2.3	Vitreous to pearly luster. Colorless.	Hardness, cleavage.
Halite	NaCl	Cubic crystals.	Perfect to give cubes.	2.5	2.2	Tastes salty. Colorless, blue.	Taste, cleavage.
Kaolinite	$Al_2Si_2O_5(OH)_4$	Soft, earthy masses. Submicroscopic crystals.	One, perfect.	2–2.5	2.6	White, yellowish. Plastic when wet; emits clayey odor. Dull luster.	Feel, plasticity, odor.
Mica: *Biotite*	$K(Mg, Fe)_3 AlSi_3O_{10}(OH)_2$	Irregular masses of flakes.	One, perfect.	2.5–3	2.8–3.2	Common in igneous and metamorphic rocks. Black, brown, dark green.	Cleavage, color. Flakes are elastic.
Mica: *Muscovite*	$KAl_3Si_3O_{10}(OH)_2$	Thin flakes.	One, perfect.	2–2.5	2.7	Common in igneous, and metamorphic rocks. Colorless, pale-green or brown.	Cleavage, color. Flakes are elastic.
Olivine	$(Mg,Fe)_2 SiO4$	Small grains. granular masses.	None Conchoidal fracture.	6.5–7	3.2–4.3	Igneous rocks. Olive green to yellowish-green.	Color, fracture, habit.
Pyroxene (A complex family of minerals. *Augite* is most common.)	$XY(SiO_3)_2$ X = Y = Ca, Mg, Fe	8-sided stubby crystals. Granular masses.	Two, perfect, nearly at right angles.	5–6	3.2–3.9	Igneous and metamorphic rocks. *Augite*, dark green to black; other varieties white to green.	Cleavage
Quartz	SiO_2	6-sided crystals, granular masses.	None. Conchoidal fracture.	7	2.6	Colorless, white, gray, but may have any color, depending on impurities Vitreous to greasy luster.	Form, fracture, striation across crystal faces at right angles to long dimension.

Chapter 1

1 (b), **2** (c), **3** (b), **4** (d), **5** (b), **6** (b), **7** (a), **8** (d), **9** (d), **10** (c), **11** (b), **12** (c), **13** (a), **14** (b), **15** (d).

Chapter 2

1 (d), **2** (see Fig. 2.1), **3** (b), **4** (c), **5** (e), **6** (a), **7** (b), **8** (a), **9** (c), **10** (e), **11** (b), **12** (a), **13** (see Fig. 2.14), **14** (c), **15** (b)

Chapter 3

1 (a), **2** (c), **3** (c), **4** (c), **5** (c), **6** (see Fig 3.3, Stage 5), **7** (d), **8** (b), **9** (b), **10** (b), **11** (see Fig 3.12), **12** (c), **13** (a), **14** (see Fig. 3.15), **15** (f)

Chapter 4

1 (a), **2** (c), **3** (d), **4** (c), **5** (see Fig. 4.6), **6** (a), **7** (see Amazing Places, Fig. D), **8** (b), **9** (c), **10** (a), **11** (d), **12** (e), **13** (c), **14** (see Fig 4.11), **15** (see Fig. 4.11).

Chapter 5

1 (c), **2** (b), **3** (d), **4** (b), **5** (a), **6** (a), **7** (see Fig. 5.10), **8** (d), **9** (d), **10** (e), **11** (a), **12** (d), **13** (see Fig. 5.15), **14** (c), **15** (b).

Chapter 6

1 (c), **2** (b), **3** (See Fig. 6.2C) (b), **4** (b), **5** (c), **6** (a), **7** (d), **8** (b), **9** (see Figs. 6.20, 6.21), **10** (a), **11** (b), **12** (1 and 2, 3 and 4, 5 and 6), **13** (c), **14** (see Fig 6.23), **15** (b).

Chapter 7

1 (See Fig. 7.1), **2** (b), **3** (d), **4** (e), **5** (c), **6** (b), **7** (see Fig. 7.13), **8** (c), **9** (d), **10** (d), **11** (b), **12** (d), **13** (c), **14** (b), **15** (see Fig. 7.18).

Chapter 8

1 (a), **2** (See Fig. 8.2C) (b), **3** (b), **4** (c), **5** (a), 6 (b), **7** (b), **8** (a), **9** (e), **10** (c), **11** (c), **12** (See Fig. 8.15)(a), **13** (b), **14** (b), **15** (c).

Chapter 9

1 (c), **2** (a), **3** (a), **4** (c), **5** (d), **6** (c), **7** (see Fig. 9.9B), **8** (a), **9** (a), **10** (d), **11** (see Fig. 9.13B), **12** (c), **13** (b), **14** (c), **15** (a).

Chapter 10

1 (b), **2** (a), **3** (see Fig. 10.2), **4** (d), **5** (c), **6** (c), **7** (c), **8** (c), **9** (d), **10** (a), **11** (c), **12** (see Fig. 10.13), **13** (a), **14** (see Fig. 10.18), **15** (b)

Chapter 11

1 (see Fig. 11.1), **2** (b), **3** (d), **4** (c), **5** (d), **6** (see Fig. 11.6A), **7** (d), **8** (d), **9** (b), **10** (c), **11** (d), **12** (c), **13** (d), **14** (a), **15** (see Fig. 11.16).

Chapter 12

1 (d), **2** (b), **3** (b), **4** (c), **5** (d), **6** (d), **7** (see What a Geologist Sees, part B), **8** (b), **9** (see Fig. 12. 9A), **10** (d), **11** (see Fig. 12.17), **12** (d), **13** (b), **14** (d), **15** (a).

Chapter 13

1 (b), **2** (d), **3** (d), **4** (see Fig. 13.7), **5** (b), **6** (c), **7** (see Fig. 13.11), **8** (a), **9** (d), **10** (a), **11** (d), **12** (a), **13** (c), **14** (b), **15** (c).

Chapter 14

1 (f), **2** (b) (Figure 14.2 is the associated figure), **3** (c), **4** (c), **5** (a), **6** (e), **7** (e), **8** (d), **9** (a), **10** (e), **11** (a), **12** (b), **13** (b) (Figure 14.12A is the associated figure), **14** (c), **15** (e).

Chapter 15

1 (a), **2** (see Figure 15B), **3** (d), **4** (c), **5** (see Figure 15.6), **6** (e), **7** (c), **8** (b), **9** (e), **10** (e), **11** (b), **12** (a), **13** (d), **14** (b), **15** (d).

Chapter 16

1 (b), **2** (a), **3** (a), **4** (see Figure 16.4), **5** (b), **6** (a), **7** (see Figure 16.13), **8** (b), **9** (d), **10** (a), **11** (e), **12** (d), **13** (c), **14** (a), **15** (see Figure 16.23).

abrasion Wind erosion in which airborne particles chip small fragments off rocks that protrude above the surface.

air A mixture of 78% nitrogen, 21% oxygen, and trace amounts of other gases, found in Earth's atmosphere.

albedo The reflectivity of a surface, as a percentage of the total reflected radiation.

alluvium Unconsolidated sediment deposited in a recent geologic time by a stream.

angiosperm A flowering, or seed-enclosed, plant.

anthropogenic greenhouse effect The portion of greenhouse warming that results from human activities rather than natural processes.

anticline A fold in the form of an arch, with the rock strata convex upward and the older rock in the core.

aphanitic An igneous rock texture with mineral grains so small they can be observed only under a magnifying lens.

aquiclude A layer of impermeable rock.

aquifer A body of rock or regolith that is water saturated, porous, and permeable.

asthenosphere A layer of weak, ductile rock in the mantle that is close to melting but not actually molten.

atmosphere The envelope of gases that surrounds Earth.

atom The smallest individual particle that retains the distinctive chemical properties of an element.

banded iron formation A type of chemical sedimentary rock rich in iron minerals and silica.

barrier island A long, narrow, sandy island lying offshore and parallel to a low-land coast.

batholith A large, irregularly shaped pluton that cuts across the layering of the rock into which it intrudes.

beach drift The movement of particles along a beach as they are driven up and down the beach slope by wave action.

beach Wave-washed sediment along a coast.

bed load Sediment that is moved along the bottom of a stream.

bedding surface The top or bottom surface of a rock stratum or bed.

bedding The layered arrangement of strata in a body of sediment or sedimentary rock.

biogenic sediment Sediment that is primarily composed of plant and animal remains or that precipitates as a result of biologic processes.

biomass energy Any form of energy that is derived more-or-less directly from plant life, including fuel wood, peat, animal dung, and agricultural wastes.

biosphere The system consisting of all living and recently dead organisms on Earth.

body wave A seismic wave that travels through Earth's interior.

bond The force that holds together the atoms in a chemical compound.

brittle deformation A permanent change in shape or volume, in which a material breaks or cracks.

burial metamorphism Metamorphism that occurs after diagenesis, as a result of the burial of sediment in deep sedimentary basins.

carbon cycle The set of processes by which carbon cycles from reservoir to reservoir through the global environment.

cave An underground open space; a cavern is a system of connected caves.

cell The basic structural and functional unit of life; a complex grouping of chemical compounds enclosed in a porous membrane.

cementation The process in which substances dissolved in pore water precipitate out and form a matrix in which grains of sediment are joined together.

channel The clearly defined natural passageway through which a stream flows.

chemical sediment Sediment formed by the precipitation of minerals dissolved in lakewater, riverwater, or seawater.

chemical weathering The decomposition of rocks and minerals by chemical and biochemical reactions.

clastic sediment Sediment formed from fragmented rock and mineral debris produced by weathering and erosion.

clay A family of hydrous alumino-silicate minerals; also, tiny mineral particles of any kind that have physical properties like those of clay minerals.

climate proxy record Records of natural events that are influenced by, and closely mimic, climate.

climate The average weather conditions of a location or region over time.

coal A combustible rock (50 to 95% carbon), formed by the compression, heating, and lithification of peat.

compaction Reduction of pore space in a sediment as a result of the weight of overlying sediment.

compound A combination of atoms of one or more elements in a specific ratio.

compression A stress that acts in a direction perpendicular to and toward a surface.

compressional wave A seismic body wave consisting of alternating pulses of compression and expansion in the direction of wave travel; also called a P wave, or primary wave.

condensation The process by which water changes from a vapor into a liquid or a solid.

conglomerate Clastic sedimentary rock with large fragments in a finer-grained matrix.

contact metamorphism Metamorphism that occurs when rock is heated and chemically changed adjacent to an intruded body of hot magma.

continental crust The older, thicker, and less dense part of Earth's crust; the bulk of Earth's land masses.

continental drift The slow lateral movement of continents across Earth's surface.

convection A form of heat transfer in which hot material circulates from hotter to colder regions, loses its heat, and then repeats the cycle.

convergent margin A boundary along which two plates come together.

core Earth's innermost compositional layer, where the magnetic field is generated and much geothermal energy resides.

Coriolis force An effect due to Earth's rotation, which causes a freely moving body to veer from a straight path.

correlation A method of equating the ages of strata that come from two or more different places.

craton A region of continental crust that has remained tectonically stable for a very long time.

creep The imperceptibly slow downslope granular flow of regolith.

crust The outermost compositional layer of the solid Earth; part of the lithosphere.

cryosphere The perennially frozen part of the hydrosphere.

crystal structure An arrangement of atoms or molecules into a regular geometric lattice. Materials that possess a crystal structure are said to be crystalline.

crystallization The process whereby mineral grains form and grow in a cooling magma (or lava).

cyclone A wind system that is circulating around a low-pressure center.

deflation Wind erosion in which loose particles of sand and dust are removed by the wind, leaving coarser particles behind.

delta A sedimentary deposit, commonly triangle-shaped, that forms where a stream enters a standing body of water.

deposition The laying down of sediment.

desert An arid land that receives less than 250 mm of rainfall or snow equivalent per year and is sparsely vegetated unless it is irrigated.

desertification Invasion of desert conditions into non-desert areas.

dip The angle between a tilted surface and a horizontal plane.

discharge (1) The amount of water passing by a point on a channel's bank during a unit of time. (2) The process by which subsurface water leaves the saturated zone and becomes surface water.

dissolution The separation of a material into ions in solution by a solvent, such as water or acid.

divergent margin A boundary along which two plates move apart from one another.

divide A topographic high that separates adjacent drainage basins.

DNA Deoxyribonucleic acid; a double-chain biopolymer that contains all the genetic information needed for an organism to grow and reproduce.

domain The broadest taxonomic category of living organisms; biologists today recognize three domains: bacteria, *Archaea*, and eukaryotes.

drainage basin The total area from which water flows into a stream.

ductile deformation A permanent but gradual change in shape or volume of a material, caused by flowing or bending.

dune A hill or ridge of sand deposited by winds.

El Niño A regional weather system that involves unusual warming of equatorial Pacific surface water.

elastic deformation A temporary change in shape or volume from which a material rebounds after the deforming stress is removed.

elastic rebound theory The theory that continuing stress along a fault results in a buildup of elastic energy in the rocks, which is abruptly released when an earthquake occurs.

element The most fundamental substance into which matter can be separated using chemical means.

eolian sediment Sediment that is carried and deposited by wind.

epicenter The point on Earth's surface directly above an earthquake's focus.

erosion The wearing-away of bedrock and transport of loosened particles by a fluid, such as water.

estuary A semi-enclosed body of coastal water, in which fresh water mixes with seawater.

eukaryote An organism composed of eukaryotic cells—that is, cells that have a well-defined nucleus and organelles.

evaporation The process by which water changes from a liquid into a vapor.

evaporite A rock formed by the evaporation of lakewater or seawater, followed by lithification of the resulting salt deposit.

evolution The theory that life on Earth has developed gradually, from one or a few simple organisms to more complex organisms.

factor of safety (FS) The balance between destabilizing forces (shear stress) and stabilizing forces (shear strength) on a slope.

fault A fracture in Earth's crust along which movement has occurred.

feedback A cycle in which the output from a process becomes an input into the same process.

flood An event in which a water body overflows its banks.

floodplain The relatively flat valley floor adjacent to a stream channel, which is inundated when the stream overflows its banks.

flow Any mass-wasting process that involves a flowing motion of regolith containing water and/or air within its pores.

focus The location where rupture commences and an earthquake's energy is first released.

fold A bend or warp in layered rock.

foliation A planar arrangement of textural features in metamorphic rock that give the rock a layered or finely banded appearance.

fossil fuel Combustible organic matter that is trapped in sediment or sedimentary rock.

fossil Remains of an organism from a past age, embedded and preserved in rock.

fractional crystallization Separation of crystals from liquids during crystallization.

fractional melt A mixture of molten and solid rock.

fractionation Separation of melted materials from the remaining solid material in the course of melting.

general circulation model A computer model of the climate system, linking processes in the atmosphere, hydrosphere, biosphere, and geosphere.

geologic column The succession of all known strata, fitted together in relative chronological order.

geologic cross section A diagram that shows geologic features that occur underground.

geologic map A map that shows the locations, kinds, and orientations of rock units, as well as structural features such as faults and folds.

geology The scientific study of Earth.

geosphere The solid Earth, as a whole.

glaciation (or **glacial period** or **ice age**) A relatively cold period, when Earth's ice cover greatly exceeded its present extent.

glacier A semi-permanent or perennially frozen body of ice, consisting largely of recrystallized snow, that moves under the pull of gravity.

global warming Present-day warming of the world's climate that most scientists believe is likely to continue and is at least partly caused by human activities.

gneiss (pronounced "nice") A coarse-grained, high-grade, strongly foliated metamorphic rock with separation of dark and light minerals into bands.

gradient The steepness of a stream channel.

greenhouse effect The absorption of long-wavelength (infrared) energy by radiatively active gases in the atmosphere, causing heat to be retained near Earth's surface.

groundwater Subsurface water contained in pore spaces in regolith and bedrock.

gymnosperm A naked-seed plant.

habit The distinctive shape of a particular mineral.

half-life The time needed for half of the parent atoms of a radioactive substance to decay into daughter atoms.

hardness A mineral's resistance to scratching.

high-grade Rock metamorphosed under temperature and pressure conditions higher than about 400°C and 400MPa.

humus Partially decayed organic matter in soil.

hydroelectric energy Electricity generated by running water.

hydrologic cycle A model that describes the movement of water through the reservoirs of the Earth system; the water cycle.

hydrosphere The system comprising all of Earth's bodies of water and ice, both on the surface and underground.

hypothesis A plausible but yet-to-be-proved explanation for how something happens.

igneous rock Rock that forms by cooling and solidification of molten rock.

infiltration The process by which water works its way into the ground through small openings in the soil.

interglaciation (or **interglacial period**) A relatively warm period, when Earth's ice cover and climate resembled those of the present day.

Intergovernmental Panel on Climate Change (IPCC) An international, interdisciplinary panel of scientists and other experts, established to keep the world community up to date on the science of the global climate system.

isostasy The flotational balance of the lithosphere on the asthenosphere.

isotopes Atoms with the same atomic number but different mass numbers.

joint A fracture in a rock, along which no appreciable movement has occurred.

kingdom The second-broadest taxonomic category. There are six recognized kingdoms, including animals and plants.

land degradation Land damage or loss of productivity caused by human activity, which may lead to the advance of desert conditions into non-desert areas.

lava Molten rock that reaches Earth's surface.

limestone A sedimentary rock that consists primarily of the mineral calcite.

lithification The group of processes by which loose sediment is transformed into sedimentary rock.

lithosphere Earth's rocky, outermost layer, comprising the crust and the uppermost part of the mantle.

load The suspended and dissolved sediment carried by a stream.

longshore current A current within the surf zone that flows parallel to the coast.

low-grade Rock metamorphosed under temperature and pressure conditions up to 400°C and 400MPa.

luster The quality and intensity of light that reflects from a mineral.

magma Molten rock that may include fragments of rock, volcanic glass and ash, or gas.

magnetic reversal A period of time in which Earth's magnetic polarity reverses itself.

mantle The middle compositional layer of Earth, between the core and the crust.

marble The product of metamorphism formed by recrystallization of limestone.

mass extinction A catastrophic episode in which a large fraction of living species become extinct within a geologically short time.

mass wasting The downslope movement of regolith and/or bedrock masses due to the pull of gravity.

mechanical weathering The breakdown of rock into solid fragments by physical processes that do not change the rock's chemical composition.

metamorphic facies The set of metamorphic mineral assemblages that form in rock of different compositions under similar temperature and stress conditions.

metamorphic rock Rock that has been altered by exposure to high temperature, high pressure, or both.

metamorphism The mineralogical, textural, chemical, and structural changes that occur in rock as a result of exposure to elevated temperatures and/or pressures.

metasomatism The process whereby the chemical composition of a rock is altered by the addition or removal of material by solution in fluids.

meteorite A fragment of extraterrestrial material that falls to Earth.

Milankovitch cycles The combined influences of astronomical-orbital factors that produce changes in Earth's climate.

mineral A naturally formed solid, inorganic substance with a characteristic crystal structure and a specific chemical composition.

molecule The smallest chemical unit that has all the properties of a particular compound.

moment magnitude scale A measure of earthquake strength that is based on the rupture size, rock properties, and amount of displacement on the fault surface.

moraine A ridge or pile of debris that has been, or is being, transported by a glacier.

mudstone A group of very fine-grained, non-fissile sedimentary rock types with differing proportions of silt- and clay-sized particles.

natural gas The gaseous form of petroleum.

natural resource A useful material that is obtained from the lithosphere, atmosphere, hydrosphere, or biosphere.

natural selection The process by which individuals that are well adapted to their environment have a survival advantage and pass on their favorable characteristics to their offspring.

nonrenewable resource A resource that cannot be replenished or regenerated on the scale of a human lifetime.

normal fault A fault in which the block of rock above the fault surface moves downward relative to the block below.

numerical age The age of a rock or geologic feature in years before the present.

oceanic crust The thinner, denser, and younger part of Earth's crust, underlying the ocean basins.

oil shale A fine-grained sedimentary rock with a high content of kerogen.

oil The liquid form of petroleum.

ore A deposit from which one or more minerals can be extracted profitably.

ore deposit A localized concentration in the crust from which one or more minerals can be profitably extracted.

orogen An elongated region of crust that has been deformed and metamorphosed through a continental collision.

ozone layer A zone in the stratosphere where ozone is concentrated.

paleoclimatology The scientific study of ancient climates.

paleomagnetism The study of rock magnetism in order to determine the intensity and direction of Earth's magnetic field in the geologic past.

paleontology The study of fossils and the record of ancient life on Earth; the use of fossils for the determination of relative ages.

paleoseismology The study of prehistoric earthquakes.

peat A biogenic sediment formed from the accumulation and compaction of plant remains from bogs and swamps, with a carbon content of about 25%.

percolation The process by which groundwater seeps downward and flows under the influence of gravity.

permafrost Ground that is perennially below the freezing point of water.

permeability A measure of how easily a solid allows fluids to pass through it.

petroleum Naturally occurring gaseous, liquid, and semisolid substances that consist chiefly of hydrocarbon compounds.

phaneritic An igneous rock texture with mineral grains large enough to be seen by the unaided eye.

photosynthesis A chemical reaction whereby plants use light energy to induce carbon dioxide to react with water, producing carbohydrates and oxygen.

phyllite A fine-grained metamorphic rock with pronounced foliation, produced by further metamorphism of slate.

plate A large fragment of rigid lithosphere bounded on all sides by faults.

plate tectonics The movement and interactions of large fragments of Earth's lithosphere, called plates.

pluton Any body of intrusive igneous rock, regardless of size or shape.

plutonic rock Igneous rock that solidifies underground, from magma.

polymerization The formation of a complex molecule by the joining of repeated simpler units.

porosity The percentage of the total volume of a body of rock or regolith that consists of open spaces (pores).

precipitation The process by which water that has condensed in the atmosphere falls back to the surface as rain, snow, or hail.

pressure A particular kind of stress in which the forces acting on a body are the same in all directions.

prokaryote A single-celled organism with no distinct nucleus—that is, no membrane separates its DNA from the rest of the cell.

pyroclastic flow Hot volcanic fragments (tephra) that flow very rapidly, buoyed by heat and volcanic gases.

quartzite The product of metamorphism formed by recrystallization of sandstone.

radioactivity A process in which an element spontaneously transforms into another isotope of the same element or into a different element.

radiometric dating The use of naturally occurring radioactive isotopes to determine the numerical age of minerals, rocks, or fossils.

recharge Replenishment of groundwater.

recrystallization The formation of new crystalline mineral grains from old ones.

reef A hard structure on a shallow ocean floor, usually but not always built by coral.

reflection The bouncing back of a wave from an interface between two different materials.

refraction The bending of a wave as it passes from one material into another material, through which it travels at a different speed.

regional metamorphism Metamorphism of an extensive area of the crust, associated with plate convergence, collision, and subduction.

regolith A loose layer of broken rock and mineral fragments that covers most of Earth's surface.

relative age The age of a rock, fossil, or other geologic feature relative to another feature.

renewable resource A resource that can be replenished or regenerated on the scale of a human lifetime.

reservoir A place in the Earth system where a material is stored for a period of time.

reverse fault A fault in which the block on top of the fault surface moves up and over the block on the bottom.

Richter magnitude scale A scale of earthquake intensity based on the recorded heights, or amplitudes, of the seismic waves recorded on a seismograph.

rift valley A linear, fault-bounded valley along a divergent plate boundary or spreading center.

rock A naturally formed, coherent aggregate of minerals and possibly other nonmineral matter.

rock cycle The set of crustal processes that form new rock, modify it, transport it, and break it down.

salinity A measure of the salt content of a solution.

saltation Sediment transport in which particles move forward in a series of short jumps along arc-shaped paths.

sand A sediment made of relatively coarse mineral grains.

sandstone Medium-grained clastic sedimentary rock in which the clasts are typically, but not necessarily, dominated by quartz grains.

schist A high-grade metamorphic rock with pronounced schistosity, in which individual mineral grains are visible.

schistosity Foliation in coarse-grained metamorphic rock.

scientific method The way a scientist approaches a problem; steps include observing, formulating a hypothesis, testing, and evaluating results.

seafloor spreading The processes through which the seafloor splits and moves apart along a midocean ridge and new oceanic crust forms along the ridge.

sediment Rock that has been fragmented, transported, and deposited.

sedimentary rock Rocks that form from sediment under conditions of low pressure and low temperature near the surface.

seismic discontinuity A boundary inside Earth where the velocities of seismic waves change abruptly.

seismic wave An elastic shock wave that travels outward in all directions from an earthquake's source.

seismogram The record made by a seismograph.

seismograph An instrument that detects and measures vibrations of Earth's surface.

seismology The scientific study of earthquakes and seismic waves.

sequestration Long-term storage of a material, in isolation from the atmosphere.

shale Very fine-grained fissile or laminated sedimentary rock, consisting primarily of silt- or clay-sized particles; a fissile mudstone.

shear A stress that acts in a direction parallel to a surface.

shear wave A seismic body wave in which rock is subjected to side-to-side or up-and-down forces, perpendicular to the wave's direction of travel; also called an S wave, or secondary wave.

shield volcano A broad, flat volcano with gently sloping sides, built of successive lava flows.

sink A reservoir that takes in more of a given material than it releases.

slate A very fine-grained, low-grade metamorphic rock with slaty cleavage; the metamorphic product of shale.

slaty cleavage Foliation in low-grade metamorphic rock, which causes the rock to break into flat, plate-like fragments.

slope failure The falling, slumping, or sliding of relatively coherent masses of rock.

soil horizon One of a succession of zones or layers within a soil profile, each with distinct physical, chemical, and biologic characteristics.

soil profile The sequence of soil horizons from the surface down to the underlying bedrock.

soil The uppermost layer of regolith, which can support rooted plants.

species A population of genetically and/or morphologically similar individuals that can interbreed and produce fertile offspring.

spring A natural outlet for ground-water that occurs where the water table intersects the land surface.

strain A change in shape or volume of rock in response to stress.

stratigraphy The science of rock layers and the process by which strata are formed.

stratosphere The layer of Earth's atmosphere above the troposphere, extending to about 50km altitude.

stratovolcano A volcano composed of solidified lava flows interlayered with pyroclastic material. Such volcanoes usually have steep sides that curve upward.

streak A thin layer of powdered mineral made by rubbing a specimen on an unglazed fragment of porcelain.

stream A body of water that flows downslope along a clearly defined natural passageway.

stress The force acting on a surface, per unit area, which may be greater in certain directions than in others.

strike The compass orientation of the line of intersection between a horizontal plane and a planar feature, such as a rock layer or fault.

strike-slip fault A fault in which the direction of the movement is mostly horizontal and parallel to the strike of the fault.

structural geology The study of stress and strain, the processes that cause them, and the deformation and rock structures that result from them.

subduction zone A boundary along which one lithospheric plate descends into the mantle beneath another plate.

surf The "broken," turbulent water found between a line of breakers and the shore.

surface creep Sediment transport in which the wind causes particles to roll along the ground.

surface runoff Precipitation that drains over the land or in stream channels.

surface wave A seismic wave that travels along Earth's surface.

suspended load Sediment that is carried in suspension by a flowing stream of water or wind.

suspension Sediment transport in which the wind carries very fine particles over long distances and periods of time.

syncline A fold in the form of a trough, with the rock strata concave upward and the younger rock in the core.

tar sand A sediment or sedimentary rock in which the pores are filled by dense, viscous, asphalt-like oil.

tectonic cycle Movements and interactions in the lithosphere by which rocks are cycled from the mantle to the crust and back; this cycle includes earthquakes, volcanism, and plate motion, driven by convection in the mantle.

tension A stress that acts in a direction perpendicular to and away from a surface.

theory A hypothesis that has been tested and is strongly supported by experimentation, observation, and scientific evidence.

thermohaline circulation The deep-ocean global "conveyor belt" circulation, driven by differences in water temperature, salinity, and density.

thrust fault A reverse fault with a shallow angle of dip.

tides A regular, daily cycle of rising and falling sea level that results from the gravitational action of the Moon, the Sun, and Earth.

till A heterogeneous mixture of crushed rock, sand, pebbles, cobbles, and boulders deposited by a glacier.

topographic map A map that shows the shape of the ground surface, as well as the location and elevation of surface features, usually by means of contour lines.

trace fossil Fossilized evidence of an organism's life processes, such as tracks, footprints, and burrows.

transform fault margin A fracture in the lithosphere where two plates slide past each other.

transpiration The process by which water taken up by plants passes directly into the atmosphere.

trend A long-term or underlying pattern in a time series of data.

troposphere The lowest layer of Earth's atmosphere, extending (variably) to about 15km in altitude.

turbidity current A turbulent, gravity-driven flow consisting of a mixture of sediment and water, which conveys sediment from the continental shelf to the deep sea.

unconformity A substantial gap in a stratigraphic sequence that marks the absence of part of the rock record.

uniformitarianism The concept that the processes governing the Earth system today have operated in a similar manner throughout geologic time.

viscosity The degree to which a substance resists flow; a less viscous liquid is runny, whereas a more viscous liquid is thick.

volcanic rock Igneous rock that solidifies on or near the surface, from lava.

volcano A vent through which lava, solid rock debris, volcanic ash, and gases erupt from Earth's crust to its surface.

water table The top surface of the saturated zone.

wave-cut cliff A coastal cliff cut by wave action at the base of a rocky coast.

weathering The chemical and physical breakdown of rock exposed to air, moisture, and living organisms.

wind Air in motion.

Table and Line Art Credits

Chapter 1

Figure 1.3 and Figure 1.4: Merali, Zeeya and Brian J. Skinner, *Visualizing Earth Science* . Copyright 2009 John Wiley & Sons, Inc. Figure 1.11: Skinner, Brian J., Stephen C. Porter and Jeffrey Park, *Dynamic Earth 5e*. Copyright 2004 John Wiley & Sons, Inc. Case Study page 17 Skinner, Brian J. and Barbara Murck. *The Blue Planet, 3e*. Copyright 2011 John Wiley & Sons, Inc. Figure 1.15: Skinner, Porter and Botkin, *The Blue Planet*. Copyright John Wiley & Sons, Inc. Where Geologists Click page 29: Used by permission of Planetary and Space Science Centre.

Chapter 2

Where Geologists Click page 51. Used by permission of the National Geographic Society.

Chapter 3

Figure 3.3: Skinner, Brian J., Stephen C. Porter and Jeffrey Park, *Dynamic Earth 5e*. Copyright 2004 John Wiley & Sons, Inc. Figure: 3.8 and Figure 3.14: Merali, Zeeya and Brian J. Skinner, *Visualizing Earth Science*. Copyright 2009 John Wiley & Sons, Inc. Amazing Places, page 87: Skinner, Brian J., Stephen C. Porter and Jeffrey Park, *Dynamic Earth 5e*. Copyright 2004 John Wiley & Sons, Inc. Where Geologists Click page 85: Courtesy of USGS.

Chapter 4

Figure 4.1, Figure 4.8 and Figure 4.14: From Skinner, Brian J. and Stephen C. Porter, *The Dynamic Earth: An Introduction to Physical Geology*. Copyright 2003 John Wiley & Sons, Inc. Figure. 4.11: Merali, Zeeya and Brian J. Skinner, *Visualizing Earth Science*. Copyright 2009 John Wiley & Sons, Inc. *Amazing Places*, page 111, Figure C: Adapted from Rubin, Ken. "Loihi Volcano—The Bathymetric Map of Loihi Seamount." Hawaii Center for Volcanology website. 1997. Retrieved October 6, 2006, from http://www.soest.hawaii.edu/GG/HCV/ loihi.html. Reprinted by permission of the Hawaii Center for Volcanology, Ken Rubin. Where Geologists Click page 108: Courtesy of USGS.

Chapter 5

Figure 5.1: Used with permission from: U.S. Geological Survey National Earthquake Information Centre. Updated March 2011. Figure 5.4B: Satake, Kenji, "2004 Sumatra Earthquake." Figure S7. December 2004. Retrieved October 3, 2006 from http://staff.aist.go.jp/kenji.satake/animation.gif. Reprinted by permission of K. Satake, Geological Survey of Japan, AIST. Table 5.2: Skinner, Brian J. and Barbara Murck. *The Blue Planet, 3e*. Copyright 2011. John Wiley & Sons, Inc Where Geologists Click page 132: Courtesy of USGS.

Chapter 6

Figure 6.10b: Adapted from Sigurdsson, Haraldur, "Volcanic pollution and climate: The 1783 Laki Eruption," *EOS*, vol. 63, pages 601–602, 1982. Copyright 1982, American Geophysical Union. Modified by permission of the American Geophysical Union. Figure 6.8: Adapted from Myers and Brantley, "United States Geological Survey Open File Report 95–231," 1995. Figure 6.19 (center): Merali, Zeeya and Brian J. Skinner, *Visualizing Earth Science*. Copyright 2009 John Wiley & Sons, Inc. Where Geologists Click page 153: Courtesy of USGS.

Chapter 7

Figure 7.2: Skinner, Brian J. and Barbara Murck. *The Blue Planet, 3e*. Copyright 2011. John Wiley & Sons, Inc. Amazing Places, page 195 Copyright © Director of National Parks (Parks Australia) www.parksaustralia.gov.au. Where Geologists Click page 204: Courtesy USDA.

Chapter 8

Where Geologists Click: page 228: The Geological Society of London.

Chapter 9

Amazing Places, page 262, Figure C: Adapted from *Earth Structure, Second Edition* by Ben A. van der Pluijm and Stephen Marshak. Copyright © 2004 by W. W. Norton & Company, Inc. Used by permission of W.W. Norton & Company, Inc. This selection may not be reproduced, stored in a retrieval system, or transmitted in any form or by any means without the prior written permission from the publisher. Where Geologists Click page 273: Courtesy USGS.

Chapter 10

Figure 10.3 (bottom) and Figure 10.17b: Skinner, Brian J. and Stephen C. Porter, *The Dynamic Earth: An Introduction to Physical Geology, 5e*. Copyright 2003 John Wiley & Sons, Inc Where Geologists Click page 289 Copyright Smithsonian Natural Museum of History.

Chapter 11

Figure 11.1a: Merali, Zeeya and Brian J. Skinner, *Visualizing Earth Science*. Copyright 2009 John Wiley & Sons, Inc. Figure 11.4 (left): Skinner, Brian J. and Stephen C. Porter, *The Dynamic Earth: An Introduction to Physical Geology, 5e*. Copyright 2003 John Wiley & Sons, Inc. Figure 11.6a: Strahler, Alan and Zeeya Merali, *Visualizing Physical Geography*. Copyright 2008 John Wiley & Sons, Inc. Figure 11.7: From The New York Times Graphics, "Coastal Defenses Are Disappearing." August 30, 2005. Used by permission of The New York Times Agency. NOTE: Any future revisions, editions thereof in print and any other format require clearance by The New York Times Agency. Where Geologists Click page 311 Courtesy of USGS.

Chapter 12

Figure 12.2: Courtesy of NASA. Figure 12.5A: Merali, Zeeya and Brian J. Skinner, *Visualizing Earth Science*. Copyright 2009 John Wiley & Sons, Inc. Figure 12.9a: Physical Geography: The Global Environment by Harm de Blij, Peter O. Muller and Richard Williams (2003), Figure 49-6 "depicting longshore drift." By permission of Oxford University Press, Inc.. Figure 12.24 Copyright University Corporation of Atmospheric Research. Figure 12.25b: Skinner, Brian J. and Barbara Murck. *The Blue Planet, 3e*. Copyright 2011. John Wiley & Sons, Inc. Case Study page 363: Courtesy of NOAA. Where Geologists Click page 371 Courtesy of NASA.

Chapter 13

Figure 13.3 a and b: Skinner, Brian J. and Stephen C. Porter, *The Dynamic Earth: An Introduction to Physical Geology, 5e*. Copyright 2003 John Wiley & Sons, Inc. Figure 13.6a: Adapted from K. Pye and L. Tsoar, *Aeolian Sand and San Dunes*, Figure 7.1, Chapman &

Hall, 1990. Reprinted with kind permission of Springer Science and Business Media. Figure 13.7a: Adapted from John T. Hack, *The Geographical Review*, vol. 31, Figure. 19, page 260 by permission of the American Geographical Society. *What a Geologist Sees*, page 410 Figures B and C: Adapted from A. H. Lachenbruch in Rhodes W. Fairbridge, ed., *The Encyclopedia of* Van Nostrand Reinhold, New York. Reprinted with kind permission of Springer Science and Business Media. Figure 13.14a: Skinner, Brian J., Stephen C. Porter and Daniel B. Botkin, *Blue Planet*. Copyright 1999 John Wiley & Sons, Inc. and adapted from K. Pye and L. Tsoar, *Aeolian Sand and San Dunes*, Where Geologists Click page 398 Copyright Food and Agriculture Organization of the United Nations.

Chapter 14

Figures 14.1, 14.3d, 14.4, 14.6b, 14.10d, 14.11 a-c, and i, 14.12, 14.15, 14.16, What a Geologist Sees Page 438: Skinner, Brian J. and Barbara Murck. *The Blue Planet, 3e*. Copyright 2011. John Wiley & Sons, Inc. Figure 14.3 a, b, c: Copyright Nigel Calder/ The Weather Machine/ BBC Publications/ London/ 1994. Figure 14.5a,b: Courtesy of NASA. Figure 14.7: Illustration by Jack Cook © Woods Hole Oceanographic Institution. Figure 14.8: Reprinted with permission from Understanding and Responding to Climate Change, 2008 by the National Academy of Sciences, Courtesy of the National Academic Press, Washington, D.C. Figure 14.9: Skinner, Brian J. and Stephen C. Porter, *The Dynamic Earth: An Introduction to Physical Geology, 5e*. Copyright 2003 John Wiley & Sons, Inc. Amazing Places page 428: Copyright The American Museum of Natural History. Case Study Page 437 graphs: Data of EPICA as shown in cdiac.gov. Where Geologists Click

page 447, Intergovernmental Panel on Climate Change, Figure 14.3b: Climate Change 2001, The Scientific Basis. Contribution of Working Group I to the Third Assessment Report of the Intergovernmental Panel on Climate Change, SPM Figure 2(a). Cambridge University Press.

Chapter 15

Figure 15.1: MacKenzie, Fred. T. *Our Changing Planet: An Introduction to Earth System Science and Global Environmental Change*, 3rd Edition, © 2003, p. 205. Reprinted by permission of Pearson Education, Inc., Upper Saddle River, NJ. Figure 15.3: From Falkowski, Paul G. et al, "The Rise of Oxygen over the Past 205 Million Years and the Evolution of Large Placental Mammals." *Science* 30 Sep. 2005, Vol. 309, no. 5744, pages 2202–2204. Reprinted with permission of AAAS. Figure 15.19f: From Nichols, D.J, D. M. Jarzen, C. J. Orth, and P. Q.Oliver. "Palynological and Iridium Anomalies at Cretaceous-Tertiary Boundary, South-Central Saskatchewan." *Science* 14 February 1986, Vol. 231, no. 4739, pages 714–717. Reprinted with permission of AAAS. Where Geologists Click page 467 Used by permissions of UC Berkeley.

Chapter 16

Figure 16.10 and 16.23: From Skinner, Brian J. and Stephen C. Porter, *The Dynamic Earth: An Introduction to Physical Geology*. Copyright 2003 John Wiley & Sons, Inc. Figure 16.13: Adapted from Bookout, J. F., in *Episodes*, vol. 12 (4), 1989, Figure 1. Used with the permission of the International Union of Geological Science. Where Geologists Click page 520 Courtesy of Natural Resources Canada.

Photo Credits

Chapter 1

Page 2: Goddard Institute for Space Studies/NASA; page 4: Jonathan Blair/NG Image Collection; page 5: (top left) Eugene Cernan/Corbis Images; page 5: (top right) Peter Carsten/NG Image Collection; page 5: (center left) Jonathan Blair/NG Image Collection; page 5: (center right) Maria Stenzel/NG Image Collection; page 5: (bottom left) Stephen Alvarez/NG Image Collection; page 5: (bottom right) Jim Richardson/NG Image Collection; page 7: (top) Gondwana Photo Art/Alamy; page 7: (bottom left) Courtesy Stephen C. Porter; page 7: (bottom right) Courtesy Stephen C. Porter; page 7: (bottom center) ©John S. Shelton/University of Washington Libraries, Special Collections; page 8: Courtesy NASA; page 10: Todd Gipstein/NG Image Collection; page 12: (right) ScapeWare3d, LLC; page 17: (left) Caltech/Nasa Jet Propulsion Laboratory; page 17: (right) NASA/Photo Researchers, Inc.; page 21: Marc Moritsch/NG Image Collection; page 22: (bottom left) Courtesy NASA; page 22: (top right) Chris Butler/Photo Researchers, Inc.; page 22: (center) Courtesy NASA; page 22: (right) Courtesy NASA; page 23: NASA/JPL-Caltech/University of Arizona/Texas A&M University; page 25: (top) John S. Shelton; page 25: (bottom) Melissa Farlow/NG Image Collection; page 26: (top right) Marvin Mattelson/NG Image Collection; page 26: NG Maps; page 26: (center) Dorling Kindersley/Getty Images; page 26: (bottom left) Publiphoto/Photo Researchers, Inc.; page 27: (center) Publiphoto/Photo Researchers, Inc.; page 27: (bottom right) Publiphoto/Photo Researchers, Inc.; page 28: Alberto Garcia/©Corbis; page 28: (inset) NG Maps; page 11: (bottom right) © John S. Shelton/University of Washington Libraries, CollectionsSpecial; page 11: (left) Yann Arthus-Bertrand/Photo Researchers, Inc.; page 11: (top right) Courtesy NASA; page 32: Thomas J. Abercrombie/NG Image Collection; page 33: (left) Marc Moritsch/NG Image Collection; page 33: (center) John S. Shelton; page 33: (right) Melissa Farlow/NG Image Collection.

Chapter 2

Page 34: Jiri Hermann, Courtesy Diavik Diamond Mines, Inc; page 34: (inset) Courtesy Diavik Diamond Mines, Inc.; page 35: NG Maps; page 37: (top) Eric Nathan/Alamy; page 37: (bottom) James P. Blair/NG Image Collection; page 37: (center) Mark A. Schneider/Photo Researchers, Inc.; page 39: (top left) C.D. Winters/Photo Researchers, Inc.; page 39: (top right) Corbis Digital Stock; page 39: (top center left) Eric Nathan/Alamy; page 39: (top center left) Victor R. Boswell, Jr. /NG Image Collection; page 39: (bottom center left) ITAR-TASS Photo Agency/Alamy; page 39: (bottom center right) Bill Curtsinger/NG Image Collection; page 39: (bottom left) Mark A. Schneider/Photo Researchers, Inc.; page 39: (bottom right) Richard Nowitz/NG Image Collection; page 41: (top left) Robert Sisson/NG Image Collection; page 41: (top right) Ed George/NG Image Collection; page 41: (top center left) Todd Gipstein/NG Image Collection; page 41: (top center right) BIANCA LAVIES/NG Image Collection; page 41: (bottom center left) James P. Blair/NG Image Collection; page 41: (bottom center right) C.D. Winters/Photo Researchers, Inc.; page 41: (bottom left) Digital Vision; page 41: (bottom right) W. Robert Moore/NG Image Collection; page 42: Courtesy Michael Hochella; page 40: (top left) Harry Taylor/Dorling Kindersley/Getty Images, Inc.; page 43: (top left) Charles D. Winters/Photo Researchers, Inc.; page 43: (bottom left) Courtesy Brian J. Skinner; page 43: (top right) Courtesy Brian J. Skinner; page 43: (bottom right) Courtesy Brian J. Skinner; page 43: (top left) William Sacco; page 44: (top) William Sacco; page 44: (bottom) Courtesy Brian J. Skinner; page 46: (bottom left) William Sacco; page 46: (top right) Mark A. Schneider/Photo Researchers, Inc.; page 46: (top left) ©Breck Kent; page 47: (left) William Sacco; page 47: (right) The Natural History Museum; page 48: William Sacco; page 49: ©Donald Stampfli/AP/Wide World Photos; page 60: (bottom) Courtesy Brian J. Skinner; page 60: (top right) Courtesy Brian J. Skinner; page 60: (top left) Courtesy Brian J. Skinner; page 52: (top) Courtesy Brian J. Skinner; page 52: (center) Joel Arem/Photo Researchers, Inc.; page 52: (bottom) ©Breck Kent; page 53: (top) ©Breck Kent; page 53: (top center) ©Breck Kent; page 53: (bottom center) ©Breck Kent; page 53: (bottom) ©Breck Kent; page 55: Courtesy Kevin Telmer; page 57: (top right) Raymond Gehman/NG Image Collection; page 57: (bottom inset) Courtesy Brian J. Skinner; page 57: (top right) Taylor S. Kennedy/NG Image Collection; page 57: (center inset) Courtesy Brian J. Skinner; page 57: (bottom left) James P. Blair/NG Image Collection; page 57: (top inset) Courtesy Brian J. Skinner; page 58: Peter Carsten/NG Image Collection; page 63: Taylor S. Kennedy/NG Image Collection; page 63: (top right) Courtesy Brian J. Skinner; page 45: (left) Javier Trueba/MSF/Photo Researchers, Inc.; page 45: (right) Javier Trueba/MSF/Photo Researchers, Inc.; page 45: (bottom) Javier Trueba/MSF/Photo Researchers, Inc.; page 61: James L. Amos/NG Image Collection; page 62: The Natural History Museum.

Chapter 3

Page 65: Loop Images/Corbis Images; page 65: (bottom inset) Steve Gschmeissner/Photo Researchers, Inc.; page 65: (top inset) NG Maps; page 67: (top left) Dale O'Dell/Alamy; page 67: (bottom left) Annie Griffiths Belt/NG Image Collection; page 67: (bottom) imagebroker.net/SuperStock; page 70: (top) David Svilar/Danita Delimont/NewsCom; page 70: (top right) NG Maps; page 70: (center) Radius Images/Corbis Images; page 68: (left) ©Marli Bryant Miller; page 68: (right) Courtesy Lee Gerhard; page 71: (top left) James L. Amos/NG Image Collection; page 71: (top center) Courtesy Brian J. Skinner; page 71: (top right) Ralph Lee Hopkins/NG Image Collection; page 71: (bottom left) James L. Amos/NG Image Collection; page 71: (bottom right) Mark Gibson/Photolibrary/Getty Images, Inc.; page 73: Jonathan Blair/NG Image Collection; page 75: (top) Mark Hallett/NG Image Collection; page 75: (center) Robert Giusty/NG Image Collection; page 75: (bottom) Kazuhiko Sano/NG Image Collection; page 76: (bottom right) O. Louis Mazzatenta/National Geographic/NG Image Collection; page 77: (H.M.S. Beagle) Mary Evans/Photo Researchers, Inc.; page 77: (The Origin of Species) Mary Evans Picture Library/Photo Researchers, Inc.; page 77: Pictorial Parade/Getty Images, Inc.; page 77: (Radioactive rock) Astrid & Hanns-Frieder Michler/Photo Researchers, Inc.; page 77: (Surfer) Patrick McFeeley/NG Image Collection; page 77: (Siccar Point) Courtesy Dr. K. Roy Gill; page 84: (bottom left) ©AP/Wide World Photos; page 84: (bottom right) Kenneth Garrett/NG Image Collection; page 85: (right) Courtesy Stephen J. Mojzsis; page 86: James L. Amos/NG Image Collection; page 87: (top left) Dr. K.Roy Gill; page 87: (top left inset) NG Maps; page 87: (top left) Photo by Gale A. Bishop; page 90: John Dunn/Arctic Light/NG Image Collection; page 89: Courtesy Stephen J. Mojzsis.

Chapter 4

Page 92: Colin Monteath/Minden Pictures/NG Image Collection; page 93: (bottom inset) Hulton Archive/Getty Images;

page 93: (top inset) NG Maps; page 97: National Library of Wales; page 98: Courtesy Willem van der Westhuizen; page 104: (top) Emory Kristof/NG Image Collection; page 104: (bottom) frans lemmens/Alamy; page 105: (top) Robert Harding/Digital Vision/Alamy; page 105: (bottom) James P. Blair/NG Image Collection; page 106: Courtesy NASA; page 110: Diane Cook and Len Jenshel/Getty Images, Inc.; page 111: (top) Chris Johns/NG Image Collection; page 111: (bottom) Courtesy Hawaii Center for Volcanology; page 113: Chris Scotese; page 116: Courtesy Tony Philpotts.

Chapter 5

Page 118: Asahi Shimbun via Getty Images; page 119: KO SA-SAKI/The New York Times/Red/Redux Pictures; page 119: NG Maps; page 121: (left) ©John S. Shelton/University of Washington Libraries, Special Collections; page 121: (right) Courtesy USGS; page 122: Peter Essick/Alamy; page 123: (left) ©George Plafker, U.S. Geological Survey; page 123: (center) FOTOSEARCH/Age Fotostock; page 123: (top right) Sacramento Bee Staff Photo/NewsCom; page 123: (bottom right) Courtesy NOAA/NGDC; page 125: (top right) DigitalGlobe/Zuma Press/NewsCom; page 125: (center right) HO/AFP/Getty Images/NewsCom; page 125: (bottom left) NG Maps; page 125: (bottom) GRONDIN EMMANUEL/Maxppp/Landov LLC; page 127: (top left) Ron Haviv/VII//©Corbis; page 127: (top right) James A. Sugar/NG Image Collection; page 127: (bottom left) ERIKO SUGITA/Reuters//©Corbis; page 129: Jim Mendenhall/NG Image Collection; page 134: (top) ©AP/Wide World Photos; page 134: (center) ©AP/Wide World Photos; page 134: (bottom) Waldemar Lindgren/NG Image Collection; page 137: Sarah Leen/NG Image Collection; page 138: (top left) Courtesy Dan Schulze; page 138: (bottom) Smithsonian Institution/©Corbis; page 139: (right) Bryan & Cherry Alexander/Photo Researchers, Inc.; page 143: (top right inset) NG Maps; page 143: (top) Courtesy Landsat.org; page 143: (bottom left) Diane Miller/Photolibrary/Getty Images, Inc.; page 143: (bottom right) Diane Miller/Photolibrary/Getty Images, Inc.; page 145: Courtesy NOAA/NGDC; page 147: Emory Kristof/NG Image Collection.

Chapter 6

Page 150: LUCAS JACKSON/Reuters/©Corbis; page 151: Arctic-Images/©Corbis; page 151: NG Maps; page 152: (left) Paul Chesley/NG Image Collection; page 152: (top right) Carsten Peter/NG Image Collection; page 152: (bottom right) ©J.D. Griggs/USGS; page 153: Photo Researchers/Getty Images, Inc.; page 154: (bottom left) ©Reuters/Corbis-Bettmann; page 154: (bottom right) Photodisc/Getty Images, Inc.; page 155: (top left) Carsten Peter/NG Image Collection; page 155: (top right) USGS; page 154: (bottom left) Frans Lanting/NG Image Collection; page 154: (bottom center) Greg Vaughn/Alamy; page 154: (bottom right) ©John S. Shelton/University of Washington Libraries, Special Collections; page 156: Greg Vaughn/Alamy; page 157: (top) Lyn Topinka/USGS; page 157: (bottom) Courtesy Stephen C. Porter; page 158: (top) Francois Gohier/Photo Researchers, Inc.; page 158: (bottom) Roger Ressmeyer/©Corbis; page 159: (top left) Photri/The Image Works; page 159: (top right) Chris Newhall//USGS; page 159: (bottom left) Chris Newhall//USGS; page 159: (top inset) NG Maps; page 161: Grant Dixon/Lonely Planet Images/Getty Images; page 162: John Stanmeyer/VII; page 163: (inset) NG Maps; page 164: (top inset) NG Maps; page 164: (top left) NASA; page 164: (top right) Robert Madden/NG Image Collection; page 164:

(bottom left) USGS; page 164: (bottom right) USGS; page 166: National Remote Sensing Centre Ltd/Photo Researchers, Inc.; page 167: (left) Charles Thatcher/Getty Images, Inc.; page 167: (right) Krafft Explorer/Photo Researchers, Inc.; page 170: (left) ©J.D. Griggs//USGS; page 170: (right) ©J.D. Griggs//USGS; page 172: (top right) OAR/National Undersea Research/Photo Researchers, Inc.; page 172: (center) Raymond Gehman/NG Image Collection; page 172: (bottom) Pablo Corral Vega/NG Image Collection; page 173: (top) Philippe Bourseiller/The Image Bank/Getty Images; page 173: (bottom) Stocktrek Images/Getty Images, Inc.; page 174: (left) Kenneth Garrett/NG Image Collection; page 174: (center) Courtesy Brian J. Skinner; page 174: (right) ©Tony Waltham; page 175: (left) Courtesy Brian J. Skinner; page 175: (right) William Sacco; page 176: Marc Moritsch/NG Image Collection; page 177: (top) Courtesy Brian J. Skinner; page 177: (center) Courtesy Brian J. Skinner; page 177: (bottom) Courtesy Brian J. Skinner; page 181: (top) Asa Thoresen/Photo Researchers, Inc.; page 181: (center) James Steinberg/Photo Researchers, Inc.; page 181: (bottom) Raymond Gehman/NG Image Collection; page 185: Winfield Parks/NG Image Collection; page 186: (top left) Frans Lanting/Alamy; page 186: (bottom right) Photodisc/Getty Images, Inc.; page 186: (bottom center) Marc Moritsch/NG Image Collection; page 186: (bottom right) ©Tony Waltham; page 187: (top) Courtesy Brian J. Skinner; page 187: (center) Courtesy Brian J. Skinner; page 187: (bottom) Courtesy Brian J. Skinner.

Chapter 7

Page 188: Frank Krahmer/Masterfile; page 189: Joe Vogan/Purestock/SuperStock; page 189: NG Maps; page 191: William E. Ferguson; page 192: Jeff Foott/Getty Images, Inc.; page 193: (top left) Prisma/SuperStock; page 193: (top right) George F. Mobley/NG Image Collection; page 193: (bottom left) Maria Stenzel/NG Image Collection; page 193: (bottom right) Paul C. Dennis/Lost Trio Hiking Trail; page 194: (bottom left) Paul Johnson/Getty Images, Inc.; page 194: (top right) Walker Howell/NG Image Collection; page 194: (bottom right) Richard Nowitz/NG Image Collection; page 195: NASA; page 196: (left) William E. Ferguson; page 196: (right) Courtesy Brian J. Skinner; page 197: (top) Courtesy Brian J. Skinner; page 197: (bottom) David Leveson; page 198: (left) Pablo Galán Cela/Age Fotostock America, Inc.; page 198: (top right) Stephanie Maze/NG Image Collection; page 198: (bottom right) Medford Taylor/NG Image Collection; page 199: (top) Courtesy Amy Larson, Smith College; page 199: (top right) Raymond Gehman/NG Image Collection; page 199: (bottom left) Frans Lanting/©Corbis; page 201: (left) Robin Siegel/NG Image Collection; page 201: (right) Hiroji Kubota/Magnum Photos, Inc.; page 202: (left) Sinclair Stammers/Photo Researchers, Inc.; page 202: (right) Photo Researchers, Inc.; page 204: (left) USDA/ Soil Conservation Service; page 204: (right) USDA/ Soil Conservation Service; page 205: (top) NG Maps; page 205: (right) Raymond Gehman/NG Image Collection; page 205: (left) Kevin Horan/Stone/Getty Images, Inc.; page 207: Michael Nichols/NG Image Collection; page 208: Courtesy NOAA; page 209: (left) Courtesy Brian J. Skinner; page 209: (top right) Courtesy Stephen C. Porter; page 209: (bottom) Hemis/Alamy; page 210: (top left) Bill Hatcher/NG Image Collection; page 210: (center right) Courtesy Stephen C. Porter; page 210: (bottom right) ©Marli Bryant Miller; page 212: W.E. Garrett R./NG Image Collection; page 215: Kevork Djansezian/©AP/Wide World Photos.

Chapter 8

Page 218: Bates Littlehales/NG Image Collection; page 218: (inset) Courtesy NASA; page 220: (far left) Fletcher & Baylis/Photo Researchers, Inc.; page 220: (left) ©Marli Bryant Miller; page 220: (right) G. R. Roberts/Photo Researchers, Inc.; page 220: (far right) Emory Kristof/NG Image Collection; page 221: (left) Martin Shields/Alamy; page 221: (right) Courtesy Stephen C. Porter; page 222: (left) Bret Webster/Photo Researchers, Inc.; page 222: (right) Courtesy Dr. Kenneth L. Finger and Chevron Corporation; page 223: (top right) STEPHEN MORRISON/EPA/NewsCom; page 223: (bottom left) Frans Lanting/©Corbis; page 223: (bottom right) Wayne G. Lawler/Photo Researchers, Inc.; page 223: (top left) Philip Smith/Getty Images, Inc.; page 225: (top right) Peter Carsten/NG Image Collection; page 225: (bottom left) James P. Blair/NG Image Collection; page 225: (bottom right) Werner Hilpert/Getty Images, Inc.; page 225: (top left) Alamy; page 226: Phil Schermeister/NG Image Collection; page 227: (top left) Raymond Gehman/NG Image Collection; page 227: (top right) FOTOSEARCH RM/Age Fotostock America, Inc.; page 227: (top right inset) Courtesy NASA; page 227: (bottom left) Bruce Dale/NG Image Collection; page 227: (bottom right) Courtesy Stephen C. Porter; page 229: Nicole Duplaix/NG Image Collection; page 230: ©Marli Bryant Miller; page 231: (top inset) NG Maps; page 231: (top left) George F. Mobley/NG Image Collection; page 231: (bottom right) Robert Sisson/NG Image Collection; page 233: (top left) Dirk Wiersma/Photo Researchers, Inc.; page 233: (top right) Phillip Hayson/Photo Researchers, Inc.; page 233: (bottom left) M.I. Walker/Photo Researchers, Inc.; page 235: Courtesy Brian J. Skinner; page 236: (top left) Courtesy Stephen C. Porter; page 236: (top right) ©Marli Bryant Miller; page 236: (bottom left) Medford Taylor/NG Image Collection; page 236: (bottom right) Courtesy Stephen C. Porter; page 237: (top left) Minden Pictures, Inc.; page 237: (top right) Minden Pictures/Getty Images, Inc.; page 237: (bottom) Kevin Schafer/Corbis Images; page 239: (top) NewsCom; page 239: (bottom) Lealisa Westerhoff/Getty Images; page 242: (top left) Creatas/Superstock, Inc; page 242: (bottom right) AgeFotostock/SuperStock, Inc; page 246: (bottom center) Courtesy Stephen C. Porter; page 246: (top) O. Louis Mazzatenta/NG Image Collection; page 246: (bottom left) Martin Shields/Alamy; page 246: (bottom right) ©Marli Bryant Miller; page 247: (top left) Dirk Wiersma/Photo Researchers, Inc.; page 247: (top right) Phillip Hayson/Photo Researchers, Inc.

Chapter 9

Page 248: Edwin Chang Photography/Flickr/Getty Images, Inc.; page 250: SuperStock; page 252: (left) Taylor Kennedy/NG Image Collection; page 252: (right) Taylor Kennedy/NG Image Collection; page 252: (left inset) Taylor Kennedy/NG Image Collection; page 253: (top left) Ted Kinsman/Photo Researchers/Getty Images; page 253: (bottom left) Sam Abell/NG Image Collection; page 253: (top right) ©Marli Bryant Miller; page 253: (bottom right) Jane Selverstone; page 254: Courtesy Mervyn Paterson; page 255: (top tthree photos) ©Cambridge Polymer Group, Boston, MA (2001). Photograph by Gavin Braithwaite; page 255: (top tthree photos) ©Cambridge Polymer Group, Boston, MA (2001). Photograph by Gavin Braithwaite; page 255: (top tthree photos) ©Cambridge Polymer Group, Boston, MA (2001). Photograph by Gavin Braithwaite; page 255: (bottom) Courtesy Dr. James Bales, Assistant Director of the MIT Edgerton Center, and Dr. Stephen Spiegelberg, Cambridge Polymer Group, Inc.; page 258: James P. Blair/NG Image Collection; page

261: Emory Kristof/NG Image Collection; page 262: (top inset) NG Maps; page 262: (top) Design Pics/Michael Interisano/Getty Images, Inc.; page 262: (right) Courtesy J. Mueller, The Harvard Structure Group; page 264: Alamy; page 265: ©Marli Bryant Miller; page 266: MDA Federal, Inc; page 267: Bernhard Edmaier/Photo Researchers, Inc.; page 269: Reproduced by permission of the British Geological Survey. © NERC. All rights reserved. IPR/106-44C; page 271: Courtesy Geological Survey of Canada, Ottawa; page 276: ©Marli Bryant Miller; page 250: SuperStock.

Chapter 10

Page 278: James L. Stanfield/NG Image Collection; page 280: (left) Alex Vasilescu/Getty Images, Inc.; page 280: (right) Courtesy Brian J. Skinner; page 282: (left) Courtesy Brian J. Skinner; page 282: (right) Courtesy Brian J. Skinner; page 283: (left) Courtesy Brian J. Skinner; page 283: (top right) Courtesy Brian J. Skinner; page 283: (bottom right) Courtesy Brian J. Skinner; page 284: Courtesy Jay Ague, Dept. of Geology and Geophysics at Yale University; page 285: (top left) Courtesy Brian J. Skinner; page 285: (top center) Courtesy Brian J. Skinner; page 285: (top right) Courtesy Brian J. Skinner; page 285: (bottom left) William Sacco; page 285: (bottom right) William Sacco; page 286: (top) Courtesy Brian J. Skinner; page 286: (bottom) ©Brenda Sirois; page 287: (left) Courtesy Brian J. Skinner; page 287: (right) Natalie Behring/NewsCom; page 288: (top) Courtesy Brian J. Skinner; page 288: (top center) Courtesy Brian J. Skinner; page 288: (bottom center) Courtesy Brian J. Skinner; page 288: (bottom) Courtesy Brian J. Skinner; page 290: (top right) Courtesy Brian J. Skinner; page 290: (top left) Courtesy Brian J. Skinner; page 290: (center right) Courtesy Brian J. Skinner; page 290: (bottom left) Courtesy Brian J. Skinner; page 290: (bottom right) Courtesy Brian J. Skinner; page 291: NG Maps; page 291: (bottom left) Courtesy George E. Harlow; page 291: (top) Universal History Archive/Getty Images, Inc.; page 291: (bottom right) American Museum of Natural History (AMNH30/10359); page 292: (top left) ©Breck Kent; page 292: (center left) Pasieka/Photo Researchers, Inc.; page 292: (top right) Joyce Photographics/Photo Researchers, Inc.; page 292: (bottom right) Dirk Wiersma/Photo Researchers, Inc.; page 293: (left) Michael P. Gadomski/Photo Researchers, Inc.; page 293: (right) Courtesy Brian J. Skinner; page 296: (left) Alamy; page 296: (right) Jonathan Blair/NG Image Collection; page 297: William Sacco; page 300: Courtesy Brian J. Skinner; page 303: SuperStock; page 301: (top left) ©Breck Kent; page 301: (bottom left) Pasieka/Photo Researchers, Inc.; page 304: (top left) ©Breck Kent; page 304: (top right) Pasieka/Photo Researchers, Inc.; page 304: (bottom left) Joyce Photographics/Photo Researchers, Inc.; page 304: (bottom right) Dirk Wiersma/Photo Researchers, Inc.

Chapter 11

Page 306: Nigel Pavitt/JAI/Corbis Images; page 306: (inset) Courtesy Brian Skinner; page 307: (inset) NG Maps; page 310: (top left) Bill Curtsinger/NG Image Collection; page 310: (bottom left) Emory Kristof/NG Image Collection; page 310: (top right) Miroslav Krob/Age Fotostock America, Inc; page 310: (bottom right) Bates Littlehales/NG Image Collection; page 311: (bottom left) Annie Griffiths Belt/NG Image Collection; page 313: (top left) George F. Mobley/NG Image Collection; page 313: (bottom left) S.C. Porter; page 313: (right) S.C. Porter; page 314: Steve Raymer/NG Image Collection; page 315: (bottom left) ©Marli Bryant Miller; page 315: (bottom right)

Courtesy Visible Earth/NASA; page 318: Courtesy NASA; page 320: Environmental Images/Age Fotostock America, Inc.; page 321: (top left) MDA Federal, Inc; page 321: (top right) MDA Federal, Inc; page 321: (bottom) ©AP/Wide World Photos; page 323: Tyrone Turner/NG Image Collection; page 326: NG Maps; page 328: NG Maps; page 338: (top right) Phil Schermeister/NG Image Collection; page 328: (bottom right) ©Greg Reis; page 328: (right) Craig Aurness/©Corbis; page 333: (top right) Edward Kinsman/Photo Researchers, Inc.; page 333: (bottom left) Courtesy Brian J. Skinner; page 335: Ted Spiegel/NG Image Collection; page 336: (left) Peter Morgan/Reuters/©Corbis; page 336: (right) ©AP/Wide World Photos; page 337: (left) USGS; page 337: (right) Bruno Barbey/Magnum Photos, Inc.; page 338: NG Maps; page 338: (top left) Michael Nichols/NG Image Collection; page 338: (top right) Michael Nichols/NG Image Collection; page 338: (bottom left) Michael Nichols/NG Image Collection; page 338: (bottom right) Michael Nichols/NG Image Collection; page 341: ©AP/Wide World Photos.

Chapter 12

Page 344: Annie Griffiths/National Geographic/Getty Images, Inc.; page 345: (inset) Robert F. Bukaty/©AP/Wide World Photos; page 347: (top) ©Don Dixon; page 347: (bottom) NG Maps/Marie Tharp; page 349: NASA; page 351: (left) Courtesy NASA; page 352: Manley, W.F., 2002, Postglacial Flooding of the Bering Land Bridge: A Geospatial Animation: INSTAAR, University of Colorado, v1, http://instaar.colorado.edu/QGISL/bering_land_bridge; page 353: Manley, W.F., 2002, Postglacial Flooding of the Bering Land Bridge: A Geospatial Animation: INSTAAR, University of Colorado, v1, http://instaar.colorado.edu/QGISL/bering_land_bridge; page 354: Richard Nowitz/NG Image Collection; page 355: Arnulf Husmo/Stone/Getty Images; page 356: Patrick McFeeley/NG Image Collection; page 357: Carl & Ann Purcell/©Corbis; page 358: James P. Blair/NG Image Collection; page 359: G.R. Roberts/The Natural Sciences Image Library; page 360: Richard Reid/NG Image Collection; page 361: (top) David Alan Harvey/NG Image Collection; page 361: (right) Cotton Coulson/NG Image Collection; page 361: (bottom) Raymond Gehman/NG Image Collection; page 363: (top) Mark Ralston/AFP/Getty Images, Inc.; page 363: (center) NASA; page 363: (bottom) Courtesy NOAA; page 362: Digital Vision/Getty Images, Inc.; page 364: NG Maps; page 364: (left) Thomas K. Gibson/Florida Keys National Marine Sanctuary; page 365: (top left) Courtesy Phillip Dustan, College of Charleston; page 365: (bottom right) Stephen Frink/Alamy; page 366: (left) Richard Bizley/Photo Researchers, Inc.; page 366: (right) Thomas and Pat Leeson/Photo Researchers, Inc.; page 367: Courtesy NASA; page 370: NASA; page 373: (left) Courtesy NASA; page 373: (right) Ginny Lloyd Photo/Shutterpoint; page 377: Courtesy NASA; page 381: NASA.

Chapter 13

Page 384: John Burcham/NG Image Collection; page 384 (inset): Carsten Peter/NG Image Collection; page 387: (top left) Carsten Peter/NG Image Collection; page 387: (top right) Dr. Cynthia M. Beall & Dr. Melvyn C. Goldstein/NG Image Collection; page 387: (bottom left) Marc Moritsch/NG Image Collection; page 387: (bottom right) Annie Griffiths Belt/NG Image Collection; page 389: (top right) Courtesy Stephen C. Porter; page 389: (bottom left) Jim Richardson/Corbis Images; page 389: (bottom right) Courtesy Stephen C. Porter; page 390: Courtesy Brian J. Skinner; page 391: Bernhard Edmaier/Photo Research-

ers, Inc.; page 392: (center right) Gerry Ellis/Minden Pictures/NG Image Collection; page 392: (center left) Courtesy Brian J. Skinner; page 392: (bottom left) George Steinmetz/NG Image Collection; page 392: (bottom right) Courtesy NASA; page 392: (top right) George Steimetz/Corbis Images; page 393: (bottom) ©John S. Shelton; page 393: (top) Marc Moritsch/NG Image Collection; page 394: (left) Courtesy Stephen C. Porter; page 394: (right) B.A.E. /Alamy; page 396: (right) G.P. Bowater/Alamy; page 396: (left) Courtesy NOAA George E. Marsh Album; page 397: Courtesy NASA; page 399: Courtesy NASA; page 400: (bottom left) Melissa Farlow/NG Image Collection; page 400: (bottom) Thomas Kitchin & Victoria Hurst/All Canada Photos/Corbis Images; page 401: (top) ©Marli Bryant Miller; page 401: (center) Gavin Hellier/Getty Images; page 401: (bottom) Harald Sund/Getty Images, Inc.; page 402: (left) North Wind Pictures Archives/Alamy; page 402: (right) Macduff Everton/Getty Images, Inc.; page 405: Courtesy Stephen C. Porter; page 406: (left) Chris Johns/NG Image Collection; page 406: (right) Chris Johns/NG Image Collection; page 408: (top) Mark Burnett/Photo Researchers, Inc.; page 408: (center) Gallo Images/Danita Delimont/Getty Images, Inc.; page 408: (bottom) James P. Blair/NG Image Collection; page 409: (top left) Raymond Gehman/NG Image Collection; page 409: (top right) Courtesy Gerald Osborn, University of Calgary; page 409: (bottom left) LOOK Die Bildagentur der Fotografen GmbH/Alamy; page 409: (bottom right) Cary Wolinsky/NG Image Collection; page 410: (top) Maria Stenzel/NG Image Collection; page 410: (bottom) Courtesy NASA; page 412: Courtesy NASA; page 413: imagebroker/Alamy; page 415: Cary Wolinsky/NG Image Collection.

Chapter 14

Page 417: SGM/NewsCom; page 418: (top) imagebroker/Alamy; page 418: (bottom) NASA; page 421: (left) Mike Hettwer; page 421: (center) age fotostock/SuperStock; page 421: (right) Mike Hettwer; page 424: Courtesy of NASA; page 425: NASA image courtesy Jeff Schmaltz, MODIS Rapid Response Team <http://rapidfire.sci.gsfc.nasa.gov/> at NASA GSFC; page 425: NASA; page 426: (left) NASA; page 426: (right) George Ostertag/SuperStock; page 426: age fotostock/SuperStock, Inc.; page 427: (top) Woods Hole Oceanographic Institution; page 427: (center) Woods Hole Oceanographic Institution; page 427: (bottom) Woods Hole Oceanographic Institution; page 428: (top left) Hemis/Alamy; page 428: (top right) Leo Hickey; page 428: (bottom left) Thuliadanta is an early Eocene (~52-53 million-year-old) tapir from Ellesmere Island, in Canada's High Arctic Archipelago. Photograph by J. Eberle (CU-Boulder).; page 428: (bottom right) Comstock Images/Getty Images, Inc.; page 434: (left) Jon Arnold Images Ltd./Alamy; page 434: (right) GYRO PHOTOGRAPHY/amanaimagesRF/Getty Images; page 435: (top left) Photo from Rob Dunbar and Glen Shen; page 435: (top center) NASA Images; page 435: R.E. Litchfield/Photo Researchers; page 436: Peter Essick/Aurora Photos Inc.; page 438: (top) ©AP/Wide World Photos; page 438: (bottom) SPL/BRITISH ANTARCTIC SURVEY/Photo Researchers, Inc.; page 444: (top left) © The New York Times/Redux Pictures; page 444: (top right) Paul Nicklen/National Geographic Creative/Getty Images, Inc.; page 444: (bottom) Ralph Lee Hopkins/NG Image Collection; page 445: courtesy IPCC; page 448: (left) James L. Stanfield/NG Image Collection; page 448: (right) Courtesy NASA; page 449: (top right) David Clapp/Photolibrary/Getty Images, Inc.; page 449: (left) David Ponton/Design Pics/NewsCom; page 449: (bottom) age fotostock/SuperStock; page 452: UniversalUclick;

page 453: SPL/BRITISH ANTARCTIC SURVEY/Photo Researchers, Inc.

Chapter 15

Page 454: Chris Jackson/Getty Images, Inc.; page 457: ©CORBIS; page 460: (left) O. Louis Mazzatenta/NG Image Collection; page 460: (right) Francois Gohier/Photo Researchers, Inc.; page 461: (top left) O. Louis Mazzatenta/NG Image Collection; page 461: (center right) Kirk Moldoff/NG Image Collection; page 461: (bottom left) AFP/NewsCom; page 461: (bottom center) Mona Lisa Production/Photo Researchers, Inc.; page 461: (bottom right) KIM JUNIPER/AFP/NewsCom; page 463: (left) ©Dr. Jeremy Burgess/ SPL/Photo Researchers, Inc.; page 463: (right) CNRI/ SPL/Photo Researchers, Inc.; page 464: (left) O. Louis Mazzatenta/NG Image Collection; page 464: (right) O. Louis Mazzatenta/NG Image Collection; page 465: (left) Photo Researchers/Getty Images, Inc.; page 465: (right) Mary Evans Picture Library/Photo Researchers, Inc.; page 466: (center map) NG Maps; page 466: (center left) Ralph Lee Hopkins/NG Image Collection; page 466: (top right) ©J. Dunning/VIREO; page 466: (top left) David Hosking/Photo Researchers, Inc.; page 466: (Just above center right) Tier und Naturfotografie/SuperStock; page 466: (center right) Richard l'Anson/Lonely Plan Lonely Planet Images/NewsCom; page 466: (bottom left) Tierbild Okapia/Photo Researchers, Inc.; page 466: (bottom right) Eric Hosking/Photo Researchers, Inc.; page 468: (top left) Paul Zahl/NG Image Collection; page 468: (bottom right) Stephen St. John/NG Image Collection; page 469: (top left) ©AP/Wide World Photos; page 469: (bottom right) Pete Oxford/Minden Pictures RM/Getty Images, Inc.; page 470: (top) NASA; page 470: (left) NASA; page 470: (right) Sinclair Stammers/Photo Researchers, Inc.; page 470: (right inset) J.W. Schopf; page 471: (bottom left) Publiphoto/Photo Researchers, Inc.; page 471: (top left) Walter Myers/Photo Researchers, Inc.; page 471: (top right) Roger Harris/Photo Researchers, Inc.; page 471: (bottom right) Mauricio Anton/Photo Researchers, Inc.; page 472: (right) Martin Versimilus/USGS; page 472: (left) Courtesy Brian J. Skinner; page 473: Michael Melford/NG Image Collection; page 474: Raymond Gehman/NG Image Collection; page 475: (top left) Scott Camazine/Photo Researchers, Inc.; page 475: (top right) Chris Johns/NG Image Collection; page 475: (bottom left) ©Breck Kent; page 475: (bottom right) Theodore Clutter/Photo Researchers, Inc.; page 476: Courtey Peabody Museum of Natural History, Yale University; page 477: (left) Hoberman Collection/Corbis Images; page 477: (right) Tom McHugh/Photo Researchers, Inc.; page 479: (top) Francois Gohier/Photo Researchers, Inc.; page 479: (bottom) Courtesy Carnegie Museum of Natural History; page 480: (top) ©Michael Rothman; page 480: (center left) Lealisa Westerhoff/AFP/Getty Images; page 480: (bottom) John Reader/ Science Photo Library/Photo Researchers, Inc.; page 482: ©Chris Butler; page 483: (bottom) Jonathan Blair/NG Image Collection; page 484: Courtesy Dr. Richard Ernst; page 485: NG Maps; page 486: Francois Gohier/Photo Researchers, Inc.; page 487: (left) Paul Zahl/NG Image Collection; page 487: (right) ©Chris Butler; page 488: (top left) Courtesy NASA; page 488: (bottom left) Chris Butler/Photo Researchers, Inc.; page 488: (center) Courtesy NASA; page 488: (right) Courtesy NASA; page 489: (top) Jonathan Blair/NG Image Collection; page 489: (bottom) ©Jonathan Blair/Corbis; page 490: (left) Dr. Jeremy Burgess/ SPL/Photo Researchers, Inc.; page 490: (right) CNRI/ SPL/Photo Researchers, Inc.

Chapter 16

Page 492: Stephanie Maze/NG Image Collection; page 493: (inset) NG Maps; page 494: NHPA/SuperStock; page 495: (left) Victor Paul Borg/Alamy; page 495: (right) ©Nick Brooks; page 495: (right inset) ©Nick Brooks; page 496: (left) G.E. Kidder Smith/Corbis Images; page 496: (right) Bill Brooks/Alamy; page 470: (left) Michael Nichols/NG Image Collection; page 470: (right) James P. Blair/NG Image Collection; page 500: Robert Sisson/NG Image Collection; page 501: (top left) Richard Olsenius/NG Image Collection; page 501: (center right) Jon McLean/Alamy; page 502: (bottom right) H. Mark Weidman Photography/Alamy; page 505: (left) Sarah Leen/NG Image Collection; page 505: (right) Jonathan P. Blair/NG Image Collection; page 508: (top) Sarah Leen/NG Image Collection; page 508: (bottom) Marc Moritsch/NG Image Collection; page 509: Lester Lefkowitz/Age Fotostock America, Inc.; page 510: (right) Roger Ressmeyer/Corbis Images; page 510: (bottom left) David Howells/Corbis Images; page 512: NG Maps; page 514: (top) Courtesy Brian J. Skinner; page 514: (center) William Sacco; page 514: (bottom) Courtesy Brian J. Skinner; page 515: (top) ©Marli Bryant Miller; page 515: (center) Courtesy Brian J. Skinner; page 515: (bottom) William E. Ferguson; page 516: ©Marli Bryant Miller; page 517: Courtesy Woods Hole Oceanographic Institution; page 521: Michael Nichols/NG Image Collection; page 523: Emory Kristof/NG Image Collection.

Index

soils and, *204*
volcanoes and, 161, *161*
weathering and, 200, *200*
Climate proxy record, 433, *434–435*
Climate system, *418*, 418–420, 433
Climate zones, *419*, 419–420
Climatologists, *5*
Closed systems, 9, *9*, 10, 12
Clouds, climate change and, 424, *424–425*
Coal, *41*, 234
Coalification, 234, *499*, 499–500
Coal seams, *501*
Coastal deserts, 387, *387*
Coastal floods, 323, *323*
Coastlines. *See also* Oceans
barrier islands and, 360–362, *361*
beaches and, 360
coral reefs and, 362–364
disappearing, *316*
lagoons and, *361*, 362
rocky coasts, *359*, 359–360
spits and, *361*, 362
wave action along, 355–357, *356*
Coelacanths, *477*
Cohesive strength, soil, 212
Color, minerals and, *47*, 47–48
Colorado:
Clear Creek, *4*
Colorado National Park, *236*
Raton Basin, *483*
Colorado National Park, Colorado, *236*
Colorado Plateau, *264*
Colorado River, *12*
Columbia Glacier, Alaska, *401*
Columbus, Christopher, 350
Comoro Islands, *477*
Compaction, 232, 334
Compounds, 37–38, *38*
Compressional stress, rock deformation and, 251, *251*
Compressional waves, 130, *131*
Computerized axial tomography (CAT) scanning, 137
Condensation, 308–309
Conduction, 108
Cone of depression, 334
Confined aquifers, 332, *332*
Confining pressure, rock deformation and, 254, *254*
Conformable strata, 68
Conglomerate rock, *220*, 233–234
Connecticut, *285*
Contact metamorphism, 294, *294*, 299
Contamination, groundwater, 334–336
Continents, adjusting height of, *257*
Continental collision zones, 106–107
Continental crust, 22, *50*, 54
Continental deserts, 387, *387*
Continental divide, 318
Continental drainage, *319*
Continental drift, 94–96, *96*, 98, *98*
Continental rift, *173*
Continental rise, *95*
Continental shelf, 95, *95*, 228
Continental slope, 95, *95*
Continent-continent collisions, 256–257
Convection, 109, *109*
Convection cells, 108, *109*, 372

Convection currents, 372
Convergent margins, *105*, 105–107, 113
Convergent plate boundaries:
collisional type, 240
subduction type, 240–243
Copernicus, Nicolaus, 94
Coprolites, 469
Corals, 362, 364, *435*
Coral barrier reef, *364*, 365
Coral bleaching, *365*
Core:
of Earth, 21, 144
of terrestrial planets, *21*
Coriolis effect, 350, 372
Correlation, fossils and, 72, *72*
Corundum, *47*
Covalent bonding, *39*
Cracks, 261
Craters:
from meteorite impact, 11, *11*
volcanoes and, *156*, 156–157
Crater Lake, Oregon, *156*, 156–157
Cratons, 256, *256*
Creep, mass wasting and, 211
Cretaceous Period, 64
Crevasses, *403*, 405
Crop irrigation, 325
Cross bedding, 230, 390
Crude oil, 363, *503*
Crust, Earth's, 20, 50–54, 140–141, *141*
Crustal faulting, lake formation and, 320
Crustal rock, 250
Cryosphere, 398–401
Crystals:
atomic structures of, *42*
faces and angles, *43*, 43–44
form and habit, 43–44
structure, 42
Crystallization, 43, 174, 178, *178*
Cubic habit, 44
Curie, Marie, 77
Curie point, 139
Currents, ocean, *350*, 350–351
Cyanobacteria, 462
Cycles, 14–15, *15*
Cyclones, 376–377, *377*

D

Dana, James Dwight, 110
Darwin, Charles, 77, 78, *465*, 465–467, 481
Daughter atoms, 79
Dawson City, Yukon Territory, Canada, *413*
The Day After Tomorrow (movie), 426
Death Valley, California, *67*, *315*, *387*, *393*, *515*
Debris avalanches, volcanoes and, 158
Debris falls, 210
Deccan Traps, India, 161, 484
Decompression melting, *168*
Deep-sea turbidites, *227*
Deepwater Horizon oil spill, 345, 363, *363*
Deflation, *389*, 389–390
Deformation, rock, *see* Rock deformation
Deforming ice, *403*, 405
Deltas, 224, 226, 316–317
Density, minerals and, 48
Density zones, ocean, *349*
Deoxyribonucleic acid (DNA), 462

Depletion, groundwater, 334–336
Deposits, streams, *315*, 315–317
Deposition, 224
glacial, 407–409, *409*
sediment and, 220
Depositional environments:
beaches, 226–227, *227*
carbonate platforms and reefs, 228
deltas and estuaries, *226*
glaciers, 224
lakes, 224
on land, 224–226, *225*
marine evaporite basins, 228
in or near oceans, 226–229, *227*
seafloors, 229
shelves, 228
streams, 224
turbidites, 228
wind and, 224, 226
Deserts and drylands:
coastal deserts, 387
continental interior deserts, 387
desertification, 395–396, *396*
desert landforms, 390–393
dunes and, 390–393, *391–393*
global belts and, 386
hot deserts, *387*
polar deserts, 387
rainfall in, 394
rainshadow deserts, 387
types of deserts, 386–387
wind-blown sediment and, 388
wind erosion and, 388–390
world's deserts, *386*
Desertification, 395–396, *396*
Desert pavements, *389*, 389–390
Devil's Tower, Wyoming, *181*
Devonian Period, *471*
Dew point, 367
Diagenesis, 233
Diamonds:
brought up by volcanism, 34
formation of, 34, *138*
Dickinsonia costata, *76*, *464*
Dickson County, Kansas, *476*
Differentiation, 20
Dikes, 180, *181*, 182
Dinosaurs, *4*, 64, 454, *469*
Diorite, 177t
Dip, structural geology and, 258–259, *259*
Direct observation, 138–139
Disappearing coastlines, *316*
Discharge:
channel, 312–313
groundwater, 331–332
Disintegration, rock, 191–195
Dissolution, chemical weathering and, 196, *196*
Dissolved load, 206
Divergent margins, *104*, 105
Divides, 318–319, *319*
DNA (deoxyribonucleic acid), 462
Dolomite, 234
Dolostone, 234
Domains, 462
Domes, 267, *267*
Downwarping, 267
Dragonfly, *476*
Drainage basins, 317–319, *317–319*

Volcanic rock *(cont.)*
  glassy texture and, *175*
  porphyritic texture and, *175*
  rapid cooling and, 175–176
  volcanic and plutonic equivalents, 177t
Volcanic tremors, 158
Volcanism, mass extinctions and, *484*
Volcanoes, 161
  basic structure of, *160*
  climate change and, *161, 423*
  craters and, 156–157
  deadly eruptions and, *160*
  debris avalanches and, 158
  eruption types and, *152,* 153–157
  establishing history and, 165
  explosive eruptions and, *152*
  flood basalts and, *155*
  hazards and, 158–161
  in Iceland, 150, *150–151*
  lava and, 152
  magma and, 153
  monitoring changes and anomalies and,
      165–166, *166*
  nonexplosive eruptions and, *152*
  Plinian eruptions, 156
  predicting eruptions and, 161–166
  primary effects and, 158
  pyroclastic flows and, 152, 153
  remote sensing and, 166, *166*
  resurgent domes and, 157
  secondary effects and, 158
  shield volcanoes, *155,* 156
  stratovolcanoes, *154,* 156
  Strombolian eruptions, 153, *155*
  tephra and, 152, *152*
  tertiary and beneficial effects and, 161
  volcanic ash, 152, *153*
  Vulcanian eruptions, 153, *154,* 156
Volcanologists, 4, *5*
Volcan Osorno, Chile, *172*
von Lauë, Max, 43
Vulcanian eruptions, 153, *154,* 156

W

Wadi Al Masilah, South Yemen, *318*
Wadi Kufra, 416, 421
Walcott, Charles, *473*
Wales, Old Red Sandstone, *265*
Wallace, Alfred Russell, 467
Walsh, Donald, 92
Waptia fidensis, 472
Washington:
  Mt. St. Helens, *153, 154,* 164, *164,* 314, *314*
  ocean–atmosphere–climate interactions, 376
  Olympic Peninsula, *227*
  Skykomish River, *324*
  Toutle River, 314, *314*
Water. *See also* Groundwater; Hydrologic cycle;
    Oceans; Streams
  classification of, *41*

as critical resource, 494
drainage basins, *318*
early origins of, 346, *346, 347,* 459
erosion by, 206
floods and, 320–325, *321*
groundwater, *329,* 329–338
heat capacity of, 375
hydrologic cycle and, *308,* 308–311
lakes and, 319–320
latent heat of, 375
overland flow and, 312
possible external sources for, *347*
residence time and, 311
springs, 332, *333*
stream deposits and, 315–317
streams and streamflow and, 312–314
stream system topography, 317–319
surface runoff and, 312
surface water resources, 325, *326–327*
world's resources and, *310*
Water cycle, *see* Hydrologic cycle
Water-scarce countries, 325
Watersheds, worldwide, 325, *326–327*
Water table, 329–330
Water vapor, 367, *368*
Waves, *355,* 355–358
  beach drift and, *357,* 358
  breakers, 356
  erosion and, 357–358
  longshore current and, 357, *357*
  sediment transport and, 357–358
  surf and, 355–357
  wave action along coastlines, 355–357, *356*
  wave refraction, 357
Wave-cut cliffs, 359, *359*
Weather:
  climate change and, 431
  defined, 374
  El Niño and La Niña, *378,* 378–379
  ocean–atmosphere–climate interactions,
      374–376
  tropical cyclones, 376–377, *377*
Weathering, *191,* 192–205
  abrasion and, 195
  biologic activity and, *199*
  chemical weathering, 196–197
  climate and, 200, *200*
  composition and, *199*
  dissolution and, 196, *196*
  factors affecting, *198–199,* 198–200
  frost wedging and, *193,* 194
  joint formation and, 192, *192, 193*
  mechanical, 192–195
  products of, 201–205
  regolith and, 191
  rock disintegration and, 191–195
  rock structure and, *198*
  root wedging and, *193,* 195
  soil and, 202–205, *203*
  soil profiles and, *203,* 203–204

tectonic setting and, *198*
topography and, *198*
vegetation and, *199*
Wegener, Alfred, *94,* 94–95, *97,* 99, 101
Wells, *334*
West Antarctic Peninsula, *444*
Wetland, 320
White Cliffs of Dover, *64*
Wildlife Conservation Society, *485*
Wilson, J. Tuzo, 113
Wilson cycle, 113
Wind:
  defined, 372
  as renewable energy source, 508–509
Wind-blown sediment, 224, 226, *388, 388*
Wind erosion, 388–390
  abrasion and, 388–390, *389–390*
  deflation and, *389,* 389–390
  desert pavement and, *389,* 389–390
  mechanisms of, 388–390
  saltation and, 388
  surface creep and, 388
  suspension and, 388
Windmills, *508,* 508–509
Windsor profile, *204*
Wind systems, 371–374, *373*
Wisconsin, Kettle-Moraine State Park, *409*
Wisconsin glacier, 408
Witwatersrand, South Africa, *515*
World Meterological Organization, 445, 446
World ocean, 346, *347,* 428
"World Series" earthquake (1989), 122
Wrangell Saint Elias National Park, Alaska, *400*
Wyoming:
  Black Thunder mine, *501*
  Devil's Tower, *181*

X

Xenoliths, *138,* 139
X-rays, seismology and, 137

Y

Yakutat Bay, Alaska, 121
Yellow River, China, 323
Yellowstone National Park, *172, 409*
Yosemite Park, California, *193*
Younger Dryas event, 426
Yucatan Peninsula, Mexico, 19, *482–483*
Yucca Mountain, Nevada, *510*
Yuma, Arizona, *25*

Z

Zion National Park, Utah, *231*
Zircon, 459
Zone of aeration, *329, 329*